CAMBRIDGE TRACTS IN MATHEMATICS

General Editors
B. BOLLOBAS, P. SARNAK, C.T.C. WALL

102   **Algebraic *L*-Theory and
Topological Manifolds**

T0269320

Algebraic *L*-Theory and Topological Manifolds

A.A. Ranicki

Reader in Mathematics
University of Edinburgh

CAMBRIDGE UNIVERSITY PRESS

CAMBRIDGE UNIVERSITY PRESS
Cambridge, New York, Melbourne, Madrid, Cape Town, Singapore, São Paulo

Cambridge University Press
The Edinburgh Building, Cambridge CB2 8RU, UK

Published in the United States of America by Cambridge University Press, New York

www.cambridge.org
Information on this title: www.cambridge.org/9780521420242

© Cambridge University Press 1992

First published 1992
This digitally printed version 2008

A catalogue record for this publication is available from the British Library

ISBN 978-0-521-42024-2 hardback
ISBN 978-0-521-05521-5 paperback

For my parents

Drawing by Tom Cheney
©1991 The New Yorker Magazine, Inc.
By Special Permission. All Rights Reserved.

Contents

Preface

The algebraic L-theory of quadratic forms relates the topology of manifolds to their homotopy types. This tract provides a reasonably self-contained account of this relationship in dimensions ≥ 5, which was established over 20 years ago by the Browder–Novikov–Sullivan–Wall surgery theory for compact differentiable and PL manifolds, and extended to topological manifolds by Kirby and Siebenmann.

The term 'algebraic L-theory' was coined by Wall, to mean the algebraic K-theory of quadratic forms, alias hermitian K-theory. In the classical theory of quadratic forms the ground ring is a field, or a ring of integers in an algebraic number field, and quadratic forms are classified up to isomorphism. In algebraic L-theory it is necessary to consider quadratic forms over more general rings, but only up to stable isomorphism. In the applications to topology the ground ring is the group ring $\mathbb{Z}[\pi]$ of the fundamental group π of a manifold.

The structure theory of high-dimensional compact differentiable and PL manifolds can be expressed in terms of the combinatorial topology of finite simplicial complexes. By contrast, the structure theory of high-dimensional compact topological manifolds involves deep geometric properties of Euclidean spaces and demands more prerequisites. For example, compare Thom's proof of the combinatorial invariance of the rational Pontrjagin classes with Novikov's proof of topological invariance. The current development of the controlled and bounded surgery theory of non-compact manifolds promises a better combinatorial understanding of these foundations, using the algebraic methods of this book and its companion on lower K- and L-theory, Ranicki [146]. The material in Appendix C is an indication of the techniques this will entail.

The book is divided into two parts, called Algebra and Topology. In principle, it is possible to start with the Introduction, and go on to the topology in Part II, referring back to Part I for novel algebraic concepts. The reader does not have to be familiar with the previous texts on surgery theory: Browder [16], Wall [176], Ranicki [143], let alone the research literature*. This book is not a replacement for any of these. Books and papers need not be read in the order in which they were written.

The text was typeset in TEX, and the diagrams in $\mathcal{LAMS}$-TEX.

Edinburgh, June 1992

---

* 'The literature on this subject is voluminous but mostly makes difficult reading'. This was Watson on integral quadratic forms, but it applies also to surgery theory.

Introduction

An *n-dimensional manifold* M is a paracompact Hausdorff topological space such that each point $x \in M$ has a neighbourhood homeomorphic to the Euclidean n-space $\mathbb{R}^n$. The homology and cohomology of a compact n-dimensional manifold M are related by the Poincaré duality isomorphisms

$$H^{n-*}(M) \cong H_*(M) ,$$

using twisted coefficients in the nonorientable case.

An *n-dimensional Poincaré space* X is a topological space such that $H^{n-*}(X) \cong H_*(X)$ with arbitrary coefficients. A Poincaré space is *finite* if it has the homotopy type of a finite CW complex. A compact n-dimensional manifold M is a finite n-dimensional Poincaré space, as is any space homotopy equivalent to M. However, a finite Poincaré space need not be homotopy equivalent to a compact manifold. The *manifold structure existence problem* is to decide if a finite Poincaré space is homotopy equivalent to a compact manifold.

A homotopy equivalence of compact manifolds need not be homotopic to a homeomorphism. The *manifold structure uniqueness problem* is to decide if a homotopy equivalence of compact manifolds is homotopic to a homeomorphism, or at least h-cobordant to one. The mapping cylinder of a homotopy equivalence of compact manifolds is a finite Poincaré h-cobordism with manifold boundary, which is homotopy equivalent rel ∂ to a compact manifold h-cobordism if and only if the homotopy equivalence is h-cobordant to a homeomorphism. The uniqueness problem is thus a relative version of the existence problem.

The Browder–Novikov–Sullivan–Wall surgery theory provides computable obstructions for deciding the manifold structure existence and uniqueness problems in dimensions ≥ 5. The obstructions use a mixture of the topological K-theory of vector bundles and the algebraic L-theory of quadratic forms. A finite Poincaré space is homotopy equivalent to a compact manifold if and only if the Spivak normal fibration admits a topological bundle reduction such that a corresponding normal map from a manifold to the Poincaré space has zero surgery obstruction. A homotopy equivalence of compact manifolds is h-cobordant to a homeomorphism if and only if it is normal bordant to the identity by a normal bordism with zero rel ∂ surgery obstruction. The theory applies in general only in dimensions ≥ 5 because it relies on the Whitney trick for removing singularities, just like the h- and s-cobordism theorems.

The algebraic theory of surgery of Ranicki [140]–[146] is extended here to a combinatorial treatment of the manifold structures existence and problems, providing an intrinsic characterization of the manifold structures in a homotopy type in terms of algebraic transversality properties on the chain level. The Poincaré duality theorem is shown to have a converse: a homotopy type contains a compact topological manifold if and only if it has sufficient local Poincaré duality. A homotopy equivalence of compact manifolds is homotopic to a homeomorphism if and only if the point inverses are algebraic Poincaré null-cobordant. The bundles and normal maps in the traditional approach are relegated from the statements of the results to the proofs.

An n-*dimensional algebraic Poincaré complex* is a chain complex C with a Poincaré duality chain equivalence $C^{n-*} \simeq C$. Algebraic Poincaré complexes are used here to define the *structure groups* $\mathbb{S}_*(X)$ of a space X. The structure groups are the value groups for the obstructions to the existence and uniqueness problems. The *total surgery obstruction* $s(X) \in \mathbb{S}_n(X)$ of an n-dimensional Poincaré space X is a homotopy invariant such that $s(X) = 0$ if (and for $n \geq 5$ only if) X is homotopy equivalent to a compact n-dimensional manifold. The *structure invariant* $s(f) \in \mathbb{S}_{n+1}(M)$ of a homotopy equivalence $f : N \longrightarrow M$ of compact n-dimensional manifolds is a homotopy invariant such that $s(f) = 0$ if (and for $n \geq 5$ only if) f is h-cobordant to a homeomorphism.

Chain homotopy theory can be used to decide if a map of spaces is a homotopy equivalence: by Whitehead's theorem a map of connected CW complexes $f : X \longrightarrow Y$ is a homotopy equivalence if and only if f induces an isomorphism of the fundamental groups $f_* : \pi_1(X) \longrightarrow \pi_1(Y)$ and a chain equivalence $\tilde{f} : C(\tilde{X}) \longrightarrow C(\tilde{Y})$ of the cellular $\mathbb{Z}[\pi_1(X)]$-module chain complexes of the universal covers $\tilde{X}, \tilde{Y}$ of X, Y. It will be shown here that the cobordism theory of algebraic Poincaré complexes can be similarly used to decide the existence and uniqueness problems in dimensions ≥ 5. A finite Poincaré space X is homotopy equivalent to a compact manifold if and only if the Poincaré duality $\mathbb{Z}[\pi_1(X)]$-module chain equivalence $[X] \cap - : C(\tilde{X})^{n-*} \longrightarrow C(\tilde{X})$ of the universal cover $\tilde{X}$ is induced up to algebraic Poincaré cobordism by a Poincaré duality of a local system of $\mathbb{Z}$-module chain complexes. A homotopy equivalence of compact manifolds f is h-cobordant to a homeomorphism if and only if the chain equivalence $\tilde{f}$ is induced up to algebraic Poincaré cobordism by an equivalence of local systems of $\mathbb{Z}$-module chain complexes. Such results are direct descendants of the h- and s-cobordism theorems, which provided necessary and sufficient cobordism-theoretic and Whitehead torsion conditions for compact manifolds of dimension ≥ 5 to be homeomorphic.

Generically, *assembly* is the passage from a local input to a global output. The input is usually topologically invariant and the output is homotopy invariant. This is the case in the original geometric assembly map of Quinn, and the *algebraic L-theory assembly* map defined here.

The passage from the topology of compact manifolds to the homotopy theory of finite Poincaré spaces is the assembly of particular interest here. In general, it is not possible to reverse the assembly process without some extra geometric hypotheses. Manifolds of a certain type are said to be *rigid* if every homotopy equivalence is homotopic to a homeomorphism, that is if the uniqueness problem has a unique affirmative solution. The classification of surfaces and their homotopy equivalences shows that compact 2-dimensional manifolds are rigid. Haken 3-dimensional manifolds are rigid, by the result of Waldhausen. The Mostow rigidity theorem for symmetric spaces and related results in hyperbolic geometry give the classic instances of higher dimensional manifolds with rigidity. The *Borel conjecture* is that every aspherical Poincaré space $B\pi$ is homotopy equivalent to a compact aspherical topological manifold, and that any homotopy equivalence of such manifolds is homotopic to a homeomorphism. Surgery theory has provided many examples of groups π with sufficient geometry to verify both this conjecture and the closely related *Novikov conjecture* on the homotopy invariance of the higher signatures. The rigidity of aspherical manifolds with fundamental group π is equivalent to the algebraic L-theory assembly map for the classifying space $B\pi$ being an isomorphism. The more complicated homotopy theory of manifolds with non-trivial higher homotopy groups is reflected in non-rigidity, with a corresponding deviation from isomorphism in the algebraic L-theory assembly map.

The Leray homology spectral sequence for a map $f\colon Y \longrightarrow X$ can be viewed as an assembly process, with input the E^2-terms

$$E^2_{p,q} \; = \; H_p(X; \{H_q(f^{-1}(x))\})$$

and output the E^∞-terms associated to $H_*(Y)$. The spectral sequence can be used to prove the Vietoris–Begle mapping theorem: if f is a map between reasonable spaces (such as paracompact polyhedra) with acyclic point inverses $f^{-1}(x)$ ($x \in X$) then f is a homology equivalence. The topologically invariant local condition of f inducing isomorphisms

$$(f|)_* \,:\, H_*(f^{-1}(x)) \xrightarrow{\;\approx\;} H_*(\{x\}) \;\; (x \in X)$$

assembles to the homotopy invariant global condition of f inducing isomorphisms

$$f_* \,:\, H_*(Y) \xrightarrow{\;\approx\;} H_*(X) \,.$$

There is also a cohomology version, with input

$$E_2^{p,q} \; = \; H^p(X; \{H^q(f^{-1}(x))\})$$

and output $H^*(Y)$. The dihomology spectral sequences of Zeeman [188] can be similarly viewed as assembly processes, piecing together the homology (resp. cohomology) of a space X from the cohomology (resp. homology) with coefficients in the local homology (resp. cohomology). The homology version has input

$$E_2^{p,q} = H^p(X; \{H_{n-q}(X, X\backslash\{x\})\})$$

and output $H_{n-*}(X)$, for any $n \in \mathbb{Z}$. The cohomology version has input

$$E_{p,q}^2 = H_p(X; \{H^{n-q}(X, X\backslash\{x\})\})$$

and output $H^{n-*}(X)$.

An n-dimensional homology manifold X is a topological space such that the local homology groups at each point $x \in X$ are the local homology groups of $\mathbb{R}^n$

$$H_*(X, X\backslash\{x\}) = H_*(\mathbb{R}^n, \mathbb{R}^n\backslash\{0\}) = \begin{cases} \mathbb{Z} & \text{if } * = n \\ 0 & \text{if } * \neq n \, . \end{cases}$$

For compact X the local fundamental classes $[X]_x \in H_n(X, X\backslash\{x\})$ assemble to a global fundamental class $[X] \in H_n(X)$, using twisted coefficients in the nonorientable case. The dihomology spectral sequences collapse for a compact homology manifold X, assembling the local Poincaré duality isomorphisms

$$[X]_x \cap - : H^{n-*}(\{x\}) \xrightarrow{\cong} H_*(X, X\backslash\{x\}) \quad (x \in X)$$

to the global Poincaré duality isomorphisms

$$[X] \cap - : H^{n-*}(X) \xrightarrow{\cong} H_*(X) \, .$$

The topologically invariant property of the local homology at each point being that of $\mathbb{R}^n$ is assembled to the homotopy invariant property of n-dimensional Poincaré duality.

The *quadratic L-groups* $L_n(R)$ $(n \geq 0)$ of Wall [176] were expressed in Ranicki [141] as the cobordism groups of quadratic Poincaré complexes (C, ψ) over a ring with involution R, with C a f.g. free R-module chain complex and ψ a quadratic structure inducing Poincaré duality isomorphisms $(1 + T)\psi_0 : H^{n-*}(C) \cong H_*(C)$.

The algebraic L-theory assembly map

$$A : H_*(X; \mathbb{L}.) \longrightarrow L_*(\mathbb{Z}[\pi_1(X)])$$

is a central feature of the combinatorial theory of surgery, with $H_*(X; \mathbb{L}.)$ the generalized homology groups of X with coefficients in the 1-connective quadratic L-theory spectrum $\mathbb{L}.$ of $\mathbb{Z}$. By construction, the structure groups $\mathbb{S}_*(X)$ of a space X are the relative homotopy groups of A, designed to fit into the *algebraic surgery exact sequence*

$$\ldots \longrightarrow H_n(X; \mathbb{L}.) \xrightarrow{A} L_n(\mathbb{Z}[\pi_1(X)]) \xrightarrow{\partial} \mathbb{S}_n(X)$$

$$\longrightarrow H_{n-1}(X; \mathbb{L}.) \longrightarrow \ldots \, .$$

The structure groups $\mathbb{S}_*(X)$ measure the extent to which the surgery obstruction groups $L_*(\mathbb{Z}[\pi_1(X)])$ fail to be a generalized homology theory, or equivalently the extent to which the algebraic L-theory assembly maps A fail to be isomorphisms. The algebraic surgery exact sequence for a compact manifold M is identified in §18 with the Sullivan–Wall surgery exact sequence for the manifold structure set of M.

The total surgery obstruction $s(X) \in \mathbb{S}_n(X)$ of an n-dimensional Poincaré space X is expressed in §17 in terms of a combinatorial formula measuring the failure on the chain level of the local homology groups $H_*(X, X\backslash\{x\})$ ($x \in X$) to be isomorphic to $H^{n-*}(\{x\}) = H_*(\mathbb{R}^n, \mathbb{R}^n\backslash\{0\})$. The condition $s(X) = 0$ is equivalent to the cellular $\mathbb{Z}[\pi_1(X)]$-module chain complex $C(\widetilde{X})$ of the universal cover $\widetilde{X}$ being algebraic Poincaré cobordant to the assembly of a local system over X of $\mathbb{Z}$-module chain complexes with Poincaré duality. The structure invariant $s(f) \in \mathbb{S}_{n+1}(M)$ of a homotopy equivalence $f: N \longrightarrow M$ of compact n-dimensional manifolds is expressed in §18 in terms of a combinatorial formula measuring the failure on the chain level of the local homology groups $H_*(f^{-1}(x))$ ($x \in M$) to be isomorphic to $H_*(\{x\})$. The condition $s(f) = 0$ is equivalent to the algebraic mapping cone $C(\widetilde{f}: C(\widetilde{N}) \longrightarrow C(\widetilde{M}))_{*+1}$ being algebraic Poincaré cobordant to the assembly of a local system over M of contractible $\mathbb{Z}$-module chain complexes.

The algebraic L-theory assembly map is constructed in §9 as a forgetful map between two algebraic Poincaré bordism theories, in which the underlying chain complexes are the same, but which differ in the duality conditions required. There is a strong 'local' condition and a weak 'global' condition, corresponding to the difference between a manifold and a Poincaré space, and between a homeomorphism and a homotopy equivalence. The assembly of a local algebraic Poincaré complex is a global algebraic Poincaré complex, by analogy with the passage from integral to rational quadratic forms in algebra, and from manifolds to Poincaré spaces in topology. The algebraic L-theory assembly maps have the advantage over the analogous topological assembly maps in that their fibres can be expressed in terms of local algebraic Poincaré complexes such that the underlying chain complexes are globally contractible.

The generalized homology groups of a simplicial complex K with L-theory coefficients are identified in §13 with the cobordism groups of local algebraic Poincaré complexes, where local means that there is a simply connected Poincaré duality condition at each simplex in K. The cobordism groups of global algebraic Poincaré complexes are the surgery obstruction groups or some symmetric analogues, where global means that there is a single non-simply connected Poincaré duality condition over the universal cover

$\widetilde{K}$. Surgery theory identifies the fibre of the assembly map from compact manifolds to finite Poincaré spaces in dimensions ≥ 5 with the fibre of the algebraic L-theory assembly map. Picture this identification as a fibre square

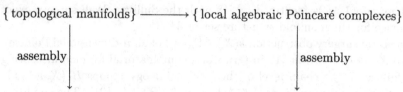

allowing the homotopy types of compact manifolds to be created out of the homotopy types of finite Poincaré spaces and some extra chain level Poincaré duality. The assembly maps forget the local structure, and the fibres of the assembly maps measure the difference between the local and global structures. The fibre square substantiates the suggestion of Siebenmann [157, §14] that 'topological manifolds bear the simplest possible relation to their underlying homotopy types'.

The *surgery obstruction* of a normal map $(f, b) \colon M \longrightarrow X$ from a compact n-dimensional manifold M to a finite n-dimensional Poincaré space X

$$\sigma_*(f, b) \in L_n(\mathbb{Z}[\pi_1(X)])$$

is such that $\sigma_*(f, b) = 0$ if (and for $n \geq 5$ only if) (f, b) is normal bordant to a homotopy equivalence. In the original construction of Wall [176] $\sigma_*(f, b)$ was defined after preliminary geometric surgeries to make (f, b) $[n/2]$-connected. In Ranicki [142] the surgery obstruction was intrepreted as the cobordism class of an n-dimensional quadratic Poincaré complex $(\mathcal{C}(f^!), \psi)$ over $\mathbb{Z}[\pi_1(X)]$ associated directly to (f, b), with

$$f^! \colon C(\widetilde{X}) \simeq C(\widetilde{X})^{n-*} \xrightarrow{\tilde{f}^*} C(\widetilde{M})^{n-*} \simeq C(\widetilde{M})$$

the Umkehr chain map.

The algebraic Poincaré cobordism approach to the quadratic L-groups $L_*(R)$ extends to n-ads, and hence to the definition of a quadratic $\mathbb{L}$-spectrum $\mathbb{L}.(R)$ with homotopy groups

$$\pi_*(\mathbb{L}.(R)) = L_*(R) .$$

In Ranicki [145] the quadratic L-groups $L_n(\mathbb{A})$ $(n \geq 0)$ of n-dimensional quadratic Poincaré complexes were defined for any additive category with involution $\mathbb{A}$, with

$$L_*(R) = L_*(\mathbb{A}(R)) , \quad \mathbb{A}(R) = \{ \text{f.g. free } R\text{-modules} \} .$$

In §1 the quadratic L-groups $L_*(\mathbb{A})$ are defined still more generally, for any additive category $\mathbb{A}$ with a *chain duality*, that is a duality involution on the chain homotopy category.

The chain complex assembly of Ranicki and Weiss [147] provides a convenient framework for dealing with the algebraic L-theory assembly over a simplicial complex K. The method can be extended to arbitrary topological spaces using nerves of open covers.

An (R, K)-*module* M is a f.g. free R-module with a direct sum decomposition

$$M = \sum_{\sigma \in K} M(\sigma)$$

with R a commutative ring. An (R, K)-module morphism $f : M \longrightarrow N$ is an R-module morphism such that

$$f(M(\sigma)) \subseteq \sum_{\tau \geq \sigma} N(\tau) \quad (\sigma \in K) .$$

An (R, K)-module chain complex C is *locally contractible* if it is contractible in the (R, K)-module category, or equivalently if each $C(\sigma)$ $(\sigma \in K)$ is a contractible f.g. R-module chain complex. The *assembly* of an (R, K)-module M is the f.g. free $R[\pi_1(K)]$-module

$$M(\widetilde{K}) = \sum_{\tilde{\sigma} \in \widetilde{K}} M(p(\tilde{\sigma})) ,$$

with $p : \widetilde{K} \longrightarrow K$ the unversal covering projection. An (R, K)-module chain complex C is *globally contractible* if the assembly $C(\widetilde{K})$ is a contractible $R[\pi_1(K)]$-module chain complex. A locally contractible complex is globally contractible, but a globally contractible complex need not be locally contractible.

An n-dimensional quadratic complex (C, ψ) in $\mathbb{A}(R, K)$ is *locally Poincaré* if the algebraic mapping cone of the (R, K)-module chain map $(1 + T)\psi_0 :$ $C^{n-*} \longrightarrow C$ is locally contractible, with each

$$(1 + T)\psi_0(\sigma) : C(\sigma)^{n-|\sigma|-*} \longrightarrow C(\sigma)/\partial C(\sigma) \quad (\sigma \in K)$$

an R-module chain equivalence. (See §5 for the construction of the chain duality on $\mathbb{A}(R, K)$.) An n-dimensional quadratic complex (C, ψ) in $\mathbb{A}(R, K)$ is *globally Poincaré* if the algebraic mapping cone of $(1 + T)\psi_0 : C^{n-*} \longrightarrow C$ is globally contractible, with

$$(1 + T)\psi_0 : C^{n-*}(\widetilde{K}) \simeq C(\widetilde{K})^{n-*} \longrightarrow C(\widetilde{K})$$

an $R[\pi_1(K)]$-module chain equivalence. Chain complexes with local (resp. global) Poincaré duality correspond to manifolds (resp. Poincaré spaces).

The generalized homology groups $H_*(K; \mathbb{L}.(R))$ are the cobordism groups of quadratic locally Poincaré complexes in $\mathbb{A}(R, K)$. The algebraic L-theory assembly map

$$A : H_n(K; \mathbb{L}.(R)) \longrightarrow L_n(R[\pi_1(K)]) ; \ (C, \psi) \longrightarrow (C(\widetilde{K}), \psi(\widetilde{K}))$$

is defined by forgetting the locally Poincaré structure. The geometric assembly map of Quinn [127], [128], [134] pieces together the non-simply connected surgery obstruction of a normal map of closed manifolds from the simply connected pieces. Similarly, the algebraic L-theory assembly map A pieces together a globally Poincaré complex over $R[\pi_1(K)]$ from a locally Poincaré complex in $\mathbb{A}(R, K)$.

The main algebraic construction of the text is the algebraic surgery exact sequence of §14

$$\ldots \longrightarrow H_n(K; \mathbb{L}.(R)) \xrightarrow{A} L_n(R[\pi_1(K)]) \xrightarrow{\partial} \mathbb{S}_n(R, K)$$
$$\longrightarrow H_{n-1}(K; \mathbb{L}.(R)) \longrightarrow \ldots \;.$$

The *quadratic structure groups* $\mathbb{S}_*(R, K)$ are the cobordism groups of quadratic complexes in $\mathbb{A}(R, K)$ which are locally Poincaré and globally contractible.

The algebraic surgery exact sequence is a generalization of the quadratic L-theory localization exact sequence of Ranicki [143, §3]

$$\ldots \longrightarrow L_n(R) \longrightarrow L_n(S^{-1}R) \longrightarrow L_n(R, S) \longrightarrow L_{n-1}(R) \longrightarrow \ldots \;,$$

for the localization $R \longrightarrow S^{-1}R$ of a ring with involution R inverting a multiplicative subset $S \subset R$ of central non-zero divisors invariant under the involution. The relative L-groups $L_*(R, S)$ are the cobordism groups of quadratic Poincaré complexes (C, ψ) over R such that C is an R-module chain complex with localization $S^{-1}C = S^{-1}R \otimes_R C$ a contractible $S^{-1}R$-module chain complex. In the classic case

$$R = \mathbb{Z} \;, \quad S = \mathbb{Z} \backslash \{0\} \;, \quad S^{-1}R = \mathbb{Q}$$

the relative L-groups $L_{2i}(R, S)$ are the Witt groups of $\mathbb{Q}/\mathbb{Z}$-valued $(-)^i$-quadratic forms on finite abelian groups, and $L_{2i+1}(R, S) = 0$.

The quadratic structure groups $\mathbb{S}_*(K)$ are defined in §15 as the 1-connective versions of $\mathbb{S}_*(\mathbb{Z}, K)$, to fit into the algebraic surgery exact sequence

$$\ldots \longrightarrow H_n(K; \mathbb{L}.) \xrightarrow{A} L_n(\mathbb{Z}[\pi_1(K)]) \xrightarrow{\partial} \mathbb{S}_n(K)$$
$$\longrightarrow H_{n-1}(K; \mathbb{L}.) \longrightarrow \ldots$$

with $\mathbb{L}.$ the 1-connective cover of $\mathbb{L}.(\mathbb{Z})$. The 0th space $\mathbb{L}_0$ of $\mathbb{L}.$ is homotopy equivalent to the homotopy fibre G/TOP of the forgetful map $BTOP \longrightarrow BG$ from the classifying space for stable topological bundles to the classifying space for stable spherical fibrations. The homotopy groups of $\mathbb{L}.$ are the simply connected surgery obstruction groups

$$\pi_n(\mathbb{L}.) = \pi_n(G/TOP) = L_n(\mathbb{Z}) = \begin{cases} \mathbb{Z} \\ 0 \\ \mathbb{Z}_2 \\ 0 \end{cases} \text{ if } n \equiv \begin{cases} 0 \\ 1 \\ 2 \\ 3 \end{cases} \pmod 4 \;.$$

The *dual cells* of a simplicial complex K are the subcomplexes of the barycentric subdivision K' defined by

$$D(\sigma, K) = \{ \hat{\sigma}_0 \hat{\sigma}_1 \ldots \hat{\sigma}_r \in K' \mid \sigma \leq \sigma_0 < \sigma_1 < \ldots < \sigma_r \} ,$$

with boundary

$$\partial D(\sigma, K) = \bigcup_{\tau > \sigma} D(\tau, K) .$$

Transversality is functorial in the PL category: Cohen [36] proved that for a simplicial map $f: M \longrightarrow K'$ from a compact n-dimensional PL manifold M the inverse images of the dual cells

$$(M(\sigma), \partial M(\sigma)) = f^{-1}(D(\sigma, K), \partial D(\sigma, K)) \quad (\sigma \in K)$$

are $(n - |\sigma|)$-dimensional PL manifolds with boundary. An abstract version of this transversality is used in §12 to express the groups $h_*(K)$ for any generalized homology theory as the cobordism groups of 'h-cycles in K', which are compatible assignments at each simplex $\sigma \in K$ of a piece of the coefficient group $h_*(\{\text{pt.}\})$. This is the combinatorial analogue of the result that every generalized homology theory is the cobordism of compact manifolds with singularities of a prescribed type (Sullivan [166], Buoncristiano, Rourke and Sanderson [21]).

A *finite n-dimensional geometric Poincaré complex* X is a finite simplicial complex such that the polyhedron is an n-dimensional Poincaré space. The total surgery obstruction of X is defined in §17 to be the cobordism class

$$s(X) = (\Gamma, \psi) \in \mathbb{S}_n(X)$$

of an $(n - 1)$-dimensional quadratic locally Poincaré globally contractible complex (Γ, ψ) in $\mathbb{A}(\mathbb{Z}, X)$ with

$$H_*(\Gamma(\sigma))$$

$$= H_{*+1}(\phi(\sigma) : C(D(\sigma, X))^{n-|\sigma|-*} \longrightarrow C(D(\sigma, X), \partial D(\sigma, X)))$$

$$= H_{*+|\sigma|+1}([X]_x \cap - : C(\{x\})^{n-*} \longrightarrow C(X, X\backslash\{x\}))$$

measuring the failure of local Poincaré duality at the barycentre $x = \hat{\sigma} \in X$ of each simplex $\sigma \in X$. The assembly $(n-1)$-dimensional quadratic Poincaré complex $(\Gamma(\widetilde{X}), \psi(\widetilde{X}))$ over $\mathbb{Z}[\pi_1(X)]$ is contractible, with

$$\Gamma(\widetilde{X}) = \mathcal{C}([X] \cap - : C(\widetilde{X})^{n-*} \longrightarrow C(\widetilde{X}))_{*+1} \simeq 0 .$$

The structure invariant $s(f) \in \mathbb{S}_{n+1}(M)$ of a homotopy equivalence $f: N \longrightarrow M$ of closed n-dimensional manifolds is defined in §18, measuring the extent up to algebraic Poincaré cobordism to which the point inverses $f^{-1}(x)$ are contractible. The invariant is such that $s(f) = 0$ if (and for $n \geq 5$ only if) f is h-cobordant to a homeomorphism. The total surgery obstruction has the following interpretation: for $n \geq 5$ a finite n-dimensional Poincaré space X is homotopy equivalent to a compact topological manifold if and only if the Poincaré duality chain equivalence has 'contractible point-inverses' up to an appropriate cobordism relation.

The *structure set* $\mathbb{S}^{TOP}(X)$ of an n-dimensional Poincaré space X is the set (possibly empty) of h-cobordism classes of pairs

(compact n-dimensional topological manifold M,

homotopy equivalence $f: M \longrightarrow X$) .

The structure set of a compact manifold M is non-empty, with base point $(M, 1) \in \mathbb{S}^{TOP}(M)$.

The structure invariant $s(f) \in \mathbb{S}_{n+1}(M)$ of a homotopy equivalence of compact n-dimensional manifolds $f: N \longrightarrow M$ is defined in §18 to be the cobordism class

$$s(f) \ = \ (\Gamma, \psi) \in \mathbb{S}_{n+1}(M)$$

of an n-dimensional quadratic locally Poincaré complex (Γ, ψ) in $\mathbb{A}(\mathbb{Z}, M)$ with contractible assembly

$$\Gamma(\widetilde{M}) \ = \ \mathcal{C}(\tilde{f}: C(\widetilde{N}) \longrightarrow C(\widetilde{M})) \ \simeq \ 0 \ .$$

The $\mathbb{Z}$-module chain complexes $\Gamma(\sigma)$ $(\sigma \in M)$ are the quadratic Poincaré kernels of the normal maps of $(n - |\sigma|)$-dimensional manifolds with boundary

$$f| \ : \ (gf)^{-1}D(\sigma, M) \ \longrightarrow \ g^{-1}D(\sigma, M) \ \ (\sigma \in M) \ .$$

(For the sake of convenience it is assumed here that M is the polyhedron of a finite simplicial complex, but this assumption is avoided in §18). The structure invariant can also be viewed as the rel ∂ total surgery obstruction

$$s(f) \ = \ s_\partial(W, N \sqcup -M) \in \mathbb{S}_{n+1}(W) \ = \ \mathbb{S}_{n+1}(M)$$

with $(W, N \sqcup -M)$ the finite $(n+1)$-dimensional Poincaré pair with manifold boundary defined by the mapping cylinder $W = N \times I \cup_f M$.

The Sullivan–Wall geometric surgery exact sequence of pointed sets for a compact n-dimensional manifold M with $n \geq 5$

$$\ldots \longrightarrow L_{n+1}(\mathbb{Z}[\pi_1(M)]) \longrightarrow \mathbb{S}^{TOP}(M)$$

$$\longrightarrow [M, G/TOP] \longrightarrow L_n(\mathbb{Z}[\pi_1(M)])$$

is shown in §18 to be isomorphic to the 1-connective algebraic surgery exact sequence of abelian groups

$$\ldots \longrightarrow L_{n+1}(\mathbb{Z}[\pi_1(M)]) \stackrel{\partial}{\longrightarrow} \mathbb{S}_{n+1}(M)$$

$$\longrightarrow H_n(M; \mathbb{L}.) \stackrel{A}{\longrightarrow} L_n(\mathbb{Z}[\pi_1(M)]) \ .$$

The function sending a homotopy equivalence of manifolds to its quadratic structure invariant defines a bijection

$$s \ : \ \mathbb{S}^{TOP}(M) \ \longrightarrow \ \mathbb{S}_{n+1}(M) \ ; \ f \ \longrightarrow \ s(f)$$

between the manifold structure set and the quadratic structure group.

The total surgery obstruction theory also has a version involving Whitehead torsion. A Poincaré space X is *simple* if it has a finite simplicial complex structure in its homotopy type with respect to which

$$\tau([X] \cap -: C(\widetilde{X})^{n-*} \longrightarrow C(\widetilde{X})) \;=\; 0 \in Wh(\pi) \quad (\pi = \pi_1(X)) \,.$$

Compact manifolds are simple Poincaré spaces, with respect to the finite structure given by the handle decomposition. The simple structure groups $\mathbb{S}_*^s(X)$ are defined to fit into the exact sequence

$$\ldots \longrightarrow H_n(X; \mathbb{L}.) \xrightarrow{\;A\;} L_n^s(\mathbb{Z}[\pi]) \xrightarrow{\;\partial\;} \mathbb{S}_n^s(X)$$
$$\longrightarrow H_{n-1}(X; \mathbb{L}.) \longrightarrow \ldots$$

with $L_*^s(\mathbb{Z}[\pi])$ the simple surgery obstruction groups of Wall [176]. The simple structure groups $\mathbb{S}_*^s(X)$ are related to the finite structure groups $\mathbb{S}_*(X)$ by an exact sequence

$$\ldots \longrightarrow \mathbb{S}_n^s(X) \longrightarrow \mathbb{S}_n(X) \longrightarrow \widehat{H}^n(\mathbb{Z}_2\,; Wh(\pi)) \longrightarrow \mathbb{S}_{n-1}^s(X) \longrightarrow \ldots$$

analogous to the Rothenberg exact sequence

$$\ldots \longrightarrow L_n^s(\mathbb{Z}[\pi]) \longrightarrow L_n(\mathbb{Z}[\pi]) \longrightarrow \widehat{H}^n(\mathbb{Z}_2\,; Wh(\pi)) \longrightarrow L_{n-1}^s(\mathbb{Z}[\pi]) \longrightarrow \ldots \,.$$

The *total simple surgery obstruction* $s(X) \in \mathbb{S}_n^s(X)$ of a simple n-dimensional Poincaré space X is such that $s(X) = 0$ if (and for $n \geq 5$ only if) X is simple homotopy equivalent to a compact n-dimensional topological manifold. The *simple structure invariant* $s(f) \in \mathbb{S}_{n+1}^s(M)$ of a simple homotopy equivalence $f: N \longrightarrow M$ of n-dimensional manifolds is such that $s(f) = 0$ if (and for $n \geq 5$ only if) f is s-cobordant to a homeomorphism. For $n \geq 5$'s-cobordant' can be replaced by 'homotopic to', by virtue of the s-cobordism theorem.

The quadratic structure group $\mathbb{S}_n(K)$ of a simplicial complex K is identified in §19 with the bordism group of finite n-dimensional Poincaré pairs $(X, \partial X)$ with a reference map $(f, \partial f): (X, \partial X) \longrightarrow K$ such that $\partial f: \partial X \longrightarrow K$ is Poincaré transverse across the dual cell decomposition of the barycentric subdivision K' of K. From this point of view, the total surgery obstruction of an n-dimensional Poincaré space X is the bordism class

$$s(X) \;=\; (X, \emptyset) \in \mathbb{S}_n(X)$$

with the identity reference map $X \longrightarrow X$. The quadratic structure group $\mathbb{S}_n(X)$ can also be identified with the bordism group of homotopy equivalences $f: N \longrightarrow M$ of compact $(n-1)$-dimensional manifolds, with a reference map $M \longrightarrow X$. The mapping cylinder W of f defines a finite n-dimensional Poincaré h-cobordism $(W, N \sqcup -M)$ with $N \sqcup -M \longrightarrow X$ Poincaré transverse by manifold transversality.

The *symmetric L-groups* $L^n(R)$ $(n \geq 0)$ of Mishchenko [112] and Ranicki [141] are the cobordism groups of n-dimensional symmetric Poincaré complexes (C, ϕ) over R, with duality isomorphisms $\phi_0 \colon H^{n-*}(C) \cong H_*(C)$. The quadratic L-groups are 4-periodic $L_*(R) = L_{*+4}(R)$. The symmetric L-groups are not 4-periodic in general, with symmetrization maps $1 + T \colon L_*(R) \longrightarrow L^*(R)$ which are isomorphisms modulo 8-torsion.

An n-dimensional Poincaré space X has a *symmetric signature*

$$\sigma^*(X) = (C(\widetilde{X}), \phi) \in L^n(\mathbb{Z}[\pi_1(X)])$$

which is homotopy invariant, with

$$\phi_0 = [X] \cap - \colon C(\widetilde{X})^{n-*} \longrightarrow C(\widetilde{X})$$

the Poincaré duality chain equivalence (Mishchenko [112], Ranicki [142]). The surgery obstruction $\sigma_*(f, b)$ of a normal map $(f, b) \colon M \longrightarrow X$ has symmetrization the difference of the symmetric signatures

$$(1 + T)\sigma_*(f, b) = \sigma^*(M) - \sigma^*(X) \in L^n(\mathbb{Z}[\pi_1(X)]) \,.$$

The symmetric L-groups are the homotopy groups of an Ω-spectrum $\mathbb{L}^{\cdot}(R)$ of symmetric Poincaré n-ads over R

$$\pi_*(\mathbb{L}^{\cdot}(R)) = L^*(R) \,.$$

The 0-connective simply connected symmetric $\mathbb{L}$-spectrum $\mathbb{L}^{\cdot} = \mathbb{L}^{\cdot}\langle 0 \rangle(\mathbb{Z})$ is a ring spectrum with homotopy groups

$$\pi_n(\mathbb{L}^{\cdot}) = L^n(\mathbb{Z}) = \begin{cases} \mathbb{Z} \\ \mathbb{Z}_2 \\ 0 \\ 0 \end{cases} \text{if } n \equiv \begin{cases} 0 \\ 1 \\ 2 \\ 3 \end{cases} (\mathrm{mod}\ 4) \,,$$

the 4-periodic symmetric L-groups of $\mathbb{Z}$. The quadratic $\mathbb{L}$-spectrum $\mathbb{L}_.$ is a module spectrum over the symmetric $\mathbb{L}$-spectrum $\mathbb{L}^{\cdot}$.

The symmetrization maps $1 + T \colon L_*(R) \longrightarrow L^*(R)$ fit into an exact sequence

$$\cdots \longrightarrow L_n(R) \xrightarrow{1+T} L^n(R) \xrightarrow{J} \widehat{L}^n(R) \xrightarrow{\partial} L_{n-1}(R) \longrightarrow \cdots$$

with $\widehat{L}^*(R)$ the exponent 8 hyperquadratic L-groups of Ranicki [143]. The 4-periodic versions of the hyperquadratic L-groups are here called the normal L-groups of R

$$NL^*(R) = \varinjlim_k \widehat{L}^{*+4k}(R) \,,$$

in accordance with the geometric theory of normal spaces of Quinn [129] and the algebraic theory of normal complexes of Weiss [182]. The normal $\mathbb{L}$-spectrum $\mathbb{NL}^{\cdot}(R)$ has homotopy groups

$$\pi_*(\mathbb{NL}^{\cdot}(R)) = NL^*(R) \,.$$

The hyperquadratic L-groups of $\mathbb{Z}$ are 4-periodic, so that the normal L-

groups of $\mathbb{Z}$ are given by

$$NL^n(\mathbb{Z}) = \widehat{L}^n(\mathbb{Z}) = \begin{cases} \mathbb{Z}_8 \\ \mathbb{Z}_2 \\ 0 \\ \mathbb{Z}_2 \end{cases} \text{if } n \equiv \begin{cases} 0 \\ 1 \\ 2 \\ 3 \end{cases} \pmod 4 .$$

The simply connected normal $\mathbb{L}$-spectrum $\mathbb{NL}^{\cdot}(\mathbb{Z})$ has a '1/2-connective' version $\widehat{\mathbb{L}}^{\cdot} = \mathbb{NL}^{\cdot}\langle 1/2 \rangle(\mathbb{Z})$, which is 0-connective and fits into a fibration sequence

$$\mathbb{L}_{\cdot} \xrightarrow{1+T} \mathbb{L}^{\cdot} \xrightarrow{J} \widehat{\mathbb{L}}^{\cdot} ,$$

with homotopy groups

$$\pi_n(\widehat{\mathbb{L}}^{\cdot}) = \begin{cases} L^0(\mathbb{Z}) = \mathbb{Z} & \text{if } n = 0 \\ \text{im}(1 + T: L_1(\mathbb{Z}) \longrightarrow L^1(\mathbb{Z})) = 0 & \text{if } n = 1 \\ \widehat{L}^n(\mathbb{Z}) & \text{if } n \geq 2 . \end{cases}$$

The normal $\mathbb{L}$-spectrum $\widehat{\mathbb{L}}^{\cdot}$ is a ring spectrum, which rationally is just the $\mathbb{Q}$-coefficient homology spectrum $\widehat{\mathbb{L}}^{\cdot} \otimes \mathbb{Q} \simeq K.(\mathbb{Q}, 0)$.

A $(k-1)$-spherical fibration $\nu: X \longrightarrow BG(k)$ has a *canonical $\widehat{\mathbb{L}}^{\cdot}$-orientation* $\widehat{U}_\nu \in \dot{H}^k(T(\nu); \widehat{\mathbb{L}}^{\cdot})$, with $T(\nu)$ the Thom space of ν and $\dot{H}$ denoting reduced cohomology with $w_1(\nu)$-twisted coefficients. The fibration sequence $\mathbb{L}_{\cdot} \longrightarrow \mathbb{L}^{\cdot} \longrightarrow \widehat{\mathbb{L}}^{\cdot}$ induces an exact sequence of cohomology groups

$$\cdots \longrightarrow \dot{H}^k(T(\nu); \mathbb{L}_{\cdot}) \xrightarrow{1+T} \dot{H}^k(T(\nu); \mathbb{L}^{\cdot}) \xrightarrow{J} \dot{H}^k(T(\nu); \widehat{\mathbb{L}}^{\cdot})$$

$$\xrightarrow{\delta} \dot{H}^{k+1}(T(\nu); \mathbb{L}_{\cdot}) \longrightarrow \cdots .$$

A topological block bundle $\tilde{\nu}: X \longrightarrow \widetilde{BTOP}(k)$ has a *canonical $\mathbb{L}^{\cdot}$-orientation* $U_{\tilde{\nu}} \in \dot{H}^k(T(\nu); \mathbb{L}^{\cdot})$, with $\nu = J\tilde{\nu}: X \longrightarrow BG(k)$. It was proved in Levitt and Ranicki [91] that $\nu: X \longrightarrow BG(k)$ admits a topological block bundle reduction $\tilde{\nu}: X \longrightarrow \widetilde{BTOP}(k)$ if and only if there exists a $\mathbb{L}^{\cdot}$-orientation $U_{\tilde{\nu}} \in \dot{H}^k(T(\nu); \mathbb{L}^{\cdot})$ such that

$$J(U_\nu) = \widehat{U}_\nu \in \text{im}(J: \dot{H}^k(T(\nu); \mathbb{L}^{\cdot}) \longrightarrow \dot{H}^k(T(\nu); \widehat{\mathbb{L}}^{\cdot}))$$

$$= \ker(\delta: \dot{H}^k(T(\nu); \widehat{\mathbb{L}}^{\cdot}) \longrightarrow \dot{H}^{k+1}(T(\nu); \mathbb{L}_{\cdot})) .$$

Thus $\delta(\widehat{U}_\nu) \in \dot{H}^{k+1}(T(\nu); \mathbb{L}_{\cdot})$ is the obstruction to the existence of a topological block bundle structure on ν. If this vanishes and $k \geq 3$ the structures are classified by the elements of the abelian group

$$\dot{H}^k(T(\nu); \mathbb{L}_{\cdot}) = H^0(X; \mathbb{L}_{\cdot}) = [X, G/TOP] = [X, G(k)/\widetilde{TOP}(k)] .$$

Rationally, the symmetric $\mathbb{L}$-spectrum of $\mathbb{Z}$ has the homotopy type of a wedge of Eilenberg-MacLane spectra

$$\mathbb{L}^{\cdot} \otimes \mathbb{Q} \simeq \bigvee_{j \geq 0} K.(\mathbb{Q}, 4j) ,$$

and the $\mathbb{L}$-orientation of an oriented topological block bundle $\tilde{\nu}\colon X \longrightarrow$ $\widetilde{BSTOP}(k)$ coincides with the inverse of the Hirzebruch $\mathcal{L}$-genus

$$U_{\tilde{\nu}} \otimes \mathbb{Q} = \mathcal{L}^{-1}(\tilde{\nu}) = \mathcal{L}(-\tilde{\nu}) \in \dot{H}^k(T(\nu); \mathbb{L}^{\cdot}) \otimes \mathbb{Q} = \sum_{j \geq 0} H^{4j}(X; \mathbb{Q}),$$

since both are determined by the signatures of submanifolds. See Taylor and Williams [169] for a general account of the homotopy theory of the algebraic $\mathbb{L}$-spectra, and for an exposition of the work of Morgan and Sullivan [116] and Wall [178] on surgery characteristic classes for manifolds and normal maps in terms of the algebraic $\mathbb{L}$-spectra.

An n-dimensional Poincaré space X has a Spivak normal structure

$$(\nu_X \colon X \longrightarrow BG(k), \rho_X \colon S^{n+k} \longrightarrow T(\nu_X))$$

with ν_X the normal $(k-1)$-spherical fibration defined by a closed regular neighbourhood $(W, \partial W)$ of an embedding $X \subset S^{n+k}$ (k large)

$$S^{k-1} \longrightarrow \partial W \longrightarrow W \simeq X$$

and ρ_X the collapsing map

$$\rho_X \colon S^{n+k} \longrightarrow S^{n+k}/\mathrm{cl}(S^{n+k} \backslash W) = W/\partial W = T(\nu_X).$$

The total surgery obstruction $s(X) \in \mathbb{S}_n(X)$ has image

$$t(X) = \delta(\widehat{U}_{\nu_X}) \in H_{n-1}(X; \mathbb{L}.) = \dot{H}^{k+1}(T(\nu_X); \mathbb{L}.),$$

the obstruction to lifting $\nu_X \colon X \longrightarrow BG(k)$ to a topological block bundle $\tilde{\nu}_X \colon X \longrightarrow \widetilde{BTOP}(k)$. A particular choice of lift $\tilde{\nu}_X$ corresponds to a bordism class of normal maps $(f, b)\colon M \longrightarrow X$ with M a closed n-dimensional manifold, by the Browder–Novikov transversality construction on $\rho_X \colon S^{n+k} \longrightarrow T(\nu_X) = T(\tilde{\nu}_X)$, with

$$f = \rho_X| \colon M = (\rho_X)^{-1}(X) \longrightarrow X, \quad b \colon \nu_M \longrightarrow \tilde{\nu}_X,$$

$$s(X) = \partial \sigma_*(f, b) \in \ker(\mathbb{S}_n(X) \longrightarrow H_{n-1}(X; \mathbb{L}.))$$

$$= \mathrm{im}(\partial \colon L_n(\mathbb{Z}[\pi_1(X)]) \longrightarrow \mathbb{S}_n(X)).$$

It follows that $s(X) = 0$ if and only if there exists a normal map $(f, b)\colon M \longrightarrow X$ with surgery obstruction

$$\sigma_*(f, b) \in \ker(\partial \colon L_n(\mathbb{Z}[\pi_1(X)]) \longrightarrow \mathbb{S}_n(X))$$

$$= \mathrm{im}(A \colon H_n(X; \mathbb{L}.) \longrightarrow L_n(\mathbb{Z}[\pi_1(X)])).$$

This is just the condition for the existence of a topological reduction $\tilde{\nu}_X$ such that the corresponding bordism class of normal maps $(f, b)\colon M \longrightarrow X$ has $\sigma_*(f, b) = 0 \in L_n(\mathbb{Z}[\pi_1(X)])$. For $n \geq 5$ this is the necessary and sufficient condition given by the Browder–Novikov–Sullivan–Wall theory for the existence of a topological manifold in the homotopy type of X. The theory has been extended to the case $n = 4$, provided the fundamental group $\pi_1(X)$ is not too large (Freedman and Quinn [53]).

A closed n-dimensional manifold M has a topologically invariant *canonical* $\mathbb{L}^{\cdot}$*-homology fundamental class* $[M]_{\mathbb{L}} \in H_n(M; \mathbb{L}^{\cdot})$ which assembles to the symmetric signature

$$A([M]_{\mathbb{L}}) = \sigma^*(M) \in L^n(\mathbb{Z}[\pi_1(M)]) .$$

Cap product with $[M]_{\mathbb{L}}$ defines the Poincaré duality isomorphism

$$[M]_{\mathbb{L}} \cap - : [M, G/TOP] = H^0(M; \mathbb{L}.) \longrightarrow H_n(M; \mathbb{L}.)$$

which is used in the identification of the algebraic and geometric surgery sequences.

A normal map $(f, b): N \longrightarrow M$ of closed n-dimensional manifolds has a *normal invariant*

$$[f, b]_{\mathbb{L}} \in H_n(M; \mathbb{L}.) = H^0(M; \mathbb{L}.) = [M, G/TOP]$$

with assembly the surgery obstruction

$$A([f, b]_{\mathbb{L}}) = \sigma_*(f, b) \in L_n(\mathbb{Z}[\pi_1(M)]) ,$$

and symmetrization the difference of the $\mathbb{L}^{\cdot}$-homology fundamental classes

$$(1 + T)[f, b]_{\mathbb{L}} = f_*[N]_{\mathbb{L}} - [M]_{\mathbb{L}} \in H_n(M; \mathbb{L}^{\cdot}) .$$

The localization away from 2 of the $\mathbb{L}^{\cdot}$-orientation $[M]_{\mathbb{L}} \in H_n(M; \mathbb{L}^{\cdot})$ of a closed n-dimensional manifold M

$$[M]_{\mathbb{L}} \otimes \mathbb{Z}[1/2] \in H_n(M; \mathbb{L}^{\cdot}) \otimes \mathbb{Z}[1/2] = KO_n(M) \otimes \mathbb{Z}[1/2]$$

is the $KO[1/2]$-orientation of Sullivan [164]. Rationally

$$[M]_{\mathbb{L}} \otimes \mathbb{Q} = [M]_{\mathbb{Q}} \cap \mathcal{L}(M) = \sum_{k \geq 0}([M]_{\mathbb{Q}} \cap \mathcal{L}_k(M))$$

$$\in H_n(M; \mathbb{L}^{\cdot}) \otimes \mathbb{Q} = \sum_{k \geq 0} H_{n-4k}(M; \mathbb{Q})$$

is the Poincaré dual of the $\mathcal{L}$-genus $\mathcal{L}(M) = \mathcal{L}(\tau_M) \in H^{4*}(M; \mathbb{Q})$ of the stable tangent bundle $\tau_M = -\nu_M: M \longrightarrow BSTOP$, with $[M]_{\mathbb{Q}} \in H_n(M; \mathbb{Q})$ the rational fundamental class. Let $(f, b): N \longrightarrow M$ be a normal map of closed n-dimensional topological manifolds, as classified by a map $c: M \longrightarrow G/TOP$ such that

$$(f^{-1})^* \nu_N - \nu_M : M \xrightarrow{c} G/TOP \longrightarrow BTOP .$$

The rational surgery obstruction of (f, b) is the assembly

$$\sigma_*(f, b) \otimes \mathbb{Q} = A([f, b]_{\mathbb{L}} \otimes \mathbb{Q}) \in L_n(\mathbb{Z}[\pi_1(M)]) \otimes \mathbb{Q}$$

of the element

$$[f, b]_{\mathbb{L}} \otimes \mathbb{Q} = f_*[N]_{\mathbb{L}} \otimes \mathbb{Q} - [M]_{\mathbb{L}} \otimes \mathbb{Q}$$

$$= [M]_{\mathbb{Q}} \cap (\mathcal{L}(M) \cup (\mathcal{L}(c) - 1))$$

$$\in H_n(M; \mathbb{L}^{\cdot}) \otimes \mathbb{Q} = \sum_{k \geq 0} H_{n-4k}(M; \mathbb{Q}) ,$$

with 0 component in $H_n(M; \mathbb{Q})$.

The *symmetric structure groups* $\mathbb{S}^*(X)$ are defined to fit into an exact sequence of abelian groups

$$\dots \longrightarrow H_n(X; \mathbb{L}^{\cdot}) \xrightarrow{A} L^n(\mathbb{Z}[\pi_1(X)]) \xrightarrow{\partial} \mathbb{S}^n(X)$$
$$\longrightarrow H_{n-1}(X; \mathbb{L}^{\cdot}) \longrightarrow \dots .$$

The symmetrization of the total surgery obstruction $s(X) \in \mathbb{S}_n(X)$ of an n-dimensional Poincaré space X is the image of the symmetric signature $\sigma^*(X) \in L^n(\mathbb{Z}[\pi_1(X)])$

$$(1 + T)s(X) = \partial\sigma^*(X) \in \mathbb{S}^n(X) .$$

Thus $(1 + T)s(X) = 0 \in \mathbb{S}^n(X)$ if and only if there exists an $\mathbb{L}^{\cdot}$-homology fundamental class $[X]_{\mathbb{L}} \in H_n(X; \mathbb{L}^{\cdot})$ with assembly the symmetric signature of X

$$A([X]_{\mathbb{L}}) = \sigma^*(X) \in L^n(\mathbb{Z}[\pi_1(X)]) .$$

The *visible symmetric L-groups* $VL^*(R[\pi])$ of Weiss [183] are defined for any commutative ring R and group π, with similar properties to $L^*(R[\pi])$. The visible analogues of the normal L-groups can be expressed as generalized homology groups of the group π with coefficients in $\mathbb{NL}^{\cdot}(R)$, so that there is defined an exact sequence

$$\dots \longrightarrow L_n(R[\pi]) \xrightarrow{1+T} VL^n(R[\pi]) \longrightarrow H_n(B\pi; \mathbb{NL}^{\cdot}(R))$$
$$\xrightarrow{\partial} L_{n-1}(R[\pi]) \longrightarrow \dots .$$

The 1/2-connective visible symmetric L-groups $VL^*(X) = VL^*\langle 1/2\rangle(\mathbb{Z}, X)$ are defined in §15 to fit into a commutative braid of exact sequences

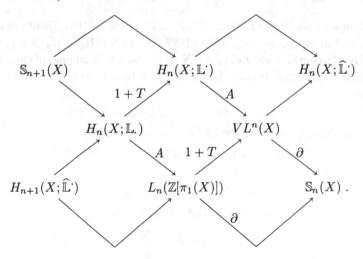

The visible symmetric L-groups $VL^*(B\pi)$ of a classifying space $B\pi$ are the versions of $VL^*(\mathbb{Z}[\pi])$ in which the chain complexes are required to be 0-connective (= positive) and the Poincaré duality chain equivalences are required to be locally 1-connected.

An n-dimensional Poincaré space X has a *1/2-connective visible symmetric signature*

$$\sigma^*(X) = (C, \phi) \in VL^n(X)$$

with assembly the symmetric signature

$$\sigma^*(X) = (C(\widetilde{X}), \phi(\widetilde{X})) \in L^n(\mathbb{Z}[\pi_1(X)]) .$$

The main geometric result of the text is the expression in §17 of the total surgery obstruction of X in terms of the 1/2-connective visible symmetric signature

$$s(X) = \partial\sigma^*(X) \in \mathbb{S}_n(X) .$$

Thus $s(X) = 0 \in \mathbb{S}_n(X)$ if and only if there exists an $\mathbb{L}^{\cdot}$-homology fundamental class $[X]_{\mathbb{L}} \in H_n(X; \mathbb{L}^{\cdot})$ with assembly the 1/2-connective visible symmetric signature

$$A([X]_{\mathbb{L}}) = \sigma^*(X) \in VL^n(X) .$$

The simply connected symmetric signature of an oriented $4k$-dimensional Poincaré space X is just the signature (alias index)

$$\sigma^*(X) = \text{signature}\,(X)$$
$$= \text{signature}\,(H^{2k}(X; \mathbb{Q}), \phi) \in L^{4k}(\mathbb{Z}) = \mathbb{Z} ,$$

with ϕ the nonsingular symmetric form

$$\phi : H^{2k}(X; \mathbb{Q}) \times H^{2k}(X; \mathbb{Q}) \longrightarrow \mathbb{Q} ; \ (x, y) \longmapsto \langle x \cup y, [X]_{\mathbb{Q}}\rangle .$$

The Hirzebruch formula expresses the signature of an oriented $4k$-dimensional manifold M as

$$\text{signature}\,(M) = \langle \mathcal{L}_k(M), [M]_{\mathbb{Q}}\rangle \in \mathbb{Z} \subset \mathbb{Q} ,$$

with $\mathcal{L}_k(M) \in H^{4k}(M; \mathbb{Q})$ the $4k$-dimensional component of the $\mathcal{L}$-genus $\mathcal{L}(M) = \mathcal{L}(\tau_M) \in H^{4*}(M; \mathbb{Q})$, and $[M]_{\mathbb{Q}} \in H_{4k}(M; \mathbb{Q})$ the rational fundamental class. This is a special case of $\sigma^*(M) = A([M]_{\mathbb{L}})$, since the signature of M in $L^{4k}(\mathbb{Z}) = \mathbb{Z}$ is the clockwise image of the fundamental $\mathbb{L}^{\cdot}$-homology class $[M]_{\mathbb{L}} \in H_{4k}(M; \mathbb{L}^{\cdot})$ in the commutative square

$$
\begin{array}{ccc}
H_{4k}(M; \mathbb{L}^{\cdot}) & \xrightarrow{\ \ A\ \ } & L^{4k}(\mathbb{Z}[\pi_1(M)]) \\
\downarrow & & \downarrow \\
H_{4k}(\{*\}; \mathbb{L}^{\cdot}) & \xrightarrow{\ \ A\ \ } & L^{4k}(\mathbb{Z})
\end{array}
$$

and the anticlockwise image is the evaluation $\langle \mathcal{L}_k(M), [M]_{\mathbb{Q}} \rangle$.

Let X be a simply connected $4k$-dimensional Poincaré space. If the Spivak normal fibration $\nu_X : X \longrightarrow BSG$ admits a topological reduction $\tilde{\nu}_X : X \longrightarrow BSTOP$ there exists a normal map $(f, b) : (M, \nu_M) \longrightarrow (X, \tilde{\nu}_X)$ from a $4k$-dimensional manifold M, with surgery obstruction the difference between the evaluation of the $\mathcal{L}$-genus of $\tilde{\nu}_X$ on $[X]_{\mathbb{Q}} \in H_{4k}(X; \mathbb{Q})$ and the signature of X

$$\sigma_*(f, b) = (\sigma^*(M) - \sigma^*(X))/8$$
$$= (\langle \mathcal{L}_k(-\tilde{\nu}_X), [X]_{\mathbb{Q}} \rangle - \text{signature}\,(X))/8 \in L_{4k}(\mathbb{Z}) = \mathbb{Z}\,.$$

There exists a manifold M^{4k} with a normal homotopy equivalence (f, b): $(M, \nu_M) \longrightarrow (X, \tilde{\nu}_X)$ if and only if there exists a topological reduction $\tilde{\nu}_X$ such that X satisfies the Hirzebruch signature formula with respect to $\tilde{\nu}_X$. The simply-connected assembly map $A : H_{4k}(X; \mathbb{L}.) \longrightarrow L_{4k}(\mathbb{Z})$ is onto, so that

$$\mathbb{S}_{4k}(X) \longrightarrow H_{4k-1}(X; \mathbb{L}.)\ ;\ s(X) \longrightarrow t(X)$$

is one-one. The total surgery obstruction of X is such that $s(X) = 0 \in \mathbb{S}_{4k}(X)$ if and only if the topological reducibility obstruction is $t(X) = 0 \in H_{4k-1}(X; \mathbb{L}.)$. Thus X is homotopy equivalent to a manifold if and only if ν_X admits a topological reduction (Browder [16] for $k \geq 2$, Freedman and Quinn [53] for $k = 1$). Moreover, it follows from the computation $L_{4k+1}(\mathbb{Z}) = 0$ that if X is homotopy equivalent to a manifold M^{4k} the structure set of M is in one-one correspondence with the set of topological reductions $\tilde{\nu}_X$ satisfying the formula, namely

$$\mathbb{S}^{TOP}(M) = \mathbb{S}_{4k+1}(X) = \ker(A : H_{4k}(X; \mathbb{L}.) \longrightarrow L_{4k}(\mathbb{Z}))$$
$$\subseteq H_{4k}(X; \mathbb{L}.) = H^0(X; \mathbb{L}.) = [X, G/TOP]\,.$$

The symmetric L-theory assembly map for any connected space M factors through the generalized homology of the fundamental group $\pi_1(M) = \pi$

$$A : H_*(M; \mathbb{L}^{\cdot}) \xrightarrow{\ f_*\ } H_*(B\pi; \mathbb{L}^{\cdot}) \xrightarrow{\ A_\pi\ } L^n(\mathbb{Z}[\pi])$$

with $f : M \longrightarrow B\pi$ the map classifying the universal cover, and A_π the assembly map for the classifying space $B\pi$. (There is a corresponding factorization of the quadratic L-theory assembly map). The $\mathbb{L}^{\cdot}$-homology fundamental class of an n-dimensional manifold M assembles to the symmetric signature

$$A([M]_{\mathbb{L}}) = A_\pi(f_*[M]_{\mathbb{L}}) = \sigma^*(M) \in \text{im}(A_\pi) \subseteq L^n(\mathbb{Z}[\pi])\,.$$

The evaluation map

$$H_{n-4*}(B\pi; \mathbb{Q}) \longrightarrow \text{Hom}_{\mathbb{Q}}(H^{n-4*}(B\pi; \mathbb{Q}), \mathbb{Q})$$

(which is an isomorphism if $H_*(B\pi; \mathbb{Q})$ is finitely generated) sends

$$f_*[M]_{\mathbb{L}} \otimes \mathbb{Q} = \sum_{k \geq 0} f_*([M]_{\mathbb{Q}} \cap \mathcal{L}_k(M))$$

$$\in H_n(B\pi; \mathbb{L}^{\cdot}) \otimes \mathbb{Q} = \sum_{k \geq 0} H_{n-4k}(B\pi; \mathbb{Q})$$

to the higher signatures of M, which are the $\mathbb{Q}$-linear morphisms defined by

$$H^{n-4*}(B\pi; \mathbb{Q}) \longrightarrow \mathbb{Q} \; ; \; x \longrightarrow \langle \mathcal{L}(M) \cup f^*x, [M]_{\mathbb{Q}} \rangle = \langle x, f_*[M]_{\mathbb{L}} \otimes \mathbb{Q} \rangle \; .$$

The assembly of $f_*[M]_{\mathbb{L}} \otimes \mathbb{Q}$ is the rational symmetric signature of M

$$A_\pi(f_*[M]_{\mathbb{L}}) \otimes \mathbb{Q} = \sigma^*(M) \otimes \mathbb{Q}$$

$$\in \mathrm{im}(A_\pi \otimes \mathbb{Q} : H_n(B\pi; \mathbb{L}^{\cdot}) \otimes \mathbb{Q} \longrightarrow L^n(\mathbb{Z}[\pi]) \otimes \mathbb{Q}) \; .$$

For finite π and $n \equiv 0 \pmod 2$ this is just the special case of the Atiyah–Singer index theorem which states that the π-signature of the free action of π on the universal cover $\widetilde{M}$ of a closed manifold M with $\pi_1(M) = \pi$ is a multiple of the regular representation of π. See §22 for the connection between the symmetric signature and the π-signature.

The Novikov conjecture on the homotopy invariance of the higher signatures of manifolds M with $\pi_1(M) = \pi$ is equivalent to the injectivity of the rational assembly map $A_\pi \otimes \mathbb{Q} : H_*(B\pi; \mathbb{L}^{\cdot}) \otimes \mathbb{Q} \longrightarrow L^*(\mathbb{Z}[\pi]) \otimes \mathbb{Q}$.

For a finitely presented group π and $n \geq 5$ every element of the $\mathbb{L}.$-homology group $H_n(B\pi; \mathbb{L}.)$ of the classifying space $B\pi$ is the image of the normal invariant $[f, b]_{\mathbb{L}} \in H_n(M; \mathbb{L}.)$ of a normal map $(f, b) : N \longrightarrow M$ of closed n-dimensional manifolds with $\pi_1(M) = \pi$. Every element of $\mathbb{S}_{n+1}(B\pi)$ is the image of the structure invariant $s(f) \in \mathbb{S}_{n+1}(M)$ of a homotopy equivalence $f : N \longrightarrow M$ of closed n-dimensional manifolds with $\pi_1(M) = \pi$. The kernel of the quadratic L-theory assembly map A_π

$$\ker(A_\pi : H_*(B\pi; \mathbb{L}.) \longrightarrow L_*(\mathbb{Z}[\pi])) = \mathrm{im}(\mathbb{S}_{*+1}(B\pi) \longrightarrow H_*(B\pi; \mathbb{L}.))$$

consists of the images of the structure invariants $s(f)$ of homotopy equivalences $f : N \longrightarrow M$ of closed manifolds with fundamental group $\pi_1(M) = \pi$. The image of the assembly map

$$\mathrm{im}(A_\pi : H_*(B\pi; \mathbb{L}.) \longrightarrow L_*(\mathbb{Z}[\pi])) = \ker(L_*(\mathbb{Z}[\pi]) \longrightarrow \mathbb{S}_*(B\pi))$$

consists of the surgery obstructions of normal maps of closed manifolds with fundamental group π. The image of A_π for finite π was determined by Hambleton, Milgram, Taylor and Williams [66] and Milgram [106].

The ultimate version of the algebraic L-theory assembly should be topologically invariant, using the language of sheaf theory to dispense with the combinatorial constructions, i.e. replacing the simplicial chain complex by the singular chain complex. From this point of view the total surgery obstruction $s(X) \in \mathbb{S}_n(X)$ of an n-dimensional Poincaré space X would measure the failure of a morphism of chain complexes of sheaves inducing the maps

$$[X] \cap - : H^{n-*}(\{x\}) \longrightarrow H_*(X, X \backslash \{x\}) \quad (x \in X)$$

to be a quasi-isomorphism, up to the appropriate sheaf cobordism relation. Although the text is primarily concerned with the applications of algebraic Poincaré complexes to the topology of manifolds and Poincaré spaces, there are also applications to the topology of singular and stratified spaces, as well as to group actions on manifolds – see Zeeman [188], Sullivan [166], McCrory [103], Goresky and MacPherson [59], [60], Siegel [159], Goresky and Siegel [61], Pardon [122], Cappell and Shaneson [27], Cappell and Weinberger [30] and Weinberger [181]. Indeed, the first version of the intersection homology theory of Goresky and MacPherson [59] used the combinatorial methods of PL topology, while the second version [60] used topologically invariant chain complexes of sheaves.

Summary

§1 develops the L-theory of algebraic Poincaré complexes in an additive category with chain duality. §2 deals with the algebraic analogue of the Spivak normal fibration. An 'algebraic bordism category' $(\mathbb{A}, \mathbb{B}, \mathbb{C})$ is an additive category with chain duality $\mathbb{A}$, together with a pair $(\mathbb{B}, \mathbb{C} \subseteq \mathbb{B})$ of subcategories of the chain homotopy category of $\mathbb{A}$. In §3 the quadratic L-groups $L_n(\mathbb{A}, \mathbb{B}, \mathbb{C})$ ($n \in \mathbb{Z}$) are defined to be the cobordism groups of finite chain complexes in $\mathbb{B}$ with an n-dimensional quadratic $\mathbb{C}$-Poincaré duality. The quadratic L-groups $L_*(R)$ of a ring with involution R are the quadratic L-groups $L_*(\Lambda(R))$ of the algebraic bordism category $\Lambda(R) = (\mathbb{A}(R), \mathbb{B}(R), \mathbb{C}(R))$ with $\mathbb{B}(R)$ the category of finite chain complexes in $\mathbb{A}(R)$, and $\mathbb{C}(R)$ the category of contractible chain complexes in $\mathbb{A}(R)$. The additive category $\mathbb{A}_*(X)$ is defined in §4, for any additive category $\mathbb{A}$ and simplicial complex X. In §5 a chain duality on $\mathbb{A}$ is extended to a chain duality on $\mathbb{A}_*(X)$. The simply connected assembly functor $\mathbb{A}_*(X) \longrightarrow \mathbb{A}$ is defined in §6. The chain duality on $\mathbb{A}_*(X)$ has a dualizing complex with respect to a derived Hom, which is obtained in §7. The chain duality on $\mathbb{A}_*(X)$ is used in §8 to extend an algebraic bordism category $(\mathbb{A}, \mathbb{B}, \mathbb{C})$ to an algebraic bordism category $(\mathbb{A}_*(X), \mathbb{B}_*(X), \mathbb{C}_*(X))$ depending co-variantly on X, as a kind of '$(\mathbb{A}, \mathbb{B}, \mathbb{C})$-coefficient algebraic bordism category of X'. The algebraic bordism category obtained in this way from $(\mathbb{A}(R), \mathbb{B}(R), \mathbb{C}(R))$ is denoted by $(\mathbb{A}(R, X), \mathbb{B}(R, X), \mathbb{C}(R)_*(X))$. The assembly functor $\mathbb{A}(R, X) \longrightarrow \mathbb{A}(R[\pi_1(X)])$ is defined in §9. In §10 this is used to define an algebraic bordism category $(\mathbb{A}(R, X), \mathbb{B}(R, X), \mathbb{C}(R, X))$, with $\mathbb{C}(R, X)$ the chain homotopy category of finite chain complexes in $\mathbb{A}(R, X)$ which assemble to a contractible chain complex in $\mathbb{A}(R[\pi_1(X)])$. An algebraic analogue of the π-π theorem of Wall [176] is used in §10 to identify the 'geometric' L-groups $L_*(\mathbb{A}(R, X), \mathbb{B}(R, X), \mathbb{C}(R, X))$ with the 'algebraic' L-groups $L_*(R[\pi_1(X)])$. The theory of Δ-sets is recalled in §11, and applied to generalized homology theory in §12. The quadratic $\mathbb{L}$-spectrum $\mathbb{L}.(\mathbb{A}, \mathbb{B}, \mathbb{C})$ of an algebraic bordism category $(\mathbb{A}, \mathbb{B}, \mathbb{C})$ is defined in §13 to be an Ω-spectrum of Kan Δ-sets with homotopy groups $\pi_*(\mathbb{L}.(\mathbb{A}, \mathbb{B}, \mathbb{C})) = L_*(\mathbb{A}, \mathbb{B}, \mathbb{C})$. The quadratic L-groups $L_*(\mathbb{A}(R, X), \mathbb{B}(R, X), \mathbb{C}(R)_*(X))$ are identified in §13 with the generalized homology groups $H_*(X; \mathbb{L}.(R))$. The braid relating the visible L-groups, the quadratic L-groups and the generalized homology with L-theory coefficients is constructed in §14, with a connective version in §15. The symmetric L-theory orientations of topological bundles and manifolds are constructed in §16. The theory developed in §1-§16 is applied in §17 to obtain the total surgery obstruction $s(X)$ and in §18 to give an algebraic description of the structure set $\mathbb{S}^{TOP}(M)$. In §19 the total surgery obstruction is identified with the obstruction to geometric Poincaré transversality. §20 deals with the simply connected case. The transfer properties of the total surgery obstruction are described in §21. The rational part of the total surgery obstruction in the case when the fundamental group is finite is computed in §22 in terms of the multisignature invariant, and this is used to construct the simplest examples of Poincaré spaces with non-zero total surgery obstruction. §23 relates the total surgery obstruction to splitting obstructions along submanifolds. §24 expresses the total surgery obstruction $s(X) \in \mathbb{S}_n(X)$ of an aspherical n-dimensional Poincaré space $X = B\pi$ satisfying the Novikov conjectures in terms of codimension n signatures. §25 deals with the 4-periodic version of the total surgery obstruction, which applies to the surgery classification of compact ANR homology manifolds. §26 considers the version of the theory appropriate to surgery with coefficients. Appendix A develops the nonorientable case of the theory. Appendix B deals with an alternative construction of assembly in L-theory, using products. Appendix C relates assembly to bounded surgery theory.

Part I

Algebra

§1. Algebraic Poincaré complexes

A chain duality (1.1) on an additive category $\mathbb{A}$ is a generalization of an involution on $\mathbb{A}$, in which the dual of an object in $\mathbb{A}$ is a chain complex in $\mathbb{A}$. A chain duality determines an involution on the derived category of chain complexes in $\mathbb{A}$ and chain homotopy classes of chain maps, allowing the definition of an n-dimensional algebraic Poincaré complex in $\mathbb{A}$ as a finite chain complex which is chain equivalent to its n-dual. The $\begin{cases} \text{symmetric} \\ \text{quadratic} \end{cases}$ L-groups $\begin{cases} L^*(\mathbb{A}) \\ L_*(\mathbb{A}) \end{cases}$ are defined to be the cobordism groups of $\begin{cases} \text{symmetric} \\ \text{quadratic} \end{cases}$ Poincaré complexes in $\mathbb{A}$. As already noted in the Introduction, geometric Poincaré complexes have a symmetric signature in $L^*(\mathbb{A})$ and normal maps have a quadratic signature (= surgery obstruction) in $L_*(\mathbb{A})$ for $\mathbb{A} = \{$ f.g. free $\mathbb{Z}[\pi]$-modules$\}$ with the standard duality involution, with π the fundamental group.

Let then $\mathbb{A}$ be an additive category. A chain complex in $\mathbb{A}$

$$C : \ldots \longrightarrow C_{r+1} \xrightarrow{d} C_r \xrightarrow{d} C_{r-1} \longrightarrow \ldots \quad (r \in \mathbb{Z})$$

is *finite* if $C_r = 0$ for all but a finite number of $r \in \mathbb{Z}$. C is *n-dimensional* if $C_r = 0$ unless $0 \leq r \leq n$.

The *algebraic mapping cone* of a chain map $f : C \longrightarrow D$ in $\mathbb{A}$ is the chain complex $C(f)$ in $\mathbb{A}$ defined by

$$d_{C(f)} = \begin{pmatrix} d_D & (-)^{r-1}f \\ 0 & d_C \end{pmatrix} :$$

$$C(f)_r = D_r \oplus C_{r-1} \longrightarrow C(f)_{r-1} = D_{r-1} \oplus C_{r-2} .$$

Inclusion and projection define chain maps

$$D \longrightarrow C(f) \ , \ C(f) \longrightarrow SC$$

with SC the *suspension* chain complex defined by

$$d_{SC} = d_C : SC_r = C_{r-1} \longrightarrow SC_{r-1} = C_{r-2} .$$

The *total complex* of a double complex $C_{*,*}$ in $\mathbb{A}$ with differentials

$$d' : C_{p,q} \longrightarrow C_{p-1,q} \ , \ d'' : C_{p,q} \longrightarrow C_{p,q-1} \quad (p, q \in \mathbb{Z})$$

such that $d'd' = 0$, $d''d'' = 0$, $d'd'' = d''d'$ is the chain complex C in $\mathbb{A}$ defined by

$$d_C = \sum_{p+q=r} (d'' + (-)^q d') : C_r = \sum_{p+q=r} C_{p,q} \longrightarrow C_{r-1} .$$

Given chain complexes C, D in $\mathbb{A}$ let $\mathrm{Hom}_{\mathbb{A}}(C, D)_{*,*}$ be the double complex of abelian groups with

$$\text{Hom}_\mathbb{A}(C,D)_{p,q} = \text{Hom}_\mathbb{A}(C_{-p}, D_q) ,$$

$$d'(f) = f d_C : C_{-p+1} \longrightarrow D_q , \quad d''(f) = d_D f : C_{-p} \longrightarrow D_{q-1} .$$

The total complex is the chain complex $\text{Hom}_\mathbb{A}(C,D)$ defined by

$$d_{\text{Hom}_\mathbb{A}(C,D)} : \text{Hom}_\mathbb{A}(C,D)_r = \sum_{p+q=r} \text{Hom}_\mathbb{A}(C_{-p}, D_q)$$

$$\longrightarrow \text{Hom}_\mathbb{A}(C,D)_{r-1} ; \quad f \longrightarrow d_D f + (-)^q f d_C .$$

Define $\Sigma^n C$ to be the chain complex in $\mathbb{A}$ with

$$d_{\Sigma^n C} = (-)^r d_C : (\Sigma^n C)_r = C_{r-n} \longrightarrow (\Sigma^n C)_{r-1} = C_{r-1-n} .$$

The nth homology group $H_n(\text{Hom}_\mathbb{A}(C,D))$ $(n \in \mathbb{Z})$ is the abelian group of chain homotopy classes of chain maps $f : \Sigma^n C \longrightarrow D$. The isomorphisms of chain objects

$$(\Sigma^n C)_r = C_{r-n} \xrightarrow{\;\cong\;} (S^n C)_r = C_{r-n} ; \quad x \longrightarrow (-)^{r(r+1)/2} x$$

define an isomorphism of chain complexes $\Sigma^n C \cong S^n C$.

Let $\mathbb{B}(\mathbb{A})$ be the additive category of finite chain complexes in $\mathbb{A}$ and chain maps. The embedding

$$1 : \mathbb{A} \longrightarrow \mathbb{B}(\mathbb{A}) ; \quad A \longrightarrow A , \quad A_r = \begin{cases} A & \text{if } r = 0 \\ 0 & \text{if } r \neq 0 \end{cases}$$

is used to identify $\mathbb{A}$ with the subcategory of $\mathbb{B}(\mathbb{A})$ consisting of 0-dimensional chain complexes.

Given a contravariant additive functor

$$T : \mathbb{A} \longrightarrow \mathbb{B}(\mathbb{A}) ; \quad A \longrightarrow T(A)$$

define an extension of T to a contravariant additive functor

$$T : \mathbb{B}(\mathbb{A}) \longrightarrow \mathbb{B}(\mathbb{A}) ; \quad C \longrightarrow T(C)$$

by sending a finite chain complex C in $\mathbb{A}$ to the total complex $T(C)$ of the double complex $T(C)_{*,*}$ in $\mathbb{A}$ defined by

$$T(C)_{p,q} = T(C_{-p})_q , \quad d' = T(d_C) , \quad d'' = d_{T(C_{-p})} ,$$

that is

$$d_{T(C)} = \sum_{p+q=r} (d_{T(C_{-p})} + (-)^q T(d_C)) :$$

$$T(C)_r = \sum_{p+q=r} T(C_{-p})_q \longrightarrow T(C)_{r-1} .$$

For any morphism $f : C \longrightarrow D$ in $\mathbb{B}(\mathbb{A})$ it is possible to identify

$$C(T(f) : T(D) \longrightarrow T(C)) = STC(f : C \longrightarrow D)$$

up to natural isomorphism in $\mathbb{B}(\mathbb{A})$.

DEFINITION 1.1 A *chain duality* (T, e) on an additive category $\mathbb{A}$ is a contravariant additive functor $T: \mathbb{A} \longrightarrow \mathbb{B}(\mathbb{A})$ together with a natural transformation

$$e : T^2 \longrightarrow 1 : \mathbb{A} \longrightarrow \mathbb{B}(\mathbb{A})$$

such that for each object A in $\mathbb{A}$
   (i) $e(T(A)) . T(e(A)) = 1 : T(A) \longrightarrow T^3(A) \longrightarrow T(A)$,
   (ii) $e(A): T^2(A) \longrightarrow A$ is a chain equivalence.
The *dual* of a chain complex C is the chain complex $T(C)$, and $\Sigma^n T(C)$ is the *n-dual* of C.

□

Note that the n-dual $\Sigma^n T(C)$ of an n-dimensional chain complex C need not be n-dimensional.

DEFINITION 1.2 A chain duality on $\mathbb{A}$ is 0-*dimensional* if for each object A in $\mathbb{A}$ the dual chain complex $T(A)$ is 0-dimensional. A 0-dimensional chain duality is an *involution* on $\mathbb{A}$.

□

In the 0-dimensional case $e(A): T^2(A) \longrightarrow A$ is an isomorphism of 0-dimensional chain complexes for each object A in $\mathbb{A}$, and the n-dual $\Sigma^n T(C)$ of an n-dimensional chain complex C is n-dimensional, with

$$\Sigma^n T(C)_r = T(C)_{r-n} = T(C_{n-r}) .$$

An involution is a contravariant additive functor $T: \mathbb{A} \longrightarrow \mathbb{A}$ together with a natural equivalence $e' = e^{-1}: 1 \longrightarrow T^2: \mathbb{A} \longrightarrow \mathbb{A}$ such that for each object A in $\mathbb{A}$

$$e'(T(A))^{-1} = T(e'(A)) : T^3(A) \longrightarrow T(A) ,$$

i.e. an involution on $\mathbb{A}$ in the sense of Ranicki [145].
   Fix an additive category $\mathbb{A}$ with a chain duality (T, e).
   For any objects M, N in $\mathbb{A}$ define the abelian group chain complex

$$M \otimes_{\mathbb{A}} N = \text{Hom}_{\mathbb{A}}(T(M), N) .$$

The construction is covariant in both variables, with morphisms $g: M \longrightarrow M'$, $h: N \longrightarrow N'$ in $\mathbb{A}$ inducing abelian group morphisms

$$g \otimes_{\mathbb{A}} h : M \otimes_{\mathbb{A}} N \longrightarrow M' \otimes_{\mathbb{A}} N' ;$$

$$(f: T(M) \longrightarrow \dot{N}) \longrightarrow (hfT(g): T(M') \longrightarrow N') .$$

The duality isomorphism of abelian group chain complexes

$$T_{M,N} : M \otimes_{\mathbb{A}} N \xrightarrow{\simeq} N \otimes_{\mathbb{A}} M$$

is defined by

$$T_{M,N} : (M \otimes_\mathbb{A} N)_n = \mathrm{Hom}_\mathbb{A}(T(M)_{-n}, N)$$

$$\xrightarrow{\simeq} (N \otimes_\mathbb{A} M)_n = \mathrm{Hom}_\mathbb{A}(T(N)_{-n}, M) ;$$

$$(f : T(M)_{-n} \longrightarrow N) \longrightarrow (T_{M,N}(f) : T(N)_{-n} \longrightarrow M)$$

with

$$T_{M,N}(f) = e(M)T(f) :$$

$$T(N)_{-n} \longrightarrow T(T(M)_{-n})_{-n} \subseteq T^2(M)_0 \longrightarrow M_0 = M .$$

The inverse of $T_{M,N}$ is

$$(T_{M,N})^{-1} = T_{N,M} : N \otimes_\mathbb{A} M \xrightarrow{\simeq} M \otimes_\mathbb{A} N ,$$

since for any $f \in M \otimes_\mathbb{A} N$

$$T_{N,M} T_{M,N}(f) = e(N) T(e(M)T(f)) = e(N) T^2(f) T(e(M))$$

$$= f e(T(M)) T(e(M)) = f \in M \otimes_\mathbb{A} N .$$

EXAMPLE 1.3 Given a ring R with an involution $R \longrightarrow R; r \longrightarrow \bar{r}$ let $\mathbb{A}^p(R)$ be the additive category of f.g. projective (left) R-modules. Define a 0-dimensional chain duality

$$T : \mathbb{A}^p(R) \longrightarrow \mathbb{A}^p(R) ; P \longrightarrow T(P) = P^* = \mathrm{Hom}_R(P, R)$$

by

$$R \times P^* \longrightarrow P^* ; (r, f) \longrightarrow (x \longrightarrow f(x).\bar{r}) ,$$

$$e(P)^{-1} : P \longrightarrow P^{**} ; x \longrightarrow (f \longrightarrow \overline{f(x)}) .$$

The tensor product of f.g. projective R-modules P,Q is the abelian group

$$P \otimes_R Q = P \otimes_\mathbb{Z} Q / \{ rx \otimes y - x \otimes \bar{r}y \,|\, x \in P, y \in Q, r \in R \} ,$$

such that the slant map defines a natural isomorphism

$$P \otimes_R Q \xrightarrow{\simeq} \mathrm{Hom}_R(P^*, Q) = P \otimes_{\mathbb{A}^p(R)} Q ; x \otimes y \longrightarrow (f \longrightarrow \overline{f(x)}.y) .$$

The duality isomorphism $T_{P,Q} : P \otimes_{\mathbb{A}^p(R)} Q \longrightarrow Q \otimes_{\mathbb{A}^p(R)} P$ corresponds to the transposition isomorphism

$$T_{P,Q} : P \otimes_R Q \xrightarrow{\simeq} Q \otimes_R P ; x \otimes y \longrightarrow y \otimes x .$$

Similarly for the full subcategory $\mathbb{A}^h(R) \subseteq \mathbb{A}^p(R)$ of f.g. free R-modules.

□

EXAMPLE 1.4 Given a commutative ring R, a group π and a group morphism $w : \pi \longrightarrow \{\pm 1\}$ let $R[\pi]^w$ denote the group ring $R[\pi]$ with the w-twisted involution

$$^- : R[\pi]^w \longrightarrow R[\pi]^w ; a = \sum_{g \in \pi} r_g g \longrightarrow \bar{a} = \sum_{g \in \pi} r_g w(g) g^{-1} \quad (r_g \in R) .$$

□

This is the example occuring most frequently in topological applications, with w an orientation character. In the orientable case $w = +1$ write $R[\pi]^w$ as $R[\pi]$. The additive category of f.g. free R-modules is written

$$\mathbb{A}^h(R) = \mathbb{A}(R) \ .$$

There is also a version of the theory for based f.g. free R-modules, with Whitehead torsion considerations.

Given a finite chain complex C in $\mathbb{A}$ write

$$C^r = T(C)_{-r} \ , \ \Sigma^n T(C) = C^{n-*} \ .$$

For a chain map $f: C \longrightarrow C'$ the components in each degree of the dual chain map $T(f): T(C') \longrightarrow T(C)$ are written

$$f^* = T(f) : C'^r = T(C')_{-r} \longrightarrow C^r = T(C)_{-r} \ .$$

Given also a finite chain complex D in $\mathbb{A}$ define the abelian group chain complex

$$C \otimes_{\mathbb{A}} D = \operatorname{Hom}_{\mathbb{A}}(T(C), D) \ .$$

The duality isomorphism

$$T_{C,D} : C \otimes_{\mathbb{A}} D \xrightarrow{\ \sim\ } D \otimes_{\mathbb{A}} C$$

is defined by

$$T_{C,D} = \Sigma(-)^{pq} T_{C_p, D_q} :$$

$$(C \otimes_{\mathbb{A}} D)_n = \sum_{p+q+r=n} (C_p \otimes_{\mathbb{A}} D_q)_r \longrightarrow (D \otimes_{\mathbb{A}} C)_n \ ,$$

with inverse

$$(T_{C,D})^{-1} = T_{D,C} : D \otimes_{\mathbb{A}} C \xrightarrow{\ \sim\ } C \otimes_{\mathbb{A}} D \ .$$

$H_n(C \otimes_{\mathbb{A}} D)$ is the abelian group of chain homotopy classes of chain maps $\phi: C^{n-*} \longrightarrow D$ in $\mathbb{A}$. The duality isomorphism for $C = D$

$$T = T_{C,C} : C \otimes_{\mathbb{A}} C \xrightarrow{\ \sim\ } C \otimes_{\mathbb{A}} C$$

is an involution ($T^2 = 1$), so that $C \otimes_{\mathbb{A}} C$ is a $\mathbb{Z}[\mathbb{Z}_2]$-module chain complex.

The algebraic theory of surgery on $\left\{ \begin{array}{l} \text{symmetric} \\ \text{quadratic} \end{array} \right.$ complexes in an additive category with involution of Ranicki [141], [145] can now be developed for an additive category $\mathbb{A}$ with chain duality.

Use the standard free $\mathbb{Z}[\mathbb{Z}_2]$-module resolution of $\mathbb{Z}$

$$W : \ \ldots \ \xrightarrow{\ 1-T\ } \mathbb{Z}[\mathbb{Z}_2] \xrightarrow{\ 1+T\ } \mathbb{Z}[\mathbb{Z}_2] \xrightarrow{\ 1-T\ } \mathbb{Z}[\mathbb{Z}_2]$$

to define for any finite chain complex C in $\mathbb{A}$ the $\mathbb{Z}$-module chain complexes

$$\begin{cases} W^\% C = \operatorname{Hom}_{\mathbb{Z}[\mathbb{Z}_2]}(W, C \otimes_{\mathbb{A}} C) = \operatorname{Hom}_{\mathbb{Z}[\mathbb{Z}_2]}(W, \operatorname{Hom}_{\mathbb{A}}(TC, C)) \\ W_\% C = W \otimes_{\mathbb{Z}[\mathbb{Z}_2]} (C \otimes_{\mathbb{A}} C) = W \otimes_{\mathbb{Z}[\mathbb{Z}_2]} \operatorname{Hom}_{\mathbb{A}}(TC, C) \ . \end{cases}$$

The boundary of the n-chain
$$\begin{cases} \phi = \{\phi_s \in \mathrm{Hom}_A(C^r, C_{n-r+s}) \mid r \in \mathbb{Z}, s \geq 0\} \in (W^\% C)_n \\ \psi = \{\psi_s \in \mathrm{Hom}_A(C^r, C_{n-r-s}) \mid r \in \mathbb{Z}, s \geq 0\} \in (W_\% C)_n \end{cases}$$
is the $(n-1)$-chain with
$$\begin{cases} (\partial\phi)_s = d_{C \otimes_A C}(\phi_s) + (-)^{n+s-1}(\phi_{s-1} + (-)^s T\phi_{s-1}) \\ (\partial\psi)_s = d_{C \otimes_A C}(\psi_s) + (-)^{n-s-1}(\psi_{s+1} + (-)^{s+1} T\psi_{s+1}) \end{cases}$$
for $s \geq 0$, with $\phi_{-1} = 0$.

DEFINITION 1.5 (i) The $\begin{cases} symmetric \\ quadratic \end{cases}$ Q-groups of a finite chain complex C in A are defined for $n \in \mathbb{Z}$ by
$$\begin{cases} Q^n(C) = H_n(W^\% C) \\ Q_n(C) = H_n(W_\% C) . \end{cases}$$
(ii) A chain map $f: C \longrightarrow D$ of finite chain complexes in A induces a $\mathbb{Z}[\mathbb{Z}_2]$-module chain map
$$f \otimes f : C \otimes_A C \longrightarrow D \otimes_A D$$
and hence $\mathbb{Z}$-module chain maps
$$\begin{cases} f^\% : W^\% C \longrightarrow W^\% D \\ f_\% : W_\% C \longrightarrow W_\% D . \end{cases}$$
□

The morphisms of Q-groups induced by a chain map $f: C \longrightarrow D$ depend only on the chain homotopy class of f, and are isomorphisms for a chain equivalence.

DEFINITION 1.6 (i) An n-dimensional $\begin{cases} symmetric \\ quadratic \end{cases}$ (Poincaré) complex in A $\begin{cases} (C, \phi) \\ (C, \psi) \end{cases}$ is a finite chain complex C in A together with an n-cycle $\begin{cases} \phi \in (W^\% C)_n \\ \psi \in (W_\% C)_n \end{cases}$ (such that the chain map $\begin{cases} \phi_0 : C^{n-*} \longrightarrow C \\ (1+T)\psi_0 : C^{n-*} \longrightarrow C \end{cases}$ is a chain equivalence in A).

(ii) A *map* of n-dimensional $\begin{cases} symmetric \\ quadratic \end{cases}$ complexes in A
$$\begin{cases} f : (C, \phi) \longrightarrow (C', \phi') \\ f : (C, \psi) \longrightarrow (C', \psi') \end{cases}$$
is a chain map $f: C \longrightarrow C'$ such that
$$\begin{cases} f^\%(\phi) = \phi' \in Q^n(C') \\ f_\%(\psi) = \psi' \in Q_n(C') . \end{cases}$$
The map is a *homotopy equivalence* if $f: C \longrightarrow C'$ is a chain equivalence.

□

Note that the chain complex C in 1.6 is only required to be finite, and not n-dimensional as in Ranicki [141].

Let $f\colon C\longrightarrow D$ be a chain map of finite chain complexes in $\mathbb{A}$. An $(n+1)$-cycle

$$\begin{cases} (\delta\phi,\phi) \in C(f^\%\colon W^\%C\longrightarrow W^\%D)_{n+1} \\ (\delta\psi,\psi) \in C(f_\%\colon W_\%C\longrightarrow W_\%D)_{n+1} \end{cases}$$

is an n-cycle $\begin{cases} \phi \in (W^\%C)_n \\ \psi \in (W_\%C)_n \end{cases}$ together with a collection

$$\begin{cases} \delta\phi = \{\delta\phi_s \in (D\otimes_{\mathbb{A}} D)_{n+1+s} \mid s \geq 0\} \\ \delta\psi = \{\delta\psi_s \in (D\otimes_{\mathbb{A}} D)_{n+1-s} \mid s \geq 0\} \end{cases}.$$

such that

$$\begin{cases} d_{D\otimes_{\mathbb{A}}D}(\delta\phi_s) + (-)^{n+s}(\delta\phi_{s-1} + (-)^s T\delta\phi_{s-1}) \\ \qquad\qquad\qquad + (-)^n (f\otimes_{\mathbb{A}} f)(\phi_s) = 0 \in (D\otimes_{\mathbb{A}} D)_{n+s} \\ d_{D\otimes_{\mathbb{A}}D}(\delta\psi_s) + (-)^{n-s}(\delta\psi_{s+1} + (-)^{s+1} T\delta\psi_{s+1}) \\ \qquad\qquad\qquad + (-)^n (f\otimes_{\mathbb{A}} f)(\psi_s) = 0 \in (D\otimes_{\mathbb{A}} D)_{n-s} \end{cases}$$

The $(n+1)$-cycle

$$\begin{cases} (\delta\phi_0,\phi_0) \in C(f\otimes_{\mathbb{A}} f\colon C\otimes_{\mathbb{A}} C\longrightarrow D\otimes_{\mathbb{A}} D)_{n+1} \\ ((1+T)\delta\psi_0,(1+T)\psi_0) \in C(f\otimes_{\mathbb{A}} f\colon C\otimes_{\mathbb{A}} C\longrightarrow D\otimes_{\mathbb{A}} D)_{n+1} \end{cases}$$

determines a chain map

$$\begin{cases} (\delta\phi_0,\phi_0)\colon D^{n+1-*} \longrightarrow C(f) \\ (1+T)(\delta\psi_0,\psi_0)\colon D^{n+1-*} \longrightarrow C(f) \end{cases}$$

with

$$\begin{cases} (\delta\phi_0,\phi_0) = \begin{pmatrix} \delta\phi_0 \\ \phi_0 f^* \end{pmatrix} : D^{n+1-r} \longrightarrow C(f)_r = D_r \oplus C_{r-1} \\[2ex] ((1+T)\delta\psi_0,(1+T)\psi_0) = \begin{pmatrix} (1+T)\delta\psi_0 \\ (1+T)\psi_0 f^* \end{pmatrix} \\ \qquad\qquad : D^{n+1-r} \longrightarrow C(f)_r = D_r \oplus C_{r-1}. \end{cases}$$

Definition 1.7 (i) An $(n+1)$-dimensional $\begin{cases} \textit{symmetric} \\ \textit{quadratic} \end{cases}$ *(Poincaré) pair* in $\mathbb{A}$

$$\begin{cases} (\,f\colon C\longrightarrow D\,,\,(\delta\phi,\phi)\,) \\ (\,f\colon C\longrightarrow D\,,\,(\delta\psi,\psi)\,) \end{cases}$$

is a chain map $f\colon C\longrightarrow D$ of finite chain complexes together with an $(n+1)$-cycle $\begin{cases} (\delta\phi,\phi) \in C(f^\%)_{n+1} \\ (\delta\psi,\psi) \in C(f_\%)_{n+1} \end{cases}$ (such that the chain map

$$\begin{cases} (\delta\phi_0,\phi_0)\colon D^{n+1-*} \longrightarrow C(f) \\ (1+T)(\delta\psi_0,\psi_0)\colon D^{n+1-*} \longrightarrow C(f) \end{cases}$$

is a chain equivalence).

(ii) A *cobordism* of n-dimensional $\begin{cases} \text{symmetric} \\ \text{quadratic} \end{cases}$ Poincaré complexes

$\begin{cases} (C,\phi) \\ (C,\psi) \end{cases}$, $\begin{cases} (C',\phi') \\ (C',\psi') \end{cases}$ is an $(n+1)$-dimensional $\begin{cases} \text{symmetric} \\ \text{quadratic} \end{cases}$ Poincaré pair

$$\begin{cases} ((f\ f'): C \oplus C' \longrightarrow D,\ (\delta\phi, \phi \oplus -\phi')) \\ ((f\ f'): C \oplus C' \longrightarrow D,\ (\delta\psi, \psi \oplus -\psi')) \end{cases}.$$

□

DEFINITION 1.8 The n-*dimensional* $\begin{cases} \text{symmetric} \\ \text{quadratic} \end{cases}$ *L-group* $\begin{cases} L^n(\mathbb{A}) \\ L_n(\mathbb{A}) \end{cases}$ $(n \in \mathbb{Z})$
an additive category with chain duality $\mathbb{A}$ is the cobordism group of n-dimensional $\begin{cases} \text{symmetric} \\ \text{quadratic} \end{cases}$ Poincaré complexes in $\mathbb{A}$.

□

DEFINITION 1.9 Given a finite chain complex C in $\mathbb{A}$ define the *double skew-suspension* isomorphism of $\mathbb{Z}$-module chain complexes

$$\begin{cases} \overline{S}^2 : S^4(W^\%C) \xrightarrow{\simeq} W^\%(S^2C) ;\ \phi \longrightarrow \overline{S}^2\phi,\ (\overline{S}^2\phi)_s = \phi_s \\ \overline{S}^2 : S^4(W_\%C) \xrightarrow{\simeq} W_\%(S^2C) ;\ \psi \longrightarrow \overline{S}^2\psi,\ (\overline{S}^2\psi)_s = \psi_s. \end{cases}$$

□

PROPOSITION 1.10 *The n-dimensional* $\begin{cases} \text{symmetric} \\ \text{quadratic} \end{cases}$ *L-groups are 4-periodic, with the double skew-suspension maps defining isomorphisms*

$$\begin{cases} \overline{S}^2 : L^n(\mathbb{A}) \xrightarrow{\simeq} L^{n+4}(\mathbb{A}) ;\ (C,\phi) \longrightarrow (S^2C, \overline{S}^2\phi) \\ \overline{S}^2 : L_n(\mathbb{A}) \xrightarrow{\simeq} L_{n+4}(\mathbb{A}) ;\ (C,\psi) \longrightarrow (S^2C, \overline{S}^2\psi) \end{cases}$$

for $n \in \mathbb{Z}$.
PROOF The functor $S^2 : \mathbb{B}(\mathbb{A}) \longrightarrow \mathbb{B}(\mathbb{A})$ is an isomorphism of additive categories.

□

EXAMPLE 1.11 Let R be a ring with involution, so that the additive categories with duality involution

$\mathbb{A}^h(R) = \{\,\text{f.g. free } R\text{-modules}\,\}$, $\mathbb{A}^p(R) = \{\,\text{f.g. projective } R\text{-modules}\,\}$

are defined as in 1.3.
(i) The quadratic L-groups of $\mathbb{A}^q(R)$ for $q = h$ (resp. p) are the free (resp. projective) versions of the 4-periodic quadratic L-groups of Wall [176]

$$L_n(\mathbb{A}^q(R)) = L_n^q(R) \quad (n \in \mathbb{Z}).$$

(ii) The symmetric L-groups of $\mathbb{A}^q(R)$ for $q = h$ (resp. p) are the 4-periodic versions of the free (resp. projective) symmetric L-groups of Mishchenko [112]

$$L^n(\mathbb{A}^q(R)) = \varinjlim_k L_q^{n+4k}(R) = L_q^{n+4*}(R) \ \ (n \in \mathbb{Z}) \ .$$

See Ranicki [141], [145] for proofs of both (i) and (ii). The 4-periodicity of the symmetric L-groups is ensured by the use of finite rather than n-dimensional chain complexes in 1.6. See 3.18 below for a further discussion.

$\square$

The 4-periodic L-groups of the additive category $\mathbb{A}^h(R)$ of a ring with involution R are written

$$L_n(\mathbb{A}^h(R)) = L_n(R) = L_{n+4*}(R) \ ,$$

$$L^n(\mathbb{A}^h(R)) = L^{n+4*}(R) \ \ (n \in \mathbb{Z}) \ .$$

DEFINITION 1.12 The n-dimensional $\begin{cases} \text{symmetric} \\ \text{quadratic} \end{cases}$ complex $\begin{cases} (C', \phi') \\ (C', \psi') \end{cases}$ in $\mathbb{A}$ obtained from an n-dimensional $\begin{cases} \text{symmetric} \\ \text{quadratic} \end{cases}$ complex $\begin{cases} (C, \phi) \\ (C, \psi) \end{cases}$ by algebraic surgery on an $(n+1)$-dimensional $\begin{cases} \text{symmetric} \\ \text{quadratic} \end{cases}$ pair

$\begin{cases} (f: C \longrightarrow D, (\delta\phi, \phi)) \\ (f: C \longrightarrow D, (\delta\psi, \psi)) \end{cases}$ is given in the symmetric case by

$$d_{C'} = \begin{pmatrix} d_C & 0 & (-)^{n+1}\phi_0 f^* \\ (-)^r f & d_D & (-)^r \delta\phi_0 f^* \\ 0 & 0 & (-)^r d_D^* \end{pmatrix} :$$

$$C'_r = C_r \oplus D_{r+1} \oplus D^{n-r-1}$$

$$\longrightarrow C'_{r-1} = C_{r-1} \oplus D_r \oplus D^{n-r+2} \ ,$$

$$\phi'_0 = \begin{pmatrix} \phi_0 & 0 & 0 \\ (-)^{n-r} fT\phi_1 & (-)^{n-r} T\delta\phi_1 & (-)^{r(n-r)} e \\ 0 & 1 & 0 \end{pmatrix} :$$

$$C'^{n-r} = C^{n-r} \oplus D^{n-r+1} \oplus (T^2 D)_{r+1}$$

$$\longrightarrow C'_r = C_r \oplus D_{r+1} \oplus D^{n-r+1}$$

$$\phi'_s = \begin{pmatrix} \phi_s & 0 & 0 \\ (-)^{n-r} fT\phi_s & (-)^{n-r+s} T\delta\phi_{s+1} & 0 \\ 0 & 0 & 0 \end{pmatrix} :$$

$$C'^{n-r+s} = C^{n-r+s} \oplus D^{n-r+s+1} \oplus (T^2 D)_{r-s+1}$$

$$\longrightarrow C'_r = C_r \oplus D_{r+1} \oplus D^{n-r+1} \ \ (s \geq 1)$$

and in the quadratic case by

$$d_{C'} = \begin{pmatrix} d_C & 0 & (-)^{n+1}(1+T)\psi_0 f^* \\ (-)^r f & d_D & (-)^r(1+T)\delta\psi_0 f^* \\ 0 & 0 & (-)^r d_D^* \end{pmatrix} :$$

$$C_r' = C_r \oplus D_{r+1} \oplus D^{n-r+1}$$

$$\longrightarrow C_{r-1}' = C_{r-1} \oplus D_r \oplus D^{n-r+2} ,$$

$$\psi_0' = \begin{pmatrix} \psi_0 & 0 & 0 \\ 0 & 0 & 0 \\ 0 & 1 & 0 \end{pmatrix} :$$

$$C'^{n-r} = C^{n-r} \oplus D^{n-r+1} \oplus (T^2 D)_{r+1}$$

$$\longrightarrow C_r' = C_r \oplus D_{r+1} \oplus D^{n-r+1}$$

$$\psi_s' = \begin{pmatrix} \psi_s & (-)^{r+s} T\psi_{s-1} f^* & 0 \\ 0 & (-)^{n-r-s+1} T\delta\psi_{s-1} & 0 \\ 0 & 0 & 0 \end{pmatrix} :$$

$$C'^{n-r-s} = C^{n-r-s} \oplus D^{n-r-s+1} \oplus (T^2 D)_{r+s+1}$$

$$\longrightarrow C_r' = C_r \oplus D_{r+1} \oplus D^{n-r+1} \quad (s \geq 1) .$$

$\square$

PROPOSITION 1.13 *Cobordism of n-dimensional* $\begin{cases} symmetric \\ quadratic \end{cases}$ *Poincaré complexes in $\mathbb{A}$ is the equivalence relation generated by homotopy equivalence and algebraic surgery.*

PROOF As for Ranicki [141, 5.1], the special case $\mathbb{A} = \mathbb{A}^p(R) = \{f.g. \text{ project-ive } R\text{-modules}\}$.

$\square$

DEFINITION 1.14 The *boundary* of an n-dimensional $\begin{cases} symmetric \\ quadratic \end{cases}$ complex

$\begin{cases} (C, \phi) \\ (C, \psi) \end{cases}$ in $\mathbb{A}$ is the $(n-1)$-dimensional $\begin{cases} symmetric \\ quadratic \end{cases}$ complex

$$\begin{cases} \partial(C, \phi) = (\partial C, \partial \phi) \\ \partial(C, \psi) = (\partial C, \partial \psi) \end{cases}$$

obtained from $(0,0)$ by surgery on the n-dimensional $\begin{cases} symmetric \\ quadratic \end{cases}$ pair

$\begin{cases} (0\colon 0 \longrightarrow C, (\phi, 0)) \\ (0\colon 0 \longrightarrow C, (\psi, 0)) \end{cases}$ In the symmetric case

$$d_{\partial C} = \begin{pmatrix} d_C & (-)^r \phi_0 \\ 0 & (-)^r d_C^* \end{pmatrix} :$$

$$\partial C_r = C_{r+1} \oplus C^{n-r} \longrightarrow \partial C_{r-1} = C_r \oplus C^{n-r-1} ,$$

$$\partial \phi_0 = \begin{pmatrix} (-)^{n-r} T\phi_1 & (-)^{r(n-r-1)} e \\ 1 & 0 \end{pmatrix} :$$

$$\partial C^{n-r-1} = C^{n-r} \oplus (T^2 C)_{r+1} \longrightarrow \partial C_r = C_{r+1} \oplus C^{n-r} ,$$

$$\partial \phi_s = \begin{pmatrix} (-)^{n-r+s} T\phi_{s+1} & 0 \\ 0 & 0 \end{pmatrix} :$$

$$\partial C^{n-r+s-1} = C^{n-r+s} \oplus (T^2 C)_{r-s+1}$$
$$\longrightarrow \partial C_r = C_{r+1} \oplus C^{n-r} \quad (s \geq 1) .$$

and in the quadratic case

$$d_{\partial C} = \begin{pmatrix} d_C & (-)^r (1+T)\psi_0 \\ 0 & (-)^r d_C^* \end{pmatrix} :$$

$$\partial C_r = C_{r+1} \oplus C^{n-r} \longrightarrow \partial C_{r-1} = C_r \oplus C^{n-r+1} ,$$

$$\partial \psi_0 = \begin{pmatrix} 0 & 0 \\ 1 & 0 \end{pmatrix} :$$

$$\partial C^{n-r-1} = C^{n-r} \oplus (T^2 C)_{r+1} \longrightarrow \partial C_r = C_{r+1} \oplus C^{n-r} ,$$

$$\partial \psi_s = \begin{pmatrix} (-)^{n-r-s-1} T\psi_{s-1} & 0 \\ 0 & 0 \end{pmatrix} :$$

$$\partial C^{n-r-s-1} = C^{n-r-s} \oplus (T^2 C)_{r+s+1}$$
$$\longrightarrow \partial C_r = C_{r+1} \oplus C^{n-r} \quad (s \geq 1) .$$

$\square$

It is immediate from the identity
$$\partial C = \begin{cases} S^{-1} C(\phi_0 : C^{n-*} \longrightarrow C) \\ S^{-1} C((1+T)\psi_0 : C^{n-*} \longrightarrow C) \end{cases}$$

that an n-dimensional $\begin{cases} \text{symmetric} \\ \text{quadratic} \end{cases}$ complex $\begin{cases} (C, \phi) \\ (C, \psi) \end{cases}$ is Poincaré if and only if the boundary $\begin{cases} \partial(C, \phi) \\ \partial(C, \psi) \end{cases}$ is contractible.

PROPOSITION 1.15 *The homotopy equivalence classes of n-dimensional* $\begin{cases} symmetric \\ quadratic \end{cases}$ *complexes in $\mathbb{A}$ are in one–one correspondence with the homotopy equivalence classes of n-dimensional* $\begin{cases} symmetric \\ quadratic \end{cases}$ *Poincaré pairs in* $\mathbb{A}$.

PROOF As for Ranicki [141, 3.4], the special case $\mathbb{A} = \mathbb{A}^p(R)$.

Given an n-dimensional $\begin{cases} \text{symmetric} \\ \text{quadratic} \end{cases}$ complex $\begin{cases} (C, \phi) \\ (C, \psi) \end{cases}$ in $\mathbb{A}$ define the n-dimensional $\begin{cases} \text{symmetric} \\ \text{quadratic} \end{cases}$ Poincaré pair

$$\begin{cases} \delta\partial(C, \phi) \\ \delta\partial(C, \psi) \end{cases} = \left(p_C = \text{projection} : \partial C \longrightarrow C^{n-*}, \begin{cases} (0, \partial\phi) \\ (0, \partial\psi) \end{cases} \right).$$

Conversely, given an n-dimensional $\begin{cases} \text{symmetric} \\ \text{quadratic} \end{cases}$ Poincaré pair in $\mathbb{A}$

$$B = \left(f : C \longrightarrow D, \begin{cases} (\delta\phi, \phi) \\ (\delta\psi, \psi) \end{cases} \right)$$

apply the algebraic Thom construction to obtain an n-dimensional $\begin{cases} \text{symmetric} \\ \text{quadratic} \end{cases}$ complex

$$B/\partial B = \begin{cases} (D, \delta\phi)/C = (C(f), \delta\phi/\phi) \\ (D, \delta\psi)/C = (C(f), \delta\psi/\psi) \end{cases}$$

with

$$(\delta\phi/\phi)_s = \begin{pmatrix} \delta\phi_s & 0 \\ (-)^{n-r-1}\phi_s f^* & (-)^{n-r+s}T\phi_{s-1} \end{pmatrix} :$$
$$C(f)^{n-r+s+1} = D^{n-r+s+1} \oplus C^{n-r+s} \longrightarrow C(f)_r = D_r \oplus C_{r-1}$$
$$(s \geq 0, \phi_{-1} = 0),$$

$$(\delta\psi/\psi)_s = \begin{pmatrix} \delta\psi_s & 0 \\ (-)^{n-r-1}\psi_s f^* & (-)^{n-r-s}T\psi_{s+1} \end{pmatrix} :$$
$$C(f)^{n-r-s+1} = D^{n-r-s+1} \oplus C^{n-r-s} \longrightarrow C(f)_r = D_r \oplus C_{r-1}$$
$$(s \geq 0),$$

which is homotopy equivalent to $\delta\partial(B/\partial B)$.

□

It follows from 1.15 that an n-dimensional $\begin{cases} \text{symmetric} \\ \text{quadratic} \end{cases}$ Poincaré complex $\begin{cases} (C, \phi) \\ (C, \psi) \end{cases}$ in $\mathbb{A}$ is such that $\begin{cases} (C, \phi) = 0 \in L^n(\mathbb{A}) \\ (C, \psi) = 0 \in L_n(\mathbb{A}) \end{cases}$ if and only if $\begin{cases} (C, \phi) \\ (C, \psi) \end{cases}$ is homotopy equivalent to the boundary $\partial(D, \theta)$ of an $(n+1)$-dimensional $\begin{cases} \text{symmetric} \\ \text{quadratic} \end{cases}$ complex (D, θ) in $\mathbb{A}$.

§2. Algebraic normal complexes

An algebraic normal complex a chain complex with the normal structure of a Poincaré complex, but not necessarily the Poincaré duality. Algebraic normal complexes are analogues of the normal spaces of Quinn [129], which have the normal structure of Poincaré spaces, but not necessarily the duality. Indeed, a normal space determines an algebraic normal complex.

The algebraic theory of normal complexes of Ranicki [142] and Weiss [182] is now generalized to an additive category $\mathbb{A}$ with a chain duality $(T: \mathbb{A} \longrightarrow \mathbb{B}(\mathbb{A}), e: T^2 \longrightarrow 1)$. Algebraic normal complexes will be used in §3 to describe the difference between symmetric and quadratic L-groups of $\mathbb{A}$.

Use the standard complete (Tate) free $\mathbb{Z}[\mathbb{Z}_2]$-module resolution of $\mathbb{Z}$

$$\widehat{W}: \ldots \longrightarrow \mathbb{Z}[\mathbb{Z}_2] \xrightarrow{1-T} \mathbb{Z}[\mathbb{Z}_2] \xrightarrow{1+T} \mathbb{Z}[\mathbb{Z}_2] \xrightarrow{1-T} \mathbb{Z}[\mathbb{Z}_2] \longrightarrow \ldots$$

to define for any finite chain complex C in $\mathbb{A}$ the $\mathbb{Z}$-module chain complex

$$\widehat{W}^\% C = \operatorname{Hom}_{\mathbb{Z}[\mathbb{Z}_2]}(\widehat{W}, C \otimes_{\mathbb{A}} C) = \operatorname{Hom}_{\mathbb{Z}[\mathbb{Z}_2]}(\widehat{W}, \operatorname{Hom}_{\mathbb{A}}(TC, C)) \ .$$

A chain $\theta \in (\widehat{W}^\% C)_n$ is a collection of morphisms

$$\theta = \{\theta_s \in \operatorname{Hom}_{\mathbb{A}}(C^{n-r+s}, C_r) \,|\, r, s \in \mathbb{Z}\} \ ,$$

with the boundary $d(\theta) \in (\widehat{W}^\% C)_{n-1}$ given by

$$d(\theta)_s = d\theta_s + (-)^r \theta_s d^* + (-)^{n+s-1}(\theta_{s-1} + (-)^s T\theta_{s-1}) :$$
$$C^{n-r+s-1} \longrightarrow C_r \quad (r, s \in \mathbb{Z}) \ .$$

DEFINITION 2.1 (i) The *hyperquadratic Q-groups* of a finite chain complex C in $\mathbb{A}$ are defined by

$$\widehat{Q}^n(C) = H_n(\widehat{W}^\% C) \quad (n \in \mathbb{Z}) \ .$$

(ii) A chain map $f: C \longrightarrow D$ of finite chain complexes in $\mathbb{A}$ induces a $\mathbb{Z}$-module chain map

$$\widehat{f}^\% : \widehat{W}^\% C \longrightarrow \widehat{W}^\% D$$

via the $\mathbb{Z}[\mathbb{Z}_2]$-module chain map $f \otimes f: C \otimes_{\mathbb{A}} C \longrightarrow D \otimes_{\mathbb{A}} D$.

□

The short exact sequence of $\mathbb{Z}$-module chain complexes

$$0 \longrightarrow W^\% C \longrightarrow \widehat{W}^\% C \longrightarrow S(W_\% C) \longrightarrow 0$$

induces the long exact sequence of Q-groups of Ranicki [141, 1.1]

$$\ldots \longrightarrow Q_n(C) \xrightarrow{1+T} Q^n(C) \xrightarrow{J} \widehat{Q}^n(C) \xrightarrow{H} Q_{n-1}(C)$$
$$\xrightarrow{1+T} Q^{n-1}(C) \longrightarrow \ldots$$

with

$$(J\phi)_s \;=\; \begin{cases} \phi_s & \text{for } s \geq 0 \\ 0 & \text{for } s < 0 \end{cases} \;, \quad ((1+T)\psi)_s \;=\; \begin{cases} (1+T)\psi_0 & \text{for } s = 0 \\ 0 & \text{for } s \geq 1 \end{cases}$$

$$(H\theta)_s \;=\; \theta_{-s-1} \text{ for } s \geq 0 \; .$$

DEFINITION 2.2 (i) A *chain bundle* (C,γ) is a chain complex C in $\mathbb{A}$ together with a 0-cycle $\gamma \in (\widehat{W}^{\%}TC)_0$.

(ii) A *map* of chain bundles in $\mathbb{A}$

$$(f,b) : (C,\gamma) \longrightarrow (C',\gamma')$$

is a chain map $f \colon C \longrightarrow C'$ together with a 1-chain $b \in (\widehat{W}^{\%}TC)_1$ such that

$$\widehat{f}^{\%}(\gamma') \;-\; \gamma \;=\; d(b) \in (\widehat{W}^{\%}TC)_0 \; .$$

□

For any chain complex C in $\mathbb{A}$ there is defined a suspension isomorphism

$$S : \widehat{W}^{\%}C \xrightarrow{\;\simeq\;} S^{-1}(\widehat{W}^{\%}SC) \; ; \; \theta \longrightarrow S\theta$$

sending an n-chain $\theta \in (\widehat{W}^{\%}C)_n$ to the $(n+1)$-chain $S\theta \in (\widehat{W}^{\%}SC)_{n+1}$ with

$$(S\theta)_t \;=\; \theta_{t-1} : (SC)^{n-r+t+1} \;=\; C^{n-r+t} \longrightarrow (SC)_r \;=\; C_{r-1} \; .$$

Hence for any $n \in \mathbb{Z}$ there is defined an n-fold suspension isomorphism

$$S^n : \widehat{W}^{\%}TC \xrightarrow{\;\simeq\;} S^{-n}(\widehat{W}^{\%}C^{n-*})$$

sending a 0-cycle $\gamma \in (\widehat{W}^{\%}TC)_0$ to the n-cycle $S^n\gamma \in (\widehat{W}^{\%}C^{n-*})_n$ with

$$(S^n\gamma)_s \;=\; \gamma_{n+s} : C_r \longrightarrow C^{-n-r-s} \quad (r,s \in \mathbb{Z}) \; .$$

DEFINITION 2.3 Given a chain bundle (C,γ) let $Q_n(C,\gamma)$ $(n \in \mathbb{Z})$ be the *twisted quadratic Q-groups* of Weiss [182], designed to fit into a long exact sequence

$$\ldots \longrightarrow Q_n(C,\gamma) \xrightarrow{\;1+T\;} Q^n(C) \xrightarrow{\;J_\gamma\;} \widehat{Q}^n(C) \xrightarrow{\;H\;} Q_{n-1}(C,\gamma)$$

$$\xrightarrow{\;1+T\;} Q^{n-1}(C) \longrightarrow \ldots$$

with

$$J_\gamma : Q^n(C) \longrightarrow \widehat{Q}^n(C) \; ; \; \phi \longrightarrow J(\phi) - \widehat{\phi}_0^{\%}(S^n\gamma) \; .$$

An element of $Q_n(C,\gamma)$ is an equivalence class of pairs

$$(\phi \in (W^{\%}C)_n \, , \, \chi \in (\widehat{W}^{\%}TC)_{n+1})$$

such that

$$d(\phi) \;=\; 0 \in (W^{\%}C)_{n-1} \, , \; J(\phi) - (\widehat{\phi}_0)^{\%}(S^n\gamma) \;=\; d(\chi) \in (\widehat{W}^{\%}C)_n \, ,$$

with
$$1 + T : Q_n(C, \gamma) \longrightarrow Q^n(C) \; ; \; (\phi, \chi) \longrightarrow \phi \; ,$$
$$H : \widehat{Q}^{n+1}(C) \longrightarrow Q_n(C, \gamma) \; ; \; \chi \longrightarrow (0, \chi) \; .$$
The addition in $Q_n(C, \gamma)$ is by
$$(\phi, \chi) + (\phi', \chi') \; = \; (\phi + \phi', \chi + \chi' + \xi) \; ,$$
with
$$\xi_s \; = \; \phi_0(\gamma_{s-n+1})\phi'_0 : C^r \longrightarrow C_{n-r+s+1} \quad (r, s \in \mathbb{Z}) \; .$$

$\square$

J_γ is induced by a morphism of the simplicial abelian groups $K(W^\% C)$ $\longrightarrow K(\widehat{W}^\% C)$ associated to the abelian group chain complexes $W^\% C, \widehat{W}^\% C$ by the Kan–Dold theorem, rather than by a chain map $W^\% C \longrightarrow \widehat{W}^\% C$. For $\gamma = 0$ $J_\gamma = J$ is induced by the chain map $J : W^\% C \longrightarrow \widehat{W}^\% C$ and $Q_*(C, 0) = Q_*(C)$.

A map of chain bundles $(f, b) : (C, \gamma) \longrightarrow (C', \gamma')$ induces morphisms of the twisted quadratic Q-groups
$$(f, b)_\% : Q_n(C, \gamma) \longrightarrow Q_n(C', \gamma') \; ; \; (\phi, \chi) \longrightarrow (f^\% \phi, \widehat{f}^\% \chi + (\widehat{f \phi_0})^\% (b)) \; .$$

DEFINITION 2.4 (i) An (algebraic) n-dimensional normal complex (C, θ) in $\mathbb{A}$ is a finite chain complex C in $\mathbb{A}$ together with a triple
$$\theta \; = \; (\phi \in (W^\% C)_n, \gamma \in (\widehat{W}^\% TC)_0, \chi \in (\widehat{W}^\% C)_{n+1})$$
such that
$$d(\phi) \; = \; 0 \in (W^\% C)_{n-1} \; , \; d(\gamma) \; = \; 0 \in (\widehat{W}^\% TC)_{-1} \; ,$$
$$J(\phi) - (\widehat{\phi}_0)^\% (S^n \gamma) \; = \; d(\chi) \in (\widehat{W}^\% C)_n \; .$$
(C, θ) is an n-dimensional symmetric complex (C, ϕ) with a *normal structure* (γ, χ).

(ii) An $(n+1)$-*dimensional normal pair* $(f : C \longrightarrow D, (\delta\theta, \theta))$ in $\mathbb{A}$ is an $(n+1)$-dimensional symmetric pair $(f : C \longrightarrow D, (\delta\phi, \phi))$ in $\mathbb{A}$ together with a map of chain bundles $(f, b) : (C, \gamma) \longrightarrow (D, \delta\gamma)$ and chains $\chi \in (\widehat{W}^\% C)_{n+1}$, $\delta\chi \in (\widehat{W}^\% D)_{n+2}$ such that
$$J(\phi) - (\widehat{\phi}_0)^\% (S^n \gamma) \; = \; d(\chi) \in (\widehat{W}^\% C)_n \; ,$$
$$J(\delta\phi) - (\delta\phi_0, \phi_0)^\% (S^n \delta\gamma) + \widehat{f}^\% (\chi - (\widehat{\phi}_0)^\% (S^n b)) \; = \; d(\delta\chi) \in (\widehat{W}^\% D)_{n+1} \; ,$$
with $(\delta\theta, \theta)$ short for $((\delta\phi, \delta\gamma, \delta\chi), (\phi, \gamma, \chi))$.

(iii) A *map* of n-dimensional normal complexes in $\mathbb{A}$
$$(f, b) : (C, \phi, \gamma, \chi) \longrightarrow (C', \phi', \gamma', \chi')$$
is a bundle map $(f, b) : (C, \gamma) \longrightarrow (C', \gamma')$ such that
$$(f, b)_\% (\phi, \chi) \; = \; (\phi', \chi') \in Q_n(C', \gamma') \; .$$

The map is a *homotopy equivalence* if $f: C \longrightarrow C'$ is a chain equivalence.

(iv) The *normal L-groups* $NL^n(\mathbb{A})$ ($n \in \mathbb{Z}$) are the cobordism groups of n-dimensional normal complexes in $\mathbb{A}$.

□

REMARK 2.5 Geometric normal (resp. Poincaré) complexes and pairs determine algebraic normal (resp. Poincaré) complexes and pairs. The methods of Ranicki [142] and Weiss [182] can be combined to associate to any $(k-1)$-spherical fibration $\nu: X \longrightarrow BG(k)$ over a finite CW complex X a chain bundle in $\mathbb{A}(\mathbb{Z}[\pi]^w)$ (cf. 1.4)

$$\widehat{\sigma}^*(\nu) = (C(\widetilde{X}), \gamma)$$

with $\widetilde{X}$ any regular covering of X such that the pullback $\widetilde{\nu}: \widetilde{X} \longrightarrow BG(k)$ is oriented, π the group of covering translations, $C(\widetilde{X})$ the cellular $\mathbb{Z}[\pi]$-module chain complex of $\widetilde{X}$, and $w: \pi \longrightarrow \{\pm 1\}$ a factorization of the orientation character

$$w_1(\nu) : \pi_1(X) \longrightarrow \pi \xrightarrow{w} \{\pm 1\} \ .$$

The hyperquadratic structure γ is unique up to equivalence (i.e. only the homology class $\gamma \in \widehat{Q}^0(C(\widetilde{X})^{-*})$ is determined), and depends only on the stable spherical fibration $\nu: X \longrightarrow BG$. Let $T(\nu)$ be the Thom space of ν, and let $U_\nu \in \dot{H}^k(T(\nu), w)$ be the w-twisted Thom class, with $\dot{H}^*$ denoting reduced cohomology. The Alexander–Whitney–Steenrod diagonal chain approximation

$$\Delta_{\widetilde{X}} : C(\widetilde{X}) \longrightarrow \mathrm{Hom}_{\mathbb{Z}[\mathbb{Z}_2]}(W, C(\widetilde{X}) \otimes_{\mathbb{Z}} C(\widetilde{X}))$$

induces the 'symmetric construction' of Ranicki [142, §1]

$$\phi_X = 1 \otimes \Delta_{\widetilde{X}} : H_n(X, w) = H_n(\mathbb{Z}^w \otimes_{\mathbb{Z}[\pi]^w} C(\widetilde{X}))$$

$$\longrightarrow Q^n(C(\widetilde{X})) = H_n(\mathrm{Hom}_{\mathbb{Z}[\mathbb{Z}_2]}(W, C(\widetilde{X}) \otimes_{\mathbb{Z}[\pi]^w} C(\widetilde{X}))) \ .$$

The composite of the Thom isomorphism and the symmetric construction

$$\dot{H}_{n+k}(T(\nu)) \xrightarrow{U_\nu \cap -} H_n(X, w) \xrightarrow{\phi_X} Q^n(C(\widetilde{X}))$$

extends to a natural transformation of exact sequences of abelian groups

$$\dots \to \Gamma_{n+k+1}(T(\nu)) \longrightarrow \pi_{n+k}(T(\nu)) \xrightarrow{h} \dot{H}_{n+k}(T(\nu)) \to \Gamma_{n+k}(T(\nu)) \to \dots$$

$$\downarrow \qquad\qquad \downarrow \qquad\qquad \downarrow \qquad\qquad \downarrow$$

$$\dots \to \widehat{Q}^{n+1}(C(\widetilde{X})) \longrightarrow Q_n(C(\widetilde{X}), \gamma) \longrightarrow Q^n(C(\widetilde{X})) \xrightarrow{J_\gamma} \widehat{Q}^n(C(\widetilde{X})) \to \dots$$

from the certain exact sequence of Whitehead [186], with h the Hurewicz map.

An n-*dimensional geometric normal complex* (X, ν_X, ρ_X) in the sense of Quinn [129] is a finite CW complex X together with a $(k-1)$-spherical fibration $\nu_X \colon X \longrightarrow BG(k)$ and a map $\rho_X \colon S^{n+k} \longrightarrow T(\nu_X)$. The algebraic normal complex of (X, ν_X, ρ_X) with respect to a covering $\widetilde{X}$ of X is defined by

$$\widehat{\sigma}^*(X, \nu_X, \rho_X) = (C(\widetilde{X}), \phi, \gamma, \chi)$$

with $(C(\widetilde{X}), \gamma) = \widehat{\sigma}^*(\nu_X)$ and $(\phi, \chi) \in Q_n(C(\widetilde{X}), \gamma)$ the image of $\rho_X \in \pi_{n+k}(T(\nu_X))$. The $\mathbb{Z}[\pi_1(X)]$-module duality chain map of $\widehat{\sigma}^*(X)$ is given by the cap product

$$\phi_0 = \phi_X([X])_0 = [X] \cap - : C(\widetilde{X})^{n-*} \longrightarrow C(\widetilde{X}),$$

with the fundamental class defined by

$$[X] = h(\rho_X) \cap U_{\nu_X} \in H_n(X, w).$$

A (finite) n-*dimensional geometric Poincaré complex* X is a (finite) CW complex together with an orientation map $w \colon \pi_1(X) \longrightarrow \mathbb{Z}_2$ and a fundamental class $[X] \in H_n(X, w)$ such that cap product defines a $\mathbb{Z}[\pi_1(X)]$-module chain equivalence

$$[X] \cap - : C(\widetilde{X})^{n-*} \xrightarrow{\simeq} C(\widetilde{X}).$$

An embedding $X \subset S^{n+k}$ (k large) determines the normal structure (ν_X, ρ_X) of Spivak [161], so that X is an n-dimensional geometric normal complex. The n-dimensional symmetric Poincaré complex in $\mathbb{A}(\mathbb{Z}[\pi]^w)$

$$\sigma^*(X) = (C(\widetilde{X}), \phi)$$

is such that $J\sigma^*(X) = \widehat{\sigma}^*(X, \nu_X, \rho_X)$.

□

The following result deals with the analogue for algebraic Poincaré complexes in any additive category with chain duality $\mathbb{A}$ of the Spivak normal structure of a geometric Poincaré complex:

PROPOSITION 2.6 (i) *An n-dimensional symmetric complex (C, ϕ) in $\mathbb{A}$ has a normal structure (γ, χ) if and only if the boundary $(n-1)$-dimensional symmetric Poincaré complex $\partial(C, \phi)$ admits a quadratic refinement.*
(ii) *There is a natural one–one correspondence between the homotopy equivalence classes of n-dimensional symmetric Poincaré complexes (C, ϕ) in $\mathbb{A}$ and those of n-dimensional normal complexes (C, ϕ, γ, χ) with $\phi_0 \colon C^{n-*} \longrightarrow C$ a chain equivalence.*
(iii) *There is a natural one–one correspondence between the homotopy equivalence classes of n-dimensional quadratic complexes (C, ψ) in $\mathbb{A}$ and those of n-dimensional normal complexes (C, ϕ, γ, χ) with $\gamma = 0$.*
PROOF (i) Write

$$\partial(C, \phi) = (\partial C, \partial \phi), \quad \partial C = S^{-1}C(\phi_0),$$

and let $e\colon C\longrightarrow S\partial C = C(\phi_0)$ be the inclusion. Consider the exact sequences of Q-groups

$$Q_{n-1}(\partial C) \xrightarrow{1+T} Q^{n-1}(\partial C) \xrightarrow{J} \widehat{Q}^{n-1}(\partial C) \ ,$$

$$\widehat{Q}^n(C^{n-*}) \xrightarrow{\widehat{\phi_0^\%}} \widehat{Q}^n(C) \xrightarrow{\widehat{e^\%}} \widehat{Q}^n(S\partial C) \ .$$

The obstruction $J(\partial\phi) \in \widehat{Q}^{n-1}(\partial C)$ to a quadratic refinement of $\partial(C,\phi)$ corresponds under the suspension isomorphism $\widehat{Q}^{n-1}(\partial C)\longrightarrow\widehat{Q}^n(S\partial C)$ to the obstruction $\hat{e}^\% J(\phi) \in \widehat{Q}^n(S\partial C)$ to a normal refinement of (C,ϕ).

(ii) An n-dimensional symmetric Poincaré complex (C,ϕ) determines an n-dimensional normal complex

$$J(C,\phi) \ = \ (C,\phi,\gamma,\chi)$$

with (γ,χ) unique up to equivalence. The class $\gamma \in \widehat{Q}^0(TC)$ is the image of $\phi \in Q^n(C)$ under the composite

$$Q^n(C) \xrightarrow{J} \widehat{Q}^n(C) \xrightarrow{(\widehat{\phi_0^\%})^{-1}} \widehat{Q}^n(C^{n-*}) \xrightarrow{S^{-n}} \widehat{Q}^0(TC) \ .$$

(iii) An n-dimensional quadratic complex (C,ψ) determines an n-dimensional normal complex with $\gamma = 0$ and

$$(1+T)(C,\psi) \ = \ (C,(1+T)\psi,0,\chi)$$

such that

$$((1+T)\psi)_s \ = \ \begin{cases} (1+T)\psi_0 & \text{if } s \geq 0 \\ 0 & \text{if } s < 0, \end{cases} \quad \chi_s \ = \ \begin{cases} 0 & \text{if } s \geq 0 \\ \psi_{-s-1} & \text{if } s < 0. \end{cases}$$

Conversely, an n-dimensional normal complex (C,ϕ,γ,χ) with $\gamma = 0$ determines an n-dimensional quadratic complex (C,ψ), by virtue of $Q_n(C,0) = Q_n(C)$.

$\square$

DEFINITION 2.7 (i) An n-dimensional (symmetric, quadratic) pair $(f\colon C\longrightarrow D, (\delta\phi,\psi))$ in $\mathbb{A}$ is an n-dimensional symmetric pair with a quadratic structure on the boundary, i.e. a chain map $f\colon C\longrightarrow D$ of finite chain complexes in $\mathbb{A}$ together with an $(n-1)$-cycle $\psi \in (W_\% C)_{n-1}$ and an n-chain $\delta\phi \in (W^\% D)_n$ such that

$$f^\%(1+T)\psi \ = \ d(\delta\phi) \in (W^\% D)_{n-1} \ .$$

(ii) The pair $(f\colon C\longrightarrow D, (\delta\phi,\psi))$ is Poincaré if the chain map

$$(\delta\phi,(1+T)\psi)_0 : D^{n-*} \longrightarrow C(f)$$

is a chain equivalence.

$\square$

PROPOSITION 2.8 (i) The homotopy equivalence classes of n-dimensional (symmetric, quadratic) pairs in $\mathbb{A}$ are in natural one–one correspondence with the homotopy equivalence classes of n-dimensional normal complexes in $\mathbb{A}$.

(ii) *The cobordism classes of n-dimensional (normal, symmetric Poincaré) pairs in $\mathbb{A}$ are in natural one–one correspondence with the cobordism classes of $(n-1)$-dimensional quadratic Poincaré complexes in $\mathbb{A}$.*

PROOF (i) An n-dimensional normal complex (C, ϕ, γ, χ) in $\mathbb{A}$ determines the n-dimensional (symmetric, quadratic) Poincaré pair in $\mathbb{A}$

$$(i_C : \partial C \longrightarrow C^{n-*}, (\delta\phi, \psi))$$

defined by

$$i_C = (0 \; 1) : \partial C_r = C_{r+1} \oplus C^{n-r} \longrightarrow C^{n-r} \, ,$$

$$d_{\partial C} = \begin{pmatrix} d_C & (-)^r \phi_0 \\ 0 & (-)^r d_C^* \end{pmatrix} :$$

$$\partial C_r = C_{r+1} \oplus C^{n-r} \longrightarrow \partial C_{r-1} = C_r \oplus C^{n-r+1} \, ,$$

$$\psi_0 = \begin{pmatrix} \chi_0 & 0 \\ 1 + \gamma_{-n}\phi_0^* & \gamma_{-n-1}^* \end{pmatrix} :$$

$$\partial C^r = C^{r+1} \oplus C_{n-r} \longrightarrow \partial C_{n-r-1} = C_{n-r} \oplus C^{r+1} \, ,$$

$$\psi_s = \begin{pmatrix} \chi_{-s} & 0 \\ \gamma_{-n-s}\phi_0^* & \gamma_{-n-s-1}^* \end{pmatrix} :$$

$$\partial C^r = C^{r+1} \oplus C_{n-r} \longrightarrow \partial C_{n-r-s-1} = C_{n-r-s} \oplus C^{r+s+1} \; (s \geq 1) \, ,$$

$$\delta\phi_s = \gamma_{-n-s} : C_r \longrightarrow C^{n-r+s} \; (s \geq 0) \, .$$

Conversely, an n-dimensional (symmetric, quadratic) Poincaré pair $(f : C \longrightarrow D, (\delta\phi, \psi))$ in $\mathbb{A}$ determines an n-dimensional normal complex $(C(f), \phi, \gamma, \chi)$ in $\mathbb{A}$ with the symmetric structure

$$\phi_s = \begin{cases} \begin{pmatrix} \delta\phi_0 & 0 \\ (1+T)\psi_0 f^* & 0 \end{pmatrix} & \text{if } s = 0 \\[2ex] \begin{pmatrix} \delta\phi_1 & 0 \\ 0 & (1+T)\psi_0 \end{pmatrix} & \text{if } s = 1 \\[2ex] \begin{pmatrix} \delta\phi_s & 0 \\ 0 & 0 \end{pmatrix} & \text{if } s \geq 2 \end{cases}$$

$$: C(f)^r = D^r \oplus C^{r-1} \longrightarrow C(f)_{n-r+s} = D_{n-r+s} \oplus C_{n-r+s-1} \, .$$

The normal structure (γ, χ) is determined up to equivalence by the Poincaré duality, with $\gamma \in \widehat{Q}^0(D^{-*})$ the image of $(\delta\phi/(1+T)\psi) \in Q^n(C(f))$ under the composite

$$Q^n(C(f)) \xrightarrow{((\delta\phi_0, (1+T)\psi_0)^\%)^{-1}} Q^n(D^{n-*}) \xrightarrow{J} \widehat{Q}^n(D^{n-*}) \xrightarrow{S^{-n}} \widehat{Q}^0(D^{-*}) \, .$$

(ii) Given an n-dimensional normal pair in $\mathbb{A}$ $(f : C \longrightarrow D, ((\delta\phi, \delta\gamma, \delta\chi), \phi))$

let (C',ϕ') be the $(n-1)$-dimensional symmetric complex obtained from (C,ϕ) by surgery on $(f:C\longrightarrow D,(\delta\phi,\phi))$. The trace of the surgery is an n-dimensional symmetric pair $((g\ g'):C\oplus C'\longrightarrow D',(\delta\phi',\phi\oplus-\phi'))$ with

$$g\ =\ \text{inclusion}: C \longrightarrow D' = C(\phi_0 f^*:D^{n-1-*}\longrightarrow C)\ ,$$

$$g'\ =\ \text{projection}: C' = S^{-1}C((\delta\phi,\phi)_0:D^{n-*}\longrightarrow C(f)) \longrightarrow D'\ .$$

The natural isomorphism

$$\widehat{Q}^{n+1}(D^{n+1-*}) \xrightarrow{\ \sim\ } \widehat{Q}^{n+1}(D'^{n-*}\longrightarrow C'^{n-*})$$

sends the chain bundle $\delta\phi \in \widehat{Q}^{n+1}(D^{n+1-*})$ to a normal structure on the trace which restricts to $0 \in \widehat{Q}^n(C'^{n-*})$, corresponding to a quadratic refinement $\psi' \in Q_{n-1}(C')$ of $\phi' \in Q^{n-1}(C')$. The symmetric complex (C,ϕ) is Poincaré if and only if the quadratic complex (C',ψ') is Poincaré.

Conversely, given an $(n-1)$-dimensional quadratic Poincaré complex (C',ψ') define an n-dimensional (normal, symmetric Poincaré) pair $(C\longrightarrow 0,(0,(1+T)\psi'))$.

$\square$

DEFINITION 2.9 The *quadratic boundary* of an n-dimensional normal complex (C,ϕ,γ,χ) is the $(n-1)$-dimensional quadratic Poincaré complex

$$\partial(C,\phi,\gamma,\chi)\ =\ (\partial C,\psi)$$

defined in 2.8 (i) above, with $\partial C = S^{-1}C(\phi_0)$ the desuspension of the algebraic mapping cone of the duality chain map $\phi_0:C^{n-*}\longrightarrow C$. This can also be viewed as the complex associated by 2.8 (ii) to the n-dimensional (normal, symmetric Poincaré) pair $(C\longrightarrow 0,(0,(1+T)\psi))$.

$\square$

DEFINITION 2.10 The *n-dimensional hyperquadratic L-group* $\widehat{L}^n(\mathbb{A})$ $(n\in\mathbb{Z})$ is the cobordism group of n-dimensional (symmetric, quadratic) Poincaré pairs in $\mathbb{A}$, designed to fit into the quadratic-symmetric exact sequence

$$\ldots \longrightarrow L_n(\mathbb{A}) \xrightarrow{1+T} L^n(\mathbb{A}) \xrightarrow{J} \widehat{L}^n(\mathbb{A}) \xrightarrow{\partial} L_{n-1}(\mathbb{A}) \longrightarrow \ldots\ .$$

$\square$

For a ring with involution R and $\mathbb{A} = \mathbb{A}^p(R)$ the hyperquadratic L-groups $\widehat{L}^*(\mathbb{A})$ of 2.10 are just the hyperquadratic L-groups $\widehat{L}^*(R)$ of Ranicki [143, p. 137].

PROPOSITION 2.11 *The hyperquadratic L-groups $\widehat{L}^*(\mathbb{A})$ are isomorphic to the cobordism groups $NL^*(\mathbb{A})$ of normal complexes in $\mathbb{A}$*

$$\widehat{L}^*(\mathbb{A})\ \cong\ NL^*(\mathbb{A})\ ,$$

so that there is defined an exact sequence

$$\ldots \longrightarrow L_n(\mathbb{A}) \xrightarrow{1+T} L^n(\mathbb{A}) \xrightarrow{J} NL^n(\mathbb{A}) \xrightarrow{\partial} L_{n-1}(\mathbb{A}) \longrightarrow \ldots$$

with

$$\partial \,:\, NL^n(\mathbb{A}) \,\longrightarrow\, L_{n-1}(\mathbb{A}) \;;\; (C,\phi,\gamma,\chi) \,\longrightarrow\, (\partial C, \psi)$$

given by the quadratic boundary (2.9) of normal complexes.

PROOF The identities $\widehat{L}^n(\mathbb{A}) = NL^n(\mathbb{A})$ $(n \in \mathbb{Z})$ are immediate from 2.8 (i) and its relative version relating (symmetric, quadratic) Poincaré triads and normal pairs. See Ranicki [143, §2.1] for algebraic Poincaré triads.

□

In the case $\mathbb{A} = \mathbb{A}^q(R)$ $(q = h, p)$ for a ring with involution R write the normal L-groups as

$$NL^*(\mathbb{A}^q(R)) \;=\; NL_q^*(R) \,.$$

EXAMPLE 2.12 The hyperquadratic L-groups $\widehat{L}_q^*(R)$ $(q = h, p)$ of Ranicki [143, p. 137] are the cobordism groups of (symmetric, quadratic) Poincaré pairs over a ring with involution R which fit into an exact sequence

$$\cdots \,\longrightarrow\, L_n^q(R) \,\overset{1+T}{\longrightarrow}\, L_q^n(R) \,\overset{J}{\longrightarrow}\, \widehat{L}_q^n(R) \,\overset{\partial}{\longrightarrow}\, L_{n-1}^q(R) \,\longrightarrow\, \cdots \,.$$

The relative terms $\widehat{H}^*(\mathbb{Z}_2\,;\widetilde{K}_0(R))$ in the Rothenberg exact sequences relating the free and projective L-groups of R are the same for the symmetric and quadratic L-groups

$$\cdots \,\longrightarrow\, L_h^n(R) \,\longrightarrow\, L_p^n(R) \,\longrightarrow\, \widehat{H}^n(\mathbb{Z}_2\,;\widetilde{K}_0(R)) \,\longrightarrow\, L_h^{n-1}(R) \,\longrightarrow\, \cdots \,,$$

$$\cdots \,\longrightarrow\, L_n^h(R) \,\longrightarrow\, L_n^p(R) \,\longrightarrow\, \widehat{H}^n(\mathbb{Z}_2\,;\widetilde{K}_0(R)) \,\longrightarrow\, L_{n-1}^h(R) \,\longrightarrow\, \cdots \,.$$

Thus the free and projective hyperquadratic L-groups of R coincide

$$\widehat{L}^*(R) \;=\; \widehat{L}_h^*(R) \;=\; \widehat{L}_p^*(R) \,.$$

Similarly, the hyperquadratic L-groups of the categories $\mathbb{A}^h(R)$ and $\mathbb{A}^p(R)$ coincide, being the 4-periodic versions of the hyperquadratic L-groups $\widehat{L}^*(R)$

$$\widehat{L}^n(\mathbb{A}^h(R)) \;=\; \widehat{L}^n(\mathbb{A}^p(R)) \;=\; \varinjlim_k \widehat{L}^{n+4k}(R) \quad (n \in \mathbb{Z}) \,,$$

the direct limits being taken with respect to the double skew-suspension maps. Use the isomorphisms given by 2.11

$$NL^*(\mathbb{A}^q(R)) \;\cong\; \widehat{L}^*(\mathbb{A}^q(R)) \quad (q = h, p)$$

to write

$$NL^*(R) \;=\; NL_h^*(R) \;=\; NL_p^*(R) \;=\; \varinjlim_k \widehat{L}^{*+4k}(R) \,.$$

□

REMARK 2.13 The exact sequence of 2.11 for $\mathbb{A} = \mathbb{A}(R) = \mathbb{A}^h(R)$ is the algebraic analogue of the exact sequence of Levitt [89], Jones [77], Quinn [129] and Hausmann and Vogel [72]

$$\cdots \,\longrightarrow\, \Omega_{n+1}^N(K) \,\longrightarrow\, L_n(\mathbb{Z}[\pi]) \,\longrightarrow\, \Omega_n^P(K) \,\longrightarrow\, \Omega_n^N(K) \,\longrightarrow\, \cdots \,,$$

with $\Omega_n^P(K)$ (resp. $\Omega_n^N(K)$) the bordism group of maps $X \longrightarrow K$ from n-dimensional geometric Poincaré (resp. normal) complexes, with $\pi = \pi_1(K)$ the fundamental group of K and $n \geq 5$. The *symmetric signature* of Mishchenko [112] and Ranicki [142, §1] defines a map from geometric to symmetric Poincaré bordism

$$\sigma^* : \Omega_n^P(K) \longrightarrow L^n(\mathbb{Z}[\pi]) \; ; \; X \longrightarrow \sigma^*(X) = (C(\tilde{X}), \phi) \, .$$

The *hyperquadratic signature* of Ranicki [143, p. 619] defines a map from geometric to algebraic normal bordism

$$\hat{\sigma}^* : \Omega_n^N(K) \longrightarrow \hat{L}^n(\mathbb{Z}[\pi]) \; ; \; X \longrightarrow \hat{\sigma}^*(X) = (C(\tilde{X}), \phi, \gamma, \chi) \, .$$

The signature maps fit together to define a map of exact sequences

$$\begin{array}{ccccccccc}
\cdots & \longrightarrow & \Omega_{n+1}^N(K) & \xrightarrow{\sigma_*} & L_n(\mathbb{Z}[\pi]) & \longrightarrow & \Omega_n^P(K) & \longrightarrow & \Omega_n^N(K) & \longrightarrow & \cdots \\
& & \hat{\sigma}^* \downarrow & & \| & & \sigma^* \downarrow & & \hat{\sigma}^* \downarrow & & \\
\cdots & \longrightarrow & \hat{L}^{n+1}(\mathbb{Z}[\pi]) & \xrightarrow{\partial} & L_n(\mathbb{Z}[\pi]) & \xrightarrow{1+T} & L^n(\mathbb{Z}[\pi]) & \xrightarrow{J} & \hat{L}^n(\mathbb{Z}[\pi]) & \longrightarrow & \cdots
\end{array}$$

The *normal signature* is the stable hyperquadratic signature

$$\hat{\sigma}^* : \Omega_n^N(K) \longrightarrow NL^n(\mathbb{Z}[\pi]) = \varinjlim_k \hat{L}^{n+4k}(\mathbb{Z}[\pi]) \, .$$

The normal signature determines the *quadratic signature*

$$\sigma_* = \partial \hat{\sigma}^* : \Omega_n^N(K) \longrightarrow \varinjlim_k L_{n+4k-1}(\mathbb{Z}[\pi]) = L_{n-1}(\mathbb{Z}[\pi]) \, .$$

There is also a twisted version for a double covering $K^w \longrightarrow K$, with the w-twisted involution on $\mathbb{Z}[\pi]$, and the bordism groups $\Omega_*(K, w)$ of maps $X \longrightarrow K$ such that the pullback $X^w \longrightarrow X$ is the orientation double cover.

$\square$

EXAMPLE 2.14 (i) Let R be a ring with involution, and let (B, β) be a chain bundle over R, with B a free R-module chain complex (not necessarily finite or finitely generated). The cobordism groups $L^n(B, \beta)$ $(n \geq 0)$ of n-dimensional symmetric Poincaré complexes (C, ϕ, γ, χ) in $\mathbb{A}(R)$ with a chain bundle map $(f, b) \colon (C, \gamma) \longrightarrow (B, \beta)$ fit into an exact sequence

$$\cdots \longrightarrow L_n(R) \longrightarrow L^n(B, \beta) \longrightarrow Q_n(B, \beta) \xrightarrow{\partial} L_{n-1}(R) \longrightarrow \cdots$$

with

$$L_n(R) \longrightarrow L^n(B, \beta) \; ; \; (C, \psi) \longrightarrow ((C, (1 + T)\psi, 0, \psi), 0) \, ,$$

$$L^n(B, \beta) \longrightarrow Q_n(B, \beta) \; ; \; ((C, \phi, \gamma, \chi), (f, b)) \longrightarrow (f, b)_\% (\gamma, \chi) \, ,$$

$$\partial : Q_n(B, \beta) \longrightarrow L_{n-1}(R) \; ; \; (\phi, \chi) \longrightarrow \partial(\bar{B}, \bar{\phi}, \bar{\beta}, \bar{\chi}) \, ,$$

where $(\bar{B}, \bar{\phi}, \bar{\beta}, \bar{\chi})$ the restriction of (B, ϕ, β, χ) to any finite subcomplex $\bar{B} \subset B$ supporting $(\phi, \chi) \in Q_n(B, \beta)$. As in Weiss [182] there is defined a universal chain bundle (B, β) over R, with $\beta \in \widehat{Q}^0(B^{-*})$ such that the algebraic Wu classes of Ranicki [143, 1.4] are isomorphisms

$$\widehat{v}^r(\beta) : H_r(B) \xrightarrow{\simeq} \widehat{H}^r(\mathbb{Z}_2 ; R) ; \quad x \longrightarrow \beta_{-2r}(x)(x) \quad (r \in \mathbb{Z}) .$$

For the universal chain bundle (B, β) and any finite chain complex C in $\mathbb{A}(R)$ there is defined an isomorphism

$$H_0(\mathrm{Hom}_R(C, B)) \xrightarrow{\simeq} \widehat{Q}^0(C^{-*}) ; \quad f \longrightarrow \widehat{f}^{\%}(\beta)$$

so that the chain bundles $(C, \gamma \in \widehat{Q}^0(C^{-*}))$ are classified up to homotopy equivalence by the chain homotopy classes of chain maps $C \longrightarrow B$. For universal (B, β) the forgetful maps define isomorphisms

$$L^n(B, \beta) \xrightarrow{\simeq} L^n(R) ; \quad (C, \phi, \gamma, \chi) \longrightarrow (C, \phi) ,$$

$$Q_n(B, \beta) \xrightarrow{\simeq} NL^n(R) ; \quad (\phi, \chi) \longrightarrow (\bar{B}^{n-*}, \partial(\bar{B}, \bar{\phi}, \bar{\beta}, \bar{\chi})) .$$

(ii) Let K be a field of characteristic 2 which is perfect, i.e. such that $K \longrightarrow K; x \longrightarrow x^2$ is an isomorphism, so that for all $n \in \mathbb{Z}$

$$\widehat{H}^n(\mathbb{Z}_2 ; K) = K , \quad K \times \widehat{H}^n(\mathbb{Z}_2 ; K) \longrightarrow \widehat{H}^n(\mathbb{Z}_2 ; K) ; \quad (x, y) \longrightarrow x^2 y$$

with the identity involution on K. The chain bundle over K

$$(B : \dots \xrightarrow{0} K \xrightarrow{0} K \xrightarrow{0} K \xrightarrow{0} K \xrightarrow{0} \dots , \ \beta = 1)$$

is universal. The quadratic Witt group $L_{2*}(K)$ is detected by the Arf invariant, and the symmetric Witt group $L^{2*}(K)$ is detected by the rank (mod 2), with isomorphisms

$$Q_{2*+1}(B, \beta) = K / \{x + x^2 \,|\, x \in K\}$$

$$\xrightarrow{\simeq} L_{2*}(K) ; \quad a \longrightarrow \left(K \oplus K, \begin{pmatrix} a & 1 \\ 0 & 1 \end{pmatrix} \right) ,$$

$$Q_{2*}(B, \beta) = \{x \in K \,|\, x + x^2 = 0\} = \mathbb{Z}_2$$

$$\xrightarrow{\simeq} NL^{2*}(K) = L^{2*}(K) ; \quad 1 \longrightarrow (K, 1)$$

and $L_{2*+1}(K) = L^{2*+1}(K) = 0$. In particular, this applies to $K = \mathbb{F}_2$.

□

By analogy with the observation of Quinn [129] that the mapping cylinder of a map of geometric normal complexes defines a cobordism, we have:

PROPOSITION 2.15 *The algebraic mapping cylinder of a map of n-dimensional normal complexes in* $\mathbb{A}$

$$(f, b) : (C', \phi', \gamma', \chi') \longrightarrow (C, \phi, \gamma, \chi)$$

is an $(n+1)$-dimensional normal pair in $\mathbb{A}$

$$M(f, b) =$$

$$((f \ 1) : C' \oplus C \to C , \ ((\delta\phi, \gamma, \delta\chi), (\phi' \oplus -\phi, \gamma' \oplus -\gamma, \chi' \oplus -\chi)), \ b \oplus 0) ,$$

which defines a cobordism between (C, ϕ, γ, χ) *and* (C, ϕ, γ, χ).

PROOF The chains $\delta\phi, \delta\chi$ are determined by a chain level representative for the identity

$$(f, b)\%(\phi', \chi') = (\phi, \chi) \in Q_n(C, \gamma) .$$

□

REMARK 2.16 (i) Let $\mathbb{A}$ be an additive category with a 0-dimensional chain duality. An *algebraic normal map* in $\mathbb{A}$ is a normal map of n-dimensional symmetric Poincaré complexes

$$(f, b) : (C', \phi', \gamma', \chi') \longrightarrow (C, \phi, \gamma, \chi) .$$

The algebraic mapping cylinder $M(f, b)$ of 2.15 is an $(n + 1)$-dimensional (normal, symmetric Poincaré) pair. The *quadratic kernel* of (f, b) is the n-dimensional quadratic Poincaré complex

$$\sigma_*(f, b) = (C(f^!), \psi)$$

obtained by applying the construction of 2.8 (ii) to $M(f, b)$, with $f^!$ the Umkehr chain map defined up to chain homotopy by the composite

$$f^! : C \xrightarrow{(\phi_0)^{-1}} C^{n-*} \xrightarrow{f^*} C'^{n-*} \xrightarrow{\phi'_0} C' .$$

The symmetrization of the quadratic kernel is an n-dimensional symmetric Poincaré complex

$$(1 + T)\sigma_*(f, b) = (C(f^!), (1 + T)\psi)$$

such that up to homotopy equivalence

$$(1 + T)\sigma_*(f, b) \oplus (C, \phi) = (C', \phi') .$$

The construction of 2.8 (ii) defines an isomorphism between the cobordism group of $(n + 1)$-dimensional (normal, symmetric Poincaré) pairs in $\mathbb{A}$ and the quadratic L-group $L_n(\mathbb{A})$. The *quadratic signature* of (f, b) is the cobordism class of the quadratic kernel

$$\sigma_*(f, b) = (C(f^!), \psi) \in L_n(\mathbb{A}) .$$

The methods of Ranicki [141], [145] show that $\sigma_*(f, b) = 0 \in L_n(\mathbb{A})$ if and only if $M(f, b)$ is algebraic normal cobordant rel ∂ to a symmetric Poincaré cobordism between (C, ϕ, γ, χ) and $(C', \phi', \gamma', \chi')$.

(ii) The quadratic kernel $\sigma_*(f, b)$ of a geometric normal map $(f, b): X' \longrightarrow X$ of n-dimensional geometric Poincaré complexes obtained in Ranicki [142] is the quadratic kernel $\sigma_*(\tilde{f}, \tilde{b})$ of an induced algebraic normal map of n-dimensional symmetric Poincaré complexes in $\mathbb{A}(\mathbb{Z}[\pi]^w)$

$$(\tilde{f}, \tilde{b}) : \sigma^*(X') = (C', \phi', \gamma', \chi') \longrightarrow \sigma^*(X) = (C, \phi, \gamma, \chi) ,$$

with $w: \pi \longrightarrow \mathbb{Z}_2$ the orientation map, and $C = C(\tilde{X})$, $C' = C(\tilde{X}')$ the cellular chain complexes of the cover $\tilde{X}$ of X and the pullback cover $\tilde{X}'$ of

X'. The quadratic signature of (f, b) is the cobordism class of $(C(f'), \psi)$

$$\sigma_*(f, b) = \sigma_*(\tilde{f}, \tilde{b}) = (C(f'), \psi) \in L_n(\mathbb{A}(\mathbb{Z}[\pi]^w)) = L_n(\mathbb{Z}[\pi]^w) ,$$

with symmetrization

$$(1 + T)\sigma_*(f, b) = \sigma^*(X') - \sigma^*(X) \in L^n(\mathbb{Z}[\pi]^w) .$$

For $X' = M$ a manifold and $(f, b): M \longrightarrow X$ a geometric normal map in the sense of Browder [16] the surgery obstruction of Wall [176] is the quadratic signature of (f, b) with $\pi = \pi_1(X)$ and $\tilde{X}$ the universal cover of X.

(iii) Geometric normal complexes can be constructed from geometric Poincaré bordisms of degree 1 normal maps of geometric Poincaré complexes, as follows. Given a normal map $(f, b): X' \longrightarrow X$ of n-dimensional geometric Poincaré complexes let $W \simeq X$ be the mapping cylinder of f, so that $(W; X, X')$ is an $(n + 1)$-dimensional normal complex cobordism. Given also a geometric Poincaré cobordism $(V; X, X')$ there is defined an $(n + 1)$-dimensional geometric normal complex

$$Y = V \cup_\partial W .$$

The normal signature of Y is the stable hyperquadratic signature

$$\hat{\sigma}^*(Y) = (C(\tilde{Y}), \phi, \gamma, \chi) \in NL^{n+1}(\mathbb{Z}[\pi_1(Y)]) = \varinjlim_k \widehat{L}^{n+4k+1}(\mathbb{Z}[\pi_1(Y)]) ,$$

with boundary the quadratic signature of (f, b) relative to $\pi_1(X) \longrightarrow \pi_1(Y)$

$$\partial \hat{\sigma}^*(Y) = \sigma_*(f, b) \in L_n(\mathbb{Z}[\pi_1(Y)]) .$$

(iv) For the mapping cylinder W of the 2-dimensional normal map

$$(f, b) : X' = S^1 \times S^1 \longrightarrow X = S^2$$

determined by the exotic framing of $S^1 \times S^1$ with Kervaire–Arf invariant 1 and for the geometric Poincaré cobordism

$$(V; X, X') = (D^3 \sqcup S^1 \times D^2; S^2, S^1 \times S^1)$$

the construction of (iii) gives a simply-connected 3-dimensional geometric normal complex $Y = V \cup_\partial W$ such that

$$\partial \hat{\sigma}^*(Y) = \sigma_*(f, b) = 1 \in L_2(\mathbb{Z}) = \mathbb{Z}_2 .$$

Thus Y is not normal bordant to a geometric Poincaré complex, and (a fortiori) the normal fibration $\nu_Y: Y \longrightarrow BSG$ is not topologically reducible, with $\nu_Y: Y \simeq S^2 \vee S^3 \longrightarrow S^3 \longrightarrow BSG$ detected by the generator

$$1 \in \pi_3(BSG) = \pi_2(G/TOP) = \pi_2^s = \Omega_2^{fr} = L_2(\mathbb{Z}) = \mathbb{Z}_2 .$$

□

From now on the normal structure (γ, χ) will be suppressed from the terminology of a normal complex (C, ϕ, γ, χ), which will be written as (C, ϕ).

§3. Algebraic bordism categories

An algebraic bordism category $\Lambda = (\mathbb{A}, \mathbb{B}, \mathbb{C})$ is a triple defined by an additive category with chain duality $\mathbb{A}$ and a pair $(\mathbb{B}, \mathbb{C} \subseteq \mathbb{B})$ of additive categories of chain complexes in $\mathbb{A}$ satisfying certain conditions. The L-groups

$$\begin{cases} L^*(\Lambda) \\ L_*(\Lambda) \\ NL^*(\Lambda) \end{cases} \text{ of } \Lambda \text{ are defined to be the cobordism groups of } \begin{cases} \text{symmetric} \\ \text{quadratic} \\ \text{normal} \end{cases}$$

complexes in $\mathbb{A}$ which are $\mathbb{B}$-contractible and $\mathbb{C}$-Poincaré. The main result of §3 is the exact sequence relating quadratic, symmetric and normal L-groups of an algebraic bordism category.

As in §§1,2 let $\mathbb{A}$ be an additive category with chain duality, and let $\mathbb{B}(\mathbb{A})$ be the additive category of finite chain complexes in $\mathbb{A}$ and chain maps.

DEFINITION 3.1 (i) A subcategory $\mathbb{C} \subseteq \mathbb{B}(\mathbb{A})$ is *closed* if it is a full additive subcategory such that the algebraic mapping cone $C(f)$ of any chain map $f \colon C \longrightarrow D$ in $\mathbb{C}$ is an object in $\mathbb{C}$.

(ii) A chain complex C in $\mathbb{A}$ is $\mathbb{C}$-*contractible* if it belongs to $\mathbb{C}$. A chain map $f \colon C \longrightarrow D$ in $\mathbb{A}$ is a $\mathbb{C}$-*equivalence* if the algebraic mapping cone $C(f)$ is $\mathbb{C}$-contractible.

(iii) An n-dimensional $\begin{cases} \text{symmetric} \\ \text{quadratic} \end{cases}$ complex $\begin{cases} (C, \phi) \\ (C, \psi) \end{cases}$ in $\mathbb{A}$ is $\mathbb{C}$-*contractible* if the chain complexes C and C^{n-*} are $\mathbb{C}$-contractible.

(iv) An n-dimensional $\begin{cases} \text{symmetric} \\ \text{quadratic} \end{cases}$ complex $\begin{cases} (C, \phi) \\ (C, \psi) \end{cases}$ in $\mathbb{A}$ is $\mathbb{C}$-*Poincaré* if the chain complex

$$\begin{cases} \partial C = S^{-1}C(\phi_0 \colon C^{n-*} \longrightarrow C) \\ \partial C = S^{-1}C((1+T)\psi_0 \colon C^{n-*} \longrightarrow C) \end{cases}$$

is $\mathbb{C}$-contractible.

□

DEFINITION 3.2 An *algebraic bordism category* $\Lambda = (\mathbb{A}, \mathbb{B}, \mathbb{C})$ is an additive category $\mathbb{A}$ with a chain duality $T \colon \mathbb{A} \longrightarrow \mathbb{B}(\mathbb{A})$, together with a pair $(\mathbb{B}, \mathbb{C} \subseteq \mathbb{B})$ of closed subcategories of $\mathbb{B}(\mathbb{A})$, such that for any object B in $\mathbb{B}$

(i) the algebraic mapping cone $C(1 \colon B \longrightarrow B)$ is an object in $\mathbb{C}$,

(ii) the chain equivalence $e(B) \colon T^2(B) \overset{\simeq}{\longrightarrow} B$ is a $\mathbb{C}$-equivalence.

□

EXAMPLE 3.3 For any additive category with chain duality $\mathbb{A}$ there is defined an algebraic bordism category

$$\Lambda(\mathbb{A}) = (\mathbb{A}, \mathbb{B}(\mathbb{A}), \mathbb{C}(\mathbb{A}))$$

with $\mathbb{B}(\mathbb{A})$ the category of finite chain complexes in $\mathbb{A}$, and $\mathbb{C}(\mathbb{A}) \subseteq \mathbb{B}(\mathbb{A})$

the subcategory of contractible complexes.

□

DEFINITION 3.4 Let $\Lambda = (\mathbb{A}, \mathbb{B}, \mathbb{C})$ be an algebraic bordism category.

(i) An n-dimensional $\left\{\begin{array}{l} symmetric \\ quadratic \\ normal \end{array}\right.$ complex $\left\{\begin{array}{l} (C, \phi) \\ (C, \psi) \\ (C, \phi) \end{array}\right.$ in Λ is an n-dimen-

sional $\left\{\begin{array}{l} symmetric \\ quadratic \\ normal \end{array}\right.$ complex in $\mathbb{A}$ which is $\mathbb{B}$-contractible and $\mathbb{C}$-Poincaré.

Similarly for pairs and cobordisms.

(ii) The $\left\{\begin{array}{l} symmetric \\ quadratic \\ normal \end{array}\right.$ L-groups $\left\{\begin{array}{l} L^n(\Lambda) \\ L_n(\Lambda) \\ NL^n(\Lambda) \end{array}\right.$ $(n \in \mathbb{Z})$ are the cobordism

groups of n-dimensional $\left\{\begin{array}{l} symmetric \\ quadratic \\ normal \end{array}\right.$ complexes in Λ.

□

PROPOSITION 3.5 *If* $\Lambda = (\mathbb{A}, \mathbb{B}, \mathbb{C})$ *is an algebraic bordism category such that* $\widehat{Q}^*(C) = 0$ *for any* $\mathbb{C}$-*contractible finite chain complex* C *in* $\mathbb{A}$ *then the forgetful maps define isomorphisms*

$$NL^n(\Lambda) \xrightarrow{\simeq} L^n(\Lambda) \; ; \; (C, \phi, \gamma, \chi) \longrightarrow (C, \phi) \; (n \in \mathbb{Z}) \; .$$

PROOF An n-dimensional symmetric complex (C, ϕ) in $\mathbb{A}$ has a normal structure if and only if

$$J(\phi) \in \text{im}(\widehat{\phi}_0^{\%} \colon \widehat{Q}^n(C^{n-*}) \longrightarrow \widehat{Q}^n(C)) \; .$$

The hyperquadratic Q-groups of C, C^{n-*} and $\partial C = S^{-1}C(\phi_0 \colon C^{n-*} \longrightarrow C)$ are related by an exact sequence

$$\ldots \longrightarrow \widehat{Q}^n(\partial C) \longrightarrow \widehat{Q}^n(C^{n-*}) \xrightarrow{\widehat{\phi}_0^{\%}} \widehat{Q}^n(C) \longrightarrow \widehat{Q}^{n-1}(\partial C) \longrightarrow \ldots \; .$$

If (C, ϕ) is $\mathbb{C}$-Poincaré then ∂C is $\mathbb{C}$-contractible, $\widehat{Q}^*(\partial C) = 0$ and there is defined an isomorphism

$$\widehat{\phi}_0^{\%} : \widehat{Q}^n(C^{n-*}) \xrightarrow{\simeq} \widehat{Q}^n(C) \; ,$$

so that (C, ϕ) has a normal structure. Similarly for pairs.

□

EXAMPLE 3.6 The algebraic bordism category $\Lambda(\mathbb{A}) = (\mathbb{A}, \mathbb{B}(\mathbb{A}), \mathbb{C}(\mathbb{A}))$ of 3.3 is such that $\widehat{Q}^*(C) = 0$ for $\mathbb{C}(\mathbb{A})$-contractible ($=$ contractible) $\mathbb{B}(\mathbb{A})$-contractible ($=$ any) finite chain complexes in $\mathbb{A}$, so that

$$NL^*(\Lambda(\mathbb{A})) = L^*(\Lambda(\mathbb{A})) = L^*(\mathbb{A}) \; .$$

□

DEFINITION 3.7 A *functor* of algebraic bordism categories
$$F : \Lambda = (\mathbb{A}, \mathbb{B}, \mathbb{C}) \longrightarrow \Lambda' = (\mathbb{A}', \mathbb{B}', \mathbb{C}')$$
is a (covariant) functor $F: \mathbb{A} \longrightarrow \mathbb{A}'$ of the additive categories, such that
(i) $F(B)$ is an object in $\mathbb{B}'$ for any object B in $\mathbb{B}$,
(ii) $F(C)$ is an object in $\mathbb{C}'$ for every object C in $\mathbb{C}$,
(iii) for every object A in $\mathbb{A}$ there is given a natural $\mathbb{C}'$-equivalence
$$G(A) : T'F(A) \xrightarrow{\simeq} FT(A)$$
with a commutative diagram

$$
\begin{array}{ccc}
T'FT(A) & \xrightarrow{GT(A)} & FT^2(A) \\
{\scriptstyle T'G(A)}\big\downarrow & & \big\downarrow{\scriptstyle Fe(A)} \\
T'^2 F(A) & \xrightarrow{e'F(A)} & F(A) .
\end{array}
$$

$\square$

PROPOSITION 3.8 *A functor of algebraic bordism categories*
$$F : \Lambda = (\mathbb{A}, \mathbb{B}, \mathbb{C}) \longrightarrow \Lambda' = (\mathbb{A}', \mathbb{B}', \mathbb{C}')$$
induces morphisms of L-groups
$$\begin{cases} F : L^*(\Lambda) \longrightarrow L^*(\Lambda') \\ F : L_*(\Lambda) \longrightarrow L_*(\Lambda') \\ F : NL^*(\Lambda) \longrightarrow NL^*(\Lambda') \end{cases}$$
and there are defined relative L-groups $\begin{cases} L^*(F) \\ L_*(F) \\ NL^*(F) \end{cases}$ *to fit into a long exact sequence*
$$\begin{cases} \dots \longrightarrow L^n(\Lambda) \xrightarrow{F} L^n(\Lambda') \longrightarrow L^n(F) \longrightarrow L^{n-1}(\Lambda) \longrightarrow \dots \\ \dots \longrightarrow L_n(\Lambda) \xrightarrow{F} L_n(\Lambda') \longrightarrow L_n(F) \longrightarrow L_{n-1}(\Lambda) \longrightarrow \dots \\ \dots \longrightarrow NL^n(\Lambda) \xrightarrow{F} NL^n(\Lambda') \longrightarrow NL^n(F) \longrightarrow NL^{n-1}(\Lambda) \longrightarrow \dots \end{cases}$$

PROOF For any objects M, N in $\mathbb{A}$ define a chain map of abelian group chain complexes
$$F(M,N) : M \otimes_\mathbb{A} N \longrightarrow F(M) \otimes_{\mathbb{A}'} F(N) ;$$
$$(\phi: T(M) \longrightarrow N) \longrightarrow (F(\phi)G(M): T'F(M) \longrightarrow FT(M) \longrightarrow F(N))$$
which is compatible with the duality equivalences. An n-dimensional symmetric complex (C, ϕ) in Λ induces an n-dimensional symmetric complex $(F(C), F(\phi))$ in Λ'. Similarly for quadratic and normal complexes, and also for pairs. Working as in Ranicki [143, §2] define the relative L-group $L^n(F)$

to be the cobordism group of pairs

$((n-1)$-dimensional symmetric complex (C, ϕ) in Λ ,

n-dimensional symmetric pair $(F(C) \longrightarrow D, (\delta\phi, F(\phi)))$ in $\Lambda')$.

Similarly for the quadratic and normal cases.

□

PROPOSITION 3.9 *Let* $\mathbb{A}$ *be an additive category with chain duality, and let* $(\mathbb{B} \subseteq \mathbb{B}(\mathbb{A}), \mathbb{C} \subseteq \mathbb{B}, \mathbb{D} \subseteq \mathbb{C})$ *be a triple of closed subcategories of* $\mathbb{B}(\mathbb{A})$. *The relative L-groups of the functor of algebraic bordism categories*

$$F : \Lambda' = (\mathbb{A}, \mathbb{B}, \mathbb{D}) \longrightarrow \Lambda = (\mathbb{A}, \mathbb{B}, \mathbb{C})$$

defined by inclusion are given up to isomorphism by the absolute L-groups of the algebraic bordism category $\Lambda'' = (\mathbb{A}, \mathbb{C}, \mathbb{D})$

(i) $L^n(F) = L^{n-1}(\Lambda'')$

(ii) $L_n(F) = L_{n-1}(\Lambda'')$

(iii) $NL^n(F) = L_{n-1}(\Lambda'')$

and there are defined exact sequences

(i) $\ldots \longrightarrow L^n(\Lambda'') \longrightarrow L^n(\Lambda') \longrightarrow L^n(\Lambda) \overset{\partial}{\longrightarrow} L^{n-1}(\Lambda'') \longrightarrow \ldots$

(ii) $\ldots \longrightarrow L_n(\Lambda'') \longrightarrow L_n(\Lambda') \longrightarrow L_n(\Lambda) \overset{\partial}{\longrightarrow} L_{n-1}(\Lambda'') \longrightarrow \ldots$

(iii) $\ldots \longrightarrow L_n(\Lambda'') \longrightarrow NL^n(\Lambda') \longrightarrow NL^n(\Lambda) \overset{\partial}{\longrightarrow} L_{n-1}(\Lambda'') \longrightarrow \ldots$

with ∂ *given by the boundary of 1.14 for* (i) *and* (ii), *and by 2.10 for* (iii).
PROOF (i) The relative symmetric L-group $L^n(F)$ is the cobordism group of n-dimensional symmetric pairs $(f : C \longrightarrow D, (\delta\phi, \phi))$ in $(\mathbb{A}, \mathbb{B}, \mathbb{C})$ with (C, ϕ) defined in $(\mathbb{A}, \mathbb{B}, \mathbb{D})$ (i.e. the pair is $\mathbb{B}$-contractible, $\mathbb{C}$-Poincaré and the boundary is $\mathbb{D}$-Poincaré). Define inverse isomorphisms

$$L^{n-1}(\mathbb{A}, \mathbb{C}, \mathbb{D}) \overset{\simeq}{\longrightarrow} L^n(F) ; \quad (C, \phi) \longrightarrow ((C, \phi), (C \longrightarrow 0, (0, \phi))) ,$$

$$L^n(F) \overset{\simeq}{\longrightarrow} L^{n-1}(\mathbb{A}, \mathbb{C}, \mathbb{D}) ; \quad (f : C \longrightarrow D, (\delta\phi, \phi)) \longrightarrow (C', \phi')$$

with (C', ϕ') the $(n-1)$-dimensional symmetric complex in $(\mathbb{A}, \mathbb{C}, \mathbb{D})$ obtained from (C, ϕ) by algebraic surgery on the n-dimensional symmetric pair $(f : C \longrightarrow D, (\delta\phi, \phi))$ in $(\mathbb{A}, \mathbb{B}, \mathbb{C})$.
(ii) As for (i), with symmetric replaced by quadratic.
(iii) As for (i), with symmetric replaced by normal, and using 2.9 (ii) to obtain a quadratic structure on the effect of surgery on a normal pair.

□

The exact sequences of 3.9 are generalizations of the localization exact sequence of Ranicki [143] (cf. 3.13 below), and of the relative L-theory exact sequences of Vogel [170].

EXAMPLE 3.10 For any algebraic bordism category $\Lambda = (\mathbb{A}, \mathbb{B}, \mathbb{C})$ the exact sequence of 3.9 (iii) for the triple $(\mathbb{B}, \mathbb{B}, \mathbb{C})$ can be written as

$$\ldots \longrightarrow L_n(\Lambda) \xrightarrow{1+T} NL^n(\Lambda) \xrightarrow{J} NL^n(\widehat{\Lambda}) \xrightarrow{\partial} L_{n-1}(\Lambda) \longrightarrow \ldots$$

with $\widehat{\Lambda} = (\mathbb{A}, \mathbb{B}, \mathbb{B})$. If Λ satisfies the hypothesis of 3.5 then $NL^*(\Lambda)$ can be replaced by $L^*(\Lambda)$. In particular, this can be done for the algebraic bordism category $\Lambda = \Lambda(\mathbb{A})$ of 3.3 (cf. 3.6), recovering the exact sequence of 2.12

$$\ldots \longrightarrow L_n(\mathbb{A}) \xrightarrow{1+T} L^n(\mathbb{A}) \xrightarrow{J} NL^n(\mathbb{A}) \xrightarrow{\partial} L_{n-1}(\mathbb{A}) \longrightarrow \ldots .$$

$\square$

EXAMPLE 3.11 Given a ring with involution R and $q = p$ (resp. h, s) define the algebraic bordism category

$$\Lambda^q(R) = (\mathbb{A}^q(R), \mathbb{B}^q(R), \mathbb{C}^q(R))$$

with $\mathbb{A}^q(R)$ the additive category of f.g. projective (resp. f.g. free, based f.g. free) R-modules with the duality involution of 1.11, $\mathbb{B}^q(R) = \mathbb{B}(\mathbb{A})^q(R)$ the category of finite chain complexes in $\mathbb{A}^q(R)$, and $\mathbb{C}^q(R) \subseteq \mathbb{B}^q(R)$ the subcategory of contractible complexes C, such that $\tau(C) = 0 \in \widetilde{K}_1(R)$ for $q = s$. The quadratic L-groups of $\Lambda^q(R)$ are the type q quadratic L-groups of R

$$L_*(\Lambda^q(R)) = L_*^q(R) .$$

Let

$$\begin{cases} * : \widetilde{K}_0(R) \xrightarrow{\simeq} \widetilde{K}_0(R) \; ; \; [P] \longrightarrow [P^*] \\ * : \widetilde{K}_1(R) \xrightarrow{\simeq} \widetilde{K}_1(R) \; ; \; \tau(f : R^n \longrightarrow R^n) \longrightarrow \tau(f^* : R^n \longrightarrow R^n) \end{cases}$$

be the induced involution of the reduced $\begin{cases} \text{projective class} \\ \text{torsion} \end{cases}$ group of R. The intermediate quadratic L-groups $L_*^X(R)$ for a $*$-invariant subgroup $X \subseteq \begin{cases} \widetilde{K}_0(R) \\ \widetilde{K}_1(R) \end{cases}$ can be expressed as the L-groups of an algebraic bordism category

$$L_*^X(R) = \begin{cases} L_*(\mathbb{A}^p(R), \mathbb{B}^X(R), \mathbb{C}^p(R)) \\ L_*(\mathbb{A}^s(R), \mathbb{B}^s(R), \mathbb{C}^X(R)) \end{cases}$$

with $\begin{cases} \mathbb{B}^X(R) \subseteq \mathbb{B}^p(R) \\ \mathbb{C}^X(R) \subseteq \mathbb{B}^s(R) \end{cases}$ the subcategory of $\begin{cases} - \\ \text{contractible} \end{cases}$ chain complexes C in the category $\begin{cases} \mathbb{A}^p(R) \\ \mathbb{A}^s(R) \end{cases}$ with $\begin{cases} \text{projective class } [C] \in X \subseteq \widetilde{K}_0(R) \\ \text{torsion } \tau(C) \in X \subseteq \widetilde{K}_1(R) . \end{cases}$ The projective, free and simple quadratic L-groups of R are the special cases

$$L_*^{\widetilde{K}_0(R)}(R) = L_*^p(R) \; , \quad L_*^{\{0\} \subseteq \widetilde{K}_1(R)}(R) = L_*^s(R) \; ,$$

$$L_*^{\{0\} \subseteq \widetilde{K}_0(R)}(R) = L_*^{\widetilde{K}_1(R)}(R) = L_*^h(R) .$$

Given $*$-invariant subgroups $Y \subseteq X \subseteq \begin{cases} \widetilde{K}_0(R) \\ \widetilde{K}_1(R) \end{cases}$ the exact sequence of

quadratic L-groups given by 3.9 (ii) for the triple $\begin{cases} (\mathbb{B}^p(R), \mathbb{B}^X(R), \mathbb{B}^Y(R)) \\ (\mathbb{C}^X(R), \mathbb{C}^Y(R), \mathbb{C}^s(R)) \end{cases}$

is isomorphic to the Rothenberg exact sequence of Ranicki [141, §9]

$$\ldots \longrightarrow L_n^Y(R) \longrightarrow L_n^X(R) \longrightarrow \widehat{H}^n(\mathbb{Z}_2 \,; X/Y) \longrightarrow L_{n-1}^Y(R) \longrightarrow \ldots ,$$

corresponding to the isomorphisms

$$\begin{cases} L_n(\mathbb{A}^p(R), \mathbb{B}^X(R), \mathbb{B}^Y(R)) \xrightarrow{\simeq} \widehat{H}^n(\mathbb{Z}_2 \,; X/Y) \,; \ (C, \psi) \longrightarrow [C] \\ L_{n-1}(\mathbb{A}^s(R), \mathbb{C}^X(R), \mathbb{C}^Y(R)) \xrightarrow{\simeq} \widehat{H}^n(\mathbb{Z}_2 \,; X/Y) \,; \\ \quad (C, \psi) \longrightarrow \tau((1+T)\psi_0 \colon C^{n-1-*} \longrightarrow C) = \tau(C) + (-)^n \tau(C)^* . \end{cases}$$

Similar considerations apply to the symmetric and normal L-groups.

□

REMARK 3.12 In dealing with the free L-theory of a ring with involution R
the terminology is abbreviated, writing

$$\Lambda^h(R) = \Lambda(R) = (\mathbb{A}(R), \mathbb{B}(R), \mathbb{C}(R)) ,$$

$$L^n(\Lambda(R)) = L_h^{n+4*}(R) = L^{n+4*}(R) ,$$

$$L_n(\Lambda(R)) = L_n^h(R) = L_n(R) ,$$

$$\widehat{\Lambda}(R) = (\mathbb{A}(R), \mathbb{B}(R), \mathbb{B}(R)) , \ NL^n(\widehat{\Lambda}(R)) = NL^n(R) .$$

□

EXAMPLE 3.13 Let R be a ring with involution, and let $S \subset R$ be a multiplicative subset of central non-zero divisors which is invariant under the
involution. The localization of R inverting S is the ring with involution

$$S^{-1}R = \{r/s \,|\, r \in R, s \in S\}$$

with

$$r/s = rt/st , \ \overline{(r/s)} = \bar{r}/\bar{s} \ (r \in R, s, t \in S) .$$

Define algebraic bordism categories

$$\Gamma(R, S) = (\mathbb{A}(R), \mathbb{B}(R), \mathbb{C}(R, S)) ,$$

$$\Lambda(R, S) = (\mathbb{A}(R), \mathbb{C}(R, S), \mathbb{C}(R))$$

with $\mathbb{C}(R, S) \subset \mathbb{B}(R)$ the closed subcategory of the finite f.g. free R-module
chain complexes C such that the localization

$$S^{-1}C = S^{-1}R \otimes_R C$$

is in $\mathbb{C}(S^{-1}R)$, i.e. a contractible finite chain complex in $\mathbb{A}(S^{-1}R)$. The
localization maps of quadratic L-groups are isomorphisms

$$L_n(\Gamma(R, S)) \xrightarrow{\simeq} L_n(\Lambda(S^{-1}R)) = L_n(S^{-1}R) \,;$$

$$(C, \psi) \longrightarrow (S^{-1}C, S^{-1}\psi) \ (n \in \mathbb{Z})$$

because

(i) for every finite chain complex C in $\mathbb{A}(R)$ localization defines isomorphisms of abelian groups

$$\varinjlim_{C \to D} Q_n(D) \xrightarrow{\simeq} \varinjlim_{C \to D} Q_n(S^{-1}D) = Q_n(S^{-1}C) \quad (n \in \mathbb{Z})$$

with the direct limits taken over all the finite chain complexes D in $\mathbb{A}(R)$ with a $\mathbb{C}(R,S)$-equivalence $C \xrightarrow{\simeq} D$,

(ii) every finite chain complex in $\mathbb{A}(S^{-1}R)$ is $\mathbb{C}(S^{-1}R)$-equivalent to $S^{-1}C$ for a finite chain complex C in $\mathbb{A}(R)$.

Let $L_n(R,S) = L_{n-1}(\Lambda(R,S))$, the cobordism group of $(n-1)$-dimensional quadratic Poincaré complexes (C, ψ) in $\mathbb{A}(R)$ with C in $\mathbb{C}(R,S)$. The localization exact sequence of Ranicki [143, §4]

$$\ldots \longrightarrow L_n(R) \longrightarrow L_n(S^{-1}R) \xrightarrow{\partial} L_n(R,S) \longrightarrow L_{n-1}(R) \longrightarrow \ldots$$

is isomorphic to the exact sequence of 3.9 (ii)

$$\ldots \longrightarrow L_n(\Lambda(R)) \longrightarrow L_n(\Gamma(R,S)) \longrightarrow L_{n-1}(\Lambda(R,S))$$
$$\longrightarrow L_{n-1}(\Lambda(R)) \longrightarrow \ldots .$$

The quadratic L-group $L_n(R,S)$ is isomorphic to the cobordism group of n-dimensional quadratic Poincaré complexes in the category of S-torsion R-modules of homological dimension 1. In particular, the boundary map for $n = 0$

$$\partial : L_0(S^{-1}R) = L_0(\Gamma(R,S)) \longrightarrow L_0(R,S) = L_{-1}(\Lambda(R,S))$$

sends the Witt class of a nonsingular quadratic form $S^{-1}(M, \lambda, \mu)$ over $S^{-1}R$ induced from a quadratic form (M, λ, μ) over R to the Witt class of a nonsingular $S^{-1}R/R$-valued quadratic linking form

$$\partial S^{-1}(M, \lambda, \mu) = (\partial M, \partial \lambda, \partial \mu) ,$$

with

$$\partial M = \operatorname{coker}(\lambda \colon M \longrightarrow M^*) ,$$

$$\partial \lambda : \partial M \times \partial M \longrightarrow S^{-1}R/R ; \; x \longrightarrow (y \longrightarrow x(z)/s)$$

$$(x, y \in M^* , \, z \in M , \, s \in S , \, \lambda(z) = sy \in M^*) .$$

Similarly for the symmetric L-groups.

$\square$

PROPOSITION 3.14 *Given an additive category with chain duality* $\mathbb{A}$ *and closed subcategories* $\mathbb{D} \subseteq \mathbb{C} \subseteq \mathbb{B} \subseteq \mathbb{B}(\mathbb{A})$ *there is defined a commutative braid of exact sequences*

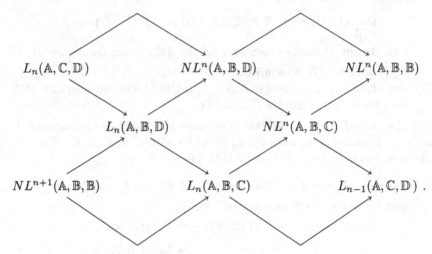

PROOF The exact sequences through $L_*(\mathbb{A}, \mathbb{C}, \mathbb{D})$ are given by 3.9 (ii), and those through $NL^*(\mathbb{A}, \mathbb{B}, \mathbb{B})$ by 3.9 (iii).   □

For any object C in a closed subcategory $\mathbb{C} \subseteq \mathbb{B}(\mathbb{A})$ the suspension $SC = C(0: C \longrightarrow 0)$ is also an object in $\mathbb{C}$.

DEFINITION 3.15 (i) A closed subcategory $\mathbb{C} \subseteq \mathbb{B}(\mathbb{A})$ is *stable* if
  (a) $\mathbb{C}$ contains the finite chain complexes C in $\mathbb{A}$ such that SC is an object in $\mathbb{C}$,
  (b) $\mathbb{C}$ contains the n-duals C^{n-*} ($n \in \mathbb{Z}$) of objects C in $\mathbb{C}$.
(ii) An algebraic bordism category $\Lambda = (\mathbb{A}, \mathbb{B}, \mathbb{C})$ is *stable* if $\mathbb{B}$ and $\mathbb{C}$ are stable closed subcategories of $\mathbb{B}(\mathbb{A})$.

□

PROPOSITION 3.16 (i) *The double skew-suspension maps of L-groups*
$$\begin{cases} \overline{S}^2 : L^n(\Lambda) \longrightarrow L^{n+4}(\Lambda) \; ; \; (C, \phi) \longrightarrow (S^2C, \phi) \\ \overline{S}^2 : L_n(\Lambda) \longrightarrow L_{n+4}(\Lambda) \; ; \; (C, \psi) \longrightarrow (S^2C, \psi) \\ \overline{S}^2 : NL^n(\Lambda) \longrightarrow NL^{n+4}(\Lambda) \; ; \; (C, \phi, \gamma, \chi) \longrightarrow (S^2C, \phi, \gamma, \chi) \end{cases}$$
are defined for any algebraic bordism category $\Lambda = (\mathbb{A}, \mathbb{B}, \mathbb{C})$ *and all* $n \in \mathbb{Z}$, *using the double skew-suspension isomorphisms of Q-groups given by 1.9.*
(ii) *The double skew-suspension maps of L-groups are isomorphisms for a stable algebraic bordism category* Λ.

PROOF (i) Trivial.

(ii) For stable Λ the double skew-suspension functor defines an isomorphism of categories

$$\overline{S}^2 : \{n\text{-dimensional symmetric complexes in } \Lambda\}$$

$$\xrightarrow{\simeq} \{(n+4)\text{-dimensional symmetric complexes in } \Lambda\}$$

for all $n \in \mathbb{Z}$ by virtue of the stability of $\mathbb{B}$ and $\mathbb{C}$. (Actually only 3.15 (i) (a) is being used here.) Similarly for quadratic and normal complexes, and also for pairs.

$\square$

EXAMPLE 3.17 (i) The algebraic bordism category $\Lambda(\mathbb{A}) = (\mathbb{A}, \mathbb{B}(\mathbb{A}), \mathbb{C}(\mathbb{A}))$ of 3.3 is stable. The $\begin{cases} \text{symmetric} \\ \text{quadratic} \end{cases}$ L-groups of $\Lambda(\mathbb{A})$ are the $\begin{cases} \text{symmetric} \\ \text{quadratic} \end{cases}$ L-groups of the additive category with chain duality $\mathbb{A}$

$$\begin{cases} L^*(\Lambda(\mathbb{A})) &= L^*(\mathbb{A}) \\ L_*(\Lambda(\mathbb{A})) &= L_*(\mathbb{A}) \ . \end{cases}$$

Also, by 3.5 the normal L-groups of $\Lambda(\mathbb{A})$ are the symmetric L-groups of $\mathbb{A}$

$$NL^*(\Lambda(\mathbb{A})) = L^*(\mathbb{A}) \ ,$$

since $\widehat{Q}^*(C) = 0$ for any $\mathbb{C}(\mathbb{A})$-contractible ($=$ contractible) finite chain complex in $\mathbb{A}$.

(ii) The normal L-groups of $\Lambda(\mathbb{A}) = (\mathbb{A}, \mathbb{B}(\mathbb{A}), \mathbb{B}(\mathbb{A}))$ are the normal L-groups of $\mathbb{A}$

$$NL^*(\Lambda(\mathbb{A})) = NL^*(\mathbb{A}) \ .$$

$\square$

EXAMPLE 3.18 Given a ring with involution R define the algebraic bordism category

$$\Lambda_+(R) = (\mathbb{A}(R), \mathbb{B}_+(R), \mathbb{C}_+(R))$$

with $\mathbb{A}(R)$ the additive category of f.g. free R-modules, $\mathbb{B}_+(R)$ the additive category of finite chain complexes C in $\mathbb{A}(R)$ which are positive (i.e. $C_r = 0$ for $r < 0$), and $\mathbb{C}_+(R) \subseteq \mathbb{B}_+(R)$ the subcategory of the contractible positive complexes. The inclusion $\Lambda_+(R) \subseteq \Lambda(R)$ in the algebraic bordism category $\Lambda(R)$ of 3.12 induces the natural maps to the 4-periodic $\begin{cases} \text{symmetric} \\ \text{quadratic} \end{cases}$ L-groups of R

$$\begin{cases} L^n(\Lambda_+(R)) &\longrightarrow L^n(\Lambda(R)) = L^n(\mathbb{A}(R)) = L^{n+4*}(R) \\ L_n(\Lambda_+(R)) &\longrightarrow L_n(\Lambda(R)) = L_n(\mathbb{A}(R)) = L_{n+4*}(R) \end{cases} \quad (n \in \mathbb{Z}) \ .$$

The symmetric L-groups of $\Lambda_+(R)$ are the symmetric L-groups of R as originally defined by Mishchenko [112]

$$L^*(\Lambda_+(R)) \ = \ L^*(R) \ .$$

It was shown in Ranicki [141] that the maps $\begin{cases} L^*(\Lambda_+(R)) \longrightarrow L^*(\Lambda(R)) \\ L_*(\Lambda_+(R)) \longrightarrow L_*(\Lambda(R)) \end{cases}$ $\begin{cases} \text{are not} \\ \text{are} \end{cases}$ isomorphisms in general, and also that

$$L_*(\Lambda_+(R)) \ = \ L_*(R)$$

with $L_*(R)$ the original 4-periodic quadratic L-groups of Wall [176].

□

Call $L^n(R)$ the *connective* symmetric L-groups of R, to distinguish them from the 4-periodic symmetric L-groups $L^{n+4*}(R)$. See §15 for the general L-theory of algebraic Poincaré complexes with connectivity conditions.

The algebraic surgery below the middle dimension used in Ranicki [141] to prove the 4-periodicity of the quadratic L-groups of rings with involution admits the following generalization for algebraic bordism categories, which is needed for §6 below.

DEFINITION 3.19 (i) An n-dimensional chain complex C in $\mathbb{A}$ is *highly connected* if there exist morphisms $\Gamma\colon C_r \longrightarrow C_{r+1}$ $(2r \geq n)$ such that

$$d\Gamma + \Gamma d \ = \ 1 : \ C_r \ \longrightarrow \ C_r \ \ (2r > n) \ .$$

(ii) An n-dimensional chain complex C in $\mathbb{A}$ is *highly $\mathbb{B}$-connected* if it is $\mathbb{B}$-equivalent to a highly connected complex.

□

EXAMPLE 3.20 Let $(\mathbb{A}, \mathbb{B}, \mathbb{C}) = (\mathbb{A}^q(R), \mathbb{B}^q(R), \mathbb{C}^q(R))$ $(q = p, h, s)$ for some ring with involution R. The following conditions on an n-dimensional chain complex C in $\mathbb{A}$ are equivalent:
  (i) C is highly connected,
  (ii) C is highly $\mathbb{B}$-connected,
  (iii) $H^r(C) = H_r(C) = 0$ for $2r > n$,
  (iv) C is $\mathbb{C}$-equivalent to an n-dimensional chain complex D in $\mathbb{A}$ such that $D_r = 0$ for $2r > n$.

□

DEFINITION 3.21 (i) An n-dimensional quadratic complex (C, ψ) (resp. pair $(f\colon C \longrightarrow D, (\delta\psi, \psi))$) in $\mathbb{A}$ is *highly $\mathbb{B}$-connected* if the chain complexes C (resp. C and D) are highly $\mathbb{B}$-connected.
(ii) Let $L_n(\Lambda)^{hc}$ $(n \in \mathbb{Z})$ be the cobordism group of highly $\mathbb{B}$-connected n-dimensional quadratic complexes in $\Lambda = (\mathbb{A}, \mathbb{B}, \mathbb{C})$.

□

DEFINITION 3.22 The algebraic bordism category $\Lambda = (\mathbb{A}, \mathbb{B}, \mathbb{C})$ is *connected* if

  (i) for each object A in $\mathbb{A}$ the dual chain complex TA is such that $(TA)_r = 0$ for $r > 0$,

  (ii) for every $\mathbb{B}$-contractible chain complex B and $k \in \mathbb{Z}$ the subcomplex $B[k] \subseteq B$ defined by

$$B[k]_r = \begin{cases} B_r & \text{if } r \geq k \\ 0 & \text{otherwise} \end{cases}$$

is $\mathbb{B}$-contractible.

□

In particular, $\Lambda = (\mathbb{A}, \mathbb{B}, \mathbb{C})$ is connected if $T: \mathbb{A} \longrightarrow \mathbb{B}(\mathbb{A})$ is 0-dimensional (1.2) and $\mathbb{B} = \mathbb{B}(\mathbb{A})$.

If C is a finite chain complex in $\mathbb{A}$ which is positive (i.e. $C_r = 0$ for $r < 0$) and $\Lambda = (\mathbb{A}, \mathbb{B}, \mathbb{C})$ is connected then $C \otimes_{\mathbb{A}} C = \text{Hom}_{\mathbb{A}}(TC, C)$ is a positive $\mathbb{Z}[\mathbb{Z}_2]$-module chain complex.

PROPOSITION 3.23 *For a connected algebraic bordism category* $\Lambda = (\mathbb{A}, \mathbb{B}, \mathbb{C})$ *the forgetful maps are isomorphisms*

$$L_n(\Lambda)^{hc} \xrightarrow{\sim} L_n(\Lambda) \; ; \; (C, \psi) \longrightarrow (C, \psi) \quad (n \in \mathbb{Z}) \; .$$

PROOF As in Ranicki [141] define inverses

$$L_n(\Lambda) \xrightarrow{\sim} L_n(\Lambda)^{hc} \; ; \; (C, \psi) \longrightarrow (C', \psi')$$

by sending an n-dimensional quadratic complex (C, ψ) in Λ to the highly $\mathbb{B}$-connected quadratic complex (C', ψ') in Λ obtained by surgery on the quadratic pair $(C \longrightarrow C[k], (0, \psi))$, with k the least integer such that $2k > n$.

□

Theorem A of Quillen [126] is an algebraic K-theory analogue of the Vietoris mapping theorem, stating that a functor $F: \mathbb{A} \longrightarrow \mathbb{A}'$ of exact categories with contractible fibres is a homotopy equivalence of categories, and so induces isomorphisms $F: K_*(\mathbb{A}) \xrightarrow{\sim} K_*(\mathbb{A}')$ in the algebraic K-groups. There is an evident algebraic L-theory analogue: a functor of algebraic bordism categories $F: \Lambda = (\mathbb{A}, \mathbb{B}, \mathbb{C}) \longrightarrow \Lambda' = (\mathbb{A}', \mathbb{B}', \mathbb{C}')$ such that

  (*) for every $n \in \mathbb{Z}$ and every $\mathbb{B}$-connected n-dimensional symmetric complex (C, ϕ) in Λ and every $\mathbb{B}'$-connected $(n+1)$-dimensional symmetric pair $E' = (f': F(C) \longrightarrow D', (\delta\phi', F(\phi)))$ in Λ' there exists an $(n+1)$-dimensional symmetric pair $E = (f: C \longrightarrow D, (\delta\phi, \phi))$ in Λ with $F(E)$ $\mathbb{B}'$-equivalent to E'

induces isomorphisms $F: L^*(\Lambda) \xrightarrow{\sim} L^*(\Lambda')$ in the symmetric L-groups, and $L^*(F: \Lambda \longrightarrow \Lambda') = 0$. Similarly for quadratic L-theory. The following highly-connected quadratic version is required for the proof of the algebraic π-π theorem in §10 below.

PROPOSITION 3.24 *A functor of connected algebraic bordism categories*

$$F : \Lambda = (\mathbb{A}, \mathbb{B}, \mathbb{C}) \longrightarrow \Lambda' = (\mathbb{A}', \mathbb{B}', \mathbb{C}')$$

such that

($*$) *for every $n \in \mathbb{Z}$ and every highly $\mathbb{B}$-connected n-dimensional quadratic complex (C, ψ) in Λ and every highly $\mathbb{B}'$-connected $(n+1)$-dimensional quadratic pair $E' = (f' \colon F(C) \longrightarrow D', (\delta\psi', F(\psi)))$ in Λ' there exists an $(n+1)$-dimensional quadratic pair $E = (f \colon C \longrightarrow D, (\delta\psi, \psi))$ in Λ with $F(E)$ $\mathbb{B}'$-equivalent to E'*

induces isomorphisms $F \colon L_*(\Lambda) \overset{\simeq}{\longrightarrow} L_*(\Lambda')$ in the quadratic L-groups, and $L_*(F \colon \Lambda \longrightarrow \Lambda') = 0$.

PROOF The induced map $F \colon L_n(\Lambda) \longrightarrow L_n(\Lambda')$ is one–one because by 3.23 an element in the kernel is represented by a highly $\mathbb{B}$-connected n-dimensional quadratic complex (C, ψ) in Λ for which there exists a highly $\mathbb{B}'$-connected $(n+1)$-dimensional quadratic pair in Λ'

$$E' = (f' \colon F(C) \longrightarrow D', (\delta\psi', F(\psi))) \ .$$

The corresponding $(n+1)$-dimensional quadratic $\mathbb{B}$-Poincaré pair $E = (f \colon C \longrightarrow D, (\delta\psi, \psi))$ in $\mathbb{A}$ with $F(E)$ $\mathbb{B}'$-equivalent to E' gives $(C, \psi) = 0 \in L_n(\Lambda)$.

The induced map $F \colon L_{n+1}(\Lambda) \longrightarrow L_{n+1}(\Lambda')$ is onto because by 3.23 every element in $L_{n+1}(\Lambda')$ is represented by a highly $\mathbb{B}'$-connected $(n+1)$-dimensional quadratic complex $(D', \delta\psi')$ in Λ', defining a highly $\mathbb{B}'$-connected $(n+1)$-dimensional quadratic pair $E' = (0 \longrightarrow D', (\delta\psi', 0))$ in Λ'. The algebraic Thom construction (Ranicki [141, 3.4]) applied to the corresponding $(n+1)$-dimensional quadratic pair $E = (f \colon C \longrightarrow D, (\delta\psi, \psi))$ in Λ with $F(E)$ $\mathbb{B}'$-equivalent to E' is an $(n+1)$-dimensional quadratic complex $(C(f), \delta\psi/\psi)$ in Λ such that

$$F(C(f), \delta\psi/\psi) = (D', \delta\psi') \in \operatorname{im}(F \colon L_{n+1}(\Lambda) \longrightarrow L_{n+1}(\Lambda')) \ .$$

$\square$

§4. Categories over complexes

An additive category $\mathbb{A}$ and a simplicial complex K are combined to define an additive category $\begin{cases} \mathbb{A}^*(K) \\ \mathbb{A}_*(K) \end{cases}$ of K-based objects in $\mathbb{A}$ which depends $\begin{cases} \text{contravariantly} \\ \text{covariantly} \end{cases}$ on K. In §5 a chain duality on $\mathbb{A}$ is extended to a chain duality on $\begin{cases} \mathbb{A}^*(K) \\ \mathbb{A}_*(K) \end{cases}$, allowing the extension of an algebraic bordism category $\Lambda = (\mathbb{A}, \mathbb{B}, \mathbb{C})$ to an algebraic bordism category $\begin{cases} \Lambda^*(K) \\ \Lambda_*(K) \end{cases}$.

DEFINITION 4.1 (i) An object M in an additive category $\mathbb{A}$ is K-*based* if it is expressed as a direct sum

$$M = \sum_{\sigma \in K} M(\sigma)$$

of objects $M(\sigma)$ in $\mathbb{A}$, such that $\{\sigma \in K \mid M(\sigma) \neq 0\}$ is finite. A morphism $f \colon M \longrightarrow N$ of K-based objects is a collection of morphisms in $\mathbb{A}$

$$f = \{f(\tau, \sigma) \colon M(\sigma) \longrightarrow N(\tau) \mid \sigma, \tau \in K\} \ .$$

(ii) Let $\begin{cases} \mathbb{A}^*(K) \\ \mathbb{A}_*(K) \end{cases}$ be the additive category of K-based objects M in $\mathbb{A}$, with morphisms $f \colon M \longrightarrow N$ such that $f(\tau, \sigma) \colon M(\sigma) \longrightarrow N(\tau)$ is 0 unless $\begin{cases} \tau \leq \sigma \\ \tau \geq \sigma \end{cases}$, so that

$$\begin{cases} f(M(\sigma)) \subseteq \sum\limits_{\tau \leq \sigma} N(\tau) \\ f(M(\sigma)) \subseteq \sum\limits_{\tau \geq \sigma} N(\tau) \ . \end{cases}$$

(iii) Forgetting the K-based structure defines the covariant *assembly* functor

$$\begin{cases} \mathbb{A}^*(K) \longrightarrow \mathbb{A} \ ; \ M \longrightarrow M^*(K) = \sum\limits_{\sigma \in K} M(\sigma) \\ \mathbb{A}_*(K) \longrightarrow \mathbb{A} \ ; \ M \longrightarrow M_*(K) = \sum\limits_{\sigma \in K} M(\sigma) \ . \end{cases}$$

□

EXAMPLE 4.2 The simplicial $\begin{cases} \text{chain} \\ \text{cochain} \end{cases}$ complex $\begin{cases} \Delta(K) \\ \Delta(K)^{-*} \end{cases}$ of K is a finite chain complex in $\begin{cases} \mathbb{A}(\mathbb{Z})^*(K) \\ \mathbb{A}(\mathbb{Z})_*(K) \end{cases}$ with

$$\begin{cases} \Delta(K)(\sigma) = S^{|\sigma|}\mathbb{Z} \\ \Delta(K)^{-*}(\sigma) = S^{-|\sigma|}\mathbb{Z} \end{cases} \quad (\sigma \in K) \ .$$

□

Regard the simplicial complex K as a category with one object for each simplex $\sigma \in K$ and one morphism $\sigma \to \tau$ for each face inclusion $\sigma \leq \tau$.

DEFINITION 4.3 Let $\begin{cases} \mathbb{A}^*[K] \\ \mathbb{A}_*[K] \end{cases}$ be the additive category with objects the $\begin{cases} \text{covariant} \\ \text{contravariant} \end{cases}$ additive functors

$$ M : K \longrightarrow \mathbb{A} \; ; \; \sigma \longrightarrow M[\sigma] $$

such that $\{\sigma \in K \,|\, M[\sigma] \neq 0\}$ is finite. The morphisms are the natural transformations of such functors.

□

Assume that the simplicial complex K is locally finite and ordered, so that for each simplex $\sigma \in K$ the set

$$ \begin{cases} K^*(\sigma) = \{\tau \in K \,|\, \tau > \sigma, \, |\tau| = |\sigma| + 1\} \\ K_*(\sigma) = \{\tau \in K \,|\, \tau < \sigma, \, |\tau| = |\sigma| - 1\} \end{cases} $$

is finite and ordered, and its elements are written

$$ \begin{cases} K^*(\sigma) = \{\delta_0\sigma, \delta_1\sigma, \delta_2\sigma, \dots\} \\ K_*(\sigma) = \{\partial_0\sigma, \partial_1\sigma, \partial_2\sigma, \dots\} \end{cases} . $$

DEFINITION 4.4 Define the covariant *assembly* functor for a simplicial complex K

$$ \begin{cases} \mathbb{B}(\mathbb{A})^*[K] = \mathbb{B}(\mathbb{A}^*[K]) \longrightarrow \mathbb{B}(\mathbb{A})_*(K) = \mathbb{B}(\mathbb{A}_*(K)) \, ; \, C \longrightarrow C^*[K] \\ \mathbb{B}(\mathbb{A})_*[K] = \mathbb{B}(\mathbb{A}_*[K]) \longrightarrow \mathbb{B}(\mathbb{A})^*(K) = \mathbb{B}(\mathbb{A}^*(K)) \, ; \, C \longrightarrow C_*[K] \end{cases} $$

by sending a finite chain complex C in $\begin{cases} \mathbb{A}^*[K] \\ \mathbb{A}_*[K] \end{cases}$ to the finite chain complex $\begin{cases} C^*[K] \\ C_*[K] \end{cases}$ in $\begin{cases} \mathbb{A}_*(K) \\ \mathbb{A}^*(K) \end{cases}$ with

$$ \begin{cases} C^*[K]_r = \displaystyle\sum_{\sigma \in K} C[\sigma]_{r+|\sigma|} \, , \quad C^*[K](\sigma) = S^{-|\sigma|}C[\sigma] \\ C_*[K]_r = \displaystyle\sum_{\sigma \in K} C[\sigma]_{r-|\sigma|} \, , \quad C_*[K](\sigma) = S^{|\sigma|}C[\sigma] \end{cases} \quad (\sigma \in K) . $$

The assembly is the total complex of the double complex in $\mathbb{A}$ defined by

$$
\begin{cases}
C^*[K]_{p,q} & = \displaystyle\sum_{\sigma \in K, |\sigma| = -p} C[\sigma]_q \\
C_*[K]_{p,q} & = \displaystyle\sum_{\sigma \in K, |\sigma| = p} C[\sigma]_q \ ,
\end{cases}
$$

$$
\begin{cases}
d' : C^*[K]_{p,q} \longrightarrow C^*[K]_{p-1,q} \ ; \ c[\sigma] \longrightarrow \displaystyle\sum_i (-)^i \delta_i c[\sigma] \\
d' : C_*[K]_{p,q} \longrightarrow C_*[K]_{p-1,q} \ ; \ c[\sigma] \longrightarrow \displaystyle\sum_i (-)^i \partial_i c[\sigma] \ ,
\end{cases}
$$

$$
\begin{cases}
d'' : C^*[K]_{p,q} \longrightarrow C^*[K]_{p,q-1} \ ; \ c[\sigma] \longrightarrow d_{C[\sigma]}(c[\sigma]) \\
d'' : C_*[K]_{p,q} \longrightarrow C_*[K]_{p,q-1} \ ; \ c[\sigma] \longrightarrow d_{C[\sigma]}(c[\sigma]) \ ,
\end{cases}
$$

the sum in d' being taken over all the elements $\begin{cases} \delta_i \sigma \in K^*(\sigma) \\ \partial_i \sigma \in K_*(\sigma) \end{cases}$ with
$\begin{cases} \delta_i : C[\sigma] \longrightarrow C[\delta_i \sigma] \\ \partial_i : C[\sigma] \longrightarrow C[\partial_i \sigma] \end{cases}$ the chain map induced by the inclusion $\begin{cases} \sigma \longrightarrow \delta_i \sigma \\ \partial_i \sigma \longrightarrow \sigma. \end{cases}$

$\square$

EXAMPLE 4.5 The assembly of the 0-dimensional chain complex $\underline{\mathbb{Z}}$ in $\begin{cases} \mathbb{A}(\mathbb{Z})^*[K] \\ \mathbb{A}(\mathbb{Z})_*[K] \end{cases}$ defined by

$$
\underline{\mathbb{Z}} : K \longrightarrow \mathbb{A}(\mathbb{Z}) \subseteq \mathbb{B}(\mathbb{A}(\mathbb{Z})) \ ; \ \sigma \longrightarrow \underline{\mathbb{Z}}[\sigma] = \mathbb{Z}
$$

with the identity structure chain maps $\underline{\mathbb{Z}}[\sigma] = \underline{\mathbb{Z}}[\tau]$ is the simplicial $\begin{cases} \text{cochain} \\ \text{chain} \end{cases}$ complex of K

$$
\begin{cases}
\underline{\mathbb{Z}}^*[K] & = \Delta(K)^{-*} \\
\underline{\mathbb{Z}}_*[K] & = \Delta(K)
\end{cases}
$$

already considered in 4.2 above as a chain complex in $\begin{cases} \mathbb{A}(\mathbb{Z})_*(K) \\ \mathbb{A}(\mathbb{Z})^*(K). \end{cases}$

$\square$

REMARK 4.6 If $\mathbb{A}$ is embedded in an abelian category the double complex $\begin{cases} C^*[K] \\ C_*[K] \end{cases}$ of 4.4 determines a spectral sequence $E(C)$ with E^2-terms

$$
E^2_{p,q} = \begin{cases} H^{-p}(K; \{H_q(C[\sigma])\}) \\ H_p(K; \{H_q(C[\sigma])\}) \ , \end{cases}
$$

which converges to $\begin{cases} H_*(C^*[K]) \\ H_*(C_*[K]) \end{cases}$ with respect to the filtration defined by

$$\begin{cases} F_p C^*[K]_q = \displaystyle\sum_{\sigma \in K, |\sigma| \geq -p} C[\sigma]_{q+|\sigma|} \subseteq C^*[K]_q \\ F_p C_*[K]_q = \displaystyle\sum_{\sigma \in K, |\sigma| \leq p} C[\sigma]_{q-|\sigma|} \subseteq C_*[K]_q \,. \end{cases}$$

$\square$

Define the covariant functors

$$\begin{cases} \mathbb{A}^*(K) \longrightarrow \mathbb{A}^*[K] \,;\, M \longrightarrow [M] \,,\, [M][\sigma] = \displaystyle\sum_{\tau \leq \sigma} M(\tau) \\[2mm] \mathbb{A}_*(K) \longrightarrow \mathbb{A}_*[K] \,;\, M \longrightarrow [M] \,,\, [M][\sigma] = \displaystyle\sum_{\tau \geq \sigma} M(\tau) \,. \end{cases}$$

For any object M in $\begin{cases} \mathbb{A}^*(K) \\ \mathbb{A}_*(K) \end{cases}$ and any object N in $\begin{cases} \mathbb{A}^*[K] \\ \mathbb{A}_*[K] \end{cases}$

$$\begin{cases} \mathrm{Hom}_{\mathbb{A}^*[K]}([M], N) = \displaystyle\sum_{\sigma \in K} \mathrm{Hom}_{\mathbb{A}}(M(\sigma), N[\sigma]) \\[2mm] \mathrm{Hom}_{\mathbb{A}_*[K]}([M], N) = \displaystyle\sum_{\sigma \in K} \mathrm{Hom}_{\mathbb{A}}(M(\sigma), N[\sigma]) \,. \end{cases}$$

A direct application of the contravariant duality functor $T: \mathbb{A} \longrightarrow \mathbb{B}(\mathbb{A})$ only gives a contravariant functor $\begin{cases} T: \mathbb{A}^*(K) \longrightarrow \mathbb{B}(\mathbb{A})_*(K) \\ T: \mathbb{A}_*(K) \longrightarrow \mathbb{B}(\mathbb{A})^*(K) \end{cases}$ and so does not define a chain duality on $\mathbb{A}_*(K)$. In §5 below the chain duality $T: \mathbb{A} \longrightarrow \mathbb{B}(\mathbb{A})$ will be extended to a chain duality $\begin{cases} T: \mathbb{A}^*(K) \longrightarrow \mathbb{B}(\mathbb{A}^*(K)) \\ T: \mathbb{A}_*(K) \longrightarrow \mathbb{B}(\mathbb{A}_*(K)) \end{cases}$ using the following embedding of $\begin{cases} \mathbb{A}^*(K) \\ \mathbb{A}_*(K) \end{cases}$ in the functor category $\begin{cases} \mathbb{A}^*[K] \\ \mathbb{A}_*[K] \end{cases}$.

PROPOSITION 4.7 (i) *A finite chain complex C in* $\begin{cases} \mathbb{A}^*(K) \\ \mathbb{A}_*(K) \end{cases}$ *is contractible if and only if each of the chain complexes $C(\sigma)$ ($\sigma \in K$) in $\mathbb{A}$ is contractible.*

(ii) *A chain map $f: C \longrightarrow D$ of finite chain complexes in* $\begin{cases} \mathbb{A}^*(K) \\ \mathbb{A}_*(K) \end{cases}$ *is a chain equivalence if and only if each of the diagonal components*

$$f(\sigma, \sigma) : C(\sigma) \longrightarrow D(\sigma) \quad (\sigma \in K)$$

is a chain equivalence in $\mathbb{A}$.

PROOF Proposition 2.9 of Ranicki and Weiss [147].

$\square$

REMARK 4.8 Given an additive category $\mathbb{A}$ let $\mathbb{D}(\mathbb{A})$ be the homotopy category of finite chain complexes in $\mathbb{A}$ and chain homotopy classes of chain maps. Let $\begin{cases} \mathbb{D}'(\mathbb{A}^*[K]) \\ \mathbb{D}'(\mathbb{A}_*[K]) \end{cases}$ be the localization of the triangulated category $\begin{cases} \mathbb{D}(\mathbb{A}^*[K]) \\ \mathbb{D}(\mathbb{A}_*[K]) \end{cases}$ inverting the chain complexes C in $\begin{cases} \mathbb{A}^*[K] \\ \mathbb{A}_*[K] \end{cases}$ such that each of the chain complexes $C[\sigma]$ $(\sigma \in K)$ in $\mathbb{A}$ is contractible. Using the methods of Ranicki and Weiss [147, §3] it can be shown that the functor $\begin{cases} \mathbb{A}^*(K) \longrightarrow \mathbb{A}^*[K] \\ \mathbb{A}_*(K) \longrightarrow \mathbb{A}_*[K] \end{cases}$ is a full embedding which determines an equivalence of the homotopy categories

$$\begin{cases} \mathbb{D}(\mathbb{A}^*(K)) \xrightarrow{\simeq} \mathbb{D}'(\mathbb{A}^*[K]) \; ; \; C \longrightarrow [C] \; , \\ \mathbb{D}(\mathbb{A}_*(K)) \xrightarrow{\simeq} \mathbb{D}'(\mathbb{A}_*[K]) \; ; \; C \longrightarrow [C] \; . \end{cases}$$

$\square$

PROPOSITION 4.9 *For any finite chain complex C in $\mathbb{A}_*(K)$ the assembly $[C]_*[K]$ of the finite chain complex $[C]$ in $\mathbb{A}_*[K]$ is naturally chain equivalent to the finite chain complex $C_*(K)$ in $\mathbb{A}$ obtained by forgetting the K-based structure.*

PROOF Define a natural chain equivalence in $\mathbb{A}$

$$\beta_C : [C]_*[K] \xrightarrow{\simeq} C_*(K)$$

by

$$\beta_C : [C]_*[K]_n = \sum_{\sigma \in K} (\Delta(\Delta^{|\sigma|}) \otimes_{\mathbb{Z}} C(\sigma))_n$$

$$\longrightarrow C_*(K)_n = \sum_{\sigma \in K} C(\sigma)_n \; ; \; a \otimes b \longrightarrow \epsilon(a)b \; ,$$

using the chain equivalences $\epsilon \colon \Delta(\Delta^{|\sigma|}) \xrightarrow{\simeq} \mathbb{Z}$ in $\mathbb{A}(\mathbb{Z})$ defined by augmentation.

$\square$

REMARK 4.10 The *star* and *link* of a simplex $\sigma \in K$ in a simplicial complex K are the subcomplexes defined by

$$\mathrm{star}_K(\sigma) = \{\tau \in K \,|\, \sigma\tau \in K\} \; ,$$

$$\mathrm{link}_K(\sigma) = \{\tau \in K \,|\, \sigma\tau \in K \, , \, \sigma \cap \tau = \emptyset\} \; .$$

The *dual cell* of σ is the contractible subcomplex of the barycentric subdivision K' defined by

$$D(\sigma, K) = \{\widehat{\sigma}_0 \widehat{\sigma}_1 \ldots \widehat{\sigma}_p \in K' \,|\, \sigma \le \sigma_0 < \sigma_1 < \ldots < \sigma_p\} \; ,$$

with *boundary*

$$\partial D(\sigma, K) = \bigcup_{\tau > \sigma} D(\tau, K) = \{\widehat{\sigma}_0 \widehat{\sigma}_1 \dots \widehat{\sigma}_p \in K' \, | \, \sigma < \sigma_0 < \sigma_1 < \dots < \sigma_p \} \, .$$

The barycentric subdivision of the link of $\sigma \in K$ is isomorphic to the boundary of the dual cell $D(\sigma, K)$

$$(\text{link}_K(\sigma))' \cong \partial D(\sigma, K) \, .$$

The star and link in K' of the barycentre $\widehat{\sigma} \in K'$ of $\sigma \in K$ are given by the joins

$$(\text{star}_{K'}(\widehat{\sigma}), \text{link}_{K'}(\widehat{\sigma})) = \partial \sigma' * (D(\sigma, K), \partial D(\sigma, K)) \, .$$

The local homology groups of $|K|$ at a point $x \in |K|$ in the interior of $\sigma \in K$ are given by

$$H_*(|K|, |K| \backslash \{x\}) \cong H_*(K, K \backslash \text{st}_K(\sigma)) \, ,$$

with

$$\text{st}_K(\sigma) = \{\tau \in K \, | \, \tau \geq \sigma\}$$

the *open star* of σ in K. Now $S^{-|\sigma|} \Delta(K, K \backslash \text{st}_K(\sigma))$ is the cellular chain complex of the relative CW pair $(|D(\sigma, K)|, |\partial D(\sigma, K)|)$, with one q-cell

$$e^q = |D(\sigma, K) \cap \tau'|$$

$$= |\bigcup \{\widehat{\tau}_0 \widehat{\tau}_1 \dots \widehat{\tau}_p \in K' \, | \, \sigma \leq \tau_0 < \tau_1 < \dots < \tau_p \leq \tau\}| \quad (q = |\tau| - |\sigma|)$$

for each $\tau \in \text{st}_K(\sigma)$. The subdivision chain equivalence

$$S^{-|\sigma|} \Delta(K, K \backslash \text{st}_K(\sigma)) = C(|D(\sigma, K)|, |\partial D(\sigma, K)|)$$

$$\xrightarrow{\simeq} \Delta(|D(\sigma, K)|, |\partial D(\sigma, K)|)$$

induces isomorphisms

$$H_*(K, K \backslash \text{st}_K(\sigma)) \cong H_{*-|\sigma|}(D(\sigma, K), \partial D(\sigma, K)) \, .$$

The following conditions on a locally finite simplicial complex K are equivalent:

(i) the polyhedron $|K|$ is an n-dimensional homology manifold, i.e. the local homology of $|K|$ at each point $x \in |K|$ is

$$H_*(|K|, |K| \backslash \{x\}) \cong H_*(\mathbb{R}^n, \mathbb{R}^n \backslash \{0\}) = \begin{cases} \mathbb{Z} & \text{if } * = n \\ 0 & \text{otherwise,} \end{cases}$$

(ii) K is a combinatorial homology n-manifold, i.e. for each simplex $\sigma \in K$

$$H_*(K, K \backslash \text{st}_K(\sigma)) \cong H_*(\mathbb{R}^n, \mathbb{R}^n \backslash \{0\}) \, ,$$

(iii) each $\text{link}_K(\sigma)$ $(\sigma \in K)$ is an $(n - |\sigma| - 1)$-dimensional homology sphere

$$H_*(\text{link}_K(\sigma)) \cong H_*(S^{n-|\sigma|-1}) = \begin{cases} \mathbb{Z} & \text{if } * = 0, \, n - |\sigma| - 1 \\ 0 & \text{otherwise} \, , \end{cases}$$

(iv) each $\partial D(\sigma, K)$ $(\sigma \in K)$ is an $(n - |\sigma| - 1)$-dimensional homology sphere,

(v) each $(D(\sigma, K), \partial D(\sigma, K))$ $(\sigma \in K)$ is an $(n - |\sigma|)$-dimensional geometric $\mathbb{Z}$-coefficient Poincaré pair

$$H^*(D(\sigma, K), \partial D(\sigma, K)) \cong H_{n-|\sigma|-*}(D(\sigma, K)) .$$

□

By contrast with 4.9, for a finite chain complex C in $\mathbb{A}^*(K)$ the assembly $[C]^*[K]$ is not chain equivalent to $C^*(K)$. If K is an oriented n-dimensional homology manifold with boundary ∂K then $[C]^*[K]$ is chain equivalent to $S^{-n}(C^*(K)/C^*(\partial K))$.

EXAMPLE 4.11 As in 4.2 regard the simplicial cochain complex $\Delta(K)^{-*}$ as a chain complex in $\mathbb{A}(\mathbb{Z})_*(K)$, with

$$\Delta(K)^{-*}(\sigma) = S^{-|\sigma|}\mathbb{Z} \ (\sigma \in K) .$$

The associated chain complex $[\Delta(K)^{-*}]$ in $\mathbb{A}(\mathbb{Z})_*[K]$ is such that

$$[\Delta(K)^{-*}][\sigma] = \Delta(K, K \backslash \mathrm{st}_K(\sigma))^{-*} \ (\sigma \in K) .$$

The spectral sequence $E([\Delta(K)^{-*}])$ of 4.6 is the dihomology spectral sequence of Zeeman [188] converging to $H^{-*}(K)$, with

$$E_{p,q}^2 = H_p(K; \{H^{-q}(K, K \backslash \mathrm{st}_K(\sigma))\}) .$$

If K is an n-dimensional homology manifold

$$H^r(K, K \backslash \mathrm{st}_K(\sigma)) = \begin{cases} \mathbb{Z} & \text{if } r = n \\ 0 & \text{otherwise} \end{cases} \ (\sigma \in K) ,$$

and the spectral sequence collapses to the Poincaré duality isomorphisms

$$H^{n-*}(K) \cong H_*(K) ,$$

using twisted coefficients in the nonorientable case. See McCrory [103] for a geometric interpretation of the Zeeman spectral sequence.

□

EXAMPLE 4.12 The simplicial chain complex $\Delta(K)$ is $\mathbb{Z}$-module chain equivalent to the assembly $B^*[K]$ of the chain complex B in $\mathbb{A}(\mathbb{Z})^*[K]$ defined by

$$B[\sigma] = \Delta(K, K \backslash \mathrm{st}_K(\sigma)) \ (\sigma \in K) ,$$

with a chain equivalence

$$\Delta(K) \xrightarrow{\simeq} B^*[K] ; \sigma \longrightarrow \hat{\sigma} .$$

For any n-cycle $[K] \in \Delta_n(K)$ let

$$[K][\sigma] \in B_n[\sigma] = \Delta_n(K, K \backslash \mathrm{st}_K(\sigma)) \ (\sigma \in K)$$

be the image n-cycles. Evaluation on $[K]$ defines a chain map in $\mathbb{A}(\mathbb{Z})^*[K]$

$$\phi = \langle [K], - \rangle : S^n\mathbb{Z} \longrightarrow B$$

with
$$\phi[\sigma] = \langle [K][\sigma], - \rangle : S^n \mathbb{Z}[\sigma] = S^n \mathbb{Z} \xrightarrow{[K]} \Delta(K)$$
$$\longrightarrow B[\sigma] = \Delta(K, K \backslash \mathrm{st}_K(\sigma)) .$$

The assembly of ϕ is the cap product $\mathbb{Z}$-module chain map
$$\phi[K] = [K] \cap - : S^n \mathbb{Z}[K] = \Delta(K)^{n-*} \longrightarrow B^*[K] \simeq \Delta(K) .$$

The following conditions on K are equivalent:

(i) K is an n-dimensional homology manifold with fundamental class $[K] \in H_n(K)$, with each $\phi[\sigma]$ ($\sigma \in K$) a $\mathbb{Z}$-module chain equivalence,

(ii) $\phi : S^n \mathbb{Z} \longrightarrow B$ is a chain equivalence in $\mathbb{A}(\mathbb{Z})^*[K]$.

For a homology manifold K the assembly $\phi[K]$ is the Poincaré duality chain equivalence.

$\square$

EXAMPLE 4.13 The simplicial chain complex $\Delta(K')$ of the barycentric subdivision K' is the assembly $C_*(K) = \Delta(K')$ of the chain complex C in $\mathbb{A}(\mathbb{Z})_*(K)$ defined by
$$C(\sigma) = \Delta(D(\sigma, K), \partial D(\sigma, K)) \quad (\sigma \in K) .$$

The $\mathbb{Z}$-module chain equivalences given by augmentation
$$\epsilon[\sigma] : [C][\sigma] = \Delta(D(\sigma, K)) \xrightarrow{\simeq} \mathbb{Z}[\sigma] = \mathbb{Z} \; ; \; \hat{\tau} \longrightarrow 1 \quad (\sigma \le \tau \in K)$$

define a chain equivalence $\epsilon : [C] \xrightarrow{\simeq} \mathbb{Z}$ in $\mathbb{A}(\mathbb{Z})_*[K]$, with $\underline{\mathbb{Z}}$ as in 4.5. C is chain equivalent in $\mathbb{A}(\mathbb{Z})_*(K)$ to the assembly $B^*[K]$ of the chain complex B in $\mathbb{A}(\mathbb{Z})^*[K]$ of 4.12, with $B[\sigma] = \Delta(K, K \backslash \mathrm{st}_K(\sigma))$.

As in McCrory [103, §5] consider the Flexner cap product $\mathbb{Z}$-module chain map
$$\Delta_F : \Delta(K) \otimes_{\mathbb{Z}} \Delta(K)^{-*} \longrightarrow \Delta(K')$$

defined by
$$\Delta_F : \Delta_p(K) \otimes_{\mathbb{Z}} \Delta^q(K) \longrightarrow \Delta_{p-q}(K') \; ; \; \sigma \otimes \tau^* \longrightarrow \begin{cases} \sum\limits_S \epsilon(S)S & \text{if } \tau \le \sigma \\ 0 & \text{otherwise} \end{cases}$$

with S running over the r-simplexes of the *dual cell* of τ in σ
$$D(\sigma, \tau) = \sigma' \cap D(\tau, K)$$
$$= \{ (\hat{\sigma}_0 \hat{\sigma}_1 \dots \hat{\sigma}_r) \in K' \,|\, \tau \le \sigma_0 < \sigma_1 < \dots < \sigma_r \le \sigma \} ,$$

with $r = p - q$ and
$$\epsilon(S) = \epsilon(\sigma_0, \sigma_1) \epsilon(\sigma_1, \sigma_2) \dots \epsilon(\sigma_{r-1}, \sigma_r) \in \{+1, -1\}$$

the product of the incidence numbers of the successive codimension 1 pairs of simplices, defined using the ordering of K. The adjoint of Δ_F is a $\mathbb{Z}$-module chain map
$$A\Delta_F : \Delta(K) \longrightarrow \mathrm{Hom}_{\mathbb{A}(\mathbb{Z})_*(K)}(\Delta(K)^{-*}, \Delta(K'))$$

which is shown to be a chain equivalence in 7.3 below. Cap product with any homology class $[K] \in H_n(K)$

$$\phi = [K] \cap - : H^{n-*}(K) \longrightarrow H_*(K') = H_*(K)$$

is induced by the chain map ϕ in $\mathbb{A}(\mathbb{Z})_*(K)$ obtained by the evaluation of $A\Delta_F$ on any representative n-cycle $[K] \in \Delta_n(K)$

$$\phi = A\Delta_F[K] = [K] \cap - : \Delta(K)^{n-*} \longrightarrow \Delta(K') .$$

The diagonal components of ϕ are the $\mathbb{Z}$-module chain maps

$$\phi(\sigma, \sigma) = \langle [K][\sigma], - \rangle :$$

$$\Delta(K)^{n-*}(\sigma) = S^{n-|\sigma|}\mathbb{Z} \longrightarrow \Delta(K')(\sigma) = \Delta(D(\sigma, K), \partial D(\sigma, K))$$

obtained by the evaluations on cycles representing the images of $[K]$

$$[K][\sigma] \in H_n(K, K \backslash \mathrm{st}_K(\sigma)) = H_{n-|\sigma|}(D(\sigma, K), \partial D(\sigma, K)) .$$

The following conditions on K are equivalent:

(i) K is an n-dimensional homology manifold with fundamental class $[K] \in H_n(K)$, with each $\phi(\sigma, \sigma)$ ($\sigma \in K$) a $\mathbb{Z}$-module chain equivalence,

(ii) $\phi = [K] \cap - : \Delta(K)^{n-*} \longrightarrow \Delta(K')$ is a chain equivalence in $\mathbb{A}(\mathbb{Z})_*(K)$.

For a homology manifold K the assembly $\phi_*(K)$ is the Poincaré duality chain equivalence.

□

DEFINITION 4.14 (i) Let X be a topological space with a covering

$$\bigcup_{v \in V} X[v] = X$$

by a collection $\{X[v] \mid v \in V\}$ of non-empty subspaces $X[v] \subseteq X$. The *nerve* of the covering is the simplicial complex K with vertex set $K^{(0)} = V$, such that distinct vertices $v_0, v_1, \ldots, v_n \in V$ span a simplex $\sigma = (v_0 v_1 \ldots v_n) \in K$ if and only if the intersection

$$X[\sigma] = X[v_0] \cap X[v_1] \cap \ldots \cap X[v_n]$$

is non-empty.

(ii) Let K be simplicial complex. A *K-dissection* of a topological space X is a collection $\{X[\sigma] \mid \sigma \in K\}$ of subspaces $X[\sigma] \subseteq X$ (some of which may be empty) indexed by the simplexes $\sigma \in K$, such that

(a) $X[\sigma] \cap X[\tau] = \begin{cases} X[\sigma\tau] & \text{if } \sigma, \tau \in K \text{ span a simplex } \sigma\tau \in K \\ \emptyset & \text{otherwise} , \end{cases}$

(b) $\bigcup_{\sigma \in K} X[\sigma] = X$.

The nerve of the covering of X is the subcomplex $\{\sigma \in K \mid X[\sigma] \neq \emptyset\} \subseteq K$.

□

Example 4.15 Let X, K be simplicial complexes. If $f : X \longrightarrow K'$ is a simplicial map then $\{X[\sigma] = f^{-1}D(\sigma, K) \mid \sigma \in K\}$ is a K-dissection of X. Conversely, any K-dissection $\{X[\sigma] \mid \sigma \in K\}$ of X determines a simplicial map $g : X' \longrightarrow K'$ with $g^{-1}D(\sigma, K) = X[\sigma]'$ $(\sigma \in K)$.

For any K-dissection $\{X[\sigma] \mid \sigma \in K\}$ of X define $\partial X[\sigma] \subseteq X[\sigma]$ to be the subcomplex

$$\partial X[\sigma] = \bigcup_{\tau > \sigma} X[\tau] \quad (\sigma \in K).$$

The simplicial chain complex of X is a chain complex $C_*(K) = \Delta(X)$ in $\mathbb{A}(\mathbb{Z})_*(K)$ with

$$C(\sigma) = \Delta(X[\sigma], \partial X[\sigma]) \ , \quad [C][\sigma] = \Delta(X[\sigma]) \ (\sigma \in K).$$

The assembly $[C]_*[K]$ is the cellular chain complex of the homotopy colimit CW complex

$$[X] = \operatorname*{hocolim}_{\sigma \in K} X[\sigma] = \left(\coprod_{\sigma \in K} \Delta^{|\sigma|} \times X[\sigma] \right) \Big/ \{(a, \partial_i b) \sim (\partial_i a, b)\}$$

with one $(p+q)$-cell for each p-simplex $\sigma \in K$ and each q-simplex in $X[\sigma]$. The projection

$$[X] \longrightarrow |X| \ ; \ (a, b) \longrightarrow b$$

is a map with contractible point inverses, inducing the chain equivalence $\beta_C : [C]_*[K] \xrightarrow{\simeq} C_*(K)$ of 4.9. Define a filtration of $[X]$ by

$$F_p[X] = \operatorname*{hocolim}_{\sigma \in K, |\sigma| \leq p} X[\sigma]$$

$$= \left(\coprod_{\sigma \in K, |\sigma| \leq p} \Delta^{|\sigma|} \times X[\sigma] \right) \Big/ \{(a, \partial_i b) \sim (\partial_i a, b)\} \ .$$

The spectral sequence determined by the corresponding filtration of $[C]_*[K]$ is the spectral sequence $E([C])$ of 4.6, namely the spectral sequence with respect to the first grading of the double complex D with

$$D_{p,q} = C_{p+q}(F_p[X]) = \sum_{\sigma \in K, |\sigma|=p} \Delta_q(X[\sigma]) \ ,$$

$$d' = \sum_{\sigma} \sum_{i} (-)^i (\partial_i \sigma \to \sigma)^* : D_{p,q} \longrightarrow D_{p-1,q} \ ,$$

$$d' = \sum_{\sigma} d_{\Delta(X[\sigma])} : D_{p,q} \longrightarrow D_{p,q-1} \ .$$

$E([C])$ is the Leray–Serre spectral sequence with E^2-terms

$$E^2_{p,q} = H_p(K; \{H_q(X[\sigma])\}) \ ,$$

converging to

$$H_*([C][K]) = H_*([X]) = H_*(X)$$

with

$$E_{p,q}^{\infty} = \mathrm{im}(H_{p+q}(F_p[X])\longrightarrow H_{p+q}(X))/\mathrm{im}(H_{p+q}(F_{p-1}[X])\longrightarrow H_{p+q}(X)) .$$

□

EXAMPLE 4.16 Given a topological space X let $Open(X)$ be the category whose objects are the open sets in X and whose morphisms are inclusions of open sets. Let K be the nerve of a finite open cover $\mathbb{U} = \{U_j \,|\, j \in J\}$ of X, and let R be a commutative ring. The Čech complex (Bott and Tu [12, p. 110]) of $\mathbb{U}$ with coefficients in a $\begin{cases} \text{contravariant} \\ \text{covariant} \end{cases}$ functor

$$F : Open(X) \longrightarrow \mathbb{B}(R) ; U \longrightarrow F(U)$$

is the assembly R-module chain complex

$$C(\mathbb{U}, F) = \begin{cases} C^*[K] \\ C_*[K] \end{cases}$$

of the $\begin{cases} \mathbb{A}(R)^*[K] \\ \mathbb{A}(R)_*[K] \end{cases}$-module chain complex C defined by

$$C[j_0 j_1 \ldots j_n] = F(U_{j_0} \cap U_{j_1} \cap \ldots \cap U_{j_n}) \ ((j_0 j_1 \ldots j_n) \in K^{(n)}) .$$

In particular, for any finite open cover $\mathbb{U}$ of a differentiable manifold X there is defined a $\begin{cases} \text{contravariant} \\ \text{covariant} \end{cases}$ functor

$$F = \begin{cases} \Omega^* \\ \Omega_c^* \end{cases} : Open(X) \longrightarrow \mathbb{B}(\mathbb{R}) ; U \longrightarrow \begin{cases} \Omega^*(U) \\ \Omega_c^*(U) \end{cases}$$

sending an open subset $U \subseteq X$ to the $\mathbb{R}$-module chain complex $\begin{cases} \Omega^*(U) \\ \Omega_c^*(U) \end{cases}$ of $\begin{cases} - \\ \text{compactly supported} \end{cases}$ differential forms on U. The assembly $\mathbb{R}$-module chain complex $C(\mathbb{U}, F)$ is the $\begin{cases} - \\ \text{compactly supported} \end{cases}$ Čech–deRham complex of X, with homology

$$H_*(C(\mathbb{U}, F)) = \begin{cases} H^{-*}(X; \mathbb{R}) \\ H_c^{-*}(X; \mathbb{R}) \end{cases}$$

the $\begin{cases} - \\ \text{compactly supported} \end{cases}$ deRham cohomology of X, as in [12, $\begin{cases} 8.5 \\ 12.12 \end{cases}$].

□

§5. Duality

An algebraic bordism category $\Lambda = (\mathbb{A}, \mathbb{B}, \mathbb{C})$ and a locally finite simplicial complex K will now be shown to determine an algebraic bordism category

$$\begin{cases} \Lambda^*(K) = (\mathbb{A}^*(K), \mathbb{B}^*(K), \mathbb{C}^*(K)) \\ \Lambda_*(K) = (\mathbb{A}_*(K), \mathbb{B}_*(K), \mathbb{C}_*(K)) \end{cases}$$

which depends $\begin{cases} \text{contravariantly} \\ \text{covariantly} \end{cases}$ on K. In §13 below the symmetric (resp. quadratic) L-groups of this category will be identified with the generalized $\begin{cases} \text{cohomology} \\ \text{homology} \end{cases}$ groups of K

$$\begin{cases} L^n(\Lambda^*(K)) = H^{-n}(K; \mathbb{L}^\cdot(\Lambda)) \\ L^n(\Lambda_*(K)) = H_n(K; \mathbb{L}^\cdot(\Lambda)) \end{cases}$$

$$\left(\text{resp.} \begin{cases} L_n(\Lambda^*(K)) = H^{-n}(K; \mathbb{L}_\cdot(\Lambda)) \\ L_n(\Lambda_*(K)) = H_n(K; \mathbb{L}_\cdot(\Lambda)) \end{cases}\right) \ (n \in \mathbb{Z})$$

with coefficients in an Ω-spectrum $\mathbb{L}^\cdot(\Lambda)$ (resp. $\mathbb{L}_\cdot(\Lambda)$) of Kan Δ-sets such that

$$\pi_n(\mathbb{L}^\cdot(\Lambda)) = L^n(\Lambda) \ (\text{resp.} \ \pi_n(\mathbb{L}_\cdot(\Lambda)) = L_n(\Lambda)) \ .$$

Algebraic Poincaré complexes in $\Lambda^*(K)$ are analogues of the 'mock bundles' over K used by Buoncristiano, Rourke and Sanderson [21] as cocycles for generalized cohomology $h^*(K)$. For PL bordism $h = \Omega^{PL}$ a $(-d)$-dimensional cocycle $p: E \longrightarrow K$ is a d-dimensional mock bundle, a PL map such that the inverse image $p^{-1}(\sigma)$ $(\sigma \in K)$ is a $(d + |\sigma|)$-dimensional PL manifold with boundary $p^{-1}(\partial\sigma)$. Dually, algebraic Poincaré complexes in $\Lambda_*(K)$ are analogues of manifold cycles for generalized homology $h_*(K)$. For PL bordism a d-dimensional cycle $p: E \longrightarrow K$ is just a PL map from a d-dimensional PL manifold E, in which case the inverse image $p^{-1}(D(\sigma, K))$ $(\sigma \in K)$ is a $(d - |\sigma|)$-dimensional manifold with boundary $p^{-1}(\partial D(\sigma, K))$.

For the additive category $\mathbb{M}(\mathbb{Z}) = \{\mathbb{Z}\text{-modules}\}$ write

$$\begin{cases} \mathbb{M}(\mathbb{Z})^*(K) = \mathbb{Z}^*(K) \\ \mathbb{M}(\mathbb{Z})_*(K) = \mathbb{Z}_*(K) \end{cases} , \quad \begin{cases} \mathbb{M}(\mathbb{Z})^*[K] = \mathbb{Z}^*[K] \\ \mathbb{M}(\mathbb{Z})_*[K] = \mathbb{Z}_*[K] \end{cases} .$$

For any finite chain complexes C, D in $\mathbb{A}$ there is defined an abelian group chain complex $C \otimes_\mathbb{A} D = \text{Hom}_\mathbb{A}(TC, D)$ as in §3. Given chain complexes C, D in $\begin{cases} \mathbb{A}^*[K] \\ \mathbb{A}_*[K] \end{cases}$ define a chain complex $C \otimes_\mathbb{A} D$ in $\begin{cases} \mathbb{Z}^*[K] \\ \mathbb{Z}_*[K] \end{cases}$ by

$$(C \otimes_\mathbb{A} D)[\sigma] = C[\sigma] \otimes_\mathbb{A} D[\sigma] \ (\sigma \in K) ,$$

and let

$$T_{C,D} : C \otimes_\mathbb{A} D \xrightarrow{\ \simeq\ } D \otimes_\mathbb{A} C$$

be the isomorphism with components

$$T_{C,D}[\sigma] = T_{C[\sigma],D[\sigma]} : C[\sigma] \otimes_{\mathbb{A}} D[\sigma] \longrightarrow D[\sigma] \otimes_{\mathbb{A}} C[\sigma] \quad (\sigma \in K) .$$

PROPOSITION 5.1 *An algebraic bordism category* $\Lambda = (\mathbb{A}, \mathbb{B}, \mathbb{C})$ *and a locally finite ordered simplicial complex* K *determine an algebraic bordism category*

$$\begin{cases} \Lambda^*(K) = (\mathbb{A}^*(K), \mathbb{B}^*(K), \mathbb{C}^*(K)) \\ \Lambda_*(K) = (\mathbb{A}_*(K), \mathbb{B}_*(K), \mathbb{C}_*(K)) . \end{cases}$$

PROOF Define a contravariant functor $\begin{cases} T : \mathbb{A}^*[K] \longrightarrow \mathbb{B}(\mathbb{A}^*(K)) \\ T : \mathbb{A}_*[K] \longrightarrow \mathbb{B}(\mathbb{A}_*(K)) \end{cases}$ by sending an object M to the chain complex TM with

$$(TM)_r(\sigma) = \begin{cases} T(M[\sigma])_{r-|\sigma|} \\ T(M[\sigma])_{r+|\sigma|} , \end{cases}$$

$$d_{TM}(\tau, \sigma) = (-)^i T(M[\tau] \longrightarrow M[\sigma]) : (TM)_r(\sigma) \longrightarrow (TM)_{r-1}(\tau)$$

$$\text{if} \begin{cases} \sigma \geq \tau, \ |\sigma| = |\tau| + 1, \ \tau = \partial_i \sigma \\ \sigma \leq \tau, \ |\sigma| = |\tau| - 1, \ \tau = \delta_i \sigma . \end{cases}$$

The contravariant functor defined by the composite

$$\begin{cases} T : \mathbb{A}^*(K) \longrightarrow \mathbb{A}^*[K] \xrightarrow{T} \mathbb{B}(\mathbb{A}^*(K)) \\ T : \mathbb{A}_*(K) \longrightarrow \mathbb{A}_*[K] \xrightarrow{T} \mathbb{B}(\mathbb{A}_*(K)) \end{cases}$$

is such that for any objects M, N in $\begin{cases} \mathbb{A}^*(K) \\ \mathbb{A}_*(K) \end{cases}$

$$\begin{cases} M \otimes_{\mathbb{A}^*(K)} N = \mathrm{Hom}_{\mathbb{A}^*(K)}(TM, N) = ([M] \otimes_{\mathbb{A}} [N])^*[K] \\ M \otimes_{\mathbb{A}_*(K)} N = \mathrm{Hom}_{\mathbb{A}_*(K)}(TM, N) = ([M] \otimes_{\mathbb{A}} [N])_*[K] . \end{cases}$$

Thus $\begin{cases} M \otimes_{\mathbb{A}^*(K)} N \\ M \otimes_{\mathbb{A}_*(K)} N \end{cases}$ is a chain complex in $\mathbb{A}$ with

$$\begin{cases} (M \otimes_{\mathbb{A}^*(K)} N)_r = \displaystyle\sum_{\sigma \in K} \sum_{\lambda, \mu \leq \sigma} (M(\lambda) \otimes_{\mathbb{A}} N(\mu))_{r+|\sigma|} \\ (M \otimes_{\mathbb{A}_*(K)} N)_r = \displaystyle\sum_{\sigma \in K} \sum_{\lambda, \mu \geq \sigma} (M(\lambda) \otimes_{\mathbb{A}} N(\mu))_{r-|\sigma|} . \end{cases}$$

The duality isomorphism of $\mathbb{Z}$-module chain complexes

$$\begin{cases} T_{M,N} : M \otimes_{\mathbb{A}^*(K)} N \xrightarrow{\simeq} N \otimes_{\mathbb{A}^*(K)} M \\ T_{M,N} : M \otimes_{\mathbb{A}_*(K)} N \xrightarrow{\simeq} N \otimes_{\mathbb{A}_*(K)} M \end{cases}$$

for $N = TM$ sends the 0-cycle

$$\begin{cases} 1 \in (M \otimes_{\mathbb{A}^*(K)} TM)_0 = \mathrm{Hom}_{\mathbb{A}^*(K)}(TM, TM)_0 \\ 1 \in (M \otimes_{\mathbb{A}_*(K)} TM)_0 = \mathrm{Hom}_{\mathbb{A}_*(K)}(TM, TM)_0 \end{cases}$$

to a 0-cycle

$$\begin{cases} e(M) \in (TM \otimes_{\mathbb{A}^*(K)} M)_0 = \mathrm{Hom}_{\mathbb{A}^*(K)}(T^2 M, M)_0 \\ e(M) \in (TM \otimes_{\mathbb{A}_*(K)} M)_0 = \mathrm{Hom}_{\mathbb{A}_*(K)}(T^2 M, M)_0 , \end{cases}$$

defining a natural transformation

$$\begin{cases} e : T^2 \longrightarrow 1 : \mathbb{A}^*(K) \longrightarrow \mathbb{B}(\mathbb{A}^*(K)) \\ e : T^2 \longrightarrow 1 : \mathbb{A}_*(K) \longrightarrow \mathbb{B}(\mathbb{A}_*(K)) \end{cases}$$

such that $e(TM) . T(e(M)) = 1$.

The additive category $\begin{cases} \mathbb{B}^*(K) \\ \mathbb{B}_*(K) \end{cases}$ is the full subcategory of $\begin{cases} \mathbb{B}(\mathbb{A}^*(K)) \\ \mathbb{B}(\mathbb{A}_*(K)) \end{cases}$

with objects the finite chain complexes C in $\begin{cases} \mathbb{A}^*(K) \\ \mathbb{A}_*(K) \end{cases}$ such that each $C(\sigma)$

($\sigma \in K$) is an object in $\mathbb{B}$. The dual chain complex TC is then also defined

in $\begin{cases} \mathbb{B}^*(K) \\ \mathbb{B}_*(K). \end{cases}$ Similarly for $\begin{cases} \mathbb{C}^*(K) \\ \mathbb{C}_*(K). \end{cases}$

□

EXAMPLE 5.2 If the chain duality on $\mathbb{A}$ is 0-dimensional (e.g. if $\mathbb{A} = \mathbb{A}(R) =$ { f.g. free R-modules}) then the dual of an object M in $\begin{cases} \mathbb{A}^*(K) \\ \mathbb{A}_*(K) \end{cases}$ is the

chain complex TM in $\begin{cases} \mathbb{A}^*(K) \\ \mathbb{A}_*(K) \end{cases}$ with

$$TM_r(\sigma) = T([M][\sigma]) \text{ if } \begin{cases} r = |\sigma| \\ r = -|\sigma| \end{cases} , = 0 \text{ otherwise .}$$

□

EXAMPLE 5.3 The chain complexes B, C in $\mathbb{A}(\mathbb{Z})_*(K)$ defined in 4.2 and 4.15 by

$$B(\sigma) = S^{-|\sigma|}\mathbb{Z} , \ B_*(K) = \Delta(K)^{-*} ,$$

$$C(\sigma) = \Delta(D(\sigma, K), \partial D(\sigma, K)) , \ C_*(K) = \Delta(K') \simeq \Delta(K)$$

are dual to each other, with the subdivision chain equivalences in $\mathbb{A}(\mathbb{Z})$

$$TB(\sigma) = S^{-|\sigma|}\Delta(K, K\backslash \text{st}_K(\sigma)) \simeq C(\sigma) = \Delta(D(\sigma, K), \partial D(\sigma, K))$$

defining a chain equivalence $TB \simeq C$ in $\mathbb{A}(\mathbb{Z})_*(K)$.

□

EXAMPLE 5.4 An m-dimensional quadratic Poincaré complex n-ad over a ring with involution R in the sense of Levitt and Ranicki [91, §3] is an $(m - n)$-dimensional quadratic Poincaré complex in $\mathbb{A}(R)^*(\Delta^n)$.

□

EXAMPLE 5.5 Let C be the chain complex in $\mathbb{A}(\mathbb{Z})_*(K)$ associated to a K-dissection $\{X[\sigma] \, | \, \sigma \in K\}$ of a simplicial complex X in 4.15, with

$$C(\sigma) = \Delta(X[\sigma], \partial X[\sigma]) \ (\sigma \in K) \ , \ C_*(K) = \Delta(X) .$$

For any $n \in \mathbb{Z}$ the n-dual of C is the chain complex $\Sigma^n TC = C^{n-*}$ in

$\mathbb{A}(\mathbb{Z})_*(K)$ with

$$C^{n-*}(\sigma) \;=\; \Delta(X[\sigma])^{n-|\sigma|-*}\,,$$

$$[C^{n-*}][\sigma] \;\simeq\; \Delta(X[\sigma],\partial X[\sigma])^{n-|\sigma|-*} \;\;(\sigma \in K)\,,$$

$$(C^{n-*})_*(K) \;=\; ([C]_*[K])^{n-*} \;\simeq\; \Delta(X)^{n-*}\,.$$

$\square$

A Δ-*map* of simplicial complexes is a simplicial map which is injective on simplexes.

PROPOSITION 5.6 *Let* $\Lambda = (\mathbb{A}, \mathbb{B}, \mathbb{C})$ *be an algebraic bordism category.*
$A \begin{cases} \Delta- \\ simplicial \end{cases}$ *map* $f\colon J \longrightarrow K$ *of finite ordered simplicial complexes induces*
$\begin{cases} contravariantly \\ covariantly \end{cases}$ *a covariant functor of algebraic bordism categories*

$$\begin{cases} f^* : \Lambda^*(K) \longrightarrow \Lambda^*(J) \\ f_* : \Lambda_*(J) \longrightarrow \Lambda_*(K)\,, \end{cases}$$

inducing morphisms of the symmetric L-groups

$$\begin{cases} f^* : L^n(\Lambda^*(K)) = H^{-n}(K; \mathbb{L}^{\cdot}(\Lambda)) \longrightarrow L^n(\Lambda^*(J)) = H^{-n}(J; \mathbb{L}^{\cdot}(\Lambda)) \\ f_* : L^n(\Lambda_*(J)) = H_n(J; \mathbb{L}^{\cdot}(\Lambda)) \longrightarrow L^n(\Lambda_*(K)) = H_n(K; \mathbb{L}^{\cdot}(\Lambda))\,. \end{cases}$$

Similarly for the quadratic L-groups.

PROOF See §13 below for the identifications of the L-groups with the generalized (co)homology groups.

(i) The functor induced by a Δ-map $f\colon J \longrightarrow K$ is defined by

$$f^* : \mathbb{A}^*(K) \longrightarrow \mathbb{A}^*(J)\,;\; M \longrightarrow f^*M\,,\; f^*M(\sigma) = M(f\sigma)\,,$$

with $T(f^*M) = f^*(TM)$.

(ii) The functor $f_*\colon \Lambda_*(J) \longrightarrow \Lambda_*(K)$ induced by a simplicial map $f\colon J \longrightarrow K$ is given by

$$f_* : \mathbb{A}_*(J) \longrightarrow \mathbb{A}_*(K)\,;\; M \longrightarrow f_*M\,,\; f_*M(\tau) = \sum_{\sigma \in J, f\sigma = \tau} M(\sigma)\,,$$

with

$$(f_*M)_*(K) \;=\; M_*(J) \;=\; \sum_{\sigma \in J} M(\sigma)\,.$$

For any object M in $\mathbb{A}_*(J)$ define a $\mathbb{C}$-equivalence in $\mathbb{A}$

$$\beta_f(M) : [M]_*[J] \xrightarrow{\;\simeq\;} [f_*M]_*[K]$$

by

$$\beta_f(M) : [M]_*[J]_r \;=\; \Big(\sum_{\sigma \in J} \Delta(\Delta^{|\sigma|}) \otimes_{\mathbb{Z}} M(\sigma)\Big)_r \longrightarrow$$

$$[f_*M]_*[K]_r \;=\; \Big(\sum_{\sigma \in J} \Delta(\Delta^{|f\sigma|}) \otimes_{\mathbb{Z}} M(\sigma)\Big)_r\,;\; a \otimes b \longrightarrow fa \otimes b\,.$$

The dual $\mathbb{C}$-equivalences determine a natural $\mathbb{C}_*(K)$-equivalence

$$G(M) \;=\; T(\beta_f(M)) : T(f_*M) \;\overset{\simeq}{\longrightarrow}\; f_*(TM) \,,$$

making $F = f_* : \Lambda_*(J) \longrightarrow \Lambda_*(K)$ a functor of algebraic bordism categories.

□

EXAMPLE 5.7 Given a simplicial complex K let $f : K \longrightarrow \{*\}$ be the unique simplicial map. The assembly of a finite chain complex C in $\mathbb{A}_*(K)$ is the finite chain complex $C_*[K] = f_*C$ in $\mathbb{A}$ induced by the functor

$$f_* : \mathbb{A}_*(K) \longrightarrow \mathbb{A}_*(\{*\}) \;=\; \mathbb{A} \,.$$

The $\mathbb{C}$-equivalence defined in the proof of 4.9 is given by

$$\beta_C \;=\; \beta_f(C) : [C]_*[K] \;\overset{\simeq}{\longrightarrow}\; C_*(K) \,.$$

□

EXAMPLE 5.8 Let X, J be simplicial complexes such that X has a J-dissection $\{X[\sigma] \,|\, \sigma \in J\}$, so that as in 4.15 there is defined a chain complex C in $\mathbb{A}(\mathbb{Z})_*(J)$ with

$$C(\sigma) \;=\; \Delta(X[\sigma], \partial X[\sigma]) \,, \quad [C][\sigma] \;=\; \Delta(X[\sigma]) \; (\sigma \in J) \,.$$

The pushforward of C with respect to a simplicial map $f : J \longrightarrow K$ is the chain complex f_*C in $\mathbb{A}(\mathbb{Z})_*(K)$ associated to the K-dissection $\{f_*X[\tau] \,|\, \tau \in K\}$ of X defined by

$$f_*X[\tau] \;=\; \bigcup_{\sigma \in J, f(\sigma) = \tau} X[\sigma] \,.$$

In particular, if $X[\sigma] = g^{-1}D(\sigma, J)$ for a simplicial map $g : X \longrightarrow J'$ then

$$f_*X[\tau] \;=\; (f'g)^{-1}D(\tau, K) \; (\tau \in K)$$

for the composite simplicial map $f'g : X \longrightarrow J' \longrightarrow K'$, since

$$f'^{-1}D(\tau, K) \;=\; \bigcup_{\sigma \in J, f(\sigma) = \tau} D(\sigma, J) \; (\tau \in K) \,.$$

□

REMARK 5.9 The method of 5.1 also applies to show that

$$\begin{cases} \Lambda^*[K] = (\mathbb{A}^*[K], \mathbb{B}^*[K], \mathbb{C}^*[K]) \\ \Lambda_*[K] = (\mathbb{A}_*[K], \mathbb{B}_*[K], \mathbb{C}_*[K]) \end{cases}$$

is an algebraic bordism category, with the chain duality

$$\begin{cases} T : \mathbb{A}^*[K] \overset{T}{\longrightarrow} \mathbb{B}(\mathbb{A}^*(K)) \longrightarrow \mathbb{B}(\mathbb{A}^*[K]) \\ T : \mathbb{A}_*[K] \overset{T}{\longrightarrow} \mathbb{B}(\mathbb{A}_*(K)) \longrightarrow \mathbb{B}(\mathbb{A}_*[K]) \end{cases}$$

such that for any objects M, N in $\begin{cases} \mathbb{A}^*[K] \\ \mathbb{A}_*[K] \end{cases}$

$$\begin{cases} M \otimes_{\mathbb{A}^*[K]} N = \mathrm{Hom}_{\mathbb{A}^*[K]}(TM, N) = (M \otimes_{\mathbb{A}} N)^*[K] \\ M \otimes_{\mathbb{A}_*[K]} N = \mathrm{Hom}_{\mathbb{A}_*[K]}(TM, N) = (M \otimes_{\mathbb{A}} N)_*[K] \, . \end{cases}$$

□

EXAMPLE 5.10 The dual in the sense of 5.9 of the object $\underline{\mathbb{Z}}$ in $\mathbb{A}(\mathbb{Z})_*[K]$ of 4.5 is the chain complex $[\Delta(K)^{-*}]$ in $\mathbb{A}(\mathbb{Z})_*[K]$ associated to the chain complex $\Delta(K)^{-*}$ in $\mathbb{A}(\mathbb{Z})_*(K)$ of 4.2

$$T\underline{\mathbb{Z}} = [\Delta(K)^{-*}] \, ,$$

with

$$T\underline{\mathbb{Z}}[\sigma] = [\Delta(K)^{-*}][\sigma] = \Delta(K, K\backslash \mathrm{st}_K(\sigma))^{-*} \quad (\sigma \in K) \, .$$

□

§6. Simply connected assembly

As in §5 let $\Lambda = (\mathbb{A}, \mathbb{B}, \mathbb{C})$ be an algebraic bordism category, and let K be a locally finite simplicial complex. The simply connected assembly functor of algebraic bordism categories $\Lambda_*(K) \longrightarrow \Lambda$ will now be defined. The simply connected assembly map for the algebraic bordism category $\Lambda = \Lambda(R)$ of a ring with involution R will be generalized in §9 to a universal assembly functor $\Lambda(R)_*(K) \longrightarrow \Lambda(R[\pi_1(K)])$.

PROPOSITION 6.1 *The assembly functor of §4*

$$\mathbb{A}_*(K) \longrightarrow \mathbb{A} \; ; \; M \longrightarrow M_*(K)$$

extends to a simply connected assembly *functor of algebraic bordism categories* $\Lambda_*(K) \longrightarrow \Lambda$ *inducing assembly maps in the* $\begin{cases} symmetric \\ quadratic \end{cases}$ *L-groups*

$$\begin{cases} L^n(\Lambda_*(K)) \longrightarrow L^n(\Lambda) \; ; \; (C, \phi) \longrightarrow (C_*(K), \phi_*(K)) \\ L_n(\Lambda_*(K)) \longrightarrow L_n(\Lambda) \; ; \; (C, \psi) \longrightarrow (C_*(K), \psi_*(K)) \; . \end{cases}$$

PROOF For any object M in $\mathbb{A}_*(K)$ use the dual of the natural chain equivalence $\beta_M \colon [M]_*[K] \xrightarrow{\simeq} M_*(K)$ given by 4.9 to define a natural $\mathbb{C}$-equivalence

$$T\beta_M \colon T(M_*(K)) \xrightarrow{\simeq} T([M]_*[K]) = (TM)_*(K) \; .$$

In particular, for any finite chain complex C in $\mathbb{A}_*(K)$ there is defined an assembly $\mathbb{Z}[\mathbb{Z}_2]$-module chain map

$$C \otimes_{\mathbb{A}_*(K)} C = \text{Hom}_{\mathbb{A}_*(K)}(TC, C) \longrightarrow$$

$$\text{Hom}_{\mathbb{A}}((TC)_*(K), C_*(K)) \simeq \text{Hom}_{\mathbb{A}}(T(C_*(K)), C_*(K))$$

$$= C_*(K) \otimes_{\mathbb{A}} C_*(K) \; .$$

$\square$

The Alexander–Whitney–Steenrod diagonal chain approximation of a simplicial complex X is a $\mathbb{Z}$-module chain map

$$\Delta_X \colon \Delta(X) \longrightarrow W^{\%}\Delta(X) = \text{Hom}_{\mathbb{Z}[\mathbb{Z}_2]}(W, \Delta(X) \otimes_{\mathbb{Z}} \Delta(X)) \; ,$$

called the symmetric construction in Ranicki [142]. The evaluation of Δ_X on any n-cycle $[X] \in \Delta_n(X)$ representing a homology class $[X] \in H_n(X)$ determines an n-dimensional symmetric complex $(\Delta(X), \phi)$ in $\mathbb{A}(\mathbb{Z})$, with $\phi = \Delta_X([X])$, such that

$$\phi_0 = [X] \cap - : \Delta(X)^{n-*} \longrightarrow \Delta(X) \; .$$

If X is an n-dimensional $\mathbb{Z}$-coefficient geometric Poincaré complex with fundamental class $[X] \in H_n(X)$, then ϕ_0 is a chain equivalence and $(\Delta(X), \phi)$ is an n-dimensional geometric Poincaré complex in $\mathbb{A}(\mathbb{Z})$.

EXAMPLE 6.2 Given a K-dissection $\{X[\sigma] \,|\, \sigma \in K\}$ of a simplicial complex X let C be the chain complex in $\mathbb{A}(\mathbb{Z})_*(K)$ defined in 4.15, with

$$C(\sigma) \;=\; \Delta(X[\sigma], \partial X[\sigma]) \;(\sigma \in K) \;,\; C_*(K) = \Delta(X) \;.$$

The symmetric constructions

$$\Delta_{X[\sigma]} : [C][\sigma] \;=\; \Delta(X[\sigma]) \;\longrightarrow\; W^{\%}\Delta(X[\sigma]) \;(\sigma \in K)$$

fit together to define a $\mathbb{Z}$-module chain map

$$\Delta_C : [C]_*[K] \;\longrightarrow\; W^{\%}C \;=\; \mathrm{Hom}_{\mathbb{Z}[\mathbb{Z}_2]}(W, ([C] \otimes_{\mathbb{Z}} [C])_*[K]) \;.$$

The evaluation of Δ_C on any n-cycle $[X] \in [C]_*[K]_n$ representing a homology class

$$[X] \in H_n([C]_*[K]) \;=\; H_n(C_*(K)) \;=\; H_n(X)$$

determines an n-dimensional symmetric complex (C, ϕ) in $\mathbb{A}(\mathbb{Z})_*(K)$ with $\phi = \Delta_C[X]$, such that the assembly is homotopy equivalent to the n-dimensional symmetric complex in $\mathbb{A}(\mathbb{Z})$

$$(C_*(K), \phi_*(K)) \;\simeq\; (\Delta(X), \Delta_X([X]))$$

considered in Ranicki [142]. Let $E = E([C])$ be the Leray–Serre spectral sequence associated to the double complex D of 4.15, with E^2-terms

$$E^2_{p,q} \;=\; H_p(K; \{H_q(X[\sigma])\}) \;,$$

converging to $H_*(X)$. For each $\sigma \in K$ let $D[\sigma]$ be the quotient double complex of D defined by

$$D[\sigma]_{p,q} \;=\; \sum_{\tau \geq \sigma, |\tau|=p} \Delta(X[\tau])_q \;,$$

and let

$$\partial_\sigma : \Delta(X) \;\simeq\; D \;\longrightarrow\; D[\sigma] \;\simeq\; S^{|\sigma|}\Delta(X[\sigma], \partial X[\sigma])$$

be the chain map determined by the projection of the total complexes. (See 8.2 below for a direct construction of ∂_σ.) The n-dimensional symmetric complex (C, ϕ) in $\mathbb{A}(\mathbb{Z})_*(K)$ is such that

$$\phi_0(\sigma) \;=\; [X(\sigma)] \cap - : C^{n-*}(\sigma) \;=\; \Delta(X[\sigma])^{n-|\sigma|-*}$$
$$\longrightarrow C(\sigma) \;=\; \Delta(X[\sigma], \partial X[\sigma]) \;,$$

with

$$[X[\sigma]] \;=\; \partial_\sigma([X]) \in H_{n-|\sigma|}(X[\sigma], \partial X[\sigma]) \;(\sigma \in K) \;.$$

The spectral sequence $\overline{E} = E([C^{n-*}])$ of 4.6 is the spectral sequence of the double complex $\overline{D}$ with

$$\overline{D}_{p,q} = \sum_{\sigma \in K, |\sigma|=p} \Delta(X[\sigma], X[\partial\sigma])^{n-p-q} \,,$$

$$\overline{d}' = \sum_{\sigma}\sum_{i}(-)^i(\sigma \to \delta_i\sigma)^* : \overline{D}_{p,q} \longrightarrow \overline{D}_{p-1,q} \,,$$

$$\overline{d}'' = \sum_{\sigma} d^*_{\Delta(X[\sigma],\partial X[\sigma])} : \overline{D}_{p,q} \longrightarrow \overline{D}_{p,q-1} \,.$$

The $\overline{E}^2$-terms are given by

$$\overline{E}^2_{p,q} = H_p(K; \{H^{n-|\sigma|-q}(X[\sigma], \partial X[\sigma])\}) \,,$$

and $\overline{E}$ converges to

$$H_*([C^{n-*}]_*[K]) = H^{n-*}(X)$$

with respect to the filtration

$$F_p H^{n-*}(X) = \ker\left(H^{n-*}(X) \longrightarrow H^{n-*}(\bigcup_{\sigma \in K, |\sigma|>p} X[\sigma])\right)\,.$$

Cap product with $[X] \in [C]_*[K]_n$ defines a map of double complexes

$$[X] \cap - : \overline{D} \longrightarrow D$$

given on the E^2-level by the cap products

$$\{[X[\sigma]] \cap -\} : \overline{E}^2_{p,q} = H_p(K; \{H^{n-|\sigma|-q}(X[\sigma], \partial X[\sigma])\})$$
$$\longrightarrow E^2_{p,q} = H_p(K; \{H_q(X[\sigma])\})$$

and converging to the cap product

$$[X] \cap - : H^{n-*}(X) \longrightarrow H_*(X)$$

on the E^∞-level. In particular, if each $(X[\sigma], \partial X[\sigma])$ $(\sigma \in K)$ is an $(n-|\sigma|)$-dimensional $\mathbb{Z}$-coefficient geometric Poincaré pair then (C, ϕ) is an n-dimensional symmetric Poincaré complex in $\mathbb{A}(\mathbb{Z})_*(K)$ and X is an n-dimensional $\mathbb{Z}$-coefficient geometric Poincaré complex. This is a generalization of the familiar result that a homology manifold is a Poincaré space.
$$\square$$

EXAMPLE 6.3 Let $\{X[\sigma] \,|\, \sigma \in K\}$ be the K-dissection of the barycentric subdivision $X = K'$ defined by the dual cells

$$X[\sigma] = D(\sigma, K) \quad (\sigma \in K)\,,$$

which are contractible. In this case the Leray–Serre spectral sequence E of 4.15 collapses, with

$$E^2_{p,q} = H_p(K; \{H_q(D(\sigma, K))\}) = \begin{cases} H_p(K) & \text{if } q = 0 \\ 0 & \text{if } q \neq 0 \end{cases}$$

and 5.5 gives the Zeeman dihomology spectral sequence $\overline{E}$ (already discussed in 4.11) converging to $H^{n-*}(K)$, with

$$\overline{E}^2_{p,q} = H_p(K; \{H^{n-|\sigma|-q}(D(\sigma, K), \partial D(\sigma, K))\}) .$$

$\square$

REMARK 6.4 The assembly functor $\mathbb{A}_*(K) \longrightarrow \mathbb{A}$ is defined in 6.1 using actual colimits, but there is also an assembly functor

$$\mathbb{A}_*[K] \longrightarrow \mathbb{B}(\mathbb{A}) ; M \longrightarrow M_*[K] ,$$

using chain homotopy colimits. By an abstract version of the Eilenberg–Zilber theorem there is defined for any chain complex C in $\mathbb{A}_*[K]$ an assembly $\mathbb{Z}$-module chain map

$$\alpha_0 : (C \otimes_{\mathbb{A}} C)_*[K] \longrightarrow C_*[K] \otimes_{\mathbb{A}} C_*[K] .$$

As for the construction of the Steenrod squares α_0 is only $\mathbb{Z}_2$-equivariant up to a chain homotopy $\alpha_1 : \alpha_0 T \simeq T\alpha_0$, with α_1 $\mathbb{Z}_2$-equivariant up to a higher chain homotopy $\alpha_2 : \alpha_1 T \simeq T\alpha_1$, and so on ..., defining a '$\mathbb{Z}_2$-isovariant chain map' $\{\alpha_s \mid s \geq 0\}$ in the sense of Ranicki [141, §1]. The simply connected assembly of an n-dimensional symmetric complex (C, ϕ) in $\mathbb{A}_*[K]$ is an n-dimensional symmetric complex in $\mathbb{A}$

$$(C, \phi)_*[K] = (C_*[K], \phi_*[K]) .$$

In particular, for any n-cycle

$$[K] = \sum_{\tau \in K, |\tau|=n} r_\tau \tau \in \mathbb{Z}_*[K]_n = \Delta(K)_n \ (r_\tau \in \mathbb{Z})$$

there is defined an n-dimensional symmetric complex $(\underline{\mathbb{Z}}, \phi)$ in $\mathbb{A}(\mathbb{Z})_*[K]$, with

$$\underline{\mathbb{Z}}_k[\sigma] = \begin{cases} \mathbb{Z} & \text{if } k = 0 \\ 0 & \text{if } k \neq 0 \end{cases} \ (\sigma \in K) \ , \ \underline{\mathbb{Z}}_*[K] = \Delta(K) ,$$

$$\phi_0 = \sum_{\tau \in K, |\tau|=n} r_\tau(1 \otimes 1) \in (\underline{\mathbb{Z}} \otimes_{\mathbb{Z}} \underline{\mathbb{Z}})_*[K]_n = \sum_{\sigma \in K} (\underline{\mathbb{Z}}[\sigma] \otimes_{\mathbb{Z}} \underline{\mathbb{Z}}[\sigma])_{n-|\sigma|} ,$$

$$\phi_s = 0 \in (\underline{\mathbb{Z}} \otimes_{\mathbb{Z}} \underline{\mathbb{Z}})_*[K]_{n+s} \ (s \geq 1)$$

such that the assembly in $\mathbb{A}(\mathbb{Z})$ is the n-dimensional symmetric complex

$$(\underline{\mathbb{Z}}, \phi)_*[K] = (\Delta(K), \phi_K([K]))$$

considered in Ranicki [142]. By 5.10 the n-dual of $\underline{\mathbb{Z}}$ is the chain complex in $\mathbb{A}(\mathbb{Z})_*[K]$

$$\underline{\mathbb{Z}}^{n-*} = S^n T\underline{\mathbb{Z}} = [C]$$

associated to the chain complex C in $\mathbb{A}(\mathbb{Z})_*(K)$ with

$$C(\sigma) = S^{n-|\sigma|}\mathbb{Z} , \ C_*(K) = \Delta(K)^{n-*} ,$$

$$\underline{\mathbb{Z}}^{n-*}[\sigma] = [C][\sigma] = \Delta(K, K \backslash \mathrm{st}_K(\sigma))^{n-*} \ (\sigma \in K) .$$

The duality chain map in $\mathbb{A}(\mathbb{Z})_*[K]$

$$\phi_0 \;=\; [K] \cap - \;:\; \underline{\mathbb{Z}}^{n-*} \longrightarrow \underline{\mathbb{Z}}$$

has components

$$\phi_0[\sigma] \;=\; \langle [K][\sigma], - \rangle \;:\; \underline{\mathbb{Z}}^{n-*}[\sigma] \;=\; \Delta(K, K\backslash \mathrm{st}_K(\sigma))^{n-*} \longrightarrow \underline{\mathbb{Z}}[\sigma] \;=\; \mathbb{Z}\;;$$

$$\tau \longrightarrow \begin{cases} r_\tau & \text{if } \tau \geq \sigma,\; |\tau| = n \\ 0 & \text{otherwise} \end{cases}$$

with $[K][\sigma]$ the image of $[K]$

$$[K][\sigma] \;=\; \sum_{\tau \geq \sigma,\, |\tau|=n} r_\tau \tau \in \Delta(K, K\backslash \mathrm{st}_K(\sigma))_n \quad (\sigma \in K)\,.$$

The assembly duality chain map $\phi_0[K] : (\underline{\mathbb{Z}}^{n-*})_*[K] \longrightarrow \underline{\mathbb{Z}}_*[K]$ in $\mathbb{A}(\mathbb{Z})$ fits into a chain homotopy commutative diagram

$$
\begin{array}{ccc}
(\underline{\mathbb{Z}}^{n-*})_*[K] \;=\; [C]_*[K] & \xrightarrow{\;\;\phi_0[K]\;\;} & \underline{\mathbb{Z}}_*[K] \;=\; \Delta(K) \\
& & \\
\beta_C \searrow {\scriptstyle\simeq} & & \nearrow [K]\cap - \\
& & \\
& C_*(K) \;=\; \Delta(K)^{n-*} &
\end{array}
$$

with β_C the chain equivalence given by 4.9. Thus K is an n-dimensional $\mathbb{Z}$-coefficient homology manifold (resp. Poincaré complex) with fundamental cycle $[K] \in \Delta(K)_n$ if and only if the chain map $\phi_0 : \underline{\mathbb{Z}}^{n-*} \longrightarrow \underline{\mathbb{Z}}$ in $\mathbb{A}(\mathbb{Z})_*[K]$ is such that each

$$\phi_0[\sigma] \;:\; \underline{\mathbb{Z}}^{n-*}[\sigma] \longrightarrow \underline{\mathbb{Z}}[\sigma] \quad (\sigma \in K)$$

is a $\mathbb{Z}$-module chain equivalence (resp. the assembly $\phi_0[K] : (\underline{\mathbb{Z}}^{n-*})_*[K] \longrightarrow \underline{\mathbb{Z}}_*[K]$ is a $\mathbb{Z}$-module chain equivalence). Identifying

$$H_*(\underline{\mathbb{Z}}^{n-*}[\sigma]) \;=\; H^{n-*}(K, K\backslash \mathrm{st}_K(\sigma)) \;=\; H^{n-*}(|K|, |K|\backslash\{\hat{\sigma}\})$$

$$=\; H^{n-|\sigma|-*}(\mathrm{star}_{K'}(\hat{\sigma}), \mathrm{link}_{K'}(\hat{\sigma})) \quad (\sigma \in K)\,,$$

we again recover the result that a homology manifold is a geometric Poincaré complex. This is the chain homotopy theoretic version of the spectral sequence argument of 6.2.

$\square$

§7. Derived product and Hom

Borel and Moore [11] defined derived duality in the category of chain complexes of sheaves of R-modules for a Dedekind ring R, using it to prove Poincaré duality for R-coefficient homology manifolds. It is a special case of the Verdier duality for chain complexes of sheaves, which plays an important role in intersection homology theory – see Goresky and MacPherson [60, 1.12]. The chain duality defined in §5 on the category of chain complexes in $\mathbb{A}(R)_*(K)$ (for any commutative ring R and finite simplicial complex K) will now be interpreted as a Verdier duality, with $\Delta(K'; R)$ as the dualizing complex.

For a Dedekind ring R with field of fractions F the *derived dual* of an R-module M is defined to be the R-module chain complex

$$TM : \ldots \longrightarrow 0 \longrightarrow \operatorname{Hom}_R(M, F) \longrightarrow \operatorname{Hom}_R(M, F/R) ,$$

using the injective resolution $F \longrightarrow F/R$ of R. The derived duality $M \longrightarrow TM$ has better homological properties than the ordinary duality $M \longrightarrow M^* = \operatorname{Hom}_R(M, R)$. The homology $H_*(TC)$ of the derived dual TC of an R-module chain complex C depends only on the homology $H_*(C)$, with universal coefficient theorem split exact sequences

$$0 \longrightarrow \operatorname{Ext}_R(H_{n-1}(C), R) \longrightarrow H_n(TC) \longrightarrow \operatorname{Hom}_R(H_n(C), R) \longrightarrow 0 .$$

For a finite f.g. free R-module chain complex C the derived dual TC is homology equivalent to the ordinary dual $C^* = \operatorname{Hom}_R(C, R)$.

Let $\mathbb{A} = \mathbb{A}(R) = \{\text{f.g. free } R\text{-modules}\}$ for a commutative ring R. From now on, the additive category $\begin{cases} \mathbb{A}(R)_*[K] \\ \mathbb{A}(R)_*(K) \end{cases}$ defined in §6 will be denoted by $\begin{cases} \mathbb{A}[R, K] \\ \mathbb{A}(R, K), \end{cases}$ and its objects will be called (f.g. free) $\begin{cases} [R, K]\text{-} \\ (R, K)\text{-} \end{cases}$ *modules*.

Given an $\begin{cases} [R, K] \\ (R, K) \end{cases}$-module chain complex C denote the corresponding R-module chain complex by $\begin{cases} C[K] \\ C(K) \end{cases}$ rather than by $\begin{cases} C_*[K] \\ C_*(K). \end{cases}$

The abelian groups $\operatorname{Hom}_R(M, N)$, $M \otimes_R N$ are R-modules, for any R-modules M, N, since the ground ring R is commutative. Thus for $\begin{cases} [R, K]\text{-} \\ (R, K)\text{-} \end{cases}$ modules M, N there are defined R-modules

$$\begin{cases} \operatorname{Hom}_{[R,K]}(M, N) = \operatorname{Hom}_{\mathbb{A}[R,K]}(M, N) \\ \operatorname{Hom}_{(R,K)}(M, N) = \operatorname{Hom}_{\mathbb{A}(R,K)}(M, N), \end{cases} \quad \begin{cases} M \otimes_{[R,K]} N = M \otimes_{\mathbb{A}[R,K]} N \\ M \otimes_{(R,K)} N = M \otimes_{\mathbb{A}(R,K)} N . \end{cases}$$

Given (R, K)-module morphisms $f : M \longrightarrow M'$, $g : N \longrightarrow N'$ there is defined an R-module morphism

$$(f^*, g_*) : \operatorname{Hom}_{(R,K)}(M', N) \longrightarrow \operatorname{Hom}_{(R,K)}(M, N') ; \quad h \longrightarrow ghf .$$

By the definition of (R, K)-module morphisms

$$\mathrm{Hom}_{(R,K)}(M, N) = \sum_{\sigma \in K} \mathrm{Hom}_R(M(\sigma), [N][\sigma]) .$$

Thus it is possible to give the R-module $\mathrm{Hom}_{(R,K)}(M, N)$ the structure of an (R, K)-module by setting

$$\mathrm{Hom}_{(R,K)}(M, N)(\sigma) = \mathrm{Hom}_R(M(\sigma), [N][\sigma]) \ (\sigma \in K) ,$$

but this is unnatural: if f is not the identity the R-module morphism (f^*, g_*) is not an (R, K)-module morphism.

The following derived products and Hom functors are modelled on the derived functors appearing in sheaf theory, and allow the resolution of $\mathrm{Hom}_{(R,K)}(M, N)$ by an (R, K)-module chain complex $\mathrm{RHom}_{(R,K)}(M, N)$ which is natural in both M and N.

DEFINITION 7.1 The *derived product* $M \boxtimes_R N$ of (R, K)-modules M, N is the (R, K)-module with

$$(M \boxtimes_R N)(K) = \sum_{\lambda, \mu \in K, \lambda \cap \mu \neq \emptyset} M(\lambda) \otimes_R N(\mu) \subseteq M(K) \otimes_R N(K) ,$$

$$(M \boxtimes_R N)(\sigma) = \sum_{\lambda \cap \mu = \sigma} M(\lambda) \otimes_R N(\mu) \ (\sigma \in K) .$$

The associated $[R, K]$-module $[M \boxtimes_R N]$ is such that

$$[M \boxtimes_R N][\sigma] = [M][\sigma] \otimes_R [N][\sigma] \ (\sigma \in K) ,$$

$$[M \boxtimes_R N][K] = M \otimes_{(R,K)} N = \mathrm{Hom}_{(R,K)}(TM, N) ,$$

with 4.9 giving an R-module chain equivalence

$$\beta_{M \boxtimes_R N} : M \otimes_{(R,K)} N \xrightarrow{\ \simeq\ } (M \boxtimes_R N)(K) .$$

$\square$

The derived product $C \boxtimes_R D$ of (R, K)-module chain complexes C, D is the (R, K)-module chain complex

$$(C \boxtimes_R D)_r = \sum_{p+q=r} C_p \boxtimes_R D_q , \ d(x \boxtimes y) = x \boxtimes dy + (-)^q dx \boxtimes y .$$

The R-module chain complex $(C \boxtimes_R D)(K)$ is a subcomplex of $C(K) \otimes_R D(K)$ such that there is defined a chain equivalence

$$\beta_{C \boxtimes_R D} : C \otimes_{(R,K)} D = [C \boxtimes_R D][K] = \mathrm{Hom}_{(R,K)}(TC, D)$$

$$\xrightarrow{\ \simeq\ } (C \boxtimes_R D)(K)$$

and

$$H_n((C \boxtimes_R D)(K)) = H_n(C \otimes_{(R,K)} D)$$

$$= H_0(\mathrm{Hom}_{(R,K)}(C^{n-*}, D)) \ (n \in \mathbb{Z}) .$$

EXAMPLE 7.2 Let $f: X \longrightarrow K'$, $g: Y \longrightarrow K'$ be simplicial maps, so that there are defined (R, K)-module chain complexes C, D as in 4.16, with

$$C = \Delta(X; R) \ , \quad [C][\sigma] = \Delta(f^{-1}D(\sigma, K); R) \ (\sigma \in K) \ ,$$

$$D = \Delta(Y; R) \ , \quad [D][\tau] = \Delta(g^{-1}D(\tau, K); R) \ (\tau \in K) \ .$$

The derived product $C \boxtimes_R D$ is chain equivalent to the (R, K)-module chain complex $\Delta(Z; R)$ associated to a simplicial map $h: Z \longrightarrow K'$, with Z a triangulation of the pullback polyhedron

$$|Z| = \{ (x, y) \in |X| \times |Y| \, | \, f(x) = g(y) \in |K'| \}$$

and h a simplicial approximation of the map

$$|Z| \longrightarrow |K'| \ ; \ (x, y) \longrightarrow f(x) = g(y) \ .$$

The R-module chain complex $\mathrm{Hom}_{(R,K)}(\Delta(X; R), \Delta(Y; R)^{-*})$ is chain equivalent to $\Delta(X \times Y, X \times Y \backslash Z; R)^{-*}$, with

$$\Delta(Y; R)^{-*}(\tau) = \Delta(g^{-1}D(\tau, K); R)^{-|\tau|-*} \ (\tau \in K) \ .$$

$\square$

EXAMPLE 7.3 The adjoint of the Flexner chain level cap product (4.13) is a $\mathbb{Z}$-module chain equivalence

$$A\Delta_F : \Delta(K) \xrightarrow{\simeq} \mathrm{Hom}_{(\mathbb{Z},K)}(\Delta(K)^{-*}, \Delta(K')) \simeq \Delta(K') \boxtimes_{\mathbb{Z}} \Delta(K') \ ,$$

by the special case $f = g = 1 : X = Y = K \longrightarrow K$ of 7.2.

$\square$

EXAMPLE 7.4 The Alexander–Whitney diagonal chain approximation for K is defined by

$$\Delta_K : \Delta(K) \longrightarrow \Delta(K) \otimes \Delta(K) \ ;$$

$$(v_0 v_1 \ldots v_n) \longrightarrow \sum_{i=0}^{n} (v_0 v_1 \ldots v_i) \otimes (v_i v_{i+1} \ldots v_n) \ .$$

Let C be the $(\mathbb{Z}, K)$-module chain complex defined as in 4.15 by

$$C(K) = \Delta(K') \ , \quad C(\sigma) = \Delta(D(\sigma, K), \partial D(\sigma, K)) \ (\sigma \in K) \ .$$

The Alexander–Whitney diagonal chain map for K' factors through a $(\mathbb{Z}, K)$-module chain equivalence

$$\Delta_{K'} : \Delta(K') \xrightarrow{\simeq} \Delta(K') \boxtimes_{\mathbb{Z}} \Delta(K') \subseteq \Delta(K') \otimes_{\mathbb{Z}} \Delta(K') \ .$$

$\square$

DEFINITION 7.5 The *derived Hom* of (R, K)-modules M, N is the (R, K)-module chain complex

$$\mathrm{RHom}_{(R,K)}(M, N) = TM \boxtimes_R N \ .$$

$\square$

The derived Hom defined for any (R, K)-module chain complexes C, D by

$$\mathrm{RHom}_{(R,K)}(C, D) = TC \boxtimes_R D \, ,$$

is such that there is defined an R-module chain equivalence

$$\beta_{TC\boxtimes_R D} : \mathrm{Hom}_{(R,K)}(C, D) = [\mathrm{RHom}_{(R,K)}(C,D)][K] = [TC \boxtimes_R D][K]$$

$$\xrightarrow{\simeq} \mathrm{RHom}_{(R,K)}(C,D)(K) = (TC \boxtimes_R D)(K) \, .$$

PROPOSITION 7.6 *The (R, K)-module chain complex $\Delta(K'; R)$ with*

$$\Delta(K'; R)(\sigma) = \Delta(D(\sigma, K), \partial D(\sigma, K); R) \quad (\sigma \in K)$$

is a dualizing complex for the chain duality $T: \mathbb{A}(R, K) \longrightarrow \mathbb{B}(\mathbb{A}(R, K))$ with respect to the derived Hom, meaning that T is naturally chain equivalent to the contravariant functor

$$T' = \mathrm{RHom}_{(R,K)}(-, \Delta(K'; R)) : \mathbb{A}(R, K) \longrightarrow \mathbb{B}(\mathbb{A}(R, K)) \, ;$$

$$M \longrightarrow T'M = \mathrm{RHom}_{(R,K)}(M, \Delta(K'; R)) \, .$$

PROOF Use the augmentation R-module chain maps $\epsilon: \Delta(K'; R)(\sigma) \longrightarrow R$ to define a natural transformation $T' \longrightarrow T$

$$T'M(\sigma) = (TM \boxtimes_R \Delta(K'; R))(\sigma) \longrightarrow TM(\sigma) \otimes_R R = TM(\sigma) \, ;$$

$$x(\lambda) \otimes y(\mu) \longrightarrow x(\lambda) \otimes \epsilon y(\mu) \, .$$

This is a natural chain equivalence, since the R-module chain maps

$$1 \otimes \epsilon : [T'M][\sigma] = [TM][\sigma] \otimes_R [\Delta(K'; R)][\sigma]$$

$$= [TM][\sigma] \otimes_R \Delta(D(\sigma, K); R) \longrightarrow [TM][\sigma] \otimes_R R = [TM][\sigma]$$

are chain equivalences.

$\square$

More generally, for any (R, K)-module chain complex C 7.6 gives a natural (R, K)-module chain equivalence

$$TC \simeq \mathrm{RHom}_{(R,K)}(C, \Delta(K'; R)) \, .$$

A simplicial map $f: K \longrightarrow L$ induces a pullback functor

$$f^* : \mathbb{A}[R, L] \longrightarrow \mathbb{A}[R, K] \, ; \quad M \longrightarrow f^*M \, , \, f^*M[\sigma] = M[f\sigma] \, .$$

EXAMPLE 7.7 The $[R, K]$-module chain complex associated to the dualizing (R, K)-module chain complex $\Delta(K'; R)$ is chain equivalent in $\mathbb{A}[R, K]$ to the pullback f^*R along the simplicial map $f: K \longrightarrow \{*\}$ of the $[R, \{*\}]$-module R

$$[\Delta(K'; R)] \simeq f^*R \, .$$

Specifically, the augmentation maps define chain equivalences

$$\epsilon[\sigma] : [\Delta(K'; R)][\sigma] = \Delta(D(\sigma, K); R) \xrightarrow{\simeq} f^*R[\sigma] = R \quad (\sigma \in K) \, .$$

$\square$

§8. Local Poincaré duality

The following notion of local Poincaré duality is an abstraction of the local Poincaré duality properties of a homology manifold, and in fact serves to characterize the geometric Poincaré complexes which are homology manifolds. The universal algebraic L-theory assembly map will be defined in §9 by passing from local Poincaré complexes to global Poincaré complexes.

Let R, K be as in §7, with R a commutative ring, K a finite simplicial complex and $\mathbb{A}(R, K)$ the additive category with chain duality defined in 5.1.

DEFINITION 8.1 An n-dimensional $\begin{cases} \text{symmetric} \\ \text{quadratic} \end{cases}$ complex $\begin{cases} (C, \phi) \\ (C, \psi) \end{cases}$ in $\mathbb{A}(R, K)$ is *locally Poincaré* if it is $\mathbb{C}(R)_*(K)$-Poincaré, i.e. if the duality is given by an (R, K)-module chain equivalence

$$\begin{cases} \phi_0 : C^{n-*} \xrightarrow{\simeq} C \\ (1 + T)\psi_0 : C^{n-*} \xrightarrow{\simeq} C . \end{cases}$$

□

The derived product $\boxtimes$ of §7 will now be used to associate to an n-dimensional $\begin{cases} \text{symmetric} \\ \text{quadratic} \end{cases}$ complex $\begin{cases} (C, \phi) \\ (C, \psi) \end{cases}$ in $\mathbb{A}(R, K)$ a collection $\begin{cases} \{(C, \phi)[\sigma] \mid \sigma \in K\} \\ \{(C, \psi)[\sigma] \mid \sigma \in K\} \end{cases}$ of $(n - |\sigma|)$-dimensional $\begin{cases} \text{symmetric} \\ \text{quadratic} \end{cases}$ pairs in $\mathbb{A}(R)$, such that $\begin{cases} (C, \phi) \\ (C, \psi) \end{cases}$ is locally Poincaré if and only if each $\begin{cases} (C, \phi)[\sigma] \\ (C, \psi)[\sigma] \end{cases}$ is a Poincaré pair in $\mathbb{A}(R)$.

By definition, an n-dimensional $\begin{cases} \text{symmetric} \\ \text{quadratic} \end{cases}$ complex $\begin{cases} (C, \phi) \\ (C, \psi) \end{cases}$ in $\mathbb{A}(R, K)$ is an n-dimensional chain complex C in $\mathbb{A}(R, K)$ together with an n-cycle

$$\begin{cases} \phi \in (W^\% C)_n = \operatorname{Hom}_{\mathbb{Z}[\mathbb{Z}_2]}(W, [C \boxtimes_R C][K])_n \\ \psi \in (W_\% C)_n = W \otimes_{\mathbb{Z}[\mathbb{Z}_2]} [C \boxtimes_R C][K]_n . \end{cases}$$

From now on, the (R, K)-module chain complex $[C \boxtimes_R C][K] = C \otimes_{(R,K)} C$ will be replaced by the (R, K)-module chain equivalent complex $(C \boxtimes_R C)(K)$.

DEFINITION 8.2 (i) Given an (R, K)-module chain complex C define the R-module chain map

$$\partial_\sigma : C(K) \longrightarrow S^{|\sigma|} C(\sigma)$$

for each simplex $\sigma = (v_0 v_1 \ldots v_{|\sigma|}) \in K$ to be the composite

$$\partial_\sigma : C(K)_n = \sum_{\tau \in K} C(\tau)_n \xrightarrow{\text{projection}} C(\sigma_0)_n$$

$$\xrightarrow{d_1} C(\sigma_1)_{n-1} \xrightarrow{d_2} C(\sigma_2)_{n-2} \longrightarrow \cdots \xrightarrow{d_{|\sigma|}} C(\sigma)_{n-|\sigma|} \, ,$$

with $\sigma_0 < \sigma_1 < \ldots < \sigma_{|\sigma|} = \sigma$ defined by $\sigma_j = (v_0 v_1 \ldots v_j)$ $(0 \le j \le |\sigma|)$ and $d_j : C(\sigma_{j-1})_{n-j+1} \longrightarrow C(\sigma_j)_{n-j}$ $(1 \le j \le |\sigma|)$ the relevant components of $d_{C(K)} : C(K)_{n-j+1} \longrightarrow C(K)_{n-j}$.

(ii) Given an n-dimensional $\left\{ \begin{array}{l} \text{symmetric} \\ \text{quadratic} \end{array} \right.$ complex $\left\{ \begin{array}{l} (C, \phi) \\ (C, \psi) \end{array} \right.$ in $\mathbb{A}(R, K)$ define for each $\sigma \in K$ an $(n - |\sigma|)$-dimensional $\left\{ \begin{array}{l} \text{symmetric} \\ \text{quadratic} \end{array} \right.$ pair in $\mathbb{A}(R)$

$$\left\{ \begin{array}{l} (C, \phi)[\sigma] = (i[\sigma]: \partial[C][\sigma] \longrightarrow [C][\sigma], (\phi[\sigma], \partial\phi[\sigma])) \\ (C, \psi)[\sigma] = (i[\sigma]: \partial[C][\sigma] \longrightarrow [C][\sigma], (\psi[\sigma], \partial\psi[\sigma])) \end{array} \right.$$

with

$$i[\sigma] = \text{inclusion} : \partial[C][\sigma]_r = \sum_{\tau > \sigma} C(\tau)_r \longrightarrow [C][\sigma]_r = \sum_{\tau \ge \sigma} C(\tau)_r \, .$$

The $\left\{ \begin{array}{l} \text{symmetric} \\ \text{quadratic} \end{array} \right.$ structure $\left\{ \begin{array}{l} (\phi[\sigma], \partial\phi[\sigma]) \\ (\psi[\sigma], \partial\psi[\sigma]) \end{array} \right.$ is the image of $\left\{ \begin{array}{l} \phi \in (W^\% C)_n \\ \psi \in (W_\% C)_n \end{array} \right.$ under the $\mathbb{Z}$-module chain map

$$\left\{ \begin{array}{l} W^\% C = \text{Hom}_{\mathbb{Z}[\mathbb{Z}_2]}(W, (C \boxtimes_R C)(K)) \\ \qquad\qquad \xrightarrow{\partial_\sigma} \text{Hom}_{\mathbb{Z}[\mathbb{Z}_2]}(W, S^{|\sigma|}(C \boxtimes_R C)(\sigma)) \\ W_\% C = W \otimes_{\mathbb{Z}[\mathbb{Z}_2]} (C \boxtimes_R C)(K) \\ \qquad\qquad \xrightarrow{\partial_\sigma} W \otimes_{\mathbb{Z}[\mathbb{Z}_2]} S^{|\sigma|}(C \boxtimes_R C)(\sigma) \, , \end{array} \right.$$

identifying

$$(C \boxtimes_R C)(\sigma) = \sum_{\lambda \cap \mu = \sigma} C(\lambda) \otimes_R C(\mu)$$

$$= \text{coker}(i[\sigma] \otimes i[\sigma]: \partial[C][\sigma] \otimes_R \partial[C][\sigma] \longrightarrow [C][\sigma] \otimes_R [C][\sigma]) \, .$$

□

EXAMPLE 8.3 Let C be the $(\mathbb{Z}, K)$-module chain complex defined as in 4.13 by

$$C(K) = \Delta(K') \, , \quad C(\sigma) = \Delta(D(\sigma, K), \partial D(\sigma, K)) \quad (\sigma \in K) \, .$$

The $\mathbb{Z}$-module chain map $\partial_\sigma : C(K) \longrightarrow S^{|\sigma|} C(\sigma)$ of 8.2 induces the natural maps passing from the global (= ordinary) homology of $|K|$ to the local homology at $\hat{\sigma} \in |K|$

$$\partial_\sigma : H_*(C(K)) = H_*(K') = H_*(K) = H_*(|K|) \xrightarrow{\text{projection}_*}$$
$$H_*(|K|, |K| \backslash \{\hat{\sigma}\}) = H_*(K, K \backslash \mathrm{st}_K(\sigma))$$
$$= H_*(\mathrm{star}_{K'}(\hat{\sigma}), \mathrm{link}_{K'}(\hat{\sigma})) = H_*(\partial\sigma' * (D(\sigma, K), \partial D(\sigma, K)))$$
$$= H_{*-|\sigma|}(D(\sigma, K), \partial D(\sigma, K)) = H_{*-|\sigma|}(C(\sigma)) .$$

If K is an n-dimensional homology manifold the images of the fundamental class $[K] \in H_n(K)$

$$\partial_\sigma([K]) = [D(\sigma, K)] \in H_{n-|\sigma|}(D(\sigma, K), \partial D(\sigma, K)) \quad (\sigma \in K)$$

are the fundamental classes of the $(n - |\sigma|)$-dimensional geometric Poincaré pairs $(D(\sigma, K), \partial D(\sigma, K))$.

$\square$

PROPOSITION 8.4 *An n-dimensional* $\begin{cases} symmetric \\ quadratic \end{cases}$ *complex* $\begin{cases} (C, \phi) \\ (C, \psi) \end{cases}$ *in*

$\mathbb{A}(R, K)$ *is locally Poincaré if and only if each* $\begin{cases} (C, \phi)[\sigma] \\ (C, \psi)[\sigma] \end{cases}$ $(\sigma \in K)$ *is*

an $(n - |\sigma|)$-dimensional $\begin{cases} symmetric \\ quadratic \end{cases}$ *Poincaré pair in* $\mathbb{A}(R)$.

PROOF By 4.7 a chain map $f: C \longrightarrow D$ in $\mathbb{A}(R, K)$ is a chain equivalence if and only if the (σ, σ)-component $f(\sigma, \sigma): C(\sigma) \longrightarrow D(\sigma)$ is a chain equivalence in $\mathbb{A}(R)$ for each $\sigma \in K$. The duality R-module chain map

$$[C][\sigma]^{n-|\sigma|-*} = C^{n-*}(\sigma) \longrightarrow [C][\sigma]/\partial[C][\sigma] = C(\sigma)$$

of $\begin{cases} (C, \phi)[\sigma] \\ (C, \psi)[\sigma] \end{cases}$ is the (σ, σ)-component of the duality (R, K)-module chain

map $\begin{cases} \phi_0: C^{n-*} \longrightarrow C \\ (1 + T)\psi_0: C^{n-*} \longrightarrow C \end{cases}$ of $\begin{cases} (C, \phi) \\ (C, \psi) \end{cases}$.

$\square$

REMARK 8.5 An *n-dimensional pseudomanifold* is a finite n-dimensional simplicial complex K such that

(i) every simplex of K is a face of an n-simplex,

(ii) every $(n - 1)$-simplex of K is a face of exactly two n-simplexes.

The result of McCrory [102] that K is a homology manifold with fundamental class $[K] \in H_n(K)$ if and only if there exists a cohomology class $U \in H^n(K \times K, K \times K \backslash \Delta)$ with the image in $H^n(K \times K)$ dual to $\Delta([K]) \in H_n(K \times K)$ can now be proved directly, using the chain duality theory of §5 and the derived product $\boxtimes$ of §7.

Assume (for simplicity) that K is oriented and connected, so that the sum of the n-simplexes is a cycle

$$[K] = \sum_{\tau \in K^{(n)}} \tau \in \ker(d \colon \Delta_n(K) \longrightarrow \Delta_{n-1}(K))$$

representing the fundamental class $[K] \in H_n(K)$. For each simplex $\sigma \in K$ the pair $(D(\sigma, K), \partial D(\sigma, K))$ is an $(n - |\sigma|)$-dimensional pseudomanifold with boundary. As in 6.2 there is defined an n-dimensional symmetric complex in $\mathbb{A}(\mathbb{Z}, K)$

$$(C, \phi) = (\Delta(K'), \Delta([K]))$$

such that

$$\phi_0(\sigma) = [D(\sigma, K)] \cap - \colon C^{n-*}(\sigma) = \Delta(D(\sigma, K))^{n-|\sigma|-*}$$
$$\longrightarrow C(\sigma) = \Delta(D(\sigma, K), \partial D(\sigma, K))$$

with assembly

$$\phi_0(K) = [K] \cap - \colon C^{n-*}(K) \simeq \Delta(K)^{n-*} \longrightarrow C(K) \simeq \Delta(K).$$

K is a homology manifold if and only if (C, ϕ) is locally Poincaré. The diagonal chain approximations are chain equivalences

$$[\Delta_0][\sigma] \colon [C][\sigma] = \Delta(D(\sigma, K))$$
$$\xrightarrow{\simeq} [C \boxtimes_{\mathbb{Z}} C][\sigma] = \Delta(D(\sigma, K)) \otimes_{\mathbb{Z}} \Delta(D(\sigma, K)),$$

so that each of the chain maps in the commutative diagram

$$
\begin{array}{ccc}
[C][K] & \xrightarrow{\ [\Delta_0]\ } & [C \boxtimes_{\mathbb{Z}} C][K] \\
{\scriptstyle \beta_C} \downarrow & & \downarrow {\scriptstyle \beta_{C \boxtimes_{\mathbb{Z}} C}} \\
C(K) & \xrightarrow{\ \Delta_0\ } & (C \boxtimes_{\mathbb{Z}} C)(K)
\end{array}
$$

is a chain equivalence, and

$$(C \otimes_{(\mathbb{Z}, K)} C)(K) \simeq \Delta(K), \quad C(K) \otimes_{\mathbb{Z}} C(K) \simeq \Delta(K \times K).$$

By 5.5 the dual $(\mathbb{Z}, K)$-module chain complex TC is such that

$$TC(\sigma) = \Delta(D(\sigma, K))^{-|\sigma|-*} \simeq S^{-|\sigma|} \mathbb{Z} \quad (\sigma \in K), \quad TC(K) \simeq \Delta(K)^{-*},$$

and

$$(TC \otimes_{(\mathbb{Z}, K)} TC)(K) \simeq \Delta(K \times K, K \times K \backslash \Delta)^{-*},$$

$$TC(K) \otimes_{\mathbb{Z}} TC(K) \simeq \Delta(K \times K)^{-*}.$$

The product $K \times K$ (or rather $K \otimes K$) is a $2n$-dimensional pseudomanifold, and the diagonal map of polyhedra

$$\Delta \colon |K| \longrightarrow |K| \times |K| \,; \; x \longrightarrow (x, x)$$

induces a diagonal map in homology

$$\Delta : H_*(K) = H_*(|K|) \longrightarrow H_*(K \times K) = H_*(|K| \times |K|) \ .$$

A *geometric Thom class* for K is an element

$$U \in H^n(K \times K, K \times K \backslash \Delta) = H_n((TC \otimes_{(\mathbb{Z},K)} TC)(K))$$
$$= H_n(\mathrm{Hom}_{(\mathbb{Z},K)}(C, TC))$$

satisfying one of the equivalent conditions:

(i) the image of U under

$$j^* = \mathrm{inclusion}^* : H^n(K \times K, K \times K \backslash \Delta)$$
$$\longrightarrow H^n(K \times K) = H_n(TC(K) \otimes_{\mathbb{Z}} TC(K))$$

is an element $j^*U \in H^n(K \times K)$ such that

$$\langle j^*U, \Delta([K]) \rangle = 1 \in \mathbb{Z} \ ,$$

(ii) the $(\mathbb{Z}, K)$-module chain map $U : C \longrightarrow C^{n-*}$ is such that

$$[\phi_0][\sigma][U][\sigma] \simeq 1 : [C][\sigma] \longrightarrow [C^{n-*}][\sigma] \longrightarrow [C][\sigma] \ (\sigma \in K) \ ,$$

with $[C][\sigma] = \Delta(D(\sigma, K)) \simeq \mathbb{Z}$.

The result of McCrory [102] is that K is a homology manifold if and only if there exists a geometric Thom class U. If K is a homology manifold then (C, ϕ) is locally Poincaré, and the inverse of the $(\mathbb{Z}, K)$-module chain equivalence $\phi_0 : C^{n-*} \longrightarrow C$ defines a geometric Thom class

$$U = (\phi_0)^{-1} \in H_n(\mathrm{Hom}_{(\mathbb{Z},K)}(C, TC)) = H^n(K \times K, K \times K \backslash \Delta) \ .$$

This is the Thom class of the homology tangent bundle of K (Spanier [160, p. 294]), the fibration

$$(K, K \backslash \{*\}) \longrightarrow (K \times K, K \times K \backslash \Delta) \longrightarrow K \ .$$

Conversely, suppose that (C, ϕ) admits a geometric Thom class U. Each $[\phi_0][\sigma]$ has a right chain homotopy inverse, and since

$$\phi_0 \simeq T\phi_0 : C^{n-*} \longrightarrow C$$

each $[\phi_0][\sigma]$ also has a left chain homotopy inverse. It follows that each $[\phi_0][\sigma]$ is a chain equivalence, so that ϕ_0 is a $(\mathbb{Z}, K)$-module chain equivalence and K is a homology manifold.

□

§9. Universal assembly

Universal assembly is the forgetful map from the L-groups of 'local' algebraic bordism categories to the L-groups of 'global' algebraic bordism categories, such as

$$A : L_*(\Lambda(R)_*(K)) = H_*(K;\mathbb{L}.(R)) \longrightarrow L_*(\Lambda(R,K)) = L_*(R[\pi_1(K)]) .$$

In §9 only the oriented case is considered; the modifications required for the nonorientable case are dealt with in Appendix A.

With R, K as in §8, let $\pi = \pi_1(K)$ be the fundamental group, and let $R[\pi]$ be the fundamental group ring. The assembly functor $\mathbb{B}(\mathbb{A}[R,K]) \longrightarrow \mathbb{B}(R)$ of 4.4 can be lifted to the universal cover $\widetilde{K}$ of K:

DEFINITION 9.1 (i) The $[R,K]$-module chain complex *universal assembly* is the functor

$$\mathbb{B}[R,K] = \mathbb{B}(\mathbb{A}[R,K]) \longrightarrow \mathbb{B}(R[\pi]) ; \; C \longrightarrow C[\widetilde{K}]$$

with

$$C[\widetilde{K}]_r = \sum_{\tilde{\sigma}\in\widetilde{K}} C[p\,\tilde{\sigma}]_{r-|\sigma|} .$$

Here, $p: \widetilde{K} \longrightarrow K$ is the covering projection.
(ii) The (R,K)-module *universal assembly* is the functor

$$\mathbb{A}(R,K) \longrightarrow \mathbb{A}(R[\pi]) ; \; M \longrightarrow M(\widetilde{K}) = \sum_{\tilde{\sigma}\in\widetilde{K}} M(p\tilde{\sigma}) ,$$

with the $R[\pi]$-module structure induced from the action of π on the universal cover $\widetilde{K}$ by covering translations. An (R,K)-module morphism $f: M \longrightarrow N$ assembles to the $R[\pi]$-module morphism $\tilde{f}: M(\widetilde{K}) \longrightarrow N(\widetilde{K})$ with components

$$\tilde{f}(\tilde{\tau},\tilde{\sigma}) = \begin{cases} f(\tau,\sigma) & \text{if } \tilde{\sigma} \leq \tilde{\tau} \\ 0 & \text{otherwise} \end{cases} : M(\tilde{\sigma}) = M(\sigma) \longrightarrow N(\tilde{\tau}) = N(\tau) .$$

$\square$

Let C be a f.g. free (R,K)-module chain complex. The R-module chain equivalence $\beta_C: [C][K] \longrightarrow C(K)$ of 4.9 lifts to an $R[\pi]$-module chain equivalence

$$\tilde{\beta}_C : [C][\widetilde{K}] \longrightarrow C(\widetilde{K}) ,$$

so that the universal assembly constructions of 9.1 (i) and (ii) agree up to chain equivalence.

PROPOSITION 9.2 *If $f: C \longrightarrow D$ is a chain map of finite $[R,K]$-module chain complexes such that each $f[\sigma]: C[\sigma] \longrightarrow D[\sigma]$ ($\sigma \in K$) is an R-module chain equivalence then the universal assembly $f[\widetilde{K}]: C[\widetilde{K}] \longrightarrow D[\widetilde{K}]$ is an $R[\pi]$-module chain equivalence.*

PROOF A chain map of finite chain complexes in an additive category is a chain equivalence if and only if the algebraic mapping cone is chain contractible. Thus it suffices to prove that a locally contractible finite $[R, K]$-module chain complex C assembles to a contractible $R[\pi]$-module chain complex $C[\widetilde{K}]$. The first quadrant spectral sequence $E(C)$ of 4.6 has E_2-terms

$$E^2_{p,q} = H_p(\widetilde{K}; \{H_q(C[\sigma])\}) ,$$

and converges to $H_*(C[\widetilde{K}])$. If C is locally contractible then $H_*(C[\sigma]) = 0$ ($\sigma \in K$), so that $H_*(C[\widetilde{K}]) = 0$ and C is globally contractible.

□

EXAMPLE 9.3 The universal assembly of the f.g. free $[R, K]$-module chain complex $\underline{R}$ defined as in 4.5 by

$$\underline{R}[\sigma] = R \ (\sigma \in K)$$

is the simplicial $R[\pi]$-module chain complex of the universal cover $\widetilde{K}$

$$\underline{R}[\widetilde{K}] = \Delta(\widetilde{K}; R) .$$

□

EXAMPLE 9.4 The Alexander–Whitney–Steenrod diagonal chain approximation for the universal cover $\widetilde{K}$

$$\Delta_{\widetilde{K}} : \Delta(\widetilde{K}; R) \longrightarrow \text{Hom}_{\mathbb{Z}[\mathbb{Z}_2]}(W, \Delta(\widetilde{K}; R) \otimes_R \Delta(\widetilde{K}; R))$$

projects to an R-module chain map

$$\tilde{\Delta}_K = 1 \otimes \Delta_{\widetilde{K}} : \Delta(K; R) = R \otimes_{R[\pi]} \Delta(\widetilde{K}; R)$$

$$\longrightarrow R \otimes_{R[\pi]} (\text{Hom}_{\mathbb{Z}[\mathbb{Z}_2]}(W, \Delta(\widetilde{K}; R) \otimes_R \Delta(\widetilde{K}; R)))$$

$$= \text{Hom}_{\mathbb{Z}[\mathbb{Z}_2]}(W, \Delta(\widetilde{K}; R) \otimes_{R[\pi]} \Delta(\widetilde{K}; R)) = W^\% \Delta(\widetilde{K}; R) ,$$

with $R[\pi]$ acting on the left of $\Delta(\widetilde{K}; R)$ via the covering translation action of π on $\widetilde{K}$, and on the right via the composition of the left action and the involution

$$R[\pi] \longrightarrow R[\pi] ; \ rg \longrightarrow rg^{-1} \ (r \in R, g \in \pi) .$$

As in Ranicki [142] for any n-cycle $[K] \in \Delta(K; R)_n$ there is defined an n-dimensional symmetric complex $(\Delta(K; R), \tilde{\Delta}_K([K]))$ in $\mathbb{A}(R[\pi])$ with

$$\tilde{\Delta}_K([K])_0 = [K] \cap - : \Delta(\widetilde{K}; R)^{n-*} \longrightarrow \Delta(\widetilde{K}; R) .$$

As in the simply connected case already considered in 6.4 the geometric nature of $\Delta_{\widetilde{K}}$ allows $(\Delta(\widetilde{K}; R), \phi)$ to be expressed as the assembly of an n-dimensional symmetric complex $(\underline{R}, \phi)$ in $\mathbb{A}[R, K]$, with $\underline{R}$ the 0-dimensional $[R, K]$-module chain complex given by

$$\underline{R}_k[\sigma] = \begin{cases} R & \text{if } k = 0 \\ 0 & \text{if } k \neq 0 \end{cases} \ (\sigma \in K) , \quad \underline{R}[K] = \Delta(K; R) .$$

By 5.6 the n-dual of $\underline{R}$ is the $[R, K]$-module chain complex
$$\underline{R}^{n-*} = \Sigma^n T\underline{R} = [C]$$
associated to the (R, K)-module chain complex C with
$$C(\sigma) = S^{n-|\sigma|}R , \; C(K) = \Delta(K; R)^{n-*} ,$$
$$\underline{R}^{n-*}[\sigma] = [C][\sigma] = \Delta(K, K\backslash \mathrm{st}_K(\sigma); R)^{n-*} \; (\sigma \in K) .$$
Write the n-cycle as
$$[K] = \sum_{\tau \in K, |\tau| = n} r_\tau \tau \in \underline{R}[K]_n = \Delta(K; R)_n \; (r_\tau \in R) .$$
The assembly of the n-dimensional symmetric complex $(\underline{R}, \phi)$ in $\mathbb{A}[R, K]$ defined by
$$\phi_0 = \sum_{\tau \in K, |\tau| = n} r_\tau (1 \otimes 1) \in (\underline{R} \otimes_R \underline{R})[K]_n = \sum_{\sigma \in K} (\underline{R}[\sigma] \otimes_R \underline{R}[\sigma])_{n - |\sigma|} ,$$
$$\phi_s = 0 \in (\underline{R} \otimes_R \underline{R})[K]_{n+s} \; (s \geq 1)$$
is the n-dimensional symmetric complex in $\mathbb{A}(R[\pi])$ defined above
$$(\underline{R}, \phi)[\tilde{K}] = (\Delta(\tilde{K}; R), \tilde{\Delta}_K([K]))$$
and there is defined a chain homotopy commutative diagram

$$\underline{R}^{n-*}[\tilde{K}] = [C][\tilde{K}] \xrightarrow{\quad \phi_0[\tilde{K}] \quad} \underline{R}[\tilde{K}] = \Delta(\tilde{K}; R)$$
$$\searrow \tilde{\beta}_C \simeq \qquad\qquad [K] \cap - \nearrow$$
$$C(\tilde{K}) = \Delta(\tilde{K}; R)^{n-*}$$

with $\tilde{\beta}_C$ the chain equivalence given by 4.9. Here, $\phi_0[K]$ is the assembly of the $[R, K]$-module chain map $\phi_0 : \underline{R}^{n-*} \longrightarrow \underline{R}$ with the components
$$\phi_0[\sigma] = \langle [K][\sigma], - \rangle :$$
$$\underline{R}^{n-*}[\sigma] = \Delta(K, K\backslash \mathrm{st}_K(\sigma); R)^{n-*} \longrightarrow \underline{R}[\sigma] = R ;$$
$$\tau \longrightarrow \begin{cases} r_\tau & \text{if } \tau \geq \sigma, \; |\tau| = n \\ 0 & \text{otherwise} \end{cases}$$
with $[K][\sigma]$ the image of $[K]$
$$[K][\sigma] = \sum_{\tau \geq \sigma, |\tau| = n} r_\tau \tau \in \Delta(K, K\backslash \mathrm{st}_K(\sigma); R)_n \; (\sigma \in K) .$$
K is an n-dimensional R-coefficient homology manifold (resp. Poincaré complex) with fundamental cycle $[K] \in \Delta(K; R)_n$ if and only if the $[R, K]$-module chain map $\phi_0 : \underline{R}^{n-*} \longrightarrow \underline{R}$ is such that each
$$\phi_0[\sigma] : \underline{R}^{n-*}[\sigma] \longrightarrow \underline{R}[\sigma] \; (\sigma \in K)$$

is an R-module chain equivalence (resp. the assembly $R[\pi]$-module chain map

$$\phi_0[K] : \underline{R}^{n-*}[K] \longrightarrow \underline{R}[K]$$

is a chain equivalence). In particular, if $(\underline{R}, \phi)$ is a Poincaré complex in $\mathbb{A}[R,K]$, then the assembly $(\underline{R}, \phi)[\widetilde{K}]$ is a Poincaré complex in $\mathbb{A}(R[\pi])$, by 9.2. The identifications

$$
\begin{aligned}
H_*(\underline{R}^{n-*}[\sigma]) &= H^{n-*}(K, K\backslash \mathrm{st}_K(\sigma); R) \\
&= H^{n-*}(\mathrm{star}_{K'}(\widehat{\sigma}), \mathrm{link}_{K'}(\widehat{\sigma}); R) \\
&= H^{n-*}(|K|, |K|\backslash\{\widehat{\sigma}\}; R) \quad (\sigma \in K)
\end{aligned}
$$

again recover the familiar result that a homology manifold is a geometric Poincaré complex. This is the chain homotopy theoretic version of the spectral sequence argument of 5.6.

□

Let $\mathbb{B}(R, K) = \mathbb{B}(\mathbb{A}(R, K))$ be the category of finite chain complexes of f.g. free (R, K)-modules.

DEFINITION 9.5 Given R, K, π as above define three algebraic bordism categories:

(i) The f.g. free $R[\pi]$-module category of 3.6

$$\Lambda(R[\pi]) = (\mathbb{A}(R[\pi]), \mathbb{B}(R[\pi]), \mathbb{C}(R[\pi])) .$$

(ii) The *local* f.g. free (R, K)-module bordism category given by 4.1

$$\Lambda(R)_*(K) = (\mathbb{A}(R, K), \mathbb{B}(R, K), \mathbb{C}(R)_*(K)) ,$$

with $\mathbb{C}(R)_*(K)$-equivalences called *local equivalences*.

(iii) The *global* f.g. free (R, K)-module bordism category

$$\Lambda(R, K) = (\mathbb{A}(R, K), \mathbb{B}(R, K), \mathbb{C}(R, K))$$

with $\mathbb{C}(R, K) \subseteq \mathbb{B}(R, K)$ the subcategory of the finite f.g. free (R, K)-module chain complexes C which assemble to contractible f.g. free $R[\pi]$-module chain complexes $C(\widetilde{K})$. $\mathbb{C}(R, K)$-equivalences are called *global equivalences*.

□

PROPOSITION 9.6 *Local equivalences are global, and inclusion defines an assembly functor of algebraic bordism categories*

$$\Lambda(R)_*(K) \longrightarrow \Lambda(R, K) .$$

PROOF The universal assembly of a finite chain complex C in $\mathbb{A}(R, K)$ is a finite chain complex $C(\widetilde{K})$ in $\mathbb{A}(R[\pi])$ which is chain equivalent (by 4.9) to the assembly $[C][\widetilde{K}]$ of the finite chain complex $[C]$ in $\mathbb{A}[R, K]$. Now apply 9.2.

□

DEFINITION 9.7

The $\begin{cases} \textit{symmetric} \\ \textit{visible symmetric} \\ \textit{quadratic} \\ \textit{normal} \end{cases}$ L-groups of (R, K) are the $\begin{cases} \text{symmetric} \\ \text{normal} \\ \text{quadratic} \\ \text{normal} \end{cases}$ L-groups

$$\begin{cases} L^n(R, K) = L^n(\Lambda(R, K)) \\ VL^n(R, K) = NL^n(\Lambda(R, K)) \\ L_n(R, K) = L_n(\Lambda(R, K)) \\ NL^n(R, K) = NL^n(\widehat{\Lambda}(R, K)) \end{cases} \quad (n \in \mathbb{Z})$$

with

$$\Lambda(R, K) = (\mathbb{A}(R, K), \mathbb{B}(R, K), \mathbb{C}(R, K)) ,$$

$$\widehat{\Lambda}(R, K) = (\mathbb{A}(R, K), \mathbb{B}(R, K), \mathbb{B}(R, K)) .$$

□

The L-groups defined in 9.7 are all 4-periodic via the double skew-suspension maps, because the underlying chain complexes are only required to be finite, allowing non-zero chain objects in negative dimensions. The (potentially) aperiodic versions defined using positive chain complexes are dealt with in §15.

The exact sequence of 3.10 can be written as

$$\ldots \longrightarrow L_n(R, K) \overset{1+T}{\longrightarrow} VL^n(R, K) \overset{J}{\longrightarrow} NL^n(R, K)$$
$$\overset{\partial}{\longrightarrow} L_{n-1}(R, K) \longrightarrow \ldots .$$

The $\begin{cases} \text{symmetric} \\ \text{visible symmetric} \\ \text{quadratic} \\ \text{normal} \end{cases}$ L-theory universal assembly maps

$\begin{cases} A\colon L^*(R, K) \longrightarrow L^*(R[\pi]) \\ A\colon VL^*(R, K) \longrightarrow VL^*(R[\pi]) \\ A\colon L_*(R, K) \longrightarrow L_*(R[\pi]) \\ A\colon NL^*(R, K) \longrightarrow NL^*(R[\pi]) \end{cases}$ are defined in 9.11 below. The quadratic L-

theory universal assembly maps are shown to be isomorphisms in §10 below, so that the quadratic L-groups of (R, K) are isomorphic to the surgery obstruction groups

$$L_*(R, K) \cong L_*(R[\pi]) .$$

(*Warning: the quadratic L-theory assembly isomorphisms* $A\colon L_*(R, K) \cong L_*(R[\pi])$ *are not to be confused with the quadratic L-theory assembly maps* $A\colon H_*(K; \mathbb{L}.(R)) \longrightarrow L_*(R[\pi])$ *defined in 14.5 below, which are not in general isomorphisms. See 9.17 below for an explicit example where the latter* A *is not an isomorphism.*)

REMARK 9.8 The *visible symmetric Q-groups* of Weiss [183] are defined for any finite f.g. free $R[\pi]$-module chain complex C to be

$$VQ^*(C) = H_*(P \otimes_{R[\pi]} (\mathrm{Hom}_{\mathbb{Z}[\mathbb{Z}_2]}(W, C \otimes_R C))) ,$$

with P a projective $R[\pi]$-module resolution of R, and there are defined natural maps

$$1 + T : Q_*(C) \longrightarrow VQ^*(C) , \quad VQ^*(C) \longrightarrow Q^*(C) .$$

In particular, the visible symmetric Q-group $VQ^0(C)$ of a 0-dimensional $R[\pi]$-module chain complex C consists of the visible symmetric forms on C^0, which are the symmetric forms $\phi = \phi^* \in \mathrm{Hom}_{R[\pi]}(C^0, C_0)$ such that

$$\phi(x)(x) \in R \subset R[\pi] \quad (x, y \in C^0) .$$

The *visible symmetric L-groups* $VL^n(R[\pi])$ ($n \in \mathbb{Z}$) of [183] are the cobordism groups of n-dimensional visible symmetric Poincaré complexes $(C, \phi \in VQ^n(C))$ over $R[\pi]$. The symmetric construction of Ranicki [142] has a visible version

$$\phi_X : H_n(X) \longrightarrow VQ^n(\Delta(\widetilde{X}))$$

for any space X with universal cover $\widetilde{X}$, so that an n-dimensional geometric Poincaré complex X has a *visible symmetric signature*

$$\sigma^*(X) = (\Delta(\widetilde{X}), \phi_X([X])) \in VL^n(\mathbb{Z}[\pi_1(X)]) .$$

By Ranicki and Weiss [147] every finite f.g. free $R[\pi]$-module chain complex is chain equivalent to the universal assembly $C(\widetilde{K})$ of a finite f.g. free (R, K)-module chain complex C, with $K = B\pi$ the classifying space of π. It is proved in [183] that for any such C the Q-group universal assembly maps are isomorphisms

$$Q^*(C) \xrightarrow{\ \sim\ } VQ^*(C(\widetilde{K})) , \quad Q_*(C) \xrightarrow{\ \sim\ } Q_*(C(\widetilde{K})) ,$$

and hence that the L-group universal assembly maps are isomorphisms

$$VL^*(R, K(\pi, 1)) \xrightarrow{\ \sim\ } VL^*(R[\pi]) , \quad L_*(R, K(\pi, 1)) \xrightarrow{\ \sim\ } L_*(R[\pi]) .$$

It is also proved in [183] that $\widehat{Q}^*(C) = 0$ for any globally contractible finite f.g. free (R, K)-module chain complex C, for any K, so that symmetric complexes in $\Lambda(R, K)$ have canonical normal structures and the forgetful maps are isomorphisms

$$VL^*(R, K) = NL^*(\Lambda(R, K)) \xrightarrow{\ \sim\ } L^*(R, K) = L^*(\Lambda(R, K))$$

(see 3.5). In the special case $K = \{*\}$ already considered in 3.6

$$VL^*(R, \{*\}) = NL^*(\Lambda(R)) = L^*(\Lambda(R)) = L^*(R) .$$

The visible symmetric L-groups $VL^*(\mathbb{Z}[\pi])$ are closely related to the R.L. symmetric L-groups $L^*_{R.L.}(\mathbb{Z}[\pi])$ of Milgram [105]. For $K = \{*\}$

$$VL^n(R, \{*\}) = \varinjlim_k L^{n+4k}(R) ,$$

the free symmetric L-groups made 4-periodic. $R.L.$ stands for Ronnie Lee, because visible symmetric forms over group rings were first used by Lee [87].

□

REMARK 9.9 It will be shown in §13 below that the L-groups of the local algebraic bordism categories are generalized homology groups

$$\begin{cases} L^n(\Lambda(R)_*(K)) = H_n(K; \mathbb{L}^{\cdot}(R)) \\ L_n(\Lambda(R)_*(K)) = H_n(K; \mathbb{L}_{\cdot}(R)) \quad (n \in \mathbb{Z}) \\ NL^n(R, K) = H_n(K; \mathbb{NL}^{\cdot}(R)) \end{cases}$$

with coefficients in algebraic L-spectra. In particular, for a classifying space $K = B\pi$ these are the generalized homology groups of the group π.

□

DEFINITION 9.10 Given (R, K)-module chain complexes C, D define the *universal assembly* $\mathbb{Z}$-module chain map

$$\alpha_{C,D} : C \otimes_{(R,K)} D = [C \boxtimes_R D][K] \xrightarrow{\beta_{C \boxtimes_R D}} (C \boxtimes_R D)(K)$$

$$\xrightarrow{\gamma_{C,D}} C(\widetilde{K}) \otimes_{R[\pi]} D(\widetilde{K}) \; ; \; \phi \longrightarrow \phi(\widetilde{K})$$

with $\beta_{C \boxtimes_R D}$ the chain equivalence given by 4.9 and

$$\gamma_{C,D} : (C \boxtimes_R D)(K) \longrightarrow C(\widetilde{K}) \otimes_{R[\pi]} D(\widetilde{K}) \; ; \; x(\lambda) \boxtimes y(\mu) \longrightarrow x(\widetilde{\lambda}) \otimes y(\widetilde{\mu})$$

the injection constructed using any lifts of the simplexes $\lambda, \mu \in K$ with $\lambda \cap \mu \neq \emptyset$ to simplexes $\widetilde{\lambda}, \widetilde{\mu} \in \widetilde{K}$ with $\widetilde{\lambda} \cap \widetilde{\mu} \neq \emptyset$.

□

The duality R-module isomorphism

$$T_{C(K),D(K)} : C(K) \otimes_R D(K) \xrightarrow{\simeq} D(K) \otimes_R C(K) \; ;$$

$$x \otimes y \longrightarrow (-)^{pq} y \otimes x \quad (x \in C(K)_p, \, y \in D(K)_q)$$

restricts to define a duality isomorphism of (R, K)-module chain complexes

$$T_{C,D} : C \boxtimes_R D \xrightarrow{\simeq} D \boxtimes_R C \; ; \; x \boxtimes y \longrightarrow (-)^{pq} y \boxtimes x \; ,$$

such that there is defined a commutative diagram

$$\begin{array}{ccc}
C \otimes_{(R,K)} D & \xrightarrow{\;\alpha_{C,D}\;} & C(\widetilde{K}) \otimes_{R[\pi]} D(\widetilde{K}) \\
\Big\downarrow{\scriptstyle T_{C,D}} & & \Big\downarrow{\scriptstyle T_{C(\widetilde{K}),D(\widetilde{K})}} \\
D \otimes_{(R,K)} C & \xrightarrow{\;\alpha_{D,C}\;} & D(\widetilde{K}) \otimes_{R[\pi]} C(\widetilde{K}) \; .
\end{array}$$

For $C = D$ universal assembly is a $\mathbb{Z}[\mathbb{Z}_2]$-module chain map
$$\alpha = \alpha_{C,C} : C \otimes_{(R,K)} C \longrightarrow C(\widetilde{K}) \otimes_{R[\pi]} C(\widetilde{K}) \; ; \; \phi \longrightarrow \phi(\widetilde{K})$$
inducing abelian group morphisms
$$\alpha^{\%} : Q^n(C) = H_n(\mathrm{Hom}_{\mathbb{Z}[\mathbb{Z}_2]}(W, (C \otimes_{(R,K)} C))) \longrightarrow$$
$$Q^n(C(\widetilde{K})) = H_n(\mathrm{Hom}_{\mathbb{Z}[\mathbb{Z}_2]}(W, C(\widetilde{K}) \otimes_{R[\pi]} C(\widetilde{K}))) \; ,$$
$$\alpha_{\%} : Q_n(C) = H_n(W \otimes_{\mathbb{Z}[\mathbb{Z}_2]} (C \otimes_{(R,K)} C)) \longrightarrow$$
$$Q_n(C(\widetilde{K})) = H_n(W \otimes_{\mathbb{Z}[\mathbb{Z}_2]} (C(\widetilde{K}) \otimes_{R[\pi]} C(\widetilde{K}))) \; (n \in \mathbb{Z}) \; .$$

PROPOSITION 9.11 *Universal assembly defines functors of algebraic bordism categories*
$$A : \Lambda(R, K) \longrightarrow \Lambda(R[\pi]) \; , \quad A : \widehat{\Lambda}(R, K) \longrightarrow \widehat{\Lambda}(R[\pi])$$

inducing universal assembly maps in the $\begin{cases} \text{symmetric} \\ \text{visible symmetric} \\ \text{quadratic} \\ \text{normal} \end{cases}$ *L-groups*

$$\begin{cases} A : L^n(R, K) \longrightarrow L^n(R[\pi]) \; ; \; (C, \phi) \longrightarrow (C, \phi)(\widetilde{K}) \\ A : VL^n(R, K) \longrightarrow VL^n(R[\pi]) \; ; \; (C, \phi) \longrightarrow (C, \phi)(\widetilde{K}) \\ A : L_n(R, K) \longrightarrow L_n(R[\pi]) \; ; \; (C, \psi) \longrightarrow (C, \psi)(\widetilde{K}) \\ A : NL^n(R, K) \longrightarrow NL^n(R[\pi]) \; ; \; (C, \phi) \longrightarrow (C, \phi)(\widetilde{K}) \; . \end{cases}$$

PROOF The universal assembly functor of the additive categories
$$A : \mathbb{A}(R, K) \longrightarrow \mathbb{A}(R[\pi]) \; ; \; M \longrightarrow M(\widetilde{K})$$
satisfies condition 3.1 (i), since $A(\mathbb{C}(R, K)) \subseteq \mathbb{C}(R[\pi])$ by the definition of $\Lambda(R, K)$. For any object M in $\mathbb{A}(R, K)$ the assembly of the 0-cycle
$$1 \in (M \otimes_{(R,K)} TM)_0 = \mathrm{Hom}_{(R,K)}(TM, TM)_0$$
is a 0-cycle
$$1(\widetilde{K}) \in (M(\widetilde{K}) \otimes_{(R,K)} (TM)(\widetilde{K}))_0 = \mathrm{Hom}_{R[\pi]}(T(M(\widetilde{K})), (TM)(\widetilde{K}))_0$$
defining a natural $\mathbb{C}(R[\pi])$-equivalence
$$B(M) = 1(\widetilde{K}) : TA(M) = T(M(\widetilde{K})) \xrightarrow{\; \simeq \;} AT(M) = (TM)(\widetilde{K})$$
satisfying condition 3.1 (ii).

For finite chain complexes C, D in $\mathbb{A}(R, K)$ an n-cycle $\phi \in (C \otimes_{(R,K)} D)_n$ is an (R, K)-module chain map $\phi \colon \Sigma^n TC \longrightarrow D$. The assembly n-cycle $\phi(\widetilde{K}) \in (C(\widetilde{K}) \otimes_{R[\pi]} D(\widetilde{K}))_n$ is the $R[\pi]$-module chain map given by the composite
$$\phi(\widetilde{K}) : C(\widetilde{K})^{n-*} = \Sigma^n T(C(\widetilde{K})) \xrightarrow{\; \Sigma^n B(C) \;}$$
$$\Sigma^n(TC)(\widetilde{K}) = (\Sigma^n TC)(\widetilde{K}) \xrightarrow{\; A(\phi) \;} D(\widetilde{K}) \; .$$

Thus $\phi: \Sigma^n TC \longrightarrow D$ is a $\mathbb{C}(R, K)$-equivalence if and only if $\phi(\widetilde{K}): C(\widetilde{K})^{n-*}$ $\longrightarrow D(\widetilde{K})$ is a $\mathbb{C}(R[\pi])$-equivalence. The universal assembly of an n-dimensional symmetric complex (C, ϕ) in $\Lambda(R, K)$ is an n-dimensional symmetric complex in $\Lambda(R[\pi])$

$$(C, \phi)(\widetilde{K}) = (C(\widetilde{K}), \phi(\widetilde{K})) ,$$

with $\phi(\widetilde{K}) \in W^\% C(\widetilde{K})_n$ the n-cycle defined by the image of the n-cycle $\phi \in (W^\% C)_n$ under the $\mathbb{Z}$-module chain map

$$\alpha^\% : W^\% C = \mathrm{Hom}_{\mathbb{Z}[\mathbb{Z}_2]}(W, C \otimes_{(R,K)} C) \longrightarrow$$

$$W^\% C(\widetilde{K}) = \mathrm{Hom}_{\mathbb{Z}[\mathbb{Z}_2]}(W, C(\widetilde{K}) \otimes_{R[\pi]} C(\widetilde{K})) .$$

Similarly for the quadratic and normal cases.

$\square$

EXAMPLE 9.12 As in 4.15 let X be a simplicial complex with a K-dissection $\{X[\sigma] \mid \sigma \in K\}$, and regard the R-coefficient simplicial chain complex $\Delta(X; R)$ as a f.g. free (R, K)-module chain complex C with

$$C(\sigma) = \Delta(X[\sigma], \partial X[\sigma]; R) , \quad [C][\sigma] = \Delta(X[\sigma]; R) ,$$

$$\partial X[\sigma] = \bigcup_{\tau > \sigma} X[\tau] \quad (\sigma \in K) .$$

The Alexander–Whitney–Steenrod diagonal chain approximation of X is an (R, K)-module chain map

$$\Delta_K = \Delta : C(K) = \Delta(X; R) \longrightarrow$$

$$(W^\% C)(K) = \mathrm{Hom}_{\mathbb{Z}[\mathbb{Z}_2]}(W, (C \boxtimes_R C)(K))$$

$$(\subseteq W^\% (C(K)) = \mathrm{Hom}_{\mathbb{Z}[\mathbb{Z}_2]}(W, \Delta(X; R) \otimes_R \Delta(X; R)))$$

with

$$\Delta_0(x) = \sum_{i=0}^{n} (x_0 x_1 \ldots x_i) \otimes (x_i x_{i+1} \ldots x_n) \in (C \boxtimes_R C)(K)_n$$

$$(x = (x_0 x_1 \ldots x_n) \in X^{(n)}) .$$

By the naturality of Δ there is defined a commutative diagram of R-module chain complexes and chain maps

$$
\begin{array}{ccc}
[C][K] & \xrightarrow{\ [\Delta]\ } & [W^\% C][K] \\
\beta_C \downarrow & & \downarrow \beta_{W^\% C} \\
C(K) & \xrightarrow{\ \ \Delta\ \ } & (W^\% C)(K) .
\end{array}
$$

Given an n-cycle
$$[X] = \sum_{\sigma \in K} x_\sigma \sigma \in C(K)_n = \Delta(X;R)_n = \sum_{\sigma \in K} \Delta(X[\sigma], \partial X[\sigma]; R)_n$$
use the chain maps $\partial_\sigma \colon \Delta(X;R) \longrightarrow S^{|\sigma|}\Delta(X[\sigma], \partial X[\sigma]; R)$ given by 9.5 to
define $(n - |\sigma|)$-cycles
$$[X(\sigma)] = \partial_\sigma([X]) \in \Delta(X[\sigma], \partial X[\sigma]; R)_{n-|\sigma|} \quad (\sigma \in K) .$$
The n-cycle
$$\phi = \Delta([X]) \in (W^\%C)(K)_n$$
defines an n-dimensional symmetric complex in $\mathbb{A}(R, K)$
$$\sigma^*(X) = (C, \phi)$$
such that
$$\sigma^*(X)[\tau] = \sigma^*(X[\tau], \partial X[\tau]) \quad (\tau \in K) .$$
The assembly of $\sigma^*(X)$ is an n-dimensional symmetric complex in $\mathbb{A}(R[\pi])$
$(C(\widetilde{K}), \phi(\widetilde{K}))$ with a chain homotopy commutative diagram

$$C(\widetilde{K})^{n-*} = \Delta(\widetilde{X};R)^{n-*} \xrightarrow{\quad [X] \cap - \quad} C(\widetilde{K}) = \Delta(\widetilde{X};R)$$

$$\widetilde{\beta}_C^* \searrow \qquad \nearrow \phi_0(\widetilde{K})$$

$$([C][\widetilde{K}])^{n-*} = C^{n-*}(\widetilde{K})$$

where $\Delta(\widetilde{X}; R)$ is the simplicial $R[\pi]$-module chain complex of the pullback
$\widetilde{X}$ to X of the universal cover $\widetilde{K}$ of K, and $\widetilde{\beta}_C^*$ is the n-dual of the $R[\pi]$-
module chain equivalence $\widetilde{\beta}_C \colon [C][\widetilde{K}] \longrightarrow C(\widetilde{K}) = \Delta(\widetilde{X}; R)$ given by 4.9. A
normal structure realizing $[X] \in H_n(X; R)$ is a pair
$$(\nu_X \colon X \longrightarrow BG(k), \rho_X \colon S^{n+k} \longrightarrow T(\nu_X)) \quad (k \gg 0)$$
such that $[X]$ is the image of the homotopy class of ρ_X under the composite
$$\pi_{n+k}(T(\nu_X)) \xrightarrow{h} \dot{H}_{n+k}(T(\nu)) \xrightarrow{t} H_n(X) \xrightarrow{c} H_n(X; R)$$
with h the Hurewicz map, t the Thom isomorphism and c the change of
rings for the morphism $\mathbb{Z} \longrightarrow R; 1 \longmapsto 1$. Such a geometric normal structure
(ν_X, ρ_X) determines an algebraic normal structure (γ, χ) for the symmetric
complex $\sigma^*(X) = (C, \phi)$, and
$$\hat{\sigma}^*(X) = (C, \phi, \gamma, \chi)$$
is an n-dimensional normal complex in $\mathbb{A}(R, K)$ with chain bundle $(C, \gamma) = \hat{\sigma}^*(\nu_X)$.

$\square$

EXAMPLE 9.13 Given a simplicial complex K set
$$X = K' \ , \quad X[\sigma] = D(\sigma, K) \ (\sigma \in K) \ , \quad R = \mathbb{Z} \ ,$$
in 9.12, so that C is the $(\mathbb{Z}, K)$-module chain complex of 4.15 with
$$C(K) = \Delta(K') \ , \quad C(\sigma) = \Delta(D(\sigma, K), \partial D(\sigma, K)) \ (\sigma \in K) \ .$$
For any n-cycle $[K] \in \Delta(K')_n$ there is defined an n-dimensional normal complex (C, ϕ) in $\mathbb{A}(\mathbb{Z}, K)$ with
$$\phi = \Delta([K]) \in H_n((W^{\%}C)(K)) \ ,$$
$$\phi_0(K) = [K] \cap - : \Sigma^n TC(K) \simeq \Delta(K')^{n-*} \longrightarrow C(K) = \Delta(K') \ ,$$
$$\phi_0(\sigma) = [D(\sigma, K)] \cap - : \Sigma^n TC(\sigma) = \Delta(D(\sigma, K))^{n-|\sigma|-*}$$
$$\longrightarrow C(\sigma) = \Delta(D(\sigma, K), \partial D(\sigma, K)) \ (\sigma \in K) \ .$$

K is an n-dimensional $\begin{cases} \text{geometric Poincaré complex} \\ \text{homology manifold} \end{cases}$ with the fundamental cycle $[K] \in \Delta(K')_n$ if and only if the symmetric complex (C, ϕ) is Poincaré. In both cases there is defined an algebraic normal structure (γ, χ), and hence a *visible symmetric signature* invariant
$$\sigma^*(K) = (C, \phi, \gamma, \chi) \in \begin{cases} VL^n(\mathbb{Z}, K) \\ L^n(\Lambda(\mathbb{Z})_*(K)) \ . \end{cases}$$
The image of (C, ϕ) under the full embedding
$$\mathbb{A}(\mathbb{Z}, K) \longrightarrow \mathbb{A}[\mathbb{Z}, K] \ ; \quad M \longrightarrow [M]$$
is homotopy equivalent to the symmetric complex $(\mathbb{Z}, \phi)$ of 9.4.

□

EXAMPLE 9.14 Let $(f, b): M \longrightarrow K'$ be a normal map from a compact n-dimensional homology manifold M to the barycentric subdivision K' of an n-dimensional $\begin{cases} \text{geometric Poincaré complex} \\ \text{homology manifold} \end{cases}$ K, so that for each $\tau \in K$ the restriction
$$(f[\tau], b[\tau]) = (f, b)| :$$
$$(M[\tau], \partial M[\tau]) = f^{-1}(D(\tau, K), \partial D(\tau, K)) \longrightarrow (D(\tau, K), \partial D(\tau, K))$$
is a normal map from an $(n - |\tau|)$-dimensional homology manifold with boundary to an $(n - |\tau|)$-dimensional geometric $\begin{cases} \text{normal} \\ \text{Poincaré} \end{cases}$ pair. The quadratic construction of Ranicki [142] associates to (f, b) an n-dimensional quadratic $\begin{cases} \text{globally} \\ \text{locally} \end{cases}$ Poincaré complex in $\mathbb{A}(\mathbb{Z}, K)$
$$\sigma_*(f, b) = (C(f^!), \psi)$$

with $C(f^!)$ the algebraic mapping cone of the Umkehr chain map in $\mathbb{A}(\mathbb{Z}, K)$

$$f^! : \Delta(K') \xrightarrow{([K']\cap -)^{-1}} \Delta(K')^{n-*} \xrightarrow{f^*} \Delta(M)^{n-*} \xrightarrow{[M]\cap -} \Delta(M) ,$$

such that

$$\sigma_*(f, b)[\tau] = \sigma_*(f[\tau], b[\tau]) \quad (\tau \in K) .$$

The *quadratic signature* of (f, b) is the cobordism class

$$\sigma_*(f, b) \in \begin{cases} L_n(\Lambda(\mathbb{Z}, K)) \\ L_n(\Lambda(\mathbb{Z})_*(K)) . \end{cases}$$

□

EXAMPLE 9.15 An n-dimensional normal complex

$$(K , \nu_K : K \longrightarrow BG(k) , \rho_K : S^{n+k} \longrightarrow T(\nu_K))$$

determines (as in 9.13) an n-dimensional normal complex $\hat{\sigma}^*(K) = (C, \phi)$ in $\mathbb{A}(\mathbb{Z}, K)$ with $C(\widetilde{K}) = \Delta(\widetilde{K}')$, and such that

$$\hat{\sigma}^*(K)[\tau] = \hat{\sigma}^*(D(\tau, K), \partial D(\tau, K)) \quad (\tau \in K) .$$

The *normal signature* of K is the cobordism class

$$\hat{\sigma}^*(K) \in NL^n(\mathbb{Z}, K) .$$

□

EXAMPLE 9.16 The assembly of the $\begin{cases} \text{visible symmetric} \\ \text{quadratic} \\ \text{normal} \end{cases}$ signature given by

$\begin{cases} 9.13 \\ 9.14 \text{ for an } n\text{-dimensional geometric} \\ 9.15 \end{cases}$ $\begin{cases} \text{Poincaré complex } K \\ \text{normal map } (f, b) : M \longrightarrow K' \text{ is the} \\ \text{normal complex } K \end{cases}$

$\begin{cases} \text{visible symmetric} \\ \text{quadratic} \\ \text{normal} \end{cases}$ signature

$$\begin{cases} \sigma^*(K) \in \text{im}(A : VL^n(\mathbb{Z}, K) \longrightarrow VL^n(\mathbb{Z}[\pi_1(K)])) \\ \sigma_*(f, b) \in \text{im}(A : L_n(\mathbb{Z}, K) \longrightarrow L_n(\mathbb{Z}[\pi_1(K)])) \\ \hat{\sigma}^*(K) \in \text{im}(A : NL^n(\mathbb{Z}, K) \longrightarrow NL^n(\mathbb{Z}[\pi_1(K)])) \end{cases}$$

of $\begin{cases} \text{Weiss [183]} \\ \text{Wall [176]} \\ \text{Ranicki [143]}. \end{cases}$ Also, $\sigma^*(K) \in L^n(\mathbb{Z}[\pi_1(K)])$ for geometric Poincaré K

is the symmetric signature of Mishchenko [112] and Ranicki [142].

□

EXAMPLE 9.17 The universal assembly maps

$$A : H_*(B\pi; \mathbb{L}_\bullet(\mathbb{Z})) \longrightarrow L_*(\mathbb{Z}[\pi]) ,$$

$$A : H_*(B\pi; \mathbb{L}^\bullet(\mathbb{Z})) \longrightarrow VL^*(\mathbb{Z}, B\pi)$$

will now be described in the special case $\pi = \mathbb{Z}_2$, $B\mathbb{Z}_2 = \mathbb{RP}^\infty$, assuming the identifications obtained in §10 and §13

$$L_*(\mathbb{Z}, B\pi) = L_*(\mathbb{Z}[\pi]) \; , \; VL^*(\mathbb{Z}, B\pi) = VL^*(\mathbb{Z}[\pi]) \; ,$$

$$L_*(\Lambda(\mathbb{Z})_*(B\pi)) = H_*(B\pi; \mathbb{L}.(\mathbb{Z})) \; , \; L^*(\Lambda(\mathbb{Z})_*(B\pi)) = H_*(B\pi; \mathbb{L}^{\cdot}(\mathbb{Z})) \; .$$

The computations have been carried out by Wall [176, §14D], Conner (Dovermann [44]) and Weiss [183, §7]. The Witt groups of the group ring

$$\mathbb{Z}[\mathbb{Z}_2] = \mathbb{Z}[T]/(T^2 - 1)$$

with the oriented involution $\overline{T} = T$ are computed using the cartesian square of rings with involution

$$\begin{array}{ccc} \mathbb{Z}[\mathbb{Z}_2] & \xrightarrow{\;j_+\;} & \mathbb{Z} \\ {\scriptstyle j_-}\Big\downarrow & & \Big\downarrow \\ \mathbb{Z} & \xrightarrow{\hspace{2cm}} & \mathbb{Z}_2 \end{array}$$

where

$$j_\pm : \mathbb{Z}[\mathbb{Z}_2] \longrightarrow \mathbb{Z} \; ; \; a + bT \longrightarrow a \pm bT \; .$$

The quadratic L-groups $L_*(\mathbb{Z}[\mathbb{Z}_2])$ fit into the Mayer–Vietoris exact sequence of Ranicki [143, 6.3.1]

$$\dots \longrightarrow L_n(\mathbb{Z}[\mathbb{Z}_2]) \xrightarrow{\;(j_+\ j_-)\;} L_n(\mathbb{Z}) \oplus L_n(\mathbb{Z})$$

$$\longrightarrow L_n(\mathbb{Z}_2) \longrightarrow L_{n-1}(\mathbb{Z}[\mathbb{Z}_2]) \longrightarrow \dots \; .$$

Although there is no such Mayer–Vietoris exact sequence for the symmetric L-groups in general (Ranicki [143, 6.4.2]) the symmetric Witt group $L^0(\mathbb{Z}[\mathbb{Z}_2])$ fits into the exact sequence

$$0 \longrightarrow L^0(\mathbb{Z}[\mathbb{Z}_2]) \xrightarrow{\;(j_+\ j_-)\;} L^0(\mathbb{Z}) \oplus L^0(\mathbb{Z}) \longrightarrow L^0(\mathbb{Z}_2) \longrightarrow 0$$

such that up to isomorphism

$$L^0(\mathbb{Z}) = \mathbb{Z} \; , \; L^0(\mathbb{Z}[\mathbb{Z}_2]) = \mathbb{Z} \oplus \mathbb{Z} \; , \; L^0(\mathbb{Z}_2) = \mathbb{Z}_2 \; .$$

The Witt group $VL^0(\mathbb{Z}[\mathbb{Z}_2], 1)$ of nonsingular visible symmetric forms over $\mathbb{Z}[\mathbb{Z}_2]$ fits into the exact sequences

$$0 \longrightarrow VL^0(\mathbb{Z}[\mathbb{Z}_2], 1) \xrightarrow{\;(j_+\ j_-)\;} L^0(\mathbb{Z}) \oplus L^0(\mathbb{Z}) \longrightarrow \widehat{L}^0(\mathbb{Z}) \longrightarrow 0$$

$$0 \longrightarrow L_0(\mathbb{Z}[\mathbb{Z}_2]) \longrightarrow VL^0(\mathbb{Z}[\mathbb{Z}_2], 1) \longrightarrow \widehat{L}^0(\mathbb{Z}) \longrightarrow 0$$

such that up to isomorphism

$$L_0(\mathbb{Z}[\mathbb{Z}_2]) = VL^0(\mathbb{Z}[\mathbb{Z}_2], 1) = \mathbb{Z} \oplus \mathbb{Z} \; , \; \widehat{L}^0(\mathbb{Z}) = NL^0(\mathbb{Z}) = \mathbb{Z}_8 \; .$$

The quadratic L-theory assembly maps are given by:

$$A : H_n(B\mathbb{Z}_2 ; \mathbb{L}.(\mathbb{Z})) = \sum_{k \in \mathbb{Z}} H_{n-k}(B\mathbb{Z}_2 ; L_k(\mathbb{Z}))$$

$$\longrightarrow H_0(B\mathbb{Z}_2 ; L_n(\mathbb{Z})) = L_n(\mathbb{Z}) = \begin{cases} \mathbb{Z} \\ 0 \\ \mathbb{Z}_2 \\ 0 \end{cases}$$

$$\xrightarrow{i_!} L_n(\mathbb{Z}[\mathbb{Z}_2]) = \begin{cases} \mathbb{Z} \oplus \mathbb{Z} \\ 0 \\ \mathbb{Z}_2 \\ \mathbb{Z}_2 \end{cases} \quad \text{if } n \equiv \begin{cases} 0 \\ 1 \\ 2 \\ 3 \end{cases} \pmod 4$$

with $i_!$ induced by the inclusion

$$i : \mathbb{Z} \longrightarrow \mathbb{Z}[\mathbb{Z}_2] \ ; \ a \longrightarrow a \ .$$

The visible symmetric L-theory assembly maps are given by

$$A : H_n(B\mathbb{Z}_2 ; \mathbb{L}^{\cdot}(\mathbb{Z})) = \sum_{k \in \mathbb{Z}} H_{n-k}(B\mathbb{Z}_2 ; L^k(\mathbb{Z})) \longrightarrow$$

$$VL^n(\mathbb{Z}[\mathbb{Z}_2]) = \begin{cases} VL^0(\mathbb{Z}[\mathbb{Z}_2],1) \oplus \sum\limits_{k \neq -1,0} H_{n-k}(B\mathbb{Z}_2 ; \widehat{L}^k(\mathbb{Z})) \\ \sum\limits_k H_{n-k}(B\mathbb{Z}_2 ; \widehat{L}^k(\mathbb{Z})) \\ \sum\limits_{k \neq 3} H_{n-k}(B\mathbb{Z}_2 ; \widehat{L}^k(\mathbb{Z})) \end{cases}$$

$$\text{if } n \equiv \begin{cases} 0 \\ 1,2 \\ 3 \end{cases} \pmod 4 \ .$$

The symmetric L-groups $L^*(\mathbb{Z}[\mathbb{Z}_2])$ are not 4-periodic.

Given a nonsingular symmetric form (M,ϕ) over $\mathbb{Z}[\mathbb{Z}_2]$ let

$$s_{\pm}(M,\phi) = \text{signature}\, j_{\pm}(M,\phi) \in L^0(\mathbb{Z}) = \mathbb{Z} \ .$$

In terms of the signatures

$$L^0(\mathbb{Z}[\mathbb{Z}_2]) = \{ (s_+,s_-) \in \mathbb{Z} \oplus \mathbb{Z} \,|\, s_+ \equiv s_- \,(\mathrm{mod}\,2) \} \ ,$$

$$VL^0(\mathbb{Z}[\mathbb{Z}_2],1) = \{ (s_+,s_-) \in \mathbb{Z} \oplus \mathbb{Z} \,|\, s_+ \equiv s_- \,(\mathrm{mod}\,8) \} \ ,$$

$$L_0(\mathbb{Z}[\mathbb{Z}_2]) = \{ (s_+,s_-) \in \mathbb{Z} \oplus \mathbb{Z} \,|\, s_+ \equiv s_- \equiv 0 \,(\mathrm{mod}\,8) \}$$

and in each case the image of the assembly map A is

$$\mathrm{im}(A) = \{ (s_+,s_-) \,|\, s_+ = s_- \in \mathbb{Z} \} \ .$$

For example

$$s_{\pm}(\mathbb{Z}[\mathbb{Z}_2],1) = 1 \ , \ (\mathbb{Z}[\mathbb{Z}_2],1) \in \mathrm{im}(A) \subset VL^0(\mathbb{Z}[\mathbb{Z}_2]) \ ,$$

$$s_{\pm}(\mathbb{Z}[\mathbb{Z}_2],T) = \pm 1 \ , \ (\mathbb{Z}[\mathbb{Z}_2],T) \notin \mathrm{im}(A) \subset L^0(\mathbb{Z}[\mathbb{Z}_2],1) \ .$$

The effect of the restriction map
$$i^! : L^0(\mathbb{Z}[\mathbb{Z}_2]) \longrightarrow L^0(\mathbb{Z}) \; ; \; (M,\phi) \longrightarrow (i^!M, i^!\phi)$$
is given by
$$i^!(s_+, s_-) = s_+ + s_- \in L^0(\mathbb{Z}) = \mathbb{Z} \, ,$$
since for any $a + bT \in \mathbb{Z}[\mathbb{Z}_2]$ the eigenvalues of
$$i^!(a + bT) = \begin{pmatrix} a & b \\ b & a \end{pmatrix} : i^!\mathbb{Z}[\mathbb{Z}_2] = \mathbb{Z} \oplus \mathbb{Z} \longrightarrow \mathbb{Z} \oplus \mathbb{Z}$$
are $j_\pm(a + bT) = a \pm b$. Thus for a nonsingular symmetric form (M, ϕ) over $\mathbb{Z}[\mathbb{Z}_2]$ the following conditions are equivalent:

    (i) $i^!(M, \phi) = 2\, j_+(M, \phi) \in \mathbb{Z}$,

    (ii) $s_+(M, \phi) = s_-(M, \phi) \in \mathbb{Z}$,

    (iii) $(M, \phi) \in \mathrm{im}(A : H_0(B\mathbb{Z}_2 ; \mathbb{L}^{\boldsymbol{\cdot}}(\mathbb{Z})) \longrightarrow L^0(\mathbb{Z}[\mathbb{Z}_2]))$

and similarly for visible symmetric and quadratic forms. For the applications to topology see Example 23.4C below.

                                                         □

§10. The algebraic π-π theorem

The geometric π-π theorem of Wall [176, 3.2] is that for $n \geq 6$ a normal map $(f, b): (M, \partial M) \longrightarrow (X, \partial X)$ from an n-dimensional manifold with boundary $(M, \partial M)$ to an n-dimensional geometric Poincaré pair $(X, \partial X)$ with $\pi_1(\partial X) \cong \pi_1(X)$ is normal bordant to a homotopy equivalence of pairs. The π-π theorem was used in Chapter 9 of [176] to identify the geometric surgery obstruction groups $L_*(K)$ with the algebraic surgery obstruction groups of the fundamental group ring $\mathbb{Z}[\pi_1(K)]$

$$L_n(K) = L_n(\mathbb{Z}[\pi_1(K)]) \ (n \geq 5) ,$$

for any connected CW complex K with a finite 2-skeleton.

An algebraic π-π theorem will now be obtained, in the form of a natural identification

$$L_n(\Lambda(R, K)) = L_n(R[\pi_1(K)]) \ (n \in \mathbb{Z})$$

for any commutative ring R and any connected ordered simplicial complex K, with $\Lambda(R, K)$ the algebraic bordism category of 9.5 (iii).

Use the base vertex $* \in K^{(0)}$ to define a f.g. free (R, K)-module Γ by

$$\Gamma_0(\sigma) = \begin{cases} R & \text{if } \sigma = * \\ 0 & \text{otherwise .} \end{cases}$$

Let $\widetilde{K}$ be the universal cover of K. Choosing a lift $\tilde{*} \in \widetilde{K}^{(0)}$ there is defined an $R[\pi_1(K)]$-module isomorphism

$$R[\pi_1(K)] \xrightarrow{\ \cong\ } \Gamma(\widetilde{K}) \ ; \ 1 \longrightarrow 1(\tilde{*}) ,$$

which will be used as an identification.

DEFINITION 10.1 The *homology assembly maps* are defined for any (R, K)-module chain complex C to be the $R[\pi_1(K)]$-module morphisms

$$H_r([C][*]) \longrightarrow H_r(C(\widetilde{K})) \ (r \in \mathbb{Z})$$

induced in homology by the chain map

$$\text{Hom}_{(R,K)}(\Gamma, C) = [C][*] \longrightarrow \text{Hom}_{R[\pi_1(K)]}(\Gamma(\widetilde{K}), C(\widetilde{K})) = C(\widetilde{K}) ;$$

$$x(\sigma) \longrightarrow x(\tilde{\sigma}) \ (* \leq \sigma, \tilde{*} \leq \tilde{\sigma}) .$$

□

The proof of the algebraic π-π theorem requires a Hurewicz theorem to represent homology classes in assembled $R[\pi]$-module chain complexes by (R, K)-module morphisms, just as the proof of the geometric π-π theorem needs the usual Hurewicz theorem to represent homology by homotopy. This requires the results of Ranicki and Weiss [147, §4] summarized in the next paragraph.

An (R, K)-module chain complex C is *homogeneous* if the inclusions define R-module chain equivalences.

$$[C][\sigma] \xrightarrow{\simeq} [C][\tau] \quad (\tau \leq \sigma \in K) .$$

The *homogeneous envelope* of a finite chain complex C in $\mathbb{A}(R, K)$ is a homogeneous (R, K)-module chain complex $V^\infty C$ with the following properties:

(i) $V^\infty C = \varinjlim_k V^k C$ is the direct limit (= union) of a sequence of inclusions of finite chain complexes in $\mathbb{A}(R, K)$
$$C = V^0 C \subseteq VC \subseteq V^2 C \subseteq \cdots$$
such that each inclusion defines a global equivalence $V^k C \longrightarrow V^{k+1} C$,

(ii) the inclusion $C \longrightarrow V^\infty C$ assembles to an $R[\pi_1(K)]$-module chain equivalence $C(\widetilde{K}) \longrightarrow V^\infty C(\widetilde{K})$,

(iii) for any finite chain complex B in $\mathbb{A}(R, K)$ and any $n \in \mathbb{Z}$ the abelian group $H_n(\mathrm{Hom}_{(R,K)}(B, V^\infty C))$ of homotopy classes of (R, K)-module chain maps $\Sigma^n B \longrightarrow V^\infty C$ is in one–one correspondence with the equivalence classes of pairs $(f: \Sigma^n B \longrightarrow D, g: C \longrightarrow D)$ of homotopy classes of (R, K)-module chain maps with D finite in $\mathbb{A}(R, K)$ and g a global equivalence, subject to the equivalence relation generated by
$$(f: \Sigma^n B \longrightarrow D, g: C \longrightarrow D) \sim (hf: \Sigma^n B \longrightarrow E, hg: D \longrightarrow E)$$
for any global equivalence $h: D \longrightarrow E$ in $\mathbb{A}(R, K)$,

(iv) the homogeneous envelope $V^\infty \Gamma$ of the 0-dimensional chain complex Γ in $\mathbb{A}(R, K)$ is chain equivalent to the (R, K)-module chain complex $\Delta(EK; R)$ associated to a triangulation EK of the pointed path space
$$E|K| = |EK| = (|K|, \{*\})^{([0,1], \{0\})}$$
and the projection
$$p : E|K| \longrightarrow |K| ; \ \omega \longrightarrow \omega(1) ,$$
and $[V^\infty \Gamma][*]$ is chain equivalent to the R-module chain complex $\Delta(\Omega K; R)$ with ΩK a triangulation of the pointed loop space
$$\Omega|K| = |\Omega K| = p^{-1}(\{*\}) = (|K|, \{*\})^{([0,1], \{0,1\})} .$$

The Hurewicz map $\pi_r(X) \longrightarrow H_r(X)$ assembles a homology class from a homotopy class. One version of the Hurewicz theorem states that if X is a space with an $(n-1)$-connected universal cover $\widetilde{X}$ and $n \geq 2$ then $\pi_r(X) = \pi_r(\widetilde{X}) \longrightarrow H_r(\widetilde{X})$ is an isomorphism for $r = n$ and an epimorphism for $r = n + 1$. Similarly:

PROPOSITION 10.2 *If C is a homogeneous (R, K)-module chain complex which is bounded below and such that*
$$H_q(C(\widetilde{K})) = 0 \ \text{for } q < n$$

then the homology assembly $R[\pi_1(K)]$-module morphism

$$H_r([C][*]) \longrightarrow H_r(C(\widetilde{K}))$$

is an isomorphism for $r = n$ and an epimorphism for $r = n + 1$.

PROOF It suffices to derive the conclusions from the hypothesis that $H_q([C]$ $[*]) = 0$ for $q < n$. By 4.9 $C(\widetilde{K})$ is chain equivalent to $[C][\widetilde{K}]$. As in 4.6 define a filtration of $[C][\widetilde{K}]$

$$F_0[C][\widetilde{K}] \subseteq F_1[C][\widetilde{K}] \subseteq F_2[C][\widetilde{K}] \subseteq \ldots \subseteq [C][\widetilde{K}]$$

by

$$F_p[C][\widetilde{K}]_q = \sum_{\tilde{\sigma} \in \widetilde{K}, |\tilde{\sigma}| \leq p} [C][\sigma]_{q-|\sigma|}$$

and consider the corresponding first quadrant spectral sequence (4.6). The E_2-terms are given by

$$E^2_{p,q} = H_p(\widetilde{K}; \{H_q([C][\sigma])\}) = H_p(\widetilde{K}; H_q([C][*])) \ (= 0 \text{ for } q < n) ,$$

using the simple connectivity of the universal cover $\widetilde{K}$ and the homogeneity of C to untwist the local coefficient systems. The spectral sequence converges to $H_*([C][\widetilde{K}]) = H_*(C(\widetilde{K}))$, with

$$E^\infty_{p,q} = \frac{\text{im}(H_{p+q}(F_p[C][\widetilde{K}]) \longrightarrow H_{p+q}(C(\widetilde{K})))}{\text{im}(H_{p+q}(F_{p-1}[C][\widetilde{K}]) \longrightarrow H_{p+q}(C(\widetilde{K})))} \ (= 0 \text{ for } q < n) .$$

The assembly map in n-dimensional homology coincides with the isomorphism defined by the edge map

$$E^2_{0,n} = H_n([C][*]) \xrightarrow{\sim} E^\infty_{0,n} = H_n(C(\widetilde{K})) .$$

A quotient of the assembly map in $(n + 1)$-dimensional homology coincides with the edge isomorphism

$$\text{coker}(d: E^2_{2,n} \longrightarrow E^2_{0,n+1}) = \text{coker}(H_2(\widetilde{K}; H_n([C][*])) \longrightarrow H_{n+1}([C][*]))$$

$$\longrightarrow E^\infty_{0,n+1} = H_{n+1}(C(\widetilde{K})) .$$

□

An application of 10.2 to the algebraic mapping cone gives that a chain map $f: C \longrightarrow D$ of homogeneous finite (R, K)-module chain complexes is a local chain equivalence if and only if it is a global chain equivalence, i.e. f is an (R, K)-module chain equivalence if and only if the assembly $f(\widetilde{K}): C(\widetilde{K}) \longrightarrow D(\widetilde{K})$ is an $R[\pi]$-module chain equivalence.

EXAMPLE 10.3 Let $f: X \longrightarrow K$ be a simplicial map with barycentric subdivision $f': X' \longrightarrow K'$, so that as in 4.15 there is defined a K-dissection $\{X[\sigma] \mid \sigma \in K\}$ of X with

$$X[\sigma] = f'^{-1}D(\sigma, K) \ (\sigma \in K) ,$$

and hence a $(\mathbb{Z}, K)$-module chain complex C with

$$C(\sigma) = \Delta(X[\sigma], \partial X[\sigma]) \ , \ [C][\sigma] = \Delta(X[\sigma]) \ (\sigma \in K) \ , \ C(K) = \Delta(K') \ .$$

The iterated mapping cylinder method of Hatcher [71, §2] shows that f is a quasifibration in the sense of Dold and Thom with fibre $F = f^{-1}(*)$ if and only if the inclusions $X[\sigma] \longrightarrow X[\tau]$ ($\tau \leq \sigma \in K$) are homotopy equivalences, in which case C is a homogeneous $(\mathbb{Z}, K)$-module chain complex with $[C][*] \simeq \Delta(F)$, and the spectral sequence of 4.6 is the Serre spectral sequence converging to $H_*(X)$ with E^2-terms

$$E^2_{p,q} = H_p(K; \{H_q(F)\}) \ .$$

The path space fibration $f: X = EK \longrightarrow K$ with fibre $F = \Omega K$ determines the homogeneous $(\mathbb{Z}, K)$-module chain complex B with

$$B(\sigma) = C(f'|: \Delta(f'^{-1}(D(\sigma, K), \partial D(\sigma, K))) \longrightarrow \Delta(D(\sigma, K), \partial D(\sigma, K)))$$
$$(\sigma \in K) \ ,$$

$$[B][*] \simeq C(\Delta(\Omega K) \longrightarrow \Delta(\{*\})) \simeq \Sigma \Delta(\Omega K, \{*\}) \ ,$$

$$B(\widetilde{K}) \simeq \Delta(\widetilde{EK} \longrightarrow \widetilde{K}) \simeq \Delta(\widetilde{K}, \pi) \ .$$

If K is $(n-1)$-connected then $H_q(B(\widetilde{K})) = 0$ for $q < n$ and 10.2 gives the usual Hurewicz theorem, with the assembly map

$$H_r([B][*]) = H_{r-1}(\Omega K, \{*\}) \longrightarrow H_r(B(\widetilde{K})) = H_r(\widetilde{K}, \pi)$$

an isomorphism for $r = n$ and an epimorphism for $r = n + 1$. (Here, $\pi = p^{-1}(\{*\}) \subset \widetilde{K}$ with $p: \widetilde{K} \longrightarrow K$ the covering projection.)

□

Identify $\Gamma = T\Gamma$ using the isomorphism

$$\Gamma_0(*) = R \xrightarrow{\ \simeq\ } T\Gamma_0(*) = \operatorname{Hom}_R(R, R) \ ; \ r \longrightarrow (s \longrightarrow sr) \ .$$

An (R, K)-module chain map $f: \Sigma^n \Gamma \longrightarrow C$ assembles to an $R[\pi_1(K)]$-module chain map

$$f(\widetilde{K}) : \Sigma^n \Gamma(\widetilde{K}) = \Sigma^n R[\pi_1(K)] \longrightarrow C(\widetilde{K}) \ ,$$

that is an n-cycle $f(\widetilde{K}) \in C(\widetilde{K})_n$. Dually, an (R, K)-module chain map $f: C \longrightarrow \Sigma^n \Gamma$ assembles to an $R[\pi_1(K)]$-module chain map

$$f(\widetilde{K}) : C(\widetilde{K}) \longrightarrow \Sigma^n \Gamma(\widetilde{K}) = \Sigma^n R[\pi_1(K)] \ ,$$

defining an n-cocycle $f(\widetilde{K}) \in C(\widetilde{K})^n$.

PROPOSITION 10.4 (i) *If C is a finite chain complex in $\mathbb{A}(R, K)$ such that $H_q(C(\widetilde{K})) = 0$ for $q < n$ then every element $x \in H_m(C(\widetilde{K}))$ for $m = n, n+1$ is represented by a pair $(f: \Sigma^m \Gamma \longrightarrow D, g: C \longrightarrow D)$ of morphisms in $\mathbb{B}(R, K)$ with g a global equivalence.*
(ii) *If C is a finite chain complex in $\mathbb{A}(R, K)$ such that $H^q(C(\widetilde{K})) = 0$ for*

$q > n$ then every element $x \in H^m(C(\widetilde{K}))$ for $m = n, n-1$ is represented by a pair $(f : C \longrightarrow D, g : \Sigma^m \Gamma \longrightarrow D)$ of morphisms in $\mathbb{B}(R, K)$ with g a global equivalence.

PROOF (i) By 10.2 the homology assembly map

$$H_m(\mathrm{Hom}_{(R,K)}(\Gamma, V^\infty C)) = H_m([V^\infty C][*])$$

$$\longrightarrow H_m(V^\infty C(\widetilde{K})) = H_m(C(\widetilde{K}))$$

is an isomorphism for $m = n$ and an epimorphism for $m = n+1$.

(ii) For any finite f.g. free (R, K)-module chain complexes B, C and $r \in \mathbb{Z}$ duality defines isomorphisms

$$H_r((B \boxtimes_R V^\infty C)(K)) = H_r(\mathrm{Hom}_{(R,K)}(TB, V^\infty C))$$

$$\xrightarrow{\simeq} H_r((C \boxtimes_R V^\infty B)(K)) = H_r(\mathrm{Hom}_{(R,K)}(TC, V^\infty B)) \ ;$$

$$(f : \Sigma^{-r} TB \longrightarrow D, g : C \longrightarrow D) \longrightarrow (f' : \Sigma^{-r} TC \longrightarrow D', g' : B \longrightarrow D')$$

with

$$D' = \Sigma^{-r} C(e(C) \oplus Th : T^2 C \longrightarrow C \oplus T(\Sigma^{-1} C(f \oplus g : \Sigma^{-r} TB \oplus C \longrightarrow D))) \ .$$

Here, f', g' are inclusions and $h : \Sigma^{-1} C(f \oplus g) \longrightarrow C$ is the projection. Let now C be such that $H^q(C(\widetilde{K})) = 0$ for $q > n$, so that the dual chain complex TC in $\mathbb{A}(R, K)$ is such that $H_q((TC)(\widetilde{K})) = H^{-q}(C(\widetilde{K})) = 0$ for $q < -n$. By the proof of (i) the assembly map

$$H_{-m}(\mathrm{Hom}_{(R,K)}(TC, V^\infty \Gamma)) = H_{-m}(\mathrm{Hom}_{(R,K)}(T\Gamma, V^\infty C))$$

$$\longrightarrow H_{-m}((TC)(\widetilde{K})) = H^m(C(\widetilde{K}))$$

is an isomorphism for $m = n$ and an epimorphism for $m = n - 1$.

$\square$

The quadratic kernel of an n-dimensional normal map of pairs

$$(f, b) : (M, \partial M) \longrightarrow (X, \partial X)$$

with a reference map $X \longrightarrow |K|$ was defined in Ranicki [142] to be an n-dimensional quadratic Poincaré pair in $\mathbb{A}(\mathbb{Z}[\pi_1(K)])$

$$\sigma_*(f, b) = (C(\partial f^!) \longrightarrow C(f^!), (\delta \psi^!, \psi^!))$$

with $f^! : C(\widetilde{X}) \longrightarrow C(\widetilde{M})$, $\partial f^! : C(\partial \widetilde{X}) \longrightarrow C(\partial \widetilde{M})$ the Umkehr chain maps between the cellular chain complexes of the covers $\widetilde{M}$, $\widetilde{X}$, $\partial \widetilde{M}$, $\partial \widetilde{X}$ of M, X, ∂M, ∂X obtained by pullback from the universal cover $\widetilde{K}$ of K. Applying the algebraic Thom construction (as in 1.15) gives an n-dimensional quadratic complex in $\mathbb{A}(\mathbb{Z}[\pi_1(K)])$

$$(C, \psi) = (C(f^!)/C(\partial f^!), \delta \psi^!/\psi^!)$$

with homology and cohomology $\mathbb{Z}[\pi_1(X)]$-modules such that

$$H_*(C) = K_*(M, \partial M) \cong K^{n-*}(M) \ ,$$

$$H^*(C) = K^*(M, \partial M) \cong K_{n-*}(M) \ .$$

If $\pi_1(\partial X) \cong \pi_1(X) \cong \pi_1(K)$, $n \geq 5$ and $(f,b):(M,\partial M)\longrightarrow(X,\partial X)$ is $(i-1)$-connected with $2i \leq n$ an element $x \in K_i(M) = H^{n-i}(C)$ can be killed by geometric surgery on a framed embedded i-sphere S^i in the interior of M with a null-homotopy in X if and only if it can be killed by an algebraic surgery on (C,ψ) using an $(n+1)$-dimensional quadratic pair $(x:C\longrightarrow\Sigma^{n-i}\mathbb{Z}[\pi_1(K)], (\delta\psi,\psi))$ (as in 1.12). The following result analogously relates algebraic surgery on a quadratic complex in $\mathbb{A}(R,K)$ to algebraic surgery on the assembly in $\mathbb{A}(R[\pi_1(K)])$. It is clear how to pass from $\mathbb{A}(R,K)$ to $\mathbb{A}(R[\pi_1(K)])$, so only the 'disassembly' of a surgery in $\mathbb{A}(R[\pi_1(K)])$ to a surgery in $\mathbb{A}(R,K)$ need be considered.

PROPOSITION 10.5 *Let* (C,ψ) *be an* n-*dimensional quadratic complex in* $\mathbb{A}(R,K)$. *For every* $(n+1)$-*dimensional quadratic pair in* $\mathbb{A}(R[\pi_1(K)])$ *of the type*

$$B' = (f':C(\widetilde{K})\longrightarrow\Sigma^{n-i}R[\pi_1(K)], (\delta\psi',\psi(\widetilde{K})))$$

with $2i \leq n$ *and* $H^q(C(\widetilde{K})) = 0$ *for* $q > n-i$ *there exists an* $(n+1)$-*dimensional quadratic pair*

$$B = (f:C\longrightarrow D, (\delta\psi,\psi))$$

in $\mathbb{A}(R,K)$ *with the assembly* $B(\widetilde{K})$ *homotopy equivalent to* B' *relative to the boundary* $(C(\widetilde{K}),\psi(\widetilde{K}))$.

PROOF By 10.4 (ii) there exists an (R,K)-module chain map $f:C\longrightarrow \Sigma^{n-i}V^\infty\Gamma$ which up to $R[\pi_1(K)]$-module chain homotopy assembles to

$$f(\widetilde{K}) = f' : C(\widetilde{K}) \longrightarrow \Sigma^{n-i}(V^\infty\Gamma)(\widetilde{K}) = \Sigma^{n-i}R[\pi_1(K)].$$

Define $\mathbb{Z}$-module chain complexes

$$E = C(1 \otimes (f \boxtimes f): W \otimes_{\mathbb{Z}[\mathbb{Z}_2]} (C \boxtimes_R C)(K) \longrightarrow$$
$$W \otimes_{\mathbb{Z}[\mathbb{Z}_2]} (\Sigma^{n-i}V^\infty\Gamma \boxtimes_R \Sigma^{n-i}V^\infty\Gamma)(K)),$$

$$E' = C(1 \otimes (f' \boxtimes f'): W \otimes_{\mathbb{Z}[\mathbb{Z}_2]} (C \boxtimes_R C)(K) \longrightarrow$$
$$W \otimes_{\mathbb{Z}[\mathbb{Z}_2]} (\Sigma^{n-i}V^\infty R[\pi_1(K)] \boxtimes_{R[\pi_1(K)]} \Sigma^{n-i}V^\infty R[\pi_1(K)])),$$

$$E'' = C(W \otimes_{\mathbb{Z}[\mathbb{Z}_2]} (\Sigma^{n-i}V^\infty\Gamma \boxtimes_R \Sigma^{n-i}V^\infty\Gamma)(K) \longrightarrow$$
$$W \otimes_{\mathbb{Z}[\mathbb{Z}_2]} (\Sigma^{n-i}V^\infty R[\pi_1(K)] \boxtimes_{R[\pi_1(K)]} \Sigma^{n-i}V^\infty R[\pi_1(K)]))$$

such that E'' is chain equivalent to the algebraic mapping cone of the assembly chain map $E\longrightarrow E'$, with an exact sequence

$$\ldots \longrightarrow H_r(E) \longrightarrow H_r(E') \longrightarrow H_r(E'') \longrightarrow H_{r-1}(E) \longrightarrow \ldots \quad (r \in \mathbb{Z}).$$

By the identification of $V^\infty\Gamma$ with $\Delta(E|K|;R)$ and by 7.2 it is possible to identify the R-module chain complex $(V^\infty\Gamma \boxtimes_R V^\infty\Gamma)(K)$ with the simplicial chain complex of a triangulation $EK \times_K EK$ of the pullback

$$|EK| \times_{|K|} |EK| = |EK \times_K EK|$$
$$= \{(\omega, \eta) \in E|K| \times E|K| \mid p(\omega) = p(\eta) \in |K|\} \cong \Omega|K| ,$$

so that up to $\mathbb{Z}[\mathbb{Z}_2]$-module chain homotopy

$$(V^\infty \Gamma \boxtimes_R V^\infty \Gamma)(K) = \Delta(EK \times_K EK; R) = \Delta(\Omega K; R) = [V^\infty \Gamma][*] .$$

The homology

$$H_*(E') = H_{*-2(n-i)}(W \otimes_{\mathbb{Z}[\mathbb{Z}_2]} \Delta(\Omega K; R) \longrightarrow W \otimes_{\mathbb{Z}[\mathbb{Z}_2]} R[\pi_1(K)])$$

is the relative R-coefficient homology of the map

$$E\mathbb{Z}_2 \times_{\mathbb{Z}_2} \Omega K \longrightarrow E\mathbb{Z}_2 \times_{\mathbb{Z}_2} \pi_1(K) ; \ (x, \omega) \longrightarrow (x, [\omega])$$

with $[\omega] \in \pi_0(\Omega K) = \pi_1(K)$ the path component of $\omega \in \Omega K$. Here $E\mathbb{Z}_2$ is a contractible space with a free $\mathbb{Z}_2$-action, the generator $T \in \mathbb{Z}_2$ acts on the pointed loop space ΩK by the reversal of loops using

$$T : [0, 1] \longrightarrow [0, 1] ; \ t \longrightarrow 1 - t$$

and on the group ring $R[\pi_1(K)]$ by the involution inverting group elements. By the usual Hurewicz theorem $H_r(E'') = 0$ for $r \leq 2(n - i) + 1$. Since $2i \leq n$ (by hypothesis) $H_{n+1}(E'') = 0$, and the assembly map

$$H_{n+1}(E) = Q_{n+1}(f : C \rightarrow \Sigma^{n-i} V^\infty \Gamma) \longrightarrow$$
$$H_{n+1}(E') = Q_{n+1}(C \rightarrow \Sigma^{n-i} R[\pi_1(K)])$$

is onto, allowing $(\delta \psi', \psi) \in H_{n+1}(E')$ to be lifted to an element $(\delta \psi, \psi) \in H_{n+1}(E)$. For sufficiently large $k \geq 0$

$$(\delta \psi, \psi) \in \mathrm{im}(Q_{n+1}(C \longrightarrow \Sigma^{n-i} V^k \Gamma) \longrightarrow Q_{n+1}(f : C \longrightarrow \Sigma^{n-i} V^\infty \Gamma))$$

with $C \longrightarrow \Sigma^{n-i} V^k \Gamma$ a restriction of $f : C \longrightarrow \Sigma^{n-i} V^\infty \Gamma$, so that $(\delta \psi', \psi)$ can be further lifted to an element $(\delta \psi, \psi) \in Q_{n+1}(C \longrightarrow \Sigma^{n-i} V^k \Gamma)$. The $(n + 1)$-dimensional quadratic pair in $\mathbb{A}(R, K)$

$$B = (C \longrightarrow \Sigma^{n-i} V^k \Gamma, (\delta \psi, \psi))$$

assembles to an $(n + 1)$-dimensional quadratic pair in $\mathbb{A}(R[\pi_1(K)])$

$$B(\widetilde{K}) = (C(\widetilde{K}) \longrightarrow \Sigma^{n-i} V^k \Gamma(\widetilde{K}), (\delta \psi(\widetilde{K}), \psi(\widetilde{K})))$$

which is homotopy equivalent to the given $(n + 1)$-dimensional quadratic pair

$$B' = (f' : C(\widetilde{K}) \longrightarrow \Sigma^{n-i} R[\pi_1(K)], (\delta \psi', \psi(\widetilde{K})))$$

relative to the boundary $(C(\widetilde{K}), \psi(\widetilde{K}))$.

$\square$

In conclusion:

ALGEBRAIC π-π THEOREM 10.6 *The global assembly maps in quadratic L-theory define isomorphisms*

$$L_n(R, K) \xrightarrow{\simeq} L_n(R[\pi_1(K)]) \; ; \; (C, \psi) \longrightarrow (C(\widetilde{K}), \psi(\widetilde{K})) \;\; (n \in \mathbb{Z}) \; .$$

PROOF Apply the criterion $(*)$ of 3.24 to the maps induced in quadratic L-theory by the global assembly functor $\Lambda(R, K) \longrightarrow \Lambda(R[\pi_1(K)])$, using 10.5 to lift surgeries in $\mathbb{A}(R[\pi_1(K)])$ to surgeries in $\mathbb{A}(R, K)$.

□

§11. Δ-sets

The semi-simplicial sets in the original theory of Kan are abstractions of the singular complex, with both face and degeneracy operations. The Δ-sets of Rourke and Sanderson [152] are 'semi-simplicial set without degeneracies'. The theory of Δ-sets is used in §12 to provide combinatorial models for generalized homology and cohomology, and in §13 to construct the algebraic $\mathbb{L}$-spectra. In §11 only the essential results of the theory are recalled – see [152] for a full exposition.

A Δ-*set* K is a sequence $K^{(n)}$ ($n \geq 0$) of sets, together with face maps

$$\partial_i : K^{(n)} \longrightarrow K^{(n-1)} \quad (0 \leq i \leq n)$$

such that

$$\partial_i \partial_j = \partial_{j-1} \partial_i \text{ for } i < j \ .$$

A Δ-set K is *locally finite* if for each $x \in K^{(n)}$ and $m \geq 1$ the set

$$\{y \in K^{(m+n)} \mid \partial_{i_1} \partial_{i_2} \ldots \partial_{i_m} y = x \text{ for some } i_1, i_2, \ldots, i_m\}$$

is finite.

The *realization* of a Δ-set K is the topological space

$$|K| = \left(\coprod_{n \geq 0} \Delta^n \times K^{(n)}\right) \Big/ \sim$$

with $\sim$ the equivalence relation generated by

$$(a, \partial_i b) \sim (\partial_i a, b) \quad (a \in \Delta^{n-1}, b \in K^{(n)}) \ ,$$

with $\partial_i : \Delta^{n-1} \longrightarrow \Delta^n$ ($0 \leq i \leq n$) the inclusion of Δ^{n-1} as the face opposite the ith vertex of Δ^n.

An ordering of a simplicial complex K is a partial ordering of the vertex set $K^{(0)}$ which restricts to a total ordering on the vertices $v_0 < v_1 < \ldots < v_n$ in any simplex $\sigma = (v_0 v_1 \ldots v_n) \in K^{(n)}$. As usual $n = |\sigma|$ is the dimension of σ, and the faces of σ are the $(n-1)$-dimensional simplexes

$$\partial_i \sigma = (v_0 v_1 \ldots v_{i-1} v_{i+1} \ldots v_n) \quad (0 \leq i \leq n)$$

and their faces. In dealing with the standard n-simplex Δ^n write the vertices as $0, 1, 2, \ldots, n$, ordering them by $0 < 1 < 2 < \ldots < n$.

A simplicial complex K is locally finite if every simplex is the face of only a finite number of simplices.

EXAMPLE 11.1 A (locally finite) ordered simplicial complex K determines a (locally finite) Δ-set K, with realization $|K|$ the polyhedron of K.

□

The product of ordered simplicial complexes K, L is the simplicial complex $K \otimes L$ with

$$(K \otimes L)^{(0)} = K^{(0)} \times L^{(0)} \ ,$$

such that the vertices $(a_0, b_0), (a_1, b_1), \ldots, (a_n, b_n)$ span an n-simplex $\sigma \in (K \otimes L)^{(n)}$ if and only if

$$a_0 \leq a_1 \leq \ldots \leq a_n \,, \; b_0 \leq b_1 \leq \ldots \leq b_n \,, \; (a_r, b_r) \neq (a_{r+1}, b_{r+1}) \; (0 \leq r < n)$$

and the sets $\{a_0, a_1, \ldots, a_n\}, \{b_0, b_1, \ldots, b_n\}$ span simplexes in K and L.

The *geometric product* of Δ-sets K, L is the Δ-set $K \otimes L$ with one p-simplex for each equivalence class of triples

$$(m\text{-simplex } \sigma \in K \,, \; n\text{-simplex } \tau \in L \,, \; p\text{-simplex } \rho \in \Delta^m \otimes \Delta^n) \,,$$

subject to the equivalence relation generated by

$$(\sigma, \tau, \rho) \sim (\sigma', \tau', \rho') \text{ if there exist } \Delta\text{-maps } f \colon \Delta^m \longrightarrow \Delta^{m'} \,,$$

$$g \colon \Delta^n \longrightarrow \Delta^{n'} \text{ such that } \sigma = f^* \sigma' \,, \; \tau = g^* \tau' \,, \; (f \otimes g)_*(\rho) = (\rho') \,.$$

EXAMPLE 11.2 The product $K \otimes L$ of ordered simplicial complexes K, L agrees with their product as Δ-sets.

□

PROPOSITION 11.3 *The realization of the geometric product $K \otimes L$ of Δ-sets K, L is homeomorphic to the product $|K| \times |L|$ of the realizations $|K|, |L|$*

$$|K \otimes L| = |K| \times |L| \,.$$

□

A Δ-*map* $f \colon K \longrightarrow L$ of Δ-sets K, L is defined in the obvious way, with realization a map of spaces $|f| \colon |K| \longrightarrow |L|$.

Let Λ_i^n be the subcomplex of Δ^n obtained by removing the n-simplex $(0, 1, \ldots, n)$ and the $(n-1)$-simplex $(0, \ldots, i-1, i+1, \ldots, n)$ opposite the ith vertex. A Δ-set K is *Kan* if it satisfies the Kan extension condition that every Δ-map $\Lambda_i^n \longrightarrow K$ extends to a Δ-map $\Delta^n \longrightarrow K$.

Given Δ-sets K, L define the *function* Δ-*set* L^K to be the Δ-set with $(L^K)^{(n)}$ the set of Δ-maps $K \otimes \Delta^n \longrightarrow L$, with ∂_i induced from $\partial_i \colon \Delta^{n-1} \longrightarrow \Delta^n$.

PROPOSITION 11.4 *For any Δ-set K and any Kan Δ-set L the function Δ-set L^K is a Kan Δ-set such that the realization $|L^K|$ is homotopy equivalent to the space $|L|^{|K|}$ of functions $|K| \longrightarrow |L|$.*

□

A *homotopy* of Δ-maps $f_0, f_1 \colon K \longrightarrow L$ is an element $g \in (L^K)^{(1)}$ with $\partial_i g = f$ $(i = 0, 1)$, that is a Δ-map $g \colon K \otimes \Delta^1 \longrightarrow L$ such that

$$g(x \otimes i) = f_i(x) \in L^{(n)} \quad (x \in K^{(n)}, \; i = 0, 1) \,.$$

PROPOSITION 11.5 *For any locally finite Δ-set K and any Kan Δ-set L homotopy is an equivalence relation on the set of Δ-maps $K \longrightarrow L$. Realization* *defines a bijection*

$$[K, L] \xrightarrow{\sim} [|K|, |L|] \; ; \; f \longrightarrow |f|$$

between the set $[K, L]$ *of homotopy classes of* Δ*-maps* $K \longrightarrow L$ *and the set* $[|K|, |L|]$ *of homotopy classes of maps* $|K| \longrightarrow |L|$.

$\square$

A Δ-set K is *finite* if there is only a finite number of pairs $(n, x \in K^{(n)})$ with $x \neq \emptyset$. A Δ-map $f \colon K \longrightarrow L$ is *compactly supported* if $\{x \in K \mid f(x) \neq \emptyset \in L\}$ is contained in a finite subobject $J \subseteq K$. Let $[K, L]_c$ denote the set of compactly supported homotopy classes of compactly supported Δ-maps $K \longrightarrow L$, and let L_c^K denote the function space of compactly supported Δ-maps $K \longrightarrow L$.

A Δ-set K is *pointed* if there is given a base n-simplex $\emptyset \in K^{(n)}$ in each dimension $n \geq 0$, with $\partial_i \emptyset = \emptyset$. In dealing with pointed Δ-sets write L^K for the function Δ-set of Δ-maps $K \otimes \Delta^n \longrightarrow L$ which preserve the base simplexes, and $[K, L]$ for the pointed homotopy classes of pointed Δ-maps. For any Δ-set K let K_+ be the pointed Δ-set with

$$(K_+)^{(n)} = K^{(n)} \cup \{\emptyset\} \quad (n \geq 0) \; .$$

The smash product of pointed Δ-sets K, L is defined by

$$K \wedge L = K \otimes L / (K \otimes \emptyset_L \cup \emptyset_K \otimes L) \; .$$

For a pointed Kan Δ-set K the pointed homotopy sets

$$\pi_n(K) = [\partial \Delta^{n+1}, K] \quad (n \geq 0)$$

can be expressed as

$$\pi_n(K) = \{x \in K^{(n)} \mid \partial_i x = \emptyset \in K^{(n-1)}, \, 0 \leq i \leq n\} / \sim \; ,$$

with the equivalence relation $\sim$ defined by $x \sim y$ if there exists $z \in K^{(n+1)}$ such that

$$\partial_i z = \begin{cases} x & \text{if } i = 0 \\ y & \text{if } i = 1 \\ \emptyset & \text{otherwise.} \end{cases}$$

For $n \geq 1$ $\pi_n(K)$ is a group, with the group law defined by

$$\pi_n(K) \times \pi_n(K) \longrightarrow \pi_n(K) \; ; \; (a, b) \longrightarrow c$$

for $a, b, c \in K^{(n)}$ such that there exists $d \in K^{(n+1)}$ with

$$\partial_i d = \begin{cases} a & \text{if } i = 0 \\ c & \text{if } i = 1 \\ b & \text{if } i = 2 \\ \emptyset & \text{otherwise.} \end{cases}$$

For $n \geq 2$, $\pi_n(K)$ is an abelian group, as usual.

The following analogue of J. H. C. Whitehead's theorem holds:

PROPOSITION 11.6 *A map of locally finite pointed Kan Δ-sets $f\colon K\longrightarrow L$ is a homotopy equivalence if and only if it induces isomorphisms of homotopy groups $f_*\colon \pi_*(K)\longrightarrow \pi_*(L)$.*

□

DEFINITION 11.7 The *mapping fibre* of a map of pointed Kan Δ-sets $f\colon K\longrightarrow L$ is the Kan Δ-set $M(f)$ with

$$M(f)^{(n)} =$$

$$\{(x,y) \in K^{(n)} \times L^{(n+1)} \,|\, \partial_0\partial_1 \ldots \partial_n y = \emptyset \in L^{(0)}, \ \partial_{n+1}y = fx \in L^{(n)}\},$$

$$\partial_i : M(f)^{(n)} \longrightarrow M(f)^{(n-1)} \,;\, (x,y) \longrightarrow (\partial_i x, \partial_i y).$$

□

The map $M(f)\longrightarrow K; (x,y)\longrightarrow x$ fits into a fibration sequence

$$M(f) \longrightarrow K \xrightarrow{\ f\ } L$$

inducing a long exact sequence of homotopy groups

$$\ldots \longrightarrow \pi_{n+1}(L) \longrightarrow \pi_n(M(f)) \longrightarrow \pi_n(K) \xrightarrow{\ f_*\ } \pi_n(L) \longrightarrow \ldots \,.$$

DEFINITION 11.8 The *loop Δ-set* of a pointed Kan Δ-set K is the pointed Kan Δ-set

$$\Omega K = K^{S^1}$$

with S^1 the pointed Δ-set defined by

$$(S^1)^{(n)} = \begin{cases} \{s, \emptyset\} & \text{if } n = 1 \\ \{\emptyset\} & \text{if } n \neq 1, \end{cases}$$

such that

$$\pi_n(\Omega K) = \pi_{n+1}(K) \ (n \geq 0).$$

□

ΩK is the mapping fibre of the unique map $\{*\}\longrightarrow K$, so that

$$\Omega K^{(n)} = \{x \in K^{(n+1)} \,|\, \partial_0\partial_1 \ldots \partial_n x = \emptyset \in K^{(0)}, \ \partial_{n+1}x = \emptyset \in K^{(n)}\}.$$

PROPOSITION 11.9 *The realization $|M(f)|$ of the mapping fibre $M(f)$ of a map $f\colon K\longrightarrow L$ of pointed Δ-sets with K locally finite and L Kan is homotopy equivalent to the mapping fibre $M(|f|)$ of the realization $|f|\colon |K|\longrightarrow|L|$. In particular, the realization $|\Omega K|$ of the loop Δ-set ΩK is homotopy equivalent to the loop space of the realization $|K|$*

$$|\Omega K| \simeq \Omega|K|.$$

□

DEFINITION 11.10 An Ω-*spectrum*

$$\mathbb{F} = \{\mathbb{F}_n, \mathbb{F}_{n+1} \xrightarrow{\simeq} \Omega\mathbb{F}_n \mid n \in \mathbb{Z}\}$$

is a sequence of pointed Kan Δ-sets $\mathbb{F}_n$ together with homotopy equivalences $\mathbb{F}_{n+1} \longrightarrow \Omega\mathbb{F}_n$. The homotopy groups of $\mathbb{F}$ are defined by

$$\pi_n(\mathbb{F}) = \pi_{n+k}(\mathbb{F}_{-k}) \quad (n, k \in \mathbb{Z}, n+k \geq 0) .$$

□

Note that the indexing of $\mathbb{F}$ is the negative of the usual terminology for an Ω-spectrum

$$\mathbb{G} = \{\mathbb{G}_n, \mathbb{G}_n \xrightarrow{\simeq} \Omega\mathbb{G}_{n+1} \mid n \in \mathbb{Z}\} .$$

DEFINITION 11.11 The *mapping cofibre* of a map $f: K \longrightarrow L$ of Ω-spectra of Kan Δ-sets is the Ω-spectrum of Kan Δ-sets

$$C(f) = \{C(f)_n = M(f: K_{n-1} \longrightarrow L_{n-1}) \mid n \in \mathbb{Z}\} .$$

□

The mapping cofibre fits into a (co)fibration sequence of Ω-spectra

$$K \xrightarrow{f} L \xrightarrow{g} C(f)$$

with

$g = $ inclusion :

$L_n = \Omega L_{n-1} = M(\{*\} \longrightarrow L_{n-1}) \longrightarrow C(f)_n = M(f: K_{n-1} \longrightarrow L_{n-1})$

inducing a long exact sequence of homotopy groups

$$\ldots \longrightarrow \pi_n(K) \xrightarrow{f_*} \pi_n(L) \xrightarrow{g_*} \pi_n(C(f)) \xrightarrow{\partial} \pi_{n-1}(K) \longrightarrow \ldots .$$

DEFINITION 11.12 The *suspension* of an Ω-spectrum $K = \{K_n \mid n \in \mathbb{Z}\}$ is the Ω-spectrum

$$\Sigma K = C(K \longrightarrow \{*\})$$

with

$$(\Sigma K)_n = K_{n-1} , \quad \pi_n(\Sigma K) = \pi_{n-1}(K) \quad (n \in \mathbb{Z}) .$$

□

The mapping cofibre of a map $f: K \longrightarrow L$ of Ω-spectra is just the suspension of the mapping fibre

$$C(f) = \Sigma M(f) .$$

§12. Generalized homology theory

The connection between generalized homology and stable homotopy theory due to G. W. Whitehead [185] and the language of Δ-sets are used to construct combinatorial models for both the cohomology and homology groups of a locally finite simplicial complex K with coefficients in an Ω-spectrum $\mathbb{F}$.

DEFINITION 12.1 Let $\mathbb{F}$ be an Ω-spectrum of Kan Δ-sets, and let K be a locally finite Δ-set. The $\begin{cases} \mathbb{F}\text{-}cohomology \\ compactly\ supported\ \mathbb{F}\text{-}cohomology\ \Omega\text{-}spectrum \\ \mathbb{F}\text{-}homology \end{cases}$ of K is defined by

$$\begin{cases} \mathbb{F}^{K_+} = \{ (\mathbb{F}_n)^{K_+} \,|\, n \in \mathbb{Z} \} \\ \mathbb{F}_c^{K_+} = \{ (\mathbb{F}_n)_c^{K_+} \,|\, n \in \mathbb{Z} \} \\ K_+ \wedge \mathbb{F} = \{ \varinjlim_j \Omega^j (K_+ \wedge \mathbb{F}_{n-j}) \,|\, n \in \mathbb{Z} \} \end{cases}$$

with homotopy groups the $\begin{cases} \mathbb{F}\text{-}cohomology \\ compactly\ supported\ \mathbb{F}\text{-}cohomology\ groups\ of \\ \mathbb{F}\text{-}homology \end{cases}$ K

$$\begin{cases} H^n(K;\mathbb{F}) = \pi_{-n}(\mathbb{F}^{K_+}) = [K_+, \mathbb{F}_{-n}] \\ H_c^n(K;\mathbb{F}) = \pi_{-n}(\mathbb{F}_c^{K_+}) = [K_+, \mathbb{F}_{-n}]_c \\ H_n(K;\mathbb{F}) = \pi_n(K_+ \wedge \mathbb{F}) = \varinjlim_j \pi_{n+j}(K_+ \wedge \mathbb{F}_{-j}) . \end{cases}$$

□

Write the $\mathbb{F}$-cohomology Ω-spectrum of K as

$$\mathbb{F}^{K_+} = \mathbb{H}^{\cdot}(K;\mathbb{F}) = \{ \mathbb{H}^n(K;\mathbb{F}) \,|\, n \in \mathbb{Z} \} ,$$

with

$$\mathbb{H}^n(K;\mathbb{F}) = (\mathbb{F}_n)^{K_+} , \quad \pi_{-n}(\mathbb{H}^{\cdot}(K;\mathbb{F})) = H^n(K;\mathbb{F}) .$$

The n-dimensional $\mathbb{F}$-cohomology group $\mathbb{F}^n(K)$ of a locally finite Δ-set K thus has a direct combinatorial description as the set of homotopy classes of Δ-maps $K_+ \longrightarrow \mathbb{F}_{-n}$, which may be called '$\mathbb{F}$-cocycles in K'. Similarly for the compactly supported $\mathbb{F}$-cohomology group $\mathbb{F}_c^n(K)$. There follows a similar description for the $\mathbb{F}$-homology group of a locally finite ordered simplicial complex K, as the set of cobordism classes of '$\mathbb{F}$-cycles in K'. On the Ω-spectrum level it is possible to replace $K_+ \wedge \mathbb{F}$ by a homotopy equivalent Ω-spectrum $\mathbb{H}_{\cdot}(K;\mathbb{F})$ which is defined directly in terms of the simplexes of K and $\mathbb{F}$.

Regard the standard n-simplex Δ^n as the simplicial complex with one k-simplex for each subset $\sigma \subseteq \{0, 1, \dots, n\}$ of $k+1$ elements. The boundary $\partial \Delta^n \subset \Delta^n$ is the subcomplex consisting of the proper subsets $\sigma \subset \{0, 1, \ldots, n\}$.

A finite ordered simplicial complex J has a canonical embedding as a subcomplex in $\partial \Delta^{m+1}$ with $m = |J^{(0)}|$, namely

$$J \longrightarrow \partial \Delta^{m+1} \; ; \; v_i \longrightarrow i \; ,$$

if $J^{(0)} = \{v_i \,|\, 0 \leq i \leq m\}$.

Let Σ^m be the simplicial complex with one k-simplex σ^* for each $(m-k)$-simplex σ in $\partial \Delta^{m+1}$, with $\sigma^* \leq \tau^* \in \Sigma^m$ if and only if $\tau \leq \sigma \in \partial \Delta^{m+1}$. The face maps in the Δ-set Σ^m are such that

$$\partial_i : (\Sigma^m)^{(k)} \longrightarrow (\Sigma^m)^{(k-1)} \; ; \; \sigma^* \longrightarrow \partial_i(\sigma^*) = (\delta_i \sigma)^* \quad (0 \leq i \leq k \leq m)$$

where

$$\delta_i : (\partial \Delta^{m+1})^{(m-k)} \longrightarrow (\partial \Delta^{m+1})^{(m-k+1)} \; ;$$

$$\sigma = \{0, 1, \ldots, m+1\} \backslash \{j_0, j_1, \ldots, j_k\} \longrightarrow \delta_i \sigma = \sigma \cup j_i \quad (0 \leq i \leq k) \; .$$

The simplicial map

$$\Sigma^m \xrightarrow{\;\cong\;} \partial \Delta^{m+1} \; ; \; \sigma^* \longrightarrow \{0, 1, \ldots, m+1\} \backslash \sigma$$

is an isomorphism of simplicial complexes. Regard Σ^m as the dual cell decomposition of the barycentric subdivision $(\partial \Delta^{m+1})'$, with σ^* the star of the barycentre $\widehat{\sigma}$ and $(\delta_i \sigma)^* \subset \partial \sigma^*$ the embedding of the star of $\delta_i \sigma$ in the link of $\widehat{\sigma}$.

DEFINITION 12.2 The *supplement* of a simplicial subcomplex $K \subseteq \partial \Delta^{m+1}$ is the subcomplex $\overline{K} \subseteq \Sigma^m$ given by

$$\overline{K} = \{\sigma^* \in \Sigma^m \,|\, \sigma \in \partial \Delta^{m+1} \backslash K\} \; .$$

$\square$

The definition of the supplement goes back to at least Blakers and Massey [10]. In particular

$$\overline{\partial \Delta^{m+1}} = \emptyset \; , \quad \overline{\emptyset} = \Sigma^m$$

and if $J \subseteq K \subseteq \partial \Delta^{m+1}$ then $\overline{K} \subseteq \overline{J} \subseteq \Sigma^m$.

DEFINITION 12.3 Let $\mathbb{F}$ be an Ω-spectrum of Kan Δ-sets.
(i) Given a finite simplicial complex J define the Ω-spectrum

$$\mathbb{H}.(J; \mathbb{F}) = \{\mathbb{H}_n(J; \mathbb{F}) \,|\, n \in \mathbb{Z}\}$$

by

$$\mathbb{H}_n(J; \mathbb{F}) = \mathbb{H}^{n-m}(\Sigma^m, \overline{J}; \mathbb{F}) \; ,$$

using the canonical embedding $J \subseteq \partial \Delta^{m+1}$ ($m = |J^{(0)}|$), with homotopy groups

$$\pi_n(\mathbb{H}.(J; \mathbb{F})) = H^{m-n}(\Sigma^m, \overline{J}; \mathbb{F}) \quad (n \in \mathbb{Z}) \; .$$

(ii) Given a locally finite ordered simplicial complex K define the Ω-spectrum

$$\mathbb{H}.(K;\mathbb{F}) = \varinjlim_J \mathbb{H}.(J;\mathbb{F})$$

with the direct limit over finite subcomplexes $J \subseteq K$. The homotopy groups are such that

$$\pi_n(\mathbb{H}.(K;\mathbb{F})) = \varinjlim_J H^{m-n}(\Sigma^m, \overline{J}; \mathbb{F}) \quad (n \in \mathbb{Z}) \ .$$

□

Given a Δ-set K let $\Delta(K)$ be the abelian group chain complex with $\Delta(K)_n$ the free abelian group generated by $K^{(n)}$, and

$$d_{\Delta(K)} : \Delta(K)_n \longrightarrow \Delta(K)_{n-1} \ ; \ x \longrightarrow \sum_{i=0}^n (-)^i \partial_i x \ .$$

PROPOSITION 12.4 *The Ω-spectrum $\mathbb{H}.(K;\mathbb{F})$ is homotopy equivalent to the $\mathbb{F}$-homology Ω-spectrum $K_+ \wedge \mathbb{F}$, with*

$$\pi_n(\mathbb{H}.(K;\mathbb{F})) = \pi_n(K_+ \wedge \mathbb{F}) = H_n(K;\mathbb{F}) \quad (n \in \mathbb{Z}) \ .$$

PROOF Since generalized homology commutes with direct limits, there is no loss of generality in assuming that K is finite, with canonical embedding $K \subseteq \partial \Delta^{m+1}$. By construction $\mathbb{H}_n(K;\mathbb{F})^{(p)}$ consists of the Δ-maps $\Sigma^m \otimes \Delta^p \longrightarrow \mathbb{F}_{n-m}$ sending $\overline{K} \otimes \Delta^p$ to $\emptyset$. Approximate the reduced diagonal map $S^m \longrightarrow |K|_+ \wedge (S^m/|\overline{K}|)$ of G.W.Whitehead [185, p. 265] by a Δ-map $\Sigma^m \longrightarrow K_+ \wedge (\Sigma^m/\overline{K})$, subdividing Σ^m if necessary – see Remark 12.5 below for an explicit construction. The Δ-map represents the m-cycle

$$\sum_{\sigma \in K} \sigma \otimes \sigma^* \in (\Delta(K) \otimes \Delta(\Sigma^m, \overline{K}))_m$$

with adjoint the isomorphism $\Delta(K)^{m-*} \xrightarrow{\simeq} \Delta(\Sigma^m, \overline{K})$ sending the elementary cochain of $\sigma \in K$ to the elementary chain of $\sigma^* \in \Sigma^m/\overline{K}$. Define a map of Ω-spectra $\mathbb{H}.(K;\mathbb{F}) \longrightarrow K_+ \wedge \mathbb{F}$ by

$$\mathbb{H}_n(K;\mathbb{F}) = (\mathbb{F}_{n-m}, \emptyset)^{(\Sigma^m, \overline{K})}$$

$$\longrightarrow (K_+ \wedge \mathbb{F}_{n-m})^{\Sigma^m} \simeq \Omega^m(K_+ \wedge \mathbb{F}_{n-m})$$

$$\longrightarrow (K_+ \wedge \mathbb{F})_n = \varinjlim_j \Omega^j(K_+ \wedge \mathbb{F}_{n-j}) \ ;$$

$$((\Sigma^m, \overline{K}) \otimes \Delta^p \longrightarrow (\mathbb{F}_{n-m}, \emptyset))$$

$$\longrightarrow (\Sigma^m \wedge \Delta^p_+ \longrightarrow K_+ \wedge (\Sigma^m/\overline{K}) \wedge \Delta^p_+ \longrightarrow K_+ \wedge \mathbb{F}_{n-m}) \ .$$

This is a homotopy equivalence by J. H. C. Whitehead's theorem, since it induces the Alexander S-duality isomorphisms

$$\pi_n(\mathbb{H}.(K;\mathbb{F})) = H^{m-n}(\Sigma^m,\overline{K};\mathbb{F}) \xrightarrow{\simeq} \pi_n(K_+\wedge\mathbb{F}) = H_n(K;\mathbb{F})\ (n \in \mathbb{Z}).$$

□

REMARK 12.5 Regard a simplicial complex K as a category with one object for each simplex $\sigma \in K$ and one morphism $\sigma\longrightarrow\tau$ for each face inclusion $\sigma \leq \tau$. The *homotopy colimit* (Bousfield and Kan [13]) of a contravariant functor

$$F : K \longrightarrow \{\text{pointed } \Delta\text{-sets}\} \ ; \ \sigma \longrightarrow F[\sigma]$$

is the pointed Δ-set

$$F[K] = (\coprod_{\sigma\in K} \Delta^{|\sigma|} \otimes F[\sigma])/\sim ,$$

with $\sim$ the equivalence relation generated by

(i) $f_*a \otimes b \sim a \otimes f^*b$ for any morphism $f{:}\sigma\longrightarrow\tau$, $a \in \Delta^{|\sigma|}$, $b \in F[\tau]$,

(ii) $\Delta^{|\sigma|} \otimes \emptyset \sim \Delta^{|\sigma'|} \otimes \emptyset'$ for any σ , $\sigma' \in K$.

Given a subcomplex $J \subseteq \partial\Delta^{m+1}$ define a contravariant functor

$$G : \partial\Delta^{m+1} \longrightarrow \{\text{pointed } \Delta\text{-sets}\} \ ; \ \sigma \longrightarrow G[\sigma] = \begin{cases} \Sigma^m/\overline{J} & \text{if } \sigma \in J \\ \emptyset & \text{otherwise} \end{cases}$$

with homotopy colimit

$$G[\partial\Delta^{m+1}] = J_+ \wedge (\Sigma^m/\overline{J}) .$$

Quinn [134] proved that the homotopy colimit $F[\partial\Delta^{m+1}]$ of the dual simplex functor

$$F : \partial\Delta^{m+1} \longrightarrow \{\text{pointed } \Delta\text{-sets}\} \ ; \ \sigma \longrightarrow \sigma^* = \Delta^{m-|\sigma|}$$

is a subdivision of Σ^m, allowing the construction of a combinatorial approximation of the reduced diagonal map $S^m\longrightarrow|J|_+ \wedge (S^m/|\overline{J}|)$ as the Δ-map

$$h[\partial\Delta^{m+1}] : F[\partial\Delta^{m+1}] \cong \Sigma^m \longrightarrow G[\partial\Delta^{m+1}] = J_+ \wedge (\Sigma^m/\overline{J})$$

induced by the natural transformation $h{:}F\longrightarrow G$ with

$$h[\sigma] = \sigma^* : F[\sigma] = \Delta^{m-|\sigma|} \longrightarrow G[\sigma] = \Sigma^m/\overline{J} \ (\sigma \in J)$$

the characteristic Δ-maps.

□

DEFINITION 12.6 An *n-dimensional* $\mathbb{F}$-*cycle* in an ordered simplicial complex K is a pair (J,x) with $J \subseteq K$ a finite subcomplex and x a 0-simplex

$$x \in \text{im}(\mathbb{H}_n(J;\mathbb{F})^{(0)}\longrightarrow\mathbb{H}_n(K;\mathbb{F})^{(0)}) ,$$

that is a collection

$$x = \{x(\sigma) \in \mathbb{F}_{n-m}^{(m-|\sigma|)} \mid \sigma \in J\}$$

defined using the canonical embedding $J \subseteq \partial \Delta^{m+1}$, such that

$$\partial_i x(\sigma) = \begin{cases} x(\delta_i \sigma) & \text{if } \delta_i \sigma \in J \\ \emptyset & \text{if } \delta_i \sigma \notin J \end{cases} \quad (0 \le i \le m - |\sigma|) .$$

□

In dealing with cycles (J, x) the finite subcomplex $J \subseteq K$ will usually be omitted from the terminology. For finite K it is always possible to take $J = K$.

DEFINITION 12.7 A *cobordism* of n-dimensional $\mathbb{F}$-cycles (J_0, x_0), (J_1, x_1) in K is a 1-simplex $y \in \mathbb{H}_n(K; \mathbb{F})^{(1)}$ such that $\partial_i y = x_i$ ($i = 0, 1$), that is a compactly supported Δ-map

$$y : (\Sigma^m, \overline{J}) \otimes \Delta^1 \longrightarrow (\mathbb{F}_{n-m}, \emptyset) \quad (J = J_0 \cup J_1)$$

such that

$$y(\sigma \otimes i) = x_i(\sigma) \in \mathbb{F}_{n-m}^{(m-|\sigma|)} \quad (\sigma \in J, i = 0, 1) .$$

□

PROPOSITION 12.8 *Cobordism is an equivalence relation on n-dimensional $\mathbb{F}$-cycles in K, such that the set of equivalence classes is the n-dimensional $\mathbb{F}$-homology group $H_n(K; \mathbb{F})$.*
PROOF Immediate from 12.4.

□

EXAMPLE 12.9 Given an abelian group π and an integer $n \ge 0$ let $K(\pi, n)$ be the Kan Δ-set defined by forgetting the degeneracies in the Eilenberg–MacLane simplicial abelian group obtained from the abelian group chain complex C with

$$C_n = \pi , \quad C_i = 0 \quad (i \ne n)$$

by the Kan–Dold construction. Let $\mathbb{F}$ be the Ω-spectrum defined by

$$\mathbb{F}_n = K(\pi, -n) \quad (n \le 0) , \quad = 0 \quad (n > 0) .$$

An n-dimensional $\mathbb{F}$-cycle (J, x) in a simplicial complex K is determined by a finite subcomplex $J \subseteq K$, with

$$x = \{x(\sigma) \in \mathbb{F}_{n-m}^{(m-|\sigma|)} \mid \sigma \in J\} \quad (m = |J^{(0)}|)$$

determined by a finite collection of group elements

$$x(\sigma) \in \mathbb{F}_{n-m}^{(m-n)} = \pi \quad (\sigma \in K^{(n)})$$

corresponding to an n-cycle

$$x = \sum_{\sigma \in K^{(n)}} x(\sigma)\sigma \in \Delta_n(K; \pi)$$

representing a homology class

$$x \in H_n(K; \mathbb{F}) \;=\; H_n(K; \pi) \,.$$

□

The cycle approach to $\mathbb{F}$-homology generalizes to the relative case. Let $(K, L \subseteq K)$ be a pair of ordered locally finite simplicial complexes. For any finite subcomplex $J \subseteq K$ with $m = |J^{(0)}|$ the supplements of J and $J \cap K$ are such that $\overline{J} \subseteq \overline{J \cap L} \subseteq \Sigma^m$, and $\mathbb{H}.(L; \mathbb{F}) \subseteq \mathbb{H}.(K; \mathbb{F})$. An n-dimensional $\mathbb{F}$-cycle in L is an n-dimensional $\mathbb{F}$-cycle (J, x) in K such that $x(\sigma) = \emptyset$ for $\sigma \in J \backslash (J \cap L)$.

DEFINITION 12.10 (i) The *relative $\mathbb{F}$-homology Ω-spectrum* of (K, L)

$$\mathbb{H}.(K, L; \mathbb{F}) \;=\; \{ \mathbb{H}_n(K, L; \mathbb{F}) \,|\, n \in \mathbb{Z} \}$$

is defined by

$$\mathbb{H}_n(K, L; \mathbb{F}) \;=\; \varinjlim_J (\mathbb{F}_{n-m}, \emptyset)^{(\overline{J \cap L}, \overline{J})} \quad (n \in \mathbb{Z}) \,,$$

with the direct limit taken over finite subcomplexes $J \subseteq K$. The relative $\mathbb{F}$-homology groups of (K, L) are the homotopy groups of $\mathbb{H}.(K, L; \mathbb{F})$

$$\pi_n(\mathbb{H}.(K, L; \mathbb{F})) \;=\; H_n(K, L; \mathbb{F}) \quad (n \in \mathbb{Z}) \,.$$

(ii) A *relative n-dimensional $\mathbb{F}$-cycle* (J, x) in (K, L) is an element of $\mathbb{H}_n(K, L; \mathbb{F})^{(0)}$, that is a finite subcomplex $J \subseteq K$ together with a collection

$$x \;=\; \{ x(\sigma) \in \mathbb{F}_{n-m}^{(m - |\sigma|)} \,|\, \sigma \in J \backslash (J \cap L) \}$$

such that

$$\partial_i x(\sigma) \;=\; \begin{cases} x(\delta_i \sigma) & \text{if } \delta_i \sigma \in J \backslash (J \cap L) \\ \emptyset & \text{if } \delta_i \sigma \notin J \end{cases} \quad (0 \leq i \leq m - |\sigma|) \,.$$

□

By analogy with 12.8:

PROPOSITION 12.11 *Cobordism of relative cycles is defined as in the absolute case, and $H_n(K, L; \mathbb{F})$ is the abelian group of cobordism classes. The fibration sequence of Ω-spectra*

$$\mathbb{H}.(L; \mathbb{F}) \longrightarrow \mathbb{H}.(K; \mathbb{F}) \longrightarrow \mathbb{H}.(K, L; \mathbb{F})$$

induces the long exact sequence of $\mathbb{F}$-homology groups

$$\ldots \longrightarrow H_n(L; \mathbb{F}) \longrightarrow H_n(K; \mathbb{F}) \longrightarrow H_n(K, L; \mathbb{F}) \longrightarrow H_{n-1}(L; \mathbb{F}) \longrightarrow \ldots \,.$$

PROOF As in the proof of 12.4 it may be assumed that K is finite, with a canonical embedding $K \subseteq \partial \Delta^{m+1}$. The homotopy equivalences $\Omega \mathbb{F}_{n-m-1} \xrightarrow{\simeq} \mathbb{F}_{n-m}$ given by the Ω-spectrum $\mathbb{F}$ and the excisive inclusion

$$(\overline{L} \otimes \Delta^1, \overline{K} \otimes \Delta^1 \cup \overline{L} \otimes \partial \Delta^1) \longrightarrow (\Sigma^m \otimes \Delta^1, \overline{K} \otimes \Delta^1 \cup \overline{L} \otimes \partial_0 \Delta^1 \cup \Sigma^m \otimes \partial_1 \Delta^1)$$

may be used to define homotopy equivalences

$$\mathbb{H}_n(K, L; \mathbb{F}) = (\mathbb{F}_{n-m}, \emptyset)^{(\overline{L}, \overline{K})}$$

$$\simeq (\Omega \mathbb{F}_{n-m-1}, \emptyset)^{(\overline{L}, \overline{K})}$$

$$= (\mathbb{F}_{n-m-1}, \emptyset)^{(\overline{L} \otimes \Delta^1, \overline{K} \otimes \Delta^1 \cup \overline{L} \otimes \partial\Delta^1)}$$

$$\simeq (\mathbb{F}_{n-m-1}, \emptyset)^{(\Sigma^m \otimes \Delta^1, \overline{K} \otimes \Delta^1 \cup \overline{L} \otimes \partial_0\Delta^1 \cup \Sigma^m \otimes \partial_1\Delta^1)}$$

$$= \text{mapping cofibre of } \mathbb{H}_n(L; \mathbb{F}) \longrightarrow \mathbb{H}_n(K; \mathbb{F}) ,$$

obtaining a homotopy equivalence between $\mathbb{H}.(K, L; \mathbb{F})$ and the mapping cofibre of the inclusion $\mathbb{H}.(L; \mathbb{F}) \longrightarrow \mathbb{H}.(K; \mathbb{F})$.

□

EXAMPLE 12.12 Consider $H_*(K, L; \mathbb{F})$ in the special case $L \subseteq K \subseteq \partial\Delta^{m+1}$ with $K = L \cup \Delta^k$ obtained from L by attaching a k-simplex along a subcomplex $\partial\Delta^k \subseteq L$. An n-dimensional $\mathbb{F}$-cycle x in (K, L) is an element $x(\Delta^k) \in \mathbb{F}_{n-m}^{(m-k)}$ such that $\partial_i x(\Delta^k) = \emptyset$ for $0 \le i \le m - k$, and the map

$$H_n(K, L; \mathbb{F}) \longrightarrow \pi_{m-k}(\mathbb{F}_{n-m}) = \pi_{n-k}(\mathbb{F}) ; \quad x \longrightarrow x(\Delta^k)$$

is an isomorphism.

□

The Kan extension condition will now be used to define the assembly map

$$A : \mathbb{H}.(K; \mathbb{F}(\{*\})) \longrightarrow \mathbb{F}.(K')$$

for any covariant functor

$$\mathbb{F} : \{\text{simplicial complexes}\} \longrightarrow \{\Omega\text{-spectra}\} ; \quad K \longrightarrow \mathbb{F}(K) .$$

Let $\Lambda^{m+1} \subset \partial\Delta^{m+1}$ be the subcomplex obtained by removing the face $\Delta^m < \Delta^{m+1}$ opposite the vertex $m + 1$, such that

$$\partial\Delta^{m+1} = \Lambda^{m+1} \cup \Delta^m , \quad \Lambda^{m+1} \cap \Delta^m = \partial\Lambda^{m+1} = \partial\Delta^m .$$

The inclusion

$$(\Lambda^{m+1}, \partial\Lambda^{m+1}) \subset (\Delta^{m+1}, \partial\Lambda^{m+1})$$

is a homotopy equivalence such that for a Kan Δ-set $\mathbb{F}$ the induced homotopy equivalence

$$(\mathbb{F}, \emptyset)^{(\Delta^{m+1}, \partial\Lambda^{m+1})} \xrightarrow{\simeq} (\mathbb{F}, \emptyset)^{(\Lambda^{m+1}, \partial\Lambda^{m+1})}$$

admits a section

$$\beta : (\mathbb{F}, \emptyset)^{(\Lambda^{m+1}, \partial\Lambda^{m+1})} \longrightarrow (\mathbb{F}, \emptyset)^{(\Delta^{m+1}, \partial\Lambda^{m+1})}$$

verifying the Kan extension condition. The inclusion

$$(\Delta^m, \partial\Delta^m) \subset (\Delta^{m+1}, \partial\Lambda^{m+1})$$

is a homotopy equivalence, inducing a homotopy equivalence

$$\gamma : (\mathbb{F}, \emptyset)^{(\Delta^{m+1}, \partial \Lambda^{m+1})} \xrightarrow{\simeq} (\mathbb{F}, \emptyset)^{(\Delta^m, \partial \Delta^m)} .$$

PROPOSITION 12.13 *For a Kan Δ-set $\mathbb{F}$ the composite Δ-map*

$$\alpha = \gamma\beta : (\mathbb{F}, \emptyset)^{(\Lambda^{m+1}, \partial \Lambda^{m+1})} \xrightarrow{\beta} (\mathbb{F}, \emptyset)^{(\Delta^{m+1}, \partial \Lambda^{m+1})}$$

$$\xrightarrow{\gamma} (\mathbb{F}, \emptyset)^{(\Delta^m, \partial \Delta^m)}$$

is a homotopy equivalence of Kan Δ-sets.
PROOF Both β and γ are homotopy equivalences.

□

The geometric realizations of Λ^{m+1} and Δ^m may be identified by means of the homeomorphism

$$|\Lambda^{m+1}| \xrightarrow{\simeq} |\Delta^m| ;$$

$$(\lambda_0, \lambda_1, \ldots, \lambda_{m+1}) \longrightarrow$$

$$(\lambda_0 + \lambda_{m+1}/(m+1), \lambda_1 + \lambda_{m+1}/(m+1), \ldots, \lambda_m + \lambda_{m+1}/(m+1))$$

$$(0 \leq \lambda_0, \lambda_1, \ldots, \lambda_{m+1} \leq 1 , \sum_{i=0}^{m+1} \lambda_i = 1 , \lambda_0 \lambda_1 \ldots \lambda_m = 0) ,$$

which maps $\partial \Lambda^{m+1}$ to $\partial \Delta^m$. This identification is used to visualize α as sending a Δ-map

$$f : (\Lambda^{m+1}, \partial \Lambda^{m+1}) \longrightarrow (\mathbb{F}, \emptyset)$$

to the Δ-map

$$\alpha(f) = \bigcup_{\sigma \in \Lambda^{m+1}} f(\sigma) : (\Delta^m, \partial \Delta^m) \longrightarrow (\mathbb{F}, \emptyset)$$

obtained by assembling together the pieces $f(\sigma) \in \mathbb{F}^{(|\sigma|)}$, glueing by the Kan extension condition.

Given an Ω-spectrum $\mathbb{F}$ let $\Theta : \mathbb{F}_n \xrightarrow{\simeq} \Omega \mathbb{F}_{n-1}$ ($n \in \mathbb{Z}$) be the given homotopy equivalences. Given a subcomplex $K \subseteq \partial \Delta^{m+1}$ define Δ-maps

$$\phi : \mathbb{H}_n(K; \mathbb{F}) = (\mathbb{F}_{n-m}, \emptyset)^{(\Sigma^m, \overline{K})}$$

$$\longrightarrow (\mathbb{F}_{n-m-1}, \emptyset)^{(\Lambda^{m+2}, \partial \Lambda^{m+2})} \quad (m \in \mathbb{Z})$$

by sending a Δ-map

$$f : (\Sigma^m, \overline{K}) \otimes \Delta^p \longrightarrow (\mathbb{F}_{n-m}, \emptyset)$$

to the Δ-map

$$\phi(f) : (\Lambda^{m+2}, \partial \Lambda^{m+2}) \otimes \Delta^p \longrightarrow (\mathbb{F}_{n-m-1}, \emptyset) ;$$

$$\tau \otimes \mu \longrightarrow \begin{cases} \Theta(f(\sigma^* \otimes \mu)) & \text{if } \sigma = \{0, 1, \ldots, m+2\} \backslash \tau \in K \\ \emptyset & \text{otherwise} . \end{cases}$$

DEFINITION 12.14 Given an Ω-spectrum $\mathbb{F}$ and a locally finite simplicial complex K define the *assembly* to be the map of Ω-spectra

$$A : \mathbb{H}.(K;\mathbb{F}) = \varinjlim_J \mathbb{H}.(J;\mathbb{F}) \longrightarrow \mathbb{F}$$

using the canonical embeddings $J \subseteq \partial\Delta^{m+1}$ of the finite subcomplexes $J \subseteq K$, with

$$A : \mathbb{H}_n(J;\mathbb{F}) \xrightarrow{\phi} (\mathbb{F}_{n-m-1},\emptyset)^{(\Lambda^{m+2},\partial\Lambda^{m+2})}$$

$$\xrightarrow{\alpha} (\mathbb{F}_{n-m-1},\emptyset)^{(\Delta^{m+1},\partial\Delta^{m+1})} = \Omega^{m+1}\mathbb{F}_{n-m-1} \xrightarrow{(\Theta^{m+1})^{-1}} \mathbb{F}_n ,$$

inducing assembly maps in the homotopy groups

$$A : \pi_n(\mathbb{H}.(K;\mathbb{F})) = H_n(K;\mathbb{F}) \longrightarrow \pi_n(\mathbb{F}) \quad (n \in \mathbb{Z}) .$$

$\square$

In terms of the homotopy equivalence $\mathbb{H}.(K;\mathbb{F}) \xrightarrow{\simeq} K_+ \wedge \mathbb{F}$ of 12.4 the assembly A is just the map of the $\mathbb{F}$-homology Ω-spectra $K_+ \wedge \mathbb{F} \longrightarrow \{*\}_+ \wedge \mathbb{F}$ induced by the unique simplicial map $K \longrightarrow \{*\}$

$$A : \mathbb{H}.(K;\mathbb{F}) \simeq K_+ \wedge \mathbb{F} \longrightarrow \{*\}_+ \wedge \mathbb{F} = \mathbb{F} .$$

An element $x \in H_n(K;\mathbb{F})$ is represented by an $\mathbb{F}$-cycle $(J \subseteq K, x)$ with

$$x = \{x(\sigma) \in \mathbb{F}_{n-m}^{(m-|\sigma|)} \,|\, \sigma \in J\} .$$

Visualize $A \colon H_n(K;\mathbb{F}) \longrightarrow \pi_n(\mathbb{F})$ as assembling the components $x(\sigma)$ to an element

$$A(x) = \bigcup_{\sigma \in J} x(\sigma) \in \mathbb{F}_n^{(0)}$$

representing

$$A(x) \in H_n(\{*\};\mathbb{F}) = \pi_0(\mathbb{F}_n) = \pi_n(\mathbb{F}) .$$

For a subcomplex $J \subseteq \partial\Delta^{m+1}$ and $\sigma \in J$ let $\overline{J}(\sigma) \subseteq \Sigma^m$ be the subcomplex consisting of the dual simplexes $\tau^* \in \Sigma^m$ of the simplexes $\tau \in \partial\Delta^{m+1}$ such that either $\tau \notin J$ or $\sigma \not\leq \tau \in J$, that is

$$\overline{J}(\sigma) = \overline{J \backslash \mathrm{st}_J(\sigma)} \subseteq \Sigma^m$$

with $\mathrm{st}_J(\sigma) = \{\rho \in J \,|\, \sigma \leq \rho\}$. If $\sigma \leq \rho \in J$ then $\overline{J}(\rho) \subseteq \overline{J}(\sigma)$, and

$$\bigcup_{\sigma \in J} \overline{J}(\sigma) = \Sigma^m .$$

The relative simplicial pair $(\overline{J}(\sigma),\overline{J})$ has one $(m-|\tau|)$-simplex τ^* for each $\tau \in \mathrm{st}_J(\sigma)$, with

$$\partial_i(\tau^*) = (\delta_i\tau)^* \in \overline{J}(\sigma) \quad (0 \leq i \leq m - |\tau|) .$$

DEFINITION 12.15 Given a covariant functor

$$\mathbb{F} : \{\text{simplicial complexes}\} \longrightarrow \{\Omega\text{-spectra}\} ; \; K \longrightarrow \mathbb{F}(K)$$

define the *local $\{\mathbb{F}\}$-coefficient homology Ω-spectrum* of a subcomplex $K \subseteq \partial\Delta^{m+1}$

$$\mathbb{H}.(K;\{\mathbb{F}\}) = \{\mathbb{H}_n(K;\{\mathbb{F}\}) \,|\, n \in \mathbb{Z}\}$$

by

$$\mathbb{H}_n(K;\{\mathbb{F}\}) = \varinjlim_{J} \varprojlim_{\sigma \in J} (\mathbb{F}_{n-m}(D(\sigma,J)),\emptyset)^{(\overline{J}(\sigma),\overline{J})} .$$

The homotopy groups of $\mathbb{H}.(K;\{\mathbb{F}\})$ are the *local $\{\mathbb{F}\}$-coefficient homology groups of K*

$$H_n(K;\{\mathbb{F}\}) = \pi_n(\mathbb{H}.(K;\{\mathbb{F}\})) \ (n \in \mathbb{Z}) ,$$

which may also be written as $H_n(K;\{\mathbb{F}(D(\sigma,K))\})$.

$\square$

EXAMPLE 12.16 If $\mathbb{F}$ is constant, with $\mathbb{F}(K) = \mathbb{F}$ for all K, then $\mathbb{H}.(K;\{\mathbb{F}\})$ is the $\mathbb{F}$-homology spectrum $\mathbb{H}.(K;\mathbb{F})$ of 12.3, with

$$\mathbb{H}_n(K;\{\mathbb{F}\}) = \varprojlim_{\sigma \in K} (\mathbb{F}_{n-m}(D(\sigma,K)),\emptyset)^{(\overline{K}(\sigma),\overline{K})}$$

$$= (\mathbb{F}_{n-m},\emptyset)^{(\Sigma^m,\overline{K})} = \mathbb{H}_n(K;\mathbb{F}) \ (n \in \mathbb{Z}) .$$

$\square$

DEFINITION 12.17 An *n-dimensional $\{\mathbb{F}\}$-cycle* in a simplicial complex K is an element of $\mathbb{H}_n(K;\{\mathbb{F}\})^{(0)}$, that is a collection

$$x = \{x(\sigma) \in \mathbb{F}_{n-m}(D(\sigma,J))^{(m-|\sigma|)} \,|\, \sigma \in J\}$$

with $J \subseteq K$ a finite subcomplex and $J \subseteq \partial\Delta^{m+1}$ the canonical embedding, such that

$$\partial_i x(\sigma) = \begin{cases} f_i x(\delta_i\sigma) & \text{if } \delta_i\sigma \in J \\ \emptyset & \text{if } \delta_i\sigma \notin J \end{cases} \ (0 \leq i \leq m - |\sigma|) ,$$

with $f_i \colon \mathbb{F}(D(\delta_i\sigma,J)) \longrightarrow \mathbb{F}(D(\sigma,J))$ the map induced by the inclusion $D(\delta_i\sigma,J) \subset D(\sigma,J)$.

$\square$

As in the constant coefficient case (12.6, 12.8) there is a corresponding notion of cobordism, such that $H_n(K;\{\mathbb{F}\})$ is the cobordism group of n-dimensional $\{\mathbb{F}\}$-cycles in K.

DEFINITION 12.18 The *local $\{\mathbb{F}\}$-coefficient assembly* is the map of Ω-spectra

$$A \colon \mathbb{H}.(K;\{\mathbb{F}\}) \longrightarrow \mathbb{F}(K')$$

given by the composite

$$A \colon \mathbb{H}.(K;\{\mathbb{F}\}) \longrightarrow \mathbb{H}.(K;\mathbb{F}(K')) \xrightarrow{A} \mathbb{F}(K')$$

of the forgetful map $\mathbb{H}.(K;\{\mathbb{F}\})\longrightarrow\mathbb{H}.(K;\mathbb{F}(K'))$ induced by all the inclusions $D(\sigma,K)\subseteq K'$ $(\sigma\in K)$ and the assembly $A\colon\mathbb{H}.(K;\mathbb{F}(K'))\longrightarrow\mathbb{F}(K')$ of 12.14.

□

A functor

$$F\ :\ \{\text{simplicial complexes}\}\ \longrightarrow\ \{\Omega\text{-spectra}\}\ ;\ K\ \longrightarrow\ \mathbb{F}(K)$$

is *homotopy invariant* if a homotopy equivalence $f\colon K\xrightarrow{\simeq} L$ induces a homotopy equivalence of Ω-spectra

$$f\ :\ \mathbb{F}(K)\ \xrightarrow{\simeq}\ \mathbb{F}(L)\ .$$

For such $\mathbb{F}$ the forgetful map from local $\mathbb{F}$-coefficient homology to constant $\mathbb{F}(\{*\})$-coefficient homology is a homotopy equivalence

$$\mathbb{H}.(K;\{\mathbb{F}\})\ \xrightarrow{\simeq}\ \mathbb{H}.(K;\mathbb{F}(\{*\}))\ ,$$

since each of the unique simplicial maps $D(\sigma,K)\longrightarrow\{*\}$ $(\sigma\in K)$ is a homotopy equivalence.

DEFINITION 12.19 The *constant $\mathbb{F}(\{*\})$-coefficient assembly* for a homotopy invariant functor $\mathbb{F}$ and a subcomplex $K\subseteq\partial\Delta^{m+1}$ is the map of Ω-spectra

$$A\ :\ \mathbb{H}.(K;\mathbb{F}(\{*\}))\ \longrightarrow\ \mathbb{F}(K)$$

given by the local $\{\mathbb{F}\}$-coefficient assembly A of 12.18, using the homotopy equivalences

$$
\begin{array}{ccc}
\mathbb{H}.(K;\{\mathbb{F}\}) & \xrightarrow{\ A\ } & \mathbb{F}(K') \\
\simeq\downarrow & & \downarrow\simeq \\
\mathbb{H}.(K;\mathbb{F}(\{*\})) & \xrightarrow{\ A\ } & \mathbb{F}(K)\ .
\end{array}
$$

□

REMARK 12.20 The assembly map $A\colon\mathbb{H}.(K;\mathbb{F}(\{*\}))\longrightarrow\mathbb{F}(K)$ of 12.19 is a combinatorial version of the assembly map of Anderson [4] and Quinn [134], which is defined as follows: a functor

$$F\ :\ \{\text{ pointed topological spaces}\}\ \longrightarrow\ \{\text{ spectra}\}$$

induces a natural transformation of function spectra

$$X\ =\ X^{\{*\}}\ \longrightarrow\ F(X)^{F(\{*\})}\ ,$$

with adjoint the assembly map

$$A\ :\ \mathbb{H}.(X;F(\{*\}))\ =\ X\wedge F(\{*\})\ \longrightarrow\ F(X)\ .$$

□

EXAMPLE 12.21 Let $\Omega^{SO}_{\cdot}(K) = \{\Omega^{SO}_{\cdot}(K)_n \,|\, n \in \mathbb{Z}\}$ be the Ω-spectrum with $\Omega^{SO}_{\cdot}(K)_n$ the Kan Δ-set defined by

$$\Omega^{SO}_{\cdot}(K)^{(k)}_n = \{ (n+k)\text{-dimensional smooth oriented manifold } k\text{-ads}$$

$$(M; \partial_0 M, \partial_1 M, \ldots, \partial_k M) \text{ such that}$$

$$\partial_0 M \cap \partial_1 M \cap \ldots \cap \partial_k M = \emptyset \, , \text{ with a map } M \longrightarrow |K|\}$$

with base simplex the empty manifold k-ad $\emptyset$. The homotopy groups

$$\pi_n(\Omega^{SO}_{\cdot}(K)) = \Omega^{SO}_n(K) \ (n \geq 0)$$

are the bordism groups of maps $M \longrightarrow |K|$ from closed smooth oriented n-dimensional manifolds. The functor

$$\Omega^{SO}_{\cdot} : \{\text{simplicial complexes}\} \longrightarrow \{\Omega\text{-spectra}\} \; ; \; K \longrightarrow \Omega^{SO}_{\cdot}(K)$$

is homotopy invariant, since for any k-simplex M in $\Omega^{SO}_{\cdot}(K)_n$ there is defined a $(k+1)$-simplex $M \otimes I$ in $\Omega^{SO}_{\cdot}(K \otimes \Delta^1)_n$, so that the two inclusions $K \longrightarrow K \otimes \Delta^1$ induce homotopic Δ-maps $\Omega^{SO}_{\cdot}(K) \longrightarrow \Omega^{SO}_{\cdot}(K \otimes \Delta^1)$. The assembly map defines a homotopy equivalence

$$A : \mathbb{H}_{\cdot}(K; \Omega^{SO}_{\cdot}(\{*\})) \overset{\simeq}{\longrightarrow} \Omega^{SO}_{\cdot}(K) \, ,$$

a combinatorial version of the Pontrjagin–Thom isomorphism and the Atiyah formulation of bordism as generalized homology. The assembly of an n-dimensional $\Omega^{SO}_{\cdot}(\{*\})$-coefficient cycle in a subcomplex $K \subseteq \partial \Delta^{m+1}$

$$x = \{M(\sigma)^{n-|\sigma|} \,|\, \sigma \in K\}$$

is a map

$$A(x) : M^n = \bigcup_{\sigma \in K} M(\sigma) \longrightarrow |K| = |K'|$$

from a closed smooth oriented n-manifold such that

$$A(x)^{-1}D(\sigma, K) = M(\sigma) \ (\sigma \in K) \, .$$

The smooth oriented bordism Ω-spectrum $\Omega^{SO}_{\cdot}(K)$ is just a combinatorial version of the Thom suspension spectrum $|K|_+ \wedge \underline{MSO}$, with

$$\underline{MSO} = \{ MSO(j), \Sigma MSO(j) \longrightarrow MSO(j+1) \,|\, j \geq 0 \} \, ,$$

$$\Omega^{SO}_{\cdot}(K)_n \simeq \mathbb{H}_n(K; \Omega^{SO}_{\cdot}(\{*\})) \simeq \varinjlim_{j} \Omega^{j+n}(|K|_+ \wedge MSO(j)) \ (n \in \mathbb{Z}) \, .$$

$\square$

§13. Algebraic $\mathbb{L}$-spectra

The algebraic $\mathbb{L}$-spectra consist of Kan Δ-sets with homotopy groups the algebraic L-groups. Given an algebraic bordism category $\Lambda = (\mathbb{A}, \mathbb{B}, \mathbb{C})$ there will now be defined an Ω-spectrum

$$\begin{cases} \mathbb{L}^{\cdot}(\Lambda) = \{\, \mathbb{L}^n(\Lambda) \,|\, n \in \mathbb{Z} \,\} \\ \mathbb{L}_{\cdot}(\Lambda) = \{\, \mathbb{L}_n(\Lambda) \,|\, n \in \mathbb{Z} \,\} \\ \mathbb{NL}^{\cdot}(\Lambda) = \{\, \mathbb{NL}^n(\Lambda) \,|\, n \in \mathbb{Z} \,\} \end{cases}$$

of Kan Δ-sets with homotopy groups the $\begin{cases} \text{symmetric} \\ \text{quadratic} \\ \text{normal} \end{cases}$ L-groups of Λ

$$\begin{cases} \pi_n(\mathbb{L}^{\cdot}(\Lambda)) = L^n(\Lambda) \\ \pi_n(\mathbb{L}_{\cdot}(\Lambda)) = L_n(\Lambda) \qquad (n \in \mathbb{Z}) \,. \\ \pi_n(\mathbb{NL}^{\cdot}(\Lambda)) = NL^n(\Lambda) \end{cases}$$

The $\begin{cases} \mathbb{L}^{\cdot}(\Lambda) \\ \mathbb{L}_{\cdot}(\Lambda) \\ \mathbb{NL}^{\cdot}(\Lambda) \end{cases}$ -cohomology (resp. homology) groups of a simplicial complex

K will be identified with the $\begin{cases} \text{symmetric} \\ \text{quadratic} \\ \text{normal} \end{cases}$ L-groups

$$\begin{cases} H^n(K; \mathbb{L}^{\cdot}(\Lambda)) = L^{-n}(\Lambda^*(K)) \\ H^n(K; \mathbb{L}_{\cdot}(\Lambda)) = L_{-n}(\Lambda^*(K)) \\ H^n(K; \mathbb{NL}^{\cdot}(\Lambda)) = NL^{-n}(\Lambda^*(K)) \end{cases}$$

$$\left(\text{resp.} \begin{cases} H_n(K; \mathbb{L}^{\cdot}(\Lambda)) = L^n(\Lambda_*(K)) \\ H_n(K; \mathbb{L}_{\cdot}(\Lambda)) = L_n(\Lambda_*(K)) \\ H_n(K; \mathbb{NL}^{\cdot}(\Lambda)) = NL^n(\Lambda_*(K)) \end{cases} \right)$$

of the algebraic bordism category $\Lambda^*(K)$ (resp. $\Lambda_*(K)$) of §5. The various algebraic $\mathbb{L}$-spectra are used in Part II to express the geometric properties of bundles and manifolds in terms of L-theory.

The algebraic surgery classifying spaces and spectra are analogues of the geometric surgery classifying spaces and spectra, which arose as follows:

Remark 13.1 (i) The classifying space G/O for fibre homotopy trivialized vector bundles and its PL analogue G/PL first appeared in the surgery classification theory of exotic spheres (Kervaire and Milnor [83], Levine [88, Appendix]). The fibration sequence

$$PL/O \longrightarrow G/O \longrightarrow G/PL$$

induces an exact sequence

$$\ldots \longrightarrow \pi_{n+1}(G/PL) \longrightarrow \pi_n(PL/O) \longrightarrow \pi_n(G/O) \longrightarrow \pi_n(G/PL) \longrightarrow \ldots$$

which for $n \geq 5$ is isomorphic to the differentiable surgery exact sequence

$$\cdots \longrightarrow L_{n+1}(\mathbb{Z}) \longrightarrow \mathbb{S}^O(S^n) \longrightarrow \pi_n(G/O) \longrightarrow L_n(\mathbb{Z})$$

with $\pi_n(PL/O) = \mathbb{S}^O(S^n) = \theta_n$ the groups of h-cobordism classes of n-dimensional exotic spheres, and $\pi_n(G/PL) = L_n(\mathbb{Z})$ the simply-connected surgery obstruction groups. An exotic sphere Σ^n is sent by $\pi_n(PL/O) \longrightarrow \pi_n(G/O)$ to the classifying map $S^n \simeq \Sigma^n \longrightarrow G/O$ for the fibre homotopy trivialization of its stable normal bundle determined by the trivial Spivak normal fibration. This is also the classifying map of the normal map $(f, b) : \Sigma^n \longrightarrow S^n$ with $f : \Sigma^n \longrightarrow S^n$ a homotopy equivalence representing the element $[f] \in \mathbb{S}^O(S^n)$ of the differentiable structure set, corresponding to $[\Sigma^n] \in \theta_n$.

(ii) The topological surgery classifying space G/TOP first appeared in the work of Casson and Sullivan in which block bundles were used to obtain the obstruction to deforming a homeomorphism $f : M \longrightarrow N$ of compact n-dimensional PL manifolds ($n \geq 5$) to a PL homeomorphism

$$\kappa(f) = \kappa(\nu_M - f^*\nu_N) \in H^3(M; \mathbb{Z}_2) \,,$$

disproving the manifold Hauptvermutung that every homeomorphism of PL manifolds is homotopic to a PL homeomorphism – see Wall [176, §17A], Armstrong, Cooke and Rourke [5]. The classifying spaces BPL, $BTOP$, BG for PL bundles, topological bundles and spherical fibrations are related by a commutative braid of fibrations

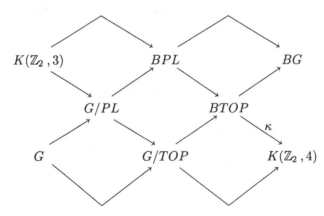

with $\kappa \in [BTOP, K(\mathbb{Z}_2, 4)] = H^4(BTOP; \mathbb{Z}_2)$ the Kirby–Siebenmann invariant.

(iii) Quinn [127] defined the geometric surgery spectrum $\mathbb{L}.(K)$ of a space K, with homotopy groups

$$\pi_*(\mathbb{L}.(K)) = L_*(\mathbb{Z}[\pi_1(K)]) \,.$$

The algebraic surgery spectrum $\mathbb{L}.(R)$ of a ring with involution R with
$$\pi_*(\mathbb{L}.(R)) = L_*(R)$$
was first constructed using forms and formations (Ranicki [135], [136]), with
$$\mathbb{L}.(K) \simeq \mathbb{L}.(\mathbb{Z}[\pi_1(K)]) .$$
The simply-connected surgery spectrum $\mathbb{L}.(\{*\}) \simeq \mathbb{L}.(\mathbb{Z})$ is the 4-periodic
delooping of G/TOP given by the characteristic variety theorem of Sullivan
[164], with
$$\mathbb{L}_0(\{*\}) = \mathbb{L}_0(\mathbb{Z}) \simeq L_0(\mathbb{Z}) \times G/TOP .$$
See Ranicki [140], [143], Levitt and Ranicki [91], Weiss and Williams [184]
for other accounts of the quadratic $\mathbb{L}$-spectra. Also, see Siegel [159], Goresky
and Siegel [61], Pardon [122], Cappell and Shaneson [27] and Weinberger
[181] for some of the connections between L-theory, the characteristic variety
theorem, intersection homology theory and stratified spaces.

$\square$

As before, let $\Lambda = (\mathbb{A}, \mathbb{B}, \mathbb{C})$ be an algebraic bordism category.

DEFINITION 13.2 Let $\begin{cases} \mathbb{L}^n(\Lambda) \\ \mathbb{L}_n(\Lambda) \\ \mathbb{NL}^n(\Lambda) \end{cases}$ $(n \in \mathbb{Z})$ be the pointed Δ-set with m-

simplexes the n-dimensional $\begin{cases} \text{symmetric} \\ \text{quadratic} \\ \text{normal} \end{cases}$ complexes in $\Lambda^*(\Delta^m)$, with the

zero complex as base m-simplex $\emptyset$. The face maps are induced from the
standard embeddings $\partial_i \colon \Delta^{m-1} \longrightarrow \Delta^m$ via the functors
$$(\partial_i)^* : \Lambda^*(\Delta^m) \longrightarrow \Lambda^*(\Delta^{m-1}) .$$

$\square$

DEFINITION 13.3 Given a pair of locally finite simplicial complexes $(K, J \subseteq K)$ let
$$\Lambda^*(K, J) = (\mathbb{A}^*(K, J), \mathbb{B}^*(K, J), \mathbb{C}^*(K, J))$$
be the algebraic bordism category defined by the full subcategory of $\Lambda^*(K)$
(5.1) with objects C such that $C(\sigma) = 0$ for $\sigma \in J$.

$\square$

PROPOSITION 13.4 $\begin{cases} \mathbb{L}^n(\Lambda) \\ \mathbb{L}_n(\Lambda) \\ \mathbb{NL}^n(\Lambda) \end{cases}$ is a Kan Δ-set with homotopy groups and

loop Δ-set
$$\begin{cases} \pi_m(\mathbb{L}^n(\Lambda)) = L^{m+n}(\Lambda) \\ \pi_m(\mathbb{L}_n(\Lambda)) = L_{m+n}(\Lambda) \\ \pi_m(\mathbb{NL}^n(\Lambda)) = NL^{m+n}(\Lambda) , \end{cases} \qquad \begin{cases} \Omega\mathbb{L}^n(\Lambda) = \mathbb{L}^{n+1}(\Lambda) \\ \Omega\mathbb{L}_n(\Lambda) = \mathbb{L}_{n+1}(\Lambda) \\ \Omega\mathbb{NL}^n(\Lambda) = \mathbb{NL}^{n+1}(\Lambda) \end{cases}$$

for $n \in \mathbb{Z}$, $m + n \geq 0$.

PROOF Only the quadratic case is considered, the symmetric and normal cases being entirely similar.

The Kan extension condition is verified using the algebraic analogues of glueing and crossing with the unit interval $I = [0, 1]$. See Ranicki [143, §1.7] for the glueing of quadratic complexes. Crossing with I corresponds to the following chain complex construction. A *pair* $(C, \partial C)$ of chain complexes in the additive category $\mathbb{A}$ is a chain complex C in $\mathbb{A}$ which is expressed as

$$d_C = \begin{pmatrix} d_{\partial C} & e_C \\ 0 & d_{\dot{C}} \end{pmatrix} : C_r = \partial C_r \oplus \dot{C}_r \longrightarrow C_{r-1} = \partial C_{r-1} \oplus \dot{C}_{r-1} ,$$

so that ∂C is a subcomplex of C and $\dot{C} = C/\partial C$ is a quotient complex. Define

$$(D, \partial D) = (C, \partial C) \otimes (I, \partial I)$$

to be the pair with

$$d_D = \begin{pmatrix} d_{\partial C} & e_C & e_C & 0 \\ 0 & d_{\dot{C}} & 0 & (-)^{r-1} \\ 0 & 0 & d_{\dot{C}} & (-)^r \\ 0 & 0 & 0 & d_{\dot{C}} \end{pmatrix} :$$

$$D_r = \partial C_r \oplus \dot{C}_r \oplus \dot{C}_r \oplus \dot{C}_{r-1}$$

$$\longrightarrow D_{r-1} = \partial C_{r-1} \oplus \dot{C}_{r-1} \oplus \dot{C}_{r-1} \oplus \dot{C}_{r-2} ,$$

$$\partial D_r = \partial C_r \oplus \dot{C}_r \oplus \dot{C}_r , \quad \dot{D}_r = \dot{C}_{r-1} .$$

Let $C \otimes \{0\}, C \otimes \{1\}$ be the subcomplexes of D defined by

$$(C \otimes \{0\})_r = \{(x, y, 0, 0) \in D_r \mid (x, y) \in C_r = \partial C_r \oplus \dot{C}_r\} ,$$

$$(C \otimes \{1\})_r = \{(x, 0, y, 0) \in D_r \mid (x, y) \in C_r = \partial C_r \oplus \dot{C}_r\} .$$

The inclusions

$$i_k : C \otimes \{k\} \longrightarrow D \quad (k = 0, 1)$$

are chain equivalences, with chain homotopy inverses $j_k : D \longrightarrow C \otimes \{k\}$ defined by

$$j_k : D_r \longrightarrow (C \otimes \{k\})_r ; \ (x, y, z, w) \longrightarrow \begin{cases} (x, y + z, 0, 0) \\ (x, 0, y + z, 0) \end{cases} \text{ if } k = \begin{cases} 0 \\ 1 \end{cases} .$$

Let $\Lambda^m \subset \Delta^m$ be the subcomplex of Δ^m obtained by removing the interiors of Δ^m and of a face $\Delta^{m-1} < \Delta^m$. Define the extension of a chain complex C in $\mathbb{A}^*(\Lambda^m)$ to a chain complex $\overline{C}$ in $\mathbb{A}^*(\Delta^m)$ by

$$(\overline{C}^*(\Delta^m), \overline{C}^*(\partial \Delta^m)) = (C^*(\Lambda^m), C^*(\partial \Lambda^m)) \otimes (I, \partial I) ,$$

with

$$\overline{C}(\sigma)_r = \begin{cases} (C(\sigma) \otimes \{0\})_r & \text{if } \sigma \in \Lambda^m \\ (\dot{C}^*(\Lambda^m) \otimes \{1\})_r & \text{if } \sigma = \Delta^{m-1} \\ \dot{C}^*(\Lambda^m)_{r-1} & \text{if } \sigma = \Delta^m . \end{cases}$$

Use the identification of pairs of abelian group chain complexes

$$((W_{\%}\overline{C})^*[\Delta^m], (W_{\%}\overline{C})^*[\partial\Delta^m]) = ((W_{\%}C)^*[\Lambda^m], (W_{\%}C)^*[\partial\Lambda^m]) \otimes (I, \partial I)$$

to define the extension of an n-dimensional quadratic complex (C, ψ) in $\Lambda^*(\Lambda^m)$ to an n-dimensional quadratic complex $(\overline{C}, \overline{\psi})$ in $\Lambda^*(\Delta^m)$ by

$$\overline{\psi} = (j_0)_{\%}(\psi) \in (W_{\%}\overline{C})^*[\Delta^m]_n .$$

The homotopy group $\pi_m(\mathbb{L}_n(\Lambda))$ is the group of equivalence classes of m-simplexes (C, ψ) in $\mathbb{L}_n(\Lambda)$ such that

$$\partial_i(C, \psi) = 0 \quad (0 \le i \le m) .$$

Such simplexes are n-dimensional quadratic complexes (C, ψ) in $\Lambda^*(\Delta^m, \partial\Delta^m)$, which are just $(m+n)$-dimensional quadratic complexes in Λ. The homotopy of simplexes corresponds to the cobordism of complexes, so that

$$\pi_m(\mathbb{L}_n(\Lambda)) = L_{m+n}(\Lambda) \quad (m \ge 0, n \in \mathbb{Z}) .$$

Let $\langle i_0, i_1, \dots, i_r \rangle$ denote the r-simplex of Δ^m with vertices $i_0, i_1, \dots, i_r$ given by a sequence $0 \le i_0 < i_1 < \dots < i_r \le m$. The standard embedding $\partial_{m+1} : \Delta^m \subset \Delta^{m+1}$ identifies Δ^m with the face of Δ^{m+1} opposite the vertex $m+1$. By definition, an m-simplex of $\Omega\mathbb{L}_n(\Lambda)$ is an n-dimensional quadratic complex (C, ψ) in $\Lambda^*(\Delta^{m+1}, \Delta^m \cup \{m+1\})$, so that

$$C(\langle m+1 \rangle) = 0 ,$$

$$C(\langle i_0, i_1, \dots, i_r \rangle) = 0 \quad (0 \le i_0 < i_1 < \dots < i_r \le m) .$$

Except for terminology this is the same as an $(n+1)$-dimensional quadratic complex (C', ψ') in $\Lambda^*(\Delta^m)$ with

$$C'(\langle i_0, i_1, \dots, i_r \rangle) = C(\langle i_0, i_1, \dots, i_r, m+1 \rangle) \quad (0 \le i_0 < i_1 < \dots < i_r \le m) .$$

This is an m-simplex of $\mathbb{L}_{n+1}(\Lambda)$, so that there is an identity of Δ-sets

$$\Omega\mathbb{L}_n(\Lambda) = \mathbb{L}_{n+1}(\Lambda) .$$

□

DEFINITION 13.5 The $\begin{cases} symmetric \\ quadratic \\ normal \end{cases}$ L-*spectrum* of an algebraic bordism category Λ is the Ω-spectrum of Kan Δ-sets given by 13.4

$$\begin{cases} \mathbb{L}^{\cdot}(\Lambda) = \{ \mathbb{L}^n(\Lambda) \,|\, n \in \mathbb{Z} \} \\ \mathbb{L}_{\cdot}(\Lambda) = \{ \mathbb{L}_n(\Lambda) \,|\, n \in \mathbb{Z} \} \\ \mathbb{NL}^{\cdot}(\Lambda) = \{ \mathbb{NL}^n(\Lambda) \,|\, n \in \mathbb{Z} \} \end{cases}$$

with homotopy groups the L-groups of Λ

$$\begin{cases} \pi_n(\mathbb{L}^{\boldsymbol{\cdot}}(\Lambda)) = \pi_{n+k}(\mathbb{L}^{-k}(\Lambda)) = L^n(\Lambda) \\ \pi_n(\mathbb{L}_{\boldsymbol{.}}(\Lambda)) = \pi_{n+k}(\mathbb{L}_{-k}(\Lambda)) = L_n(\Lambda) \qquad (n, k \in \mathbb{Z}, n + k \geq 0) \, . \\ \pi_n(\mathbb{NL}^{\boldsymbol{\cdot}}(\Lambda)) = \pi_{n+k}(\mathbb{NL}^{-k}(\Lambda)) = NL^n(\Lambda) \end{cases}$$

□

EXAMPLE 13.6 For any additive category with chain duality $\mathbb{A}$ and the algebraic bordism category of 3.3

$$\Lambda(\mathbb{A}) = (\mathbb{A}, \mathbb{B}(\mathbb{A}), \mathbb{C}(\mathbb{A}))$$

the $\begin{cases} \text{symmetric} \\ \text{quadratic} \\ \text{normal} \end{cases}$ $\mathbb{L}$-spectrum of $\Lambda(\mathbb{A})$ has homotopy groups the $\begin{cases} \text{symmetric} \\ \text{quadratic} \\ \text{normal} \end{cases}$

L-groups of $\mathbb{A}$

$$\begin{cases} \pi_*(\mathbb{L}^{\boldsymbol{\cdot}}(\Lambda(\mathbb{A}))) = L^*(\mathbb{A}) \\ \pi_*(\mathbb{L}_{\boldsymbol{.}}(\Lambda(\mathbb{A}))) = L_*(\mathbb{A}) \\ \pi_*(\mathbb{NL}^{\boldsymbol{\cdot}}(\Lambda(\mathbb{A}))) = NL^*(\mathbb{A}) \, . \end{cases}$$

Also, by 3.6

$$\pi_*(\mathbb{L}^{\boldsymbol{\cdot}}(\Lambda(\mathbb{A}))) = L^*(\mathbb{A}) = NL^*(\mathbb{A}) = \pi_*(\mathbb{NL}^{\boldsymbol{\cdot}}(\Lambda(\mathbb{A}))) \, ,$$

so that the forgetful map defines a homotopy equivalence

$$\mathbb{NL}^{\boldsymbol{\cdot}}(\Lambda(\mathbb{A})) \xrightarrow{\simeq} \mathbb{L}^{\boldsymbol{\cdot}}(\Lambda(\mathbb{A})) \, .$$

□

PROPOSITION 13.7 *The* $\begin{cases} symmetric \\ quadratic \\ normal \end{cases}$ L-*spectrum of* $\Lambda^*(K)$ *(resp.* $\Lambda_*(K)$*)*

is the $\begin{cases} \mathbb{L}^{\boldsymbol{\cdot}}(\Lambda) \\ \mathbb{L}_{\boldsymbol{.}}(\Lambda) \\ \mathbb{NL}^{\boldsymbol{\cdot}}(\Lambda) \end{cases}$ *-cohomology (resp. homology) spectrum of the locally finite simplicial complex* K

$$\begin{cases} \mathbb{L}^{\boldsymbol{\cdot}}(\Lambda^*(K)) = \mathbb{H}^{\boldsymbol{\cdot}}(K; \mathbb{L}^{\boldsymbol{\cdot}}(\Lambda)) \\ \mathbb{L}_{\boldsymbol{.}}(\Lambda^*(K)) = \mathbb{H}^{\boldsymbol{\cdot}}(K; \mathbb{L}_{\boldsymbol{.}}(\Lambda)) \\ \mathbb{NL}^{\boldsymbol{\cdot}}(\Lambda^*(K)) = \mathbb{H}^{\boldsymbol{\cdot}}(K; \mathbb{NL}^{\boldsymbol{\cdot}}(\Lambda)) \, , \end{cases}$$

$$\left(resp. \begin{cases} \mathbb{L}^{\boldsymbol{\cdot}}(\Lambda_*(K)) = \mathbb{H}_{\boldsymbol{.}}(K; \mathbb{L}^{\boldsymbol{\cdot}}(\Lambda)) \\ \mathbb{L}_{\boldsymbol{.}}(\Lambda_*(K)) = \mathbb{H}_{\boldsymbol{.}}(K; \mathbb{L}_{\boldsymbol{.}}(\Lambda)) \\ \mathbb{NL}^{\boldsymbol{\cdot}}(\Lambda_*(K)) = \mathbb{H}_{\boldsymbol{.}}(K; \mathbb{NL}^{\boldsymbol{\cdot}}(\Lambda)) \end{cases} \right)$$

so that on the level of homotopy groups

$$\begin{cases} L^n(\Lambda^*(K)) = H^{-n}(K; \mathbb{L}^{\cdot}(\Lambda)) \\ L_n(\Lambda^*(K)) = H^{-n}(K; \mathbb{L}.(\Lambda)) \\ NL^n(\Lambda^*(K)) = H^{-n}(K; \mathbb{NL}^{\cdot}(\Lambda)), \end{cases}$$

$$\left(resp. \begin{cases} L^n(\Lambda_*(K)) = H_n(K; \mathbb{L}^{\cdot}(\Lambda)) \\ L_n(\Lambda_*(K)) = H_n(K; \mathbb{L}.(\Lambda)) \\ NL^n(\Lambda_*(K)) = H_n(K; \mathbb{NL}^{\cdot}(\Lambda)). \end{cases} \right)$$

PROOF As in 13.4 consider only the quadratic case, the symmetric and normal cases being entirely similar. An n-dimensional quadratic complex in $\Lambda^*(K)$ is a collection of n-dimensional quadratic complexes in $\Lambda^*(\Delta^m)$, one for each m-simplex of K, with the common faces in K corresponding to common faces of the quadratic complexes. Thus the Δ-maps $K_+ \longrightarrow \mathbb{L}_n(\Lambda)$ are just the n-dimensional quadratic complexes in $\Lambda^*(K)$. For each $p \geq 0$ identify

$$\mathbb{L}_n(\Lambda^*(K))^{(p)} = \{\, n\text{-dimensional quadratic complexes in } \Lambda^*(K)^*(\Delta^p) \,\}$$

$$= \{\, n\text{-dimensional quadratic complexes in } \Lambda^*(K \otimes \Delta^p) \,\}$$

$$= \{\, \Delta\text{-maps } (K \otimes \Delta^p)_+ \longrightarrow \mathbb{L}_n(\Lambda) \,\} = \mathbb{H}^n(K; \mathbb{L}.(\Lambda))^{(p)}$$

and so

$$\mathbb{L}_n(\Lambda^*(K)) = \mathbb{L}_n(\Lambda)^{K_+} = \mathbb{H}^n(K; \mathbb{L}.(\Lambda)).$$

As in 12.4 there is no loss of generality in taking K to be finite, so that there is an embedding $K \subset \partial \Delta^{m+1}$ for some $m \geq 0$, and the supplement $\overline{K} \subseteq \Sigma^m$ is defined (12.2). There is a natural one–one correspondence between chain complexes C in $\mathbb{A}_*(K)$ and chain complexes D in $\mathbb{A}^*(\Sigma^m, \overline{K})$, with

$$C(\sigma) = D(\sigma^*) \quad (\sigma \in K), \quad [C]_*[K] = S^m[D]^*[\Sigma^m, \overline{K}].$$

For each $p \geq 0$ identify

$$\mathbb{L}_n(\Lambda_*(K))^{(p)} = \{\, n\text{-dimensional quadratic complexes in } \Lambda_*(K)^*(\Delta^p) \,\}$$

$$= \{\, (n-m)\text{-dimensional quadratic complexes in}$$

$$\Lambda^*(\Sigma^m \otimes \Delta^p, \overline{K} \otimes \Delta^p) \,\},$$

$$= \mathbb{H}^{n-m}(\Sigma^m, \overline{K}; \mathbb{L}.(\Lambda))^{(p)} = \mathbb{L}_{n-m}(\Lambda^*(\Sigma^m, \overline{K}))^{(p)}$$

$$= \{\, \Delta\text{-maps } (\Sigma^m, \overline{K}) \otimes \Delta^p \longrightarrow \mathbb{L}_{n-m}(\Lambda) \,\}$$

$$= \mathbb{H}_n(K; \mathbb{L}.(\Lambda))^{(p)},$$

and so

$$\mathbb{L}_n(\Lambda_*(K)) = \mathbb{H}_n(K; \mathbb{L}.(\Lambda)).$$

$\square$

REMARK 13.8 The identification $H^0(K; \mathbb{L}^{\cdot}(\mathbb{A})) = L^0(\mathbb{A}^*(K))$ is an analogue of the identification due to Gelfand and Mishchenko [57] (cf. Mishchenko [113, 4.2])

$$K(X) = L^0(\mathcal{C}(X, \mathbb{C}))$$

of the topological K-group of complex vector bundles over a topological space X with the symmetric Witt group of the ring $\mathcal{C}(X, \mathbb{C})$ of continuous functions $X \longrightarrow \mathbb{C}$ with respect to the involution determined by complex conjugation $z \longrightarrow \bar{z}$. See Milnor and Husemoller [110, p. 106] for the corresponding identification of the real K-group

$$KO(M) = L^0(\mathcal{C}^\infty(M, \mathbb{R}))$$

with M a differentiable manifold and $\mathcal{C}^\infty(M, \mathbb{R})$ the ring of differentiable functions $M \longrightarrow \mathbb{R}$ with the identity involution.

□

PROPOSITION 13.9 *Given an algebraic bordism category* $\Lambda = (\mathbb{A}, \mathbb{B}, \mathbb{C})$ *let* $\widehat{\Lambda} = (\mathbb{A}, \mathbb{B}, \mathbb{B})$.
(i) *The exact sequence of 3.10*

$$\ldots \longrightarrow L_n(\Lambda) \xrightarrow{1+T} NL^n(\Lambda) \xrightarrow{J} NL^n(\widehat{\Lambda}) \xrightarrow{\partial} L_{n-1}(\Lambda) \longrightarrow \ldots$$

is the exact sequence of homotopy groups of a fibration sequence of Ω-*spectra*

$$\mathbb{L}_{\cdot}(\Lambda) \xrightarrow{1+T} \mathbb{NL}^{\cdot}(\Lambda) \xrightarrow{J} \mathbb{NL}^{\cdot}(\widehat{\Lambda}) \ .$$

(ii) *If* $\widehat{Q}^*(C) = 0$ *for* $\mathbb{C}$-*contractible* C *then the forgetful map defines a homotopy equivalence of* $\mathbb{L}$-*spectra*

$$\mathbb{NL}^{\cdot}(\Lambda) \xrightarrow{\simeq} \mathbb{L}^{\cdot}(\Lambda) \ .$$

PROOF (i) The one–one correspondence between the $\mathbb{C}^*(\Delta^m)$-equivalence classes of (normal, quadratic) pairs in $\Lambda^*(\Delta^m)$ and the $\mathbb{B}^*(\Delta^m)$-equivalence classes of normal complexes in $\widehat{\Lambda}^*(\Delta^m)$ given for any $n \geq 0$ by 2.8 (i) defines a homotopy equivalence of Ω-spectra

$$(\text{mapping cofibre of } 1 + T \colon \mathbb{L}_{\cdot}(\Lambda) \longrightarrow \mathbb{NL}^{\cdot}(\Lambda)) \xrightarrow{\simeq} \mathbb{NL}^{\cdot}(\widehat{\Lambda}) \ .$$

(ii) Immediate from 3.5.

□

PROPOSITION 13.10 *The relative symmetric L-theory exact sequence of 3.8 for a functor* $F \colon \Lambda \longrightarrow \Lambda'$ *of algebraic bordism categories*

$$\ldots \longrightarrow L^{n+1}(F) \longrightarrow L^n(\Lambda) \xrightarrow{F} L^n(\Lambda') \longrightarrow L^n(F) \longrightarrow L^{n-1}(\Lambda) \longrightarrow \ldots$$

is the exact sequence of homotopy groups of a fibration sequence of Ω-*spectra*

$$\mathbb{L}^{\cdot}(\Lambda) \xrightarrow{F} \mathbb{L}^{\cdot}(\Lambda') \longrightarrow \mathbb{L}^{\cdot}(F)$$

and similarly for quadratic and normal L-theory.

PROOF Let $\mathbb{L}^{\cdot}(F) = \{\mathbb{L}^n(F) \,|\, n \in \mathbb{Z}\}$ be the Ω-spectrum of Kan Δ-sets with homotopy groups $\pi_*(\mathbb{L}^{\cdot}(F)) = L^*(F)$ defined by

$$\mathbb{L}^n(F) = \text{mapping cofibre of } F \colon \mathbb{L}^n(\Lambda) \longrightarrow \mathbb{L}^n(\Lambda') \ .$$

$\square$

PROPOSITION 13.11 *Let* $\mathbb{A}$ *be an additive category with chain duality, and let* $(\mathbb{B} \subseteq \mathbb{B}(\mathbb{A}),\ \mathbb{C} \subseteq \mathbb{B},\ \mathbb{D} \subseteq \mathbb{C})$ *be a triple of closed subcategories of* $\mathbb{B}(\mathbb{A})$.
(i) *The exact sequence of 3.9* (i)

$$\cdots \longrightarrow L^n(\mathbb{A}, \mathbb{C}, \mathbb{D}) \longrightarrow L^n(\mathbb{A}, \mathbb{B}, \mathbb{D}) \longrightarrow L^n(\mathbb{A}, \mathbb{B}, \mathbb{C})$$
$$\overset{\partial}{\longrightarrow} L^{n-1}(\mathbb{A}, \mathbb{C}, \mathbb{D}) \longrightarrow \cdots$$

is the exact sequence of the homotopy groups of symmetric $\mathbb{L}$-*spectra in a fibration sequence*

$$\mathbb{L}^{\cdot}(\mathbb{A}, \mathbb{C}, \mathbb{D}) \longrightarrow \mathbb{L}^{\cdot}(\mathbb{A}, \mathbb{B}, \mathbb{D}) \longrightarrow \mathbb{L}^{\cdot}(\mathbb{A}, \mathbb{B}, \mathbb{C}) \ .$$

Similarly in the quadratic and normal cases.
(ii) *The braid of exact sequences of algebraic L-groups of 3.13*

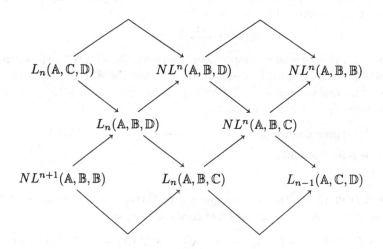

consists of the exact sequences of homotopy groups of algebraic $\mathbb{L}$-*spectra in a braid of fibration sequences*

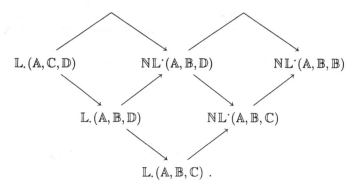

PROOF (i) Inclusion defines a functor $F: (\mathbb{A}, \mathbb{B}, \mathbb{D}) \longrightarrow (\mathbb{A}, \mathbb{B}, \mathbb{C})$, so that $\mathbb{L}.(F)$ is defined as in 13.10. The inverse isomorphisms of quadratic L-groups defined in 3.9 (ii)

$$L_{*-1}(\mathbb{A}, \mathbb{C}, \mathbb{D}) \; \underset{\longleftarrow}{\overset{\longrightarrow}{\quad}} \; L_*(F)$$

are induced by inverse homotopy equivalences of quadratic $\mathbb{L}$-spectra

$$\Sigma \mathbb{L}.(\mathbb{A}, \mathbb{C}, \mathbb{D}) \; \underset{\longleftarrow}{\overset{\longrightarrow}{\quad}} \; \mathbb{L}.(F)$$

defined by

$\Sigma \mathbb{L}_n(\mathbb{A}, \mathbb{C}, \mathbb{D})^{(m)}$

$$= \mathbb{L}_{n-1}(\mathbb{A}, \mathbb{C}, \mathbb{D})^{(m)} = \mathbb{L}_{n-1}(\mathbb{A}^*(\Delta^m), \mathbb{C}^*(\Delta^m), \mathbb{D}^*(\Delta^m))^{(0)}$$

$$\longrightarrow \mathbb{L}_n(F)^{(m)} = \mathbb{L}_n(F^*(\Delta^m))^{(0)} ;$$

$(C, \psi) \longrightarrow$ algebraic mapping cylinder of $(C \longrightarrow 0, (0, \psi))$,

$\mathbb{L}_n(F)^{(m)} = \mathbb{L}_n(F^*(\Delta^m))^{(0)}$

$$\longrightarrow \Sigma \mathbb{L}_n(\mathbb{A}, \mathbb{C}, \mathbb{D})^{(m)} = \mathbb{L}_{n-1}(\mathbb{A}^*(\Delta^m), \mathbb{C}^*(\Delta^m), \mathbb{D}^*(\Delta^m))^{(0)} ;$$

$(f: C \longrightarrow D, (\delta\psi, \psi)) \longrightarrow (C', \psi')$,

with (C', ψ') the quadratic complex obtained from (C, ψ) by algebraic surgery on the quadratic pair $(f: C \longrightarrow D, (\delta\psi, \psi))$, and $F^*(\Delta^m)$ the functor of algebraic bordism categories

$F^*(\Delta^m) : (\mathbb{A}^*(\Delta^m), \mathbb{B}^*(\Delta^m), \mathbb{D}^*(\Delta^m)) \longrightarrow (\mathbb{A}^*(\Delta^m), \mathbb{B}^*(\Delta^m), \mathbb{C}^*(\Delta^m))$.

(ii) The fibration sequences through $\mathbb{NL}^{\cdot}(\mathbb{A}, \mathbb{B}, \mathbb{B})$ are given by 13.9, and those through $\mathbb{L}.(\mathbb{A}, \mathbb{C}, \mathbb{D})$ by (i).

□

§14. The algebraic surgery exact sequence

Given a commutative ring R and a simplicial complex K the visible symmetric L-groups $VL^*(R,K)$, the generalized homology groups of K with coefficients in the various L-theories of R and the quadratic L-groups $L_*(R[\pi_1(K)])$ are related by a commutative braid of exact sequences

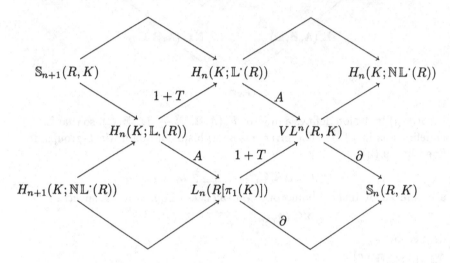

with the 'quadratic structure group'

$$\mathbb{S}_n(R,K) \;=\; L_{n-1}(\mathbb{A}(R,K),\mathbb{C}(R,K),\mathbb{C}(R)_*(K))$$

defined to be the cobordism group of $(n-1)$-dimensional quadratic complexes in $\mathbb{A}(R,K)$ which are globally contractible and locally Poincaré.

The 'algebraic surgery exact sequence' is the exact sequence
$$\ldots \longrightarrow H_n(K;\mathbb{L}_\cdot(R)) \longrightarrow L_n(R[\pi_1(K)]) \longrightarrow \mathbb{S}_n(R,K)$$
$$\longrightarrow H_{n-1}(K;\mathbb{L}_\cdot(R)) \longrightarrow \ldots$$
relating the generalized homology groups $H_*(K;\mathbb{L}_\cdot(R))$, the surgery obstruction groups $L_*(R[\pi_1(K)])$ and the quadratic structure groups $\mathbb{S}_*(R,K)$. The algebraic characterization in §18 of the topological manifold structure sets actually requires the '1/2-connective' version of the algebraic surgery exact sequence for $R=\mathbb{Z}$, and this will be developed in §15.

DEFINITION 14.1 The $\begin{cases} symmetric \\ quadratic \\ normal \end{cases}$ L-spectrum of a ring with involution
R

$$\begin{cases} \mathbb{L}^{\cdot}(R) = \{\mathbb{L}^n(R) \,|\, n \in \mathbb{Z}\} \\ \mathbb{L}_{\cdot}(R) = \{\mathbb{L}_n(R) \,|\, n \in \mathbb{Z}\} \\ \mathbb{NL}^{\cdot}(R) = \{\mathbb{NL}^n(R) \,|\, n \in \mathbb{Z}\} \end{cases}$$

is the $\begin{cases} symmetric \\ quadratic \\ normal \end{cases}$ L-spectrum $\begin{cases} \mathbb{L}^{\cdot}(\Lambda(R)) = \{\mathbb{L}^n(\Lambda(R)) \,|\, n \in \mathbb{Z}\} \\ \mathbb{L}_{\cdot}(\Lambda(R)) = \{\mathbb{L}_n(\Lambda(R)) \,|\, n \in \mathbb{Z}\} \\ \mathbb{NL}^{\cdot}(\Lambda(R)) = \{\mathbb{NL}^n(\Lambda(R)) \,|\, n \in \mathbb{Z}\} \end{cases}$ of
13.5 with

$$\Lambda(R) = (\mathbb{A}(R), \mathbb{B}(R), \mathbb{C}(R)) \,.$$

The homotopy groups are the $\begin{cases} symmetric \\ quadratic \\ normal \end{cases}$ L-groups of R

$$\begin{cases} \pi_*(\mathbb{L}^{\cdot}(R)) = L^*(R) \\ \pi_*(\mathbb{L}_{\cdot}(R)) = L_*(R) \\ \pi_*(\mathbb{NL}^{\cdot}(R)) = NL^*(R) \,. \end{cases}$$

□

The algebraic L-spectra of 14.1 are the special case $K = \{*\}$ of:

DEFINITION 14.2 The $\begin{cases} symmetric \\ visible\ symmetric \\ quadratic \\ normal \end{cases}$ L-spectrum of a pair (R, K)

with R a commutative ring and K a simplicial complex is the algebraic
L-spectrum

$$\begin{cases} \mathbb{L}^{\cdot}(R, K) = \{\mathbb{L}^n(R, K) \,|\, n \in \mathbb{Z}\} = \mathbb{L}^{\cdot}(\Lambda(R, K)) \\ \mathbb{VL}^{\cdot}(R, K) = \{\mathbb{VL}^n(R, K) \,|\, n \in \mathbb{Z}\} = \mathbb{NL}^{\cdot}(\Lambda(R, K)) \\ \mathbb{L}_{\cdot}(R, K) = \{\mathbb{L}_n(R, K) \,|\, n \in \mathbb{Z}\} = \mathbb{L}_{\cdot}(\Lambda(R, K)) \\ \mathbb{NL}^{\cdot}(R, K) = \{\mathbb{NL}^n(R, K) \,|\, n \in \mathbb{Z}\} = \mathbb{NL}^{\cdot}(\widehat{\Lambda}(R, K)) \end{cases}$$

of 13.4 for the algebraic bordism categories

$$\Lambda(R, K) = (\mathbb{A}(R, K), \mathbb{B}(R, K), \mathbb{C}(R, K)) \,,$$
$$\widehat{\Lambda}(R, K) = (\mathbb{A}(R, K), \mathbb{B}(R, K), \mathbb{B}(R, K)) \,.$$

The homotopy groups are the $\begin{cases} \text{symmetric} \\ \text{visible symmetric} \\ \text{quadratic} \\ \text{normal} \end{cases}$ L-groups of (R, K)

$$\begin{cases} \pi_*(\mathbb{L}^{\cdot}(R, K)) = L^*(R, K) \\ \pi_*(\mathbb{VL}^{\cdot}(R, K)) = VL^*(R, K) \\ \pi_*(\mathbb{L}.(R, K)) = L_*(R, K) \\ \pi_*(\mathbb{NL}^{\cdot}(R, K)) = NL^*(R, K) \end{cases}$$

defined in 9.7.

$\square$

REMARK 14.3 It follows from 9.8 that the forgetful map defines a homotopy equivalence

$$\mathbb{VL}^{\cdot}(R, K) \xrightarrow{\simeq} \mathbb{L}^{\cdot}(R, K) .$$

In the special case $K = \{*\}$ already considered in 3.6 this is

$$\mathbb{VL}^{\cdot}(R, \{*\}) = \mathbb{NL}^{\cdot}(\Lambda(R)) \simeq \mathbb{L}^{\cdot}(\Lambda(R)) = \mathbb{L}^{\cdot}(R) .$$

$\square$

A functor of algebraic bordism categories $F: \Lambda \longrightarrow \Lambda'$ induces a map of algebraic $\mathbb{L}$-spectra

$$\begin{cases} F : \mathbb{L}^{\cdot}(\Lambda) \longrightarrow \mathbb{L}^{\cdot}(\Lambda') \\ F : \mathbb{L}.(\Lambda) \longrightarrow \mathbb{L}.(\Lambda') \\ F : \mathbb{NL}^{\cdot}(\Lambda) \longrightarrow \mathbb{NL}^{\cdot}(\Lambda') . \end{cases}$$

PROPOSITION 14.4 *The universal assembly functor of §9*

$$A : \Lambda(R, K) \longrightarrow \Lambda(R[\pi_1(K)])$$

induces maps of the algebraic $\mathbb{L}$-spectra

$$\begin{cases} A : \mathbb{L}^{\cdot}(R, K) \longrightarrow \mathbb{L}^{\cdot}(R[\pi_1(K)]) \\ A : \mathbb{VL}^{\cdot}(R, K) \longrightarrow \mathbb{L}^{\cdot}(R[\pi_1(K)]) \\ A : \mathbb{L}.(R, K) \longrightarrow \mathbb{L}.(R[\pi_1(K)]) \\ A : \mathbb{NL}^{\cdot}(R, K) \longrightarrow \mathbb{NL}^{\cdot}(R[\pi_1(K)]) \end{cases}$$

which is a homotopy equivalence $\mathbb{L}.(R, K) \simeq \mathbb{L}.(R[\pi_1(K)])$ *in the quadratic case.*

PROOF The universal assembly maps in quadratic L-theory define isomorphisms $A: L_*(R, K) \longrightarrow L_*(R[\pi_1(K)])$ by the algebraic π-π theorem (10.6).

$\square$

Recall from §9 the local algebraic bordism category of (R, K)

$$\Lambda(R)_*(K) = (\mathbb{A}(R, K), \mathbb{B}(R, K), \mathbb{C}(R)_*(K)) .$$

An object in $\mathbb{C}(R)_*(K)$ is a finite f.g. free (R, K)-module chain complex C such that each $[C][\sigma]$ ($\sigma \in K$) is a contractible finite f.g. free R-module chain complex. The assembly $C(\widetilde{K})$ over the universal cover $\widetilde{K}$ is a contractible finite f.g. free $R[\pi_1(K)]$-module chain complex, by the algebraic analogue of the Vietoris theorem (which may be proved using the algebraic Leray–Serre spectral sequence of the proof of 10.2). Thus $\mathbb{C}(R)_*(K)$ is a subcategory of $\mathbb{C}(R,K)$, and there is defined a forgetful functor of algebraic bordism categories $\Lambda(R)_*(K) \longrightarrow \Lambda(R,K)$.

PROPOSITION 14.5 (i) The $\left\{ \begin{array}{l} symmetric \\ quadratic \end{array} \right.$ L-theory homology Ω-spectrum of (R,K) is the $\left\{ \begin{array}{l} symmetric \\ quadratic \end{array} \right.$ L-spectrum of the algebraic bordism category $\Lambda_*(R,K)$

$$\left\{ \begin{array}{l} \mathbb{H}.(K;\mathbb{L}^{\cdot}(R)) \;=\; \mathbb{L}^{\cdot}(\Lambda(R)_*(K)) \\ \mathbb{H}.(K;\mathbb{L}.(R)) \;=\; \mathbb{L}.(\Lambda(R)_*(K)) \;, \end{array} \right.$$

with homotopy groups

$$\left\{ \begin{array}{l} \pi_*(\mathbb{H}.(K;\mathbb{L}^{\cdot}(R))) \;=\; H_*(K;\mathbb{L}^{\cdot}(R)) \;=\; L^*(\Lambda(R)_*(K)) \\ \pi_*(\mathbb{H}.(K;\mathbb{L}.(R))) \;=\; H_*(K;\mathbb{L}.(R)) \;=\; L_*(\Lambda(R)_*(K)) \;. \end{array} \right.$$

The assembly maps given by 12.19

$$\left\{ \begin{array}{l} A \,:\, H_*(K;\mathbb{L}^{\cdot}(R)) \;\longrightarrow\; L^*(R,K) \;=\; L^*(\Lambda(R,K)) \\ A \,:\, H_*(K;\mathbb{L}.(R)) \;\longrightarrow\; L_*(R,K) \;=\; L_*(\Lambda(R,K)) \end{array} \right.$$

coincide with the maps induced by the forgetful functor $\Lambda(R)_*(K) \longrightarrow \Lambda(R,K)$.

(ii) The normal L-theory homology Ω-spectrum of (R,K) is the normal $\mathbb{L}$-spectrum of the algebraic bordism category $\widehat{\Lambda}(R,K)$

$$\mathbb{H}.(K;\mathbb{NL}^{\cdot}(R)) \;=\; \mathbb{NL}^{\cdot}(\widehat{\Lambda}(R,K)) \;=\; \mathbb{NL}^{\cdot}(R,K) \;,$$

with homotopy groups

$$\pi_*(\mathbb{H}.(K;\mathbb{NL}^{\cdot}(R))) \;=\; H_*(K;\mathbb{NL}^{\cdot}(R)) \;=\; NL^*(R,K) \;.$$

The assembly maps given by 12.19

$$A \,:\, H_*(K;\mathbb{NL}^{\cdot}(R)) \;\longrightarrow\; NL^*(R,K) \;=\; L^*(\widehat{\Lambda}(R,K))$$

are isomorphisms.

(iii) The $\mathbb{L}$-homology spectra of (i) and (ii) fit into a fibration sequence

$$\mathbb{H}.(K;\mathbb{L}.(R)) \;\longrightarrow\; \mathbb{H}.(K;\mathbb{L}^{\cdot}(R)) \;\longrightarrow\; \mathbb{H}.(K;\mathbb{NL}^{\cdot}(R)) \;=\; \mathbb{NL}^{\cdot}(R,K) \;.$$

PROOF (i) Only the quadratic case is considered, the symmetric case being entirely similar. The identification of the quadratic L-theory of $\Lambda_*(R,K)$ with the $\mathbb{L}.(R)$-homology of K is the quadratic case of 13.7, with $\Lambda = \Lambda(R)$. The covariant functor

$$\mathbb{L}.(R,-) \,:\, \{\,\text{simplicial complexes}\,\} \;\longrightarrow\; \{\,\Omega\text{-spectra}\,\} \;;$$

$$K \;\longrightarrow\; \mathbb{L}.(R,K) \;=\; \mathbb{L}.(\Lambda(R,K))$$

is homotopy invariant, since for any quadratic Poincaré complex (C, ψ) in $\Lambda(R, K)$ there is defined a quadratic Poincaré cobordism $(C, \psi) \otimes I$ in $\Lambda(R, K \otimes \Delta^1)$ (as in the verification of the Kan extension condition in 14.3), so that the two inclusions $K \longrightarrow K \otimes \Delta^1$ induce homotopic Δ-maps $\mathbb{L}.(R, K) \longrightarrow \mathbb{L}.(R, K \otimes \Delta^1)$. Also, there is defined a commutative diagram of Ω-spectra

$$
\begin{array}{ccc}
\mathbb{H}.(K; \{\mathbb{L}.(R, D(\sigma, K))\}) & \xrightarrow{\quad A \quad} & \mathbb{L}.(R, K') \\
\simeq \Big\downarrow & & \Big\downarrow \simeq \\
\mathbb{H}.(K; \mathbb{L}.(R)) & \xrightarrow{\quad A \quad} & \mathbb{L}.(R, K)
\end{array}
$$

with A the local $\{\mathbb{L}.(R, -)\}$-coefficient assembly of 12.18.
(ii) This is the normal case of 13.7 with $\Lambda = \widehat{\Lambda}(R, K)$.
(iii) This is the special case of 13.9 (i) with $\Lambda = \Lambda(R)_*(K)$.

□

The forgetful map $\mathbb{H}.(K; \mathbb{L}.(R)) \longrightarrow \mathbb{L}.(R, K)$ may be composed with the homotopy equivalence of 14.4 $\mathbb{L}.(R, K) \simeq \mathbb{L}.(R[\pi_1(K)])$ to define an assembly map

$$A : \mathbb{H}.(K; \mathbb{L}.(R)) \longrightarrow \mathbb{L}.(R[\pi_1(K)]) .$$

DEFINITION 14.6 (i) The *quadratic structure groups* of (R, K) are the cobordism groups

$$\mathbb{S}_n(R, K) = L_{n-1}(\mathbb{A}(R, K), \mathbb{C}(R, K), \mathbb{C}(R)_*(K)) \quad (n \in \mathbb{Z})$$

of $(n-1)$-dimensional quadratic complexes in $\mathbb{A}(R, K)$ which are globally contractible and locally Poincaré.
(ii) The *quadratic structure spectrum* of (R, K) is the quadratic $\mathbb{L}$-spectrum

$$\mathbb{S}.(R, K) = \Sigma \mathbb{L}.(\mathbb{A}(R, K), \mathbb{C}(R, K), \mathbb{C}(R)_*(K))$$

with homotopy groups

$$\pi_*(\mathbb{S}.(R, K)) = \mathbb{S}_*(R, K) .$$

(iii) The *algebraic surgery exact sequence* is the exact sequence of homotopy groups

$$\ldots \longrightarrow H_n(K; \mathbb{L}.(R)) \xrightarrow{\quad A \quad} L_n(R[\pi_1(K)])$$

$$\xrightarrow{\quad \partial \quad} \mathbb{S}_n(R, K) \longrightarrow H_{n-1}(K; \mathbb{L}.(R)) \longrightarrow \ldots$$

induced by the fibration sequence of spectra

$$\mathbb{H}.(K; \mathbb{L}.(R)) \longrightarrow \mathbb{L}.(R[\pi_1(K)]) \longrightarrow \mathbb{S}.(R, K) .$$

□

The *symmetric structure groups* $\mathbb{S}^*(R, K)$ and the *symmetric structure spectrum* $\mathbb{S}^{\cdot}(R, K)$ are defined entirely similarly, using symmetric L-theory.

PROPOSITION 14.7 *For any commutative ring R and simplicial complex K there is defined a commutative braid of exact sequences of algebraic L-groups*

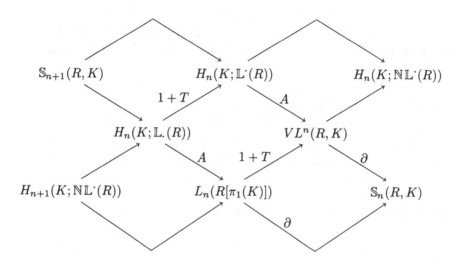

which are the exact sequences of homotopy groups of algebraic $\mathbb{L}$-spectra in a braid of fibration sequences

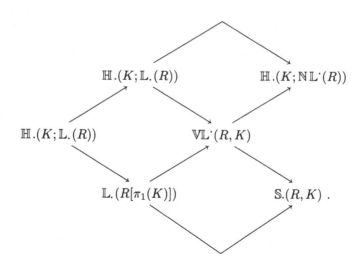

PROOF These are just the braids of 13.11 for

$$\mathbb{A} = \mathbb{A}(R,K) \ , \ \mathbb{B} = \mathbb{B}(R,K) \ , \ \mathbb{C} = \mathbb{C}(R,K) \ , \ \mathbb{D} = \mathbb{C}(R)_*(K) \ ,$$

using 14.4 to replace $\mathbb{L}.(\mathbb{A},\mathbb{B},\mathbb{C})$ by $\mathbb{L}.(R[\pi_1(K)])$ and 14.5 (ii) to replace $\mathbb{NL}^{\cdot}(\mathbb{A},\mathbb{B},\mathbb{B})$ by $\mathbb{NL}^{\cdot}(R,K)$. The universal assembly map

$$A : H_n(K;\mathbb{L}.(R)) \longrightarrow L_n(R[\pi_1(K)]) \ ; \ (C,\psi) \longrightarrow (C(\widetilde{K}),\psi(\widetilde{K}))$$

is defined in 9.10. The map

$$\partial : VL^n(R,K) \longrightarrow \mathbb{S}_n(R,K) \ ; \ (C,\phi) \longrightarrow (\partial C,\psi)$$

sends an n-dimensional globally Poincaré normal complex in $\mathbb{A}(R,K)$ to the boundary $(n-1)$-dimensional globally contractible locally Poincaré quadratic complex in $\mathbb{A}(R,K)$ defined in 2.10.

$\square$

PROPOSITION 14.8 *The visible symmetrization maps*

$$1+T : L_n(R[\pi_1(K)]) \longrightarrow VL^n(R,K) \ \ (n \in \mathbb{Z})$$

are isomorphisms modulo 8-torsion.
PROOF The relative homotopy groups $H_*(K;\mathbb{NL}^{\cdot}(R))$ are 8-torsion, since

$$\pi_n(\mathbb{NL}^{\cdot}(R)) = NL^n(R) = \varinjlim_k \widehat{L}^{n+4k}(R) \ \ (n \in \mathbb{Z}) \ ,$$

and the hyperquadratic L-groups $\widehat{L}^*(R)$ of Ranicki [143, p.137] are 8-torsion.

$\square$

§15. Connective L-theory

Let $q \in \mathbb{Z}$. An Ω-spectrum $\mathbb{F}$ is *q-connective* if $\pi_n(\mathbb{F}) = 0$ for $n < q$. A *q-connective cover* of an Ω-spectrum $\mathbb{F}$ is a q-connective Ω-spectrum $\mathbb{F}\langle q \rangle$ together with a map $\mathbb{F}\langle q \rangle \longrightarrow \mathbb{F}$ inducing isomorphisms $\pi_n(\mathbb{F}\langle q \rangle) \cong \pi_n(\mathbb{F})$ for $n \geq q$. In general, $\mathbb{F}\langle q \rangle$ is obtained from $\mathbb{F}$ by killing the homotopy groups $\pi_n(\mathbb{F})$ for $n < q$, using Postnikov decompositions and Eilenberg–MacLane spectra.

The q-connective L-theory required for the applications to topology will now be developed. The q-connective covers of the $\mathbb{L}$-spectra are explicitly constructed using algebraic Poincaré complexes of the appropriate connectivity, rather than by killing the homotopy groups using the general machinery.

Let $\Lambda = (\mathbb{A}, \mathbb{B}, \mathbb{C})$ be an algebraic bordism category.

DEFINITION 15.1 The q-connective $\begin{cases} symmetric \\ quadratic \\ normal \end{cases}$ L-groups $\begin{cases} L^*\langle q \rangle(\Lambda) \\ L_*\langle q \rangle(\Lambda) \\ NL^*\langle q \rangle(\Lambda) \end{cases}$ of Λ are defined by

$$\begin{cases} L^n\langle q \rangle(\Lambda) = L^n(\Lambda) & \text{if } n \geq q, \ 0 \text{ if } n < q \\ L_n\langle q \rangle(\Lambda) = L_n(\Lambda) & \text{if } n \geq q, \ 0 \text{ if } n < q \\ NL^n\langle q \rangle(\Lambda) = NL^n(\Lambda) & \text{if } n \geq q, \ 0 \text{ if } n < q. \end{cases}$$

$\square$

Write the p-skeleton of a simplicial complex K as $K^{[p]}$. Similarly, the p-skeleton of a pointed Δ-set K is the pointed Δ-set $K^{[p]}$ with

$$(K^{[p]})^{(q)} = \begin{cases} K^{(q)} & \text{if } q \leq p \\ \{\emptyset\} & \text{otherwise .} \end{cases}$$

DEFINITION 15.2 The q-connective $\begin{cases} symmetric \\ quadratic \\ normal \end{cases}$ L-spectrum of Λ is the Ω-spectrum of Kan Δ-sets

$$\begin{cases} \mathbb{L}^{\cdot}\langle q \rangle(\Lambda) = \{\mathbb{L}^n\langle q \rangle(\Lambda) \mid n \in \mathbb{Z}\} \\ \mathbb{L}_{\cdot}\langle q \rangle(\Lambda) = \{\mathbb{L}_n\langle q \rangle(\Lambda) \mid n \in \mathbb{Z}\} \\ \mathbb{NL}^{\cdot}\langle q \rangle(\Lambda) = \{\mathbb{NL}^n\langle q \rangle(\Lambda) \mid n \in \mathbb{Z}\} \end{cases}$$

with

$$\begin{cases} \mathbb{L}^n\langle q \rangle(\Lambda)^{(m)} \\ \mathbb{L}_n\langle q \rangle(\Lambda)^{(m)} \\ \mathbb{NL}^n\langle q \rangle(\Lambda)^{(m)} \end{cases} = \{ n\text{-dimensional} \begin{cases} symmetric \\ quadratic \\ normal \end{cases} \text{complexes} \begin{cases} (C, \phi) \\ (C, \psi) \\ (C, \phi) \end{cases}$$

in $\Lambda^*(\Delta^m)$ such that C is $\mathbb{C}^*((\Delta^m)^{[q-n-1]})$-contractible,

i.e. $C(\sigma)$ is in $\mathbb{C}$ for $\sigma \in \Delta^m$ with $|\sigma| \leq q - n - 1 \}$, such that

$$\begin{cases} \Omega\,\mathbb{L}^n\langle q\rangle(\Lambda) = \mathbb{L}^{n+1}\langle q\rangle(\Lambda) \\ \Omega\,\mathbb{L}_n\langle q\rangle(\Lambda) = \mathbb{L}_{n+1}\langle q\rangle(\Lambda) \\ \Omega\mathbb{NL}^n\langle q\rangle(\Lambda) = \mathbb{NL}^{n+1}\langle q\rangle(\Lambda), \end{cases} \quad \begin{cases} \pi_m(\mathbb{L}^n\langle q\rangle(\Lambda)) = L^{m+n}\langle q\rangle(\Lambda) \\ \pi_m(\mathbb{L}_n\langle q\rangle(\Lambda)) = L_{m+n}\langle q\rangle(\Lambda) \\ \pi_m(\mathbb{NL}^n\langle q\rangle(\Lambda)) = NL^{m+n}\langle q\rangle(\Lambda), \end{cases}$$

$$\begin{cases} \pi_*(\mathbb{L}^{\cdot}\langle q\rangle(\Lambda)) = L^*\langle q\rangle(\Lambda) \\ \pi_*(\mathbb{L}_{\cdot}\langle q\rangle(\Lambda)) = L_*\langle q\rangle(\Lambda) \\ \pi_*(\mathbb{NL}^{\cdot}\langle q\rangle(\Lambda)) = NL^*\langle q\rangle(\Lambda), \end{cases}$$

with $\begin{cases} \mathbb{L}^n\langle q\rangle(\Lambda) \\ \mathbb{L}_n\langle q\rangle(\Lambda) \\ \mathbb{NL}^n\langle q\rangle(\Lambda) \end{cases}$ $(q-n-1)$-connected.

$\square$

For a ring with involution R and the algebraic bordism category

$$\Lambda(R) = (\mathbb{A}(R), \mathbb{B}(R), \mathbb{C}(R))$$

of 3.12 write

$$\begin{cases} \mathbb{L}^{\cdot}\langle q\rangle(\Lambda(R)) = \mathbb{L}^{\cdot}\langle q\rangle(R) \\ \mathbb{L}_{\cdot}\langle q\rangle(\Lambda(R)) = \mathbb{L}_{\cdot}\langle q\rangle(R) \\ \mathbb{NL}^{\cdot}\langle q\rangle(\Lambda(R)) = \mathbb{NL}^{\cdot}\langle q\rangle(R), \end{cases} \quad \begin{cases} L^*\langle q\rangle(\Lambda(R)) = L^*\langle q\rangle(R) \\ L_*\langle q\rangle(\Lambda(R)) = L_*\langle q\rangle(R) \\ NL^*\langle q\rangle(\Lambda(R)) = NL^*\langle q\rangle(R). \end{cases}$$

Given a simplicial complex K and an abelian group A let $\Delta^*(K; A)$ be the A-coefficient simplicial cochain complex of K.

The following results hold in symmetric, normal and quadratic L-theory, although they are only stated in the symmetric case:

PROPOSITION 15.3 (i) *The* $\mathbb{L}^{\cdot}\langle q\rangle(\Lambda)$-$\begin{cases} cohomology \\ homology \end{cases}$ *of a simplicial complex*

K *is expressed in terms of the* $\mathbb{L}^{\cdot}(\Lambda)$-$\begin{cases} cohomology \\ homology \end{cases}$ *and the simplicial*

$L^q(\Lambda)$-*coefficient* $\begin{cases} cochain \\ chain \end{cases}$ *groups of* K *by*

$$\begin{cases} H^{-n}(K; \mathbb{L}^{\cdot}\langle q\rangle(\Lambda)) \\ \quad = \mathrm{im}(H^{-n}(K, K^{[q-n-1]}; \mathbb{L}^{\cdot}(\Lambda)) \longrightarrow H^{-n}(K, K^{[q-n-2]}; \mathbb{L}^{\cdot}(\Lambda))) \\ \quad = \mathrm{coker}(\delta\colon \Delta^{q-n-1}(K; L^q(\Lambda)) \longrightarrow H^{-n}(K, K^{[q-n-1]}; \mathbb{L}^{\cdot}(\Lambda))), \\ H_n(K; \mathbb{L}^{\cdot}\langle q\rangle(\Lambda)) \\ \quad = \mathrm{im}(H_n(K^{[n-q]}; \mathbb{L}^{\cdot}(\Lambda)) \longrightarrow H_n(K^{[n-q+1]}; \mathbb{L}^{\cdot}(\Lambda))) \\ \quad = \mathrm{coker}(\partial\colon \Delta_{n-q+1}(K; L^q(\Lambda)) \longrightarrow H_n(K^{[n-q]}; \mathbb{L}^{\cdot}(\Lambda))). \end{cases}$$

(ii) *The* $\mathbb{L}^{\cdot}\langle q\rangle(\Lambda)$- *and* $\mathbb{L}^{\cdot}\langle q+1\rangle(\Lambda)$-$\begin{cases} cohomology \\ homology \end{cases}$ *groups are related by an exact sequence*

$$\left\{ \begin{array}{l} \cdots \longrightarrow H^{-n}(K; \mathbb{L}^{\cdot}\langle q+1\rangle(\Lambda)) \longrightarrow H^{-n}(K; \mathbb{L}^{\cdot}\langle q\rangle(\Lambda)) \\ \qquad \longrightarrow H^{q-n}(K; L^q(\Lambda)) \longrightarrow H^{-n-1}(K; \mathbb{L}^{\cdot}\langle q+1\rangle(\Lambda)) \longrightarrow \cdots , \\ \cdots \longrightarrow H_n(K; \mathbb{L}^{\cdot}\langle q+1\rangle(\Lambda)) \longrightarrow H_n(K; \mathbb{L}^{\cdot}\langle q\rangle(\Lambda)) \\ \qquad \longrightarrow H_{n-q}(K; L^q(\Lambda)) \longrightarrow H_{n-1}(K; \mathbb{L}^{\cdot}\langle q+1\rangle(\Lambda)) \longrightarrow \cdots \end{array} \right.$$

with

$$\left\{ \begin{array}{l} H^{-n}(K; \mathbb{L}^{\cdot}\langle q\rangle(\Lambda)) \longrightarrow H^{q-n}(K; L^q(\Lambda)) ; \\ \qquad (C,\phi) \longrightarrow \displaystyle\sum_{\sigma \in K^{(q-n)}} (C(\sigma), \phi(\sigma))\sigma , \\ H_n(K; \mathbb{L}^{\cdot}\langle q\rangle(\Lambda)) \longrightarrow H_{n-q}(K; L^q(\Lambda)) ; \\ \qquad (C,\phi) \longrightarrow \displaystyle\sum_{\sigma \in K^{(n-q)}} (C(\sigma), \phi(\sigma))\sigma . \end{array} \right.$$

(iii) *If* $\left\{ \begin{array}{l} n - q \geq 1 \\ n - q \geq \dim(K) \end{array} \right.$ *then the natural map defines an isomorphism*

$$\left\{ \begin{array}{l} H^{-n}(K; \mathbb{L}^{\cdot}\langle q\rangle(\Lambda)) \xrightarrow{\;\simeq\;} H^{-n}(K; \mathbb{L}^{\cdot}(\Lambda)) \\ H_n(K; \mathbb{L}^{\cdot}\langle q\rangle(\Lambda)) \xrightarrow{\;\simeq\;} H_n(K; \mathbb{L}^{\cdot}(\Lambda)) . \end{array} \right.$$

PROOF It is convenient to replace $\mathbb{L}^{\cdot}\langle q\rangle(\Lambda)$ by the deformation retract

$$\mathbb{L}^{\cdot}(q)(\Lambda) = \{ \mathbb{L}^n(q)(\Lambda) \,|\, n \in \mathbb{Z} \} ,$$

with $\mathbb{L}^n(q)(\Lambda)$ the Kan Δ-set defined by

$\mathbb{L}^n(q)(\Lambda)^{(m)}$

$\qquad = \{ n\text{-dimensional symmetric complexes in } \Lambda^*(\Delta^m, (\Delta^m)^{[q-n-1]}) \}$

$\qquad = \{ n\text{-dimensional symmetric complexes } (C, \phi) \text{ in } \Lambda^*(\Delta^m)$

$\qquad\qquad \text{such that } C(\sigma) = 0 \text{ for } \sigma \in \Delta^m \text{ with } |\sigma| \leq q - n - 1 \} ,$

such that

$$\Omega \mathbb{L}^n(q)(\Lambda) = \mathbb{L}^{n+1}(q)(\Lambda) , \quad \pi_m(\mathbb{L}^n(q)(\Lambda)) = L^{m+n}\langle q\rangle(\Lambda) .$$

Define an embedding

$$\left\{ \begin{array}{l} \mathbb{H}^{\cdot}(K; \mathbb{L}^{\cdot}(q)(\Lambda)) \longrightarrow \mathbb{L}^{\cdot}(q)(\Lambda^*(K)) \\ \mathbb{H}_{\cdot}(K; \mathbb{L}^{\cdot}(q)(\Lambda)) \longrightarrow \mathbb{L}^{\cdot}(q)(\Lambda_*(K)) \end{array} \right.$$

of the $\mathbb{L}^{\cdot}(q)(\Lambda)$-$\left\{ \begin{array}{l} \text{cohomology} \\ \text{homology} \end{array} \right.$ spectrum in the q-connective symmetric $\mathbb{L}$-spectrum as follows. For cohomology use the embeddings of Δ-sets

$$\mathbb{H}^n(K; \mathbb{L}^{\cdot}(q)(\Lambda)) \longrightarrow \mathbb{L}^n(q)(\Lambda_*(K)) \quad (n \in \mathbb{Z})$$

by means of the identifications

$$\mathbb{H}^n(K;\mathbb{L}^{\cdot}(q)(\Lambda))^{(p)} =$$

$\{\,n\text{-dimensional symmetric complexes in } \Lambda^*(K \otimes \Delta^p, (K \otimes \Delta^p)^{[q-n-1]})\,\}$

$$\mathbb{L}^n(q)(\Lambda^*(K))^{(p)} =$$

$\{\,n\text{-dimensional symmetric complexes in } \Lambda^*(K \otimes \Delta^p, K \otimes (\Delta^p)^{[q-n-1]})\,\}$

and the inclusion

$$(K \otimes \Delta^p, (K \otimes \Delta^p)^{[q-n-1]}) \longrightarrow (K \otimes \Delta^p, K \otimes (\Delta^p)^{[q-n-1]})\,.$$

For homology use an embedding $K \subseteq \partial\Delta^{m+1}$ to define embeddings of Δ-sets

$$\mathbb{H}_n(K;\mathbb{L}^{\cdot}(q)(\Lambda)) \longrightarrow \mathbb{L}^n(q)(\Lambda_*(K)) \quad (n \in \mathbb{Z})$$

by means of the identifications

$$\mathbb{H}_n(K;\mathbb{L}^{\cdot}(q)(\Lambda))^{(p)} = \mathbb{H}^{n-m}(\Sigma^m,\overline{K};\mathbb{L}^{\cdot}(q)(\Lambda))^{(p)}$$

$$= \{\,(n-m)\text{-dimensional symmetric}$$

$$\text{complexes in } \Lambda^*(\Sigma^m \otimes \Delta^p, (\Sigma^m \otimes \Delta^p)^{[q-n+m-1]} \cup \overline{K} \otimes \Delta^p)\,\}\,,$$

$$\mathbb{L}^n(q)(\Lambda_*(K))^{(p)} = \mathbb{L}^{n-m}(q)(\mathbb{A}^*(\Sigma^m,\overline{K}),\mathbb{B}^*(\Sigma^m,\overline{K}),\mathbb{C}^*(\Sigma^m,\overline{K}))^{(p)}$$

$$= \{\,n\text{-dimensional symmetric complexes in } \Lambda_*(K)^*(\Delta^p,(\Delta^p)^{[q-n-1]})\,\}$$

$$= \{\,(n-m)\text{-dimensional symmetric}$$

$$\text{complexes in } \Lambda^*((\Sigma^m,\overline{K}) \otimes (\Delta^p,(\Delta^p)^{[q-n-1]}))\,\}$$

and the inclusion

$$(\Sigma^m \otimes \Delta^p, (\Sigma^m \otimes \Delta^p)^{[q-n+m-1]} \cup \overline{K} \otimes \Delta^p) \longrightarrow$$

$$(\Sigma^m,\overline{K}) \otimes (\Delta^p,(\Delta^p)^{[q-n-1]}) = (\Sigma^m \otimes \Delta^p, \Sigma^m \otimes (\Delta^p)^{[q-n-1]} \cup \overline{K} \otimes \Delta^p)\,.$$

(i) Consider the two cases separately, starting with cohomology. Use the identifications

$$\{\,\Delta\text{-maps } K_+ \longrightarrow \mathbb{L}^n(q)(\Lambda)\,\} = \{\,\Delta\text{-maps } (K, K^{[q-n-1]}) \longrightarrow \mathbb{L}^n(\Lambda)\,\}$$

$$= \{\,n\text{-dimensional symmetric complexes in } \mathbb{A}^*(K, K^{[q-n-1]})\,\}$$

to define a surjection of homotopy groups

$$H^{-n}(K, K^{[q-n-1]};\mathbb{L}^{\cdot}(\Lambda)) = [K, K^{[q-n-1]};\mathbb{L}^n(\Lambda),\emptyset]$$

$$\longrightarrow H^{-n}(K;\mathbb{L}^{\cdot}(q)(\Lambda)) = [K_+, \mathbb{L}^n(q)(\Lambda)]\,.$$

An element in the kernel is represented by a Δ-map

$$(K, K^{[q-n-1]}) \otimes \{0\} \longrightarrow (\mathbb{L}^n(\Lambda),\emptyset)$$

which extends to a Δ-map

$$(K \otimes \Delta^1, K^{[q-n-1]} \otimes \{0\} \cup K^{[q-n-2]} \otimes \Delta^1 \cup K \otimes \{1\}) \longrightarrow (\mathbb{L}^n(\Lambda),\emptyset)\,.$$

The first map in the exact sequence
$$H^{-n}(K \otimes \Delta^1, K^{[q-n-1]} \otimes \{0\} \cup K^{[q-n-2]} \otimes \Delta^1 \cup K \otimes \{1\}; \mathbb{L}^{\cdot}(\Lambda))$$
$$\xrightarrow{\text{inclusion}^*} H^{-n}(K \otimes \{0\}, K^{[q-n-1]} \otimes \{0\}; \mathbb{L}^{\cdot}(\Lambda))$$
$$\longrightarrow H^{-n}(K; \mathbb{L}^{\cdot}(q)(\Lambda)) \longrightarrow 0$$
is isomorphic to the first map in the exact sequence
$$H^{-n-1}(K^{[q-n-1]}, K^{[q-n-2]}; \mathbb{L}^{\cdot}(\Lambda)) \xrightarrow{\delta}$$
$$H^{-n}(K, K^{[q-n-1]}; \mathbb{L}^{\cdot}(\Lambda)) \longrightarrow H^{-n}(K, K^{[q-n-2]}; \mathbb{L}^{\cdot}(\Lambda)) \ .$$
This gives an identification
$$H^{-n}(K; \mathbb{L}^{\cdot}(q)(\Lambda)) \ = \ \mathrm{coker}(\delta) \ ,$$
and the domain of δ can be expressed as a cochain group
$$H^{-n-1}(K^{[q-n-1]}, K^{[q-n-2]}; \mathbb{L}^{\cdot}(\Lambda)) \ = \ \Delta^{q-n-1}(K; L^q(\Lambda)) \ .$$
The result for homology may now be deduced from the cohomology result. Embed $K \subseteq \partial\Delta^{m+1}$ for some $m \geq 0$, and note that the supplement of the p-skeleton $K^{[p]}$ in $\partial\Delta^{m+1}$ is given by
$$\overline{K^{[p]}} \ = \ \overline{K} \cup (\Sigma^m)^{[m-p-1]} \subseteq \Sigma^m \quad (p \geq 0) \ .$$
By duality and the cohomology result
$$H_n(K; \mathbb{L}^{\cdot}(q)(\Lambda)) \ = \ H^{m-n}(\Sigma^m, \overline{K}; \mathbb{L}^{\cdot}(q)(\Lambda))$$
$$= \ \mathrm{coker}(\delta \colon H^{m-n-1}(\overline{K} \cup (\Sigma^m)^{[q-n+m-1]}, \overline{K} \cup (\Sigma^m)^{[q-n+m-2]}; \mathbb{L}^{\cdot}(\Lambda))$$
$$\longrightarrow H^{m-n}(\Sigma^m, \overline{K} \cup (\Sigma^m)^{[q-n+m-1]}; \mathbb{L}^{\cdot}(\Lambda)))$$
$$= \ \mathrm{coker}(\partial \colon H_{n+1}(K^{[n-q+1]}, K^{[n-q]}; \mathbb{L}^{\cdot}(\Lambda)) \longrightarrow H_n(K^{[n-q]}; \mathbb{L}^{\cdot}(\Lambda)))$$
$$= \ \mathrm{coker}(\partial \colon \Delta_{n-q+1}(K; L^q(\Lambda)) \longrightarrow H_n(K^{[n-q]}; \mathbb{L}^{\cdot}(\Lambda))) \ .$$
(ii) The relative homotopy groups of the pair $(\mathbb{L}^{\cdot}(q)(\Lambda), \mathbb{L}^{\cdot}(q+1)(\Lambda))$ are given by
$$\pi_m(\mathbb{L}^{\cdot}(q)(\Lambda), \mathbb{L}^{\cdot}(q+1)(\Lambda)) \ = \ \begin{cases} L^q(\Lambda) & \text{if } m = q \\ 0 & \text{otherwise} \end{cases}$$
so that there is defined a fibration sequence of Ω-spectra
$$\mathbb{L}^{\cdot}(q+1)(\Lambda) \longrightarrow \mathbb{L}^{\cdot}(q)(\Lambda) \longrightarrow K.(L^q(\Lambda), q) \ .$$
Here, $K.(L^q(\Lambda), q)$ is the Ω-spectrum of Eilenberg–MacLane spaces with
$$K.(L^q(\Lambda), q)_n \ = \ K(L^q(\Lambda), q - n) \quad (n \leq q) \ .$$
(iii) This follows from
$$\begin{cases} H^{-n}(K; \mathbb{L}^{\cdot}(\Lambda)) \ = \ \lim\limits_{q \to -\infty} H^{-n}(K; \mathbb{L}^{\cdot}\langle q \rangle(\Lambda)) \\ H_n(K; \mathbb{L}^{\cdot}(\Lambda)) \ = \ \lim\limits_{q \to -\infty} H_n(K; \mathbb{L}^{\cdot}\langle q \rangle(\Lambda)) \ , \end{cases}$$

and the identifications given by (ii)

$$\begin{cases} H^{-n}(K;\mathbb{L}^{\cdot}\langle n-1\rangle(\Lambda)) = H^{-n}(\mathbb{L}^{\cdot}\langle n-2\rangle(\Lambda)) = \ldots = H^{-n}(K;\mathbb{L}^{\cdot}(\Lambda)) \\ H_n(K;\mathbb{L}^{\cdot}\langle n-k\rangle(\Lambda)) = H_n(\mathbb{L}^{\cdot}\langle n-k-1\rangle(\Lambda)) = \ldots = H_n(K;\mathbb{L}^{\cdot}(\Lambda)), \end{cases}$$

with $k = \dim(K)$.

$\square$

DEFINITION 15.4 (i) A finite chain complex C in $\mathbb{A}(R)$ is q-*connective* if

$$H_r(C) = 0 \text{ for } r < q,$$

or equivalently if C is chain equivalent to a complex D with $D_r = 0$ for $r < q$.

(ii) A finite chain complex C in $\mathbb{A}(R,K)$ is q-*connective* if each $C(\sigma)$ ($\sigma \in K$) is q-connective, or equivalently if each $[C][\sigma]$ ($\sigma \in K$) is q-connective.

(iii) An n-dimensional symmetric complex (C,ϕ) in $\mathbb{A}(R,K)$ is q-*connective* if C and C^{n-*} are q-connective.

(iv) An n-dimensional symmetric complex (C,ϕ) in $\mathbb{A}(R,K)$ is *locally* q-*Poincaré* if $\partial C = S^{-1}C(\phi_0: C^{n-*} \longrightarrow C)$ is q-connective.

Similarly for normal and quadratic complexes.

$\square$

Note that the assembly of a q-connective chain complex C in $\mathbb{A}(R,K)$ is a q-connective chain complex $C(\widetilde{K})$ in $\mathbb{A}(R[\pi_1(K)])$.

EXAMPLE 15.5 Given a simplicial complex K and any homology class $[K] \in H_n(K)$ let (C,ϕ) be the n-dimensional symmetric complex defined as in 9.13, with

$$C(\sigma) = \Delta(D(\sigma,K), \partial D(\sigma,K)) \ (\sigma \in K),$$

$$\phi_0(K) = [K] \cap - : C^{n-*}(K) \simeq \Delta(K')^{n-*} \longrightarrow C(K) = \Delta(K').$$

If K is n-dimensional then (C,ϕ) is 0-connective. (C,ϕ) is locally q-Poincaré if and only if

$$H_r([D(\sigma,K)] \cap -: \Delta(D(\sigma,K))^{n-|\sigma|-*} \longrightarrow \Delta(D(\sigma,K), \partial D(\sigma,K))) = 0$$

$$(\sigma \in K, r \leq q),$$

in which case

$$H_r(\partial D(\sigma,K)) = H_r(\text{link}_K(\sigma)) = H_r(S^{n-|\sigma|-1}) \ (r \leq q-1).$$

$\square$

The following conditions on an n-dimensional symmetric complex (C,ϕ) in $\mathbb{A}(R,K)$ are equivalent:

(i) (C,ϕ) is locally q-Poincaré,

(ii) the R-module chain complexes

$$\partial C(\sigma) = S^{-1}C(\phi_0(\sigma): [C][\sigma]^{n-|\sigma|-*} \longrightarrow C(\sigma)) \ (\sigma \in K)$$

are q-connective,

(iii) the R-module chain complexes
$$[\partial C][\sigma] = S^{-1}C([\phi_0][\sigma]:C(\sigma)^{n-|\sigma|-*}\longrightarrow[C][\sigma]) \quad (\sigma \in K)$$
are q-connective.

A simplicial complex K is *locally q-Poincaré* with respect to a homology class $[K] \in H_n(K)$ if the n-dimensional symmetric complex (C, ϕ) in $\mathbb{A}(\mathbb{Z}, K)$ defined in 9.13 (with $C(K) = \Delta(K')$ etc.) is locally q-Poincaré.

REMARK 15.6 The following conditions on a simplicial complex K with a homology class $[K] \in H_n(K)$ are equivalent:
(i) K is locally q-Poincaré,
(ii) $H_r([D(\sigma, K)]\cap-:\Delta(D(\sigma, K))^{n-|\sigma|-*}\longrightarrow\Delta(D(\sigma, K), \partial D(\sigma, K))) = 0$ for all $\sigma \in K$, $r \leq q$,
(iii) $H_r([D(\sigma, K)]\cap-:\Delta(D(\sigma, K), \partial D(\sigma, K))^{n-|\sigma|-*}\longrightarrow\Delta(D(\sigma, K))) = 0$ for all $\sigma \in K$, $r \leq q$.

$\square$

DEFINITION 15.7 (i) The *q-connective algebraic bordism categories* of a ring with involution R are
$$\Lambda\langle q\rangle(R) = (\mathbb{A}(R), \mathbb{B}\langle q\rangle(R), \mathbb{C}\langle q\rangle(R)),$$
$$\widehat{\Lambda}\langle q\rangle(R) = (\mathbb{A}(R), \mathbb{B}\langle q\rangle(R), \mathbb{B}\langle q\rangle(R))$$
with $\mathbb{B}\langle q\rangle(R)$ the category of q-connective finite chain complexes C in $\mathbb{A}(R)$, and $\mathbb{C}\langle q\rangle(R) = \mathbb{C}(R) \subseteq \mathbb{B}\langle q\rangle(R)$ the subcategory of contractible complexes.
(ii) The *q-connective algebraic bordism categories* of a commutative ring R and a simplicial complex K are
$$\Lambda\langle q\rangle(R, K) = (\mathbb{A}(R, K), \mathbb{B}\langle q\rangle(R, K), \mathbb{C}\langle q\rangle(R, K)),$$
$$\widehat{\Lambda}\langle q\rangle(R, K) = (\mathbb{A}(R, K), \mathbb{B}\langle q\rangle(R, K), \mathbb{B}\langle q\rangle(R, K))$$
with $\mathbb{B}\langle q\rangle(R, K) = \mathbb{B}\langle q\rangle(R)_*(K)$ the category of q-connective finite chain complexes C in $\mathbb{A}(R, K)$ and $\mathbb{C}\langle q\rangle(R, K) \subseteq \mathbb{B}\langle q\rangle(R, K)$ the subcategory of the globally contractible complexes.

$\square$

In the special case $K = \{*\}$ write the q-connective algebraic bordism categories as
$$\Lambda\langle q\rangle(R, \{*\}) = \Lambda\langle q\rangle(R),$$
$$\widehat{\Lambda}\langle q\rangle(R, \{*\}) = \widehat{\Lambda}\langle q\rangle(R).$$

It should be noted that the symmetric L-groups $L^*(\Lambda\langle q\rangle(R))$ of the q-connective algebraic bordism category $\Lambda\langle q\rangle(R)$ need not be the same as the q-connective symmetric L-groups $L^*\langle q\rangle(R)$ of R. Likewise for the other categories.

EXAMPLE 15.8 (i) The quadratic L-groups of $\Lambda\langle 0\rangle(R)$ coincide with the 0-connective quadratic L-groups of R

$$L_n(\Lambda\langle 0\rangle(R)) = L_n\langle 0\rangle(R) = L_n(R) \ (n \geq 0) \,,$$

by virtue of the 4-periodicity of the quadratic L-groups, and the map of quadratic $\mathbb{L}$-spectra

$$\mathbb{L}.(\Lambda\langle 0\rangle(R)) \longrightarrow \mathbb{L}.\langle 0\rangle(R)$$

is a homotopy equivalence.
(ii) The symmetric L-groups of $\Lambda\langle 0\rangle(R)$ are the connective symmetric L-groups of 3.18

$$L^n(\Lambda\langle 0\rangle(R)) = L^n(R) \ (n \geq 0) \,,$$

while the 0-connective symmetric L-groups of R are the 4-periodic symmetric L-groups of 3.12

$$L^n\langle 0\rangle(R) = L^{n+4*}(R) \ (n \geq 0) \,.$$

If R is a ring such that the symmetric L-groups $L^*(R)$ are 4-periodic (such as $R = \mathbb{Z}$) then the map of symmetric $\mathbb{L}$-spectra

$$\mathbb{L}^{\cdot}(\Lambda\langle 0\rangle(R)) \longrightarrow \mathbb{L}^{\cdot}\langle 0\rangle(R)$$

is a homotopy equivalence. If also $L^0(R) \longrightarrow NL^0(R)$ is onto then the map of normal $\mathbb{L}$-spectra

$$\mathbb{NL}^{\cdot}(\Lambda\langle 0\rangle(R)) \longrightarrow \mathbb{NL}^{\cdot}\langle 0\rangle(R)$$

is a homotopy equivalence.

□

PROPOSITION 15.9 *For any commutative ring R and a simplicial complex K the* $\begin{cases} symmetric \\ quadratic \\ normal \end{cases}$ $\mathbb{L}$*-spectrum of the algebraic bordism category* $\Lambda\langle q\rangle(R)_*(K)$ *(given by 4.1) is the homology spectrum of K with coefficients in the corresponding q-connective $\mathbb{L}$-spectrum of R*

$$\begin{cases} \mathbb{L}^{\cdot}(\Lambda\langle q\rangle(R)_*(K)) = \mathbb{H}.(K; \mathbb{L}^{\cdot}(\Lambda\langle q\rangle(R))) \\ \mathbb{L}.(\Lambda\langle q\rangle(R)_*(K)) = \mathbb{H}.(K; \mathbb{L}.(\Lambda\langle q\rangle(R))) \\ \mathbb{NL}^{\cdot}(\widehat{\Lambda}\langle q\rangle(R)_*(K)) = \mathbb{H}.(K; \mathbb{NL}^{\cdot}(\widehat{\Lambda}\langle q\rangle(R))) \,. \end{cases}$$

PROOF Exactly as for 13.7, which is the special case $q = -\infty$.

□

By analogy with 15.6:

DEFINITION 15.10 (i) The *q-connective quadratic structure groups* of (R, K) are the cobordism groups

$$\mathbb{S}_n\langle q\rangle(R, K) = L_{n-1}(\mathbb{A}(R, K), \mathbb{C}\langle q\rangle(R, K), \mathbb{C}\langle q\rangle(R)_*(K)) \ (n \in \mathbb{Z})$$

of q-connective $(n-1)$-dimensional quadratic complexes (C,ψ) in $\mathbb{A}(R,K)$ which are globally contractible and locally Poincaré.

(ii) The *q-connective quadratic structure spectrum* of (R,K) is the quadratic $\mathbb{L}$-spectrum

$$\mathbb{S}.\langle q\rangle(R,K) \;=\; \Sigma\mathbb{L}.(\mathbb{A}(R,K),\mathbb{C}\langle q\rangle(R,K),\mathbb{C}\langle q\rangle(R)_*(K))$$

with homotopy groups

$$\pi_*(\mathbb{S}.\langle q\rangle(R,K)) \;=\; \mathbb{S}_*\langle q\rangle(R,K) \;.$$

(iii) The *q-connective algebraic surgery exact sequence* is the exact sequence of homotopy groups

$$\cdots \longrightarrow H_n(K;\mathbb{L}.\langle q\rangle(R)) \xrightarrow{\;A\;} L_n(\Lambda\langle q\rangle(R,K)) \xrightarrow{\;\partial\;}$$
$$\mathbb{S}_n\langle q\rangle(R,K) \longrightarrow H_{n-1}(K;\mathbb{L}.\langle q\rangle(R)) \longrightarrow \cdots$$

induced by the fibration sequence of spectra

$$\mathbb{H}.(K;\mathbb{L}.\langle q\rangle(R)) \longrightarrow \mathbb{L}.(\Lambda\langle q\rangle(R,K)) \longrightarrow \mathbb{S}.\langle q\rangle(R,K) \;.$$

□

The *q-connective symmetric structure groups* $\mathbb{S}^*\langle q\rangle(R,K)$ and the *q-connective symmetric structure spectrum* $\mathbb{S}^{\cdot}\langle q\rangle(R,K)$ are defined entirely similarly, using symmetric L-theory.

PROPOSITION 15.11 (i) *The assembly map*

$$L_n(\Lambda\langle q\rangle(R,K)) \longrightarrow L_n(R[\pi_1(K)])$$

is an isomorphism if $n \geq 2q$.

(ii) *For $n \geq \max(2q+1, q+2)$*

$$\mathbb{S}_n\langle q\rangle(R,K) \;=\; \ker(\mathbb{S}_n(R,K^{[n-q]}) \longrightarrow \Delta_{n-q}(K;L_{q-1}(R)))$$

and for $n \geq \max(2q+1, q+3)$

$$\mathbb{S}_n\langle q\rangle(R,K) \;=\; \operatorname{im}(\mathbb{S}_n(R,K^{[n-q-1]}) \longrightarrow \mathbb{S}_n(R,K^{[n-q]})) \;.$$

(iii) *For $n \geq 2q+4$ the q-connective and $(q+1)$-connective quadratic $\mathbb{S}$-groups are related by an exact sequence*

$$\cdots \longrightarrow H_{n-q}(K;L_q(R)) \longrightarrow \mathbb{S}_n\langle q+1\rangle(R,K) \longrightarrow$$
$$\mathbb{S}_n\langle q\rangle(R,K) \longrightarrow H_{n-q-1}(K;L_q(R)) \longrightarrow \cdots \;.$$

(iv) *If K is k-dimensional and $n \geq \max(q+k+1, 2q+4)$ then*

$$\mathbb{S}_n\langle q\rangle(R,K) = \mathbb{S}_n(R,K) \;.$$

(v) *If K is k-dimensional and $n \geq \max(q+k, 2q+4)$ then there are defined* *exact sequences*

$$0 \longrightarrow \mathbb{S}_n\langle q\rangle(R,K) \longrightarrow \mathbb{S}_n\langle q-1\rangle(R,K) \longrightarrow H_{n-q}(K;L_{q-1}(R))$$
$$\longrightarrow \mathbb{S}_{n-1}\langle q\rangle(R,K) \longrightarrow \mathbb{S}_{n-1}\langle q-1\rangle(R,K) \longrightarrow \cdots ,$$
$$0 \longrightarrow \mathbb{S}_{n-1}\langle q-1\rangle(R,K) \longrightarrow \mathbb{S}_{n-1}\langle q-2\rangle(R,K) \longrightarrow H_{n-q}(K;L_{q-2}(R))$$
$$\longrightarrow \mathbb{S}_{n-2}\langle q-1\rangle(R,K) \longrightarrow \mathbb{S}_{n-2}\langle q-2\rangle(R,K) \longrightarrow \cdots$$

with

$$\mathbb{S}_n\langle q-1\rangle(R,K) = \mathbb{S}_n(R,K) \ , \quad \mathbb{S}_{n-1}\langle q-2\rangle(R,K) = \mathbb{S}_{n-1}(R,K) \ .$$

PROOF (i) The assembly map $L_n(R,K)\longrightarrow L_n(R[\pi_1(K)])$ is an isomorphism
by the algebraic π-π theorem (10.6). The forgetful map $L_n(\Lambda\langle q\rangle(R,K))\longrightarrow$
$L_n(R,K)$ is an isomorphism for $n \geq 2q$, with the inverse

$$L_n(R,K) \longrightarrow L_n(\Lambda\langle q\rangle(R,K)) \ ; \ (C,\psi) \longrightarrow (C',\psi')$$

defined by sending an n-dimensional quadratic complex (C,ψ) in $\Lambda(R,K)$
to the n-dimensional quadratic complex (C',ψ') in $\Lambda\langle q\rangle(R,K)$ obtained by
surgery below the middle dimension using the quadratic pair $(C\longrightarrow D,(0,\psi))$
with

$$D_r = \begin{cases} C_r & \text{if } 2r > n+1 \\ 0 & \text{otherwise.} \end{cases}$$

(ii) Consider the map of exact sequences

$$
\begin{array}{ccc}
H_n(K^{[n-q]};\mathbb{L}.\langle q\rangle(R)) & \longrightarrow & H_n(K;\mathbb{L}.\langle q\rangle(R)) \\
\downarrow & & \downarrow \\
L_n(\Lambda\langle q\rangle(R,K^{[n-q]})) & \longrightarrow & L_n(\Lambda\langle q\rangle(R,K)) \\
\downarrow & & \downarrow \\
\mathbb{S}_n\langle q\rangle(R,K^{[n-q]}) & \longrightarrow & \mathbb{S}_n\langle q\rangle(R,K) \\
\downarrow & & \downarrow \\
H_{n-1}(K^{[n-q]};\mathbb{L}.\langle q\rangle(R)) & \longrightarrow & H_{n-1}(K;\mathbb{L}.\langle q\rangle(R)) \\
\downarrow & & \downarrow \\
L_{n-1}(\Lambda\langle q\rangle(R,K^{[n-q]})) & \longrightarrow & L_{n-1}(\Lambda\langle q\rangle(R,K)) \ .
\end{array}
$$

The condition $n - q \geq 2$ is used to identify

$$\pi_1(K^{[n-q]}) = \pi_1(K) \ ,$$

and since $n - 1 \geq 2q$ (i) applies to show that up to isomorphism

$$L_m(\Lambda\langle q\rangle(R, K^{[n-q]})) = L_m(R[\pi_1(K^{[n-q]})])$$

$$= L_m(R[\pi_1(K)]) = L_m(\Lambda\langle q\rangle(R, K))$$

for $m = n, n-1$. By 15.3 (i) the map

$$H_n(K^{[n-q]}; \mathbb{L}.\langle q\rangle(R)) = H_n(K^{[n-q]}; \mathbb{L}.(R)) \longrightarrow$$

$$H_n(K; \mathbb{L}.\langle q\rangle(R)) = \mathrm{im}(H_n(K^{[n-q]}; \mathbb{L}.(R)) \longrightarrow H_n(K^{[n-q+1]}; \mathbb{L}.(R)))$$

is a surjection, so that there is defined an isomorphism

$$H_{n-1}(K^{[n-q]}; \mathbb{L}.\langle q\rangle(R)) \xrightarrow{\cong}$$

$$H_{n-1}(K; \mathbb{L}.\langle q\rangle(R)) = \mathrm{im}(H_{n-1}(K^{[n-q-1]}; \mathbb{L}.(R)) \to H_{n-1}(K^{[n-q]}; \mathbb{L}.(R))) .$$

An application of the 5-lemma gives an isomorphism

$$\mathbb{S}_n\langle q\rangle(R, K^{[n-q]}) \xrightarrow{\cong} \mathbb{S}_n\langle q\rangle(R, K) .$$

Consider the map of exact sequences

$$
\begin{array}{ccc}
H_n(K^{[n-q]}; \mathbb{L}.\langle q\rangle(R)) & \longrightarrow & H_n(K^{[n-q]}; \mathbb{L}.(R)) \\
\downarrow & & \downarrow \\
L_n(\Lambda\langle q\rangle(R, K^{[n-q]})) & \longrightarrow & L_n(\Lambda(R, K^{[n-q]})) \\
\downarrow & & \downarrow \\
\mathbb{S}_n\langle q\rangle(R, K^{[n-q]}) & \longrightarrow & \mathbb{S}_n\langle q\rangle(R, K^{[n-q]}) \\
\downarrow & & \downarrow \\
H_{n-1}(K^{[n-q]}; \mathbb{L}.\langle q\rangle(R)) & \longrightarrow & H_{n-1}(K^{[n-q]}; \mathbb{L}.(R)) \\
\downarrow & & \downarrow \\
L_{n-1}(\Lambda\langle q\rangle(R, K^{[n-q]})) & \longrightarrow & L_{n-1}(\Lambda(R, K^{[n-q]})) .
\end{array}
$$

Again, (i) applies to show that up to isomorphism

$$L_m(\Lambda\langle q\rangle(R, K^{[n-q]})) = L_m(R[\pi_1(K^{[n-q]})])$$

$$= L_m(R[\pi_1(K)]) = L_m(\Lambda(R, K))$$

for $m = n, n-1$. By 15.3 (i) there is defined an isomorphism

$$H_n(K^{[n-q]}; \mathbb{L}.\langle q\rangle(R)) \xrightarrow{\cong} H_n(K^{[n-q]}; \mathbb{L}.(R)) ,$$

and there is also defined an exact sequence

$$0 \longrightarrow H_{n-1}(K^{[n-q]}; \mathbb{L}.\langle q \rangle(R))$$
$$\longrightarrow H_{n-1}(K^{[n-q]}; \mathbb{L}.(R)) \longrightarrow \Delta_{n-q}(K; L_{q-1}(R)) \ .$$

It follows that

$$\mathbb{S}_n\langle q \rangle(R, K) = \mathbb{S}_n\langle q \rangle(R, K^{[n-q]})$$
$$= \ker(\mathbb{S}_n(R, K^{[n-q]}) \longrightarrow \Delta_{n-q}(K; L_{q-1}(R))) \ .$$

If $n - q \geq 3$ there is defined a map of (co)fibration sequences of Ω-spectra

$$\mathbb{H}.(K^{[n-q-1]}; \mathbb{L}.(R)) \longrightarrow \mathbb{L}.(R, K^{[n-q-1]}) \longrightarrow \mathbb{S}.(R, K^{[n-q-1]})$$
$$\downarrow \qquad\qquad\qquad \downarrow \qquad\qquad\qquad \downarrow$$
$$\mathbb{H}.(K^{[n-q]}; \mathbb{L}.(R)) \longrightarrow \mathbb{L}.(R, K^{[n-q]}) \longrightarrow \mathbb{S}.(R, K^{[n-q]})$$

with $\mathbb{L}.(R, K^{[n-q-1]}) \longrightarrow \mathbb{L}.(R, K^{[n-q]})$ a homotopy equivalence, giving rise to a homotopy equivalence

homotopy fibre of $\mathbb{S}.(R, K^{[n-q-1]}) \longrightarrow \mathbb{S}.(R, K^{[n-q]})$

$\simeq$ homotopy cofibre of $\mathbb{H}.(K^{[n-q-1]}; \mathbb{L}.(R)) \longrightarrow \mathbb{H}.(K^{[n-q]}; \mathbb{L}.(R))$.

Thus there is defined an exact sequence

$$\mathbb{S}_n(R, K^{[n-q-1]}) \longrightarrow \mathbb{S}_n(R, K^{[n-q]}) \longrightarrow H_{n-1}(K^{[n-q]}, K^{[n-q-1]}; \mathbb{L}.(R))$$
$$(= \Delta_{n-q}(K; L_{q-1}(R)))$$

and

$$\mathbb{S}_n\langle q \rangle(R, K) = \operatorname{im}(\mathbb{S}_n(R, K^{[n-q-1]}) \longrightarrow \mathbb{S}_n(R, K^{[n-q]})) \ .$$

(iv) There is defined a map of (co)fibration sequences of Ω-spectra

$$\mathbb{H}.(K; \mathbb{L}.\langle q+1 \rangle(R)) \longrightarrow \mathbb{L}.(\Lambda\langle q+1 \rangle(R, K)) \longrightarrow \mathbb{S}.\langle q+1 \rangle(R, K)$$
$$\downarrow \alpha \qquad\qquad\qquad\qquad \downarrow \beta \qquad\qquad\qquad\qquad \downarrow \gamma$$
$$\mathbb{H}.(K; \mathbb{L}.\langle q \rangle(R)) \longrightarrow \mathbb{L}.(\Lambda\langle q \rangle(R, K)) \longrightarrow \mathbb{S}.\langle q \rangle(R, K)$$

inducing an exact sequence of relative homotopy groups

$$\dots \longrightarrow \pi_n(\alpha) \longrightarrow \pi_n(\beta) \longrightarrow \pi_n(\gamma) \longrightarrow \pi_{n-1}(\alpha) \longrightarrow \dots \ .$$

By (i) $\pi_n(\beta) = 0$ for $n \geq 2q + 3$, so that for $n \geq 2q + 4$

$$\pi_n(\gamma) = \pi_{n-1}(\alpha) = H_{n-q-1}(K; L_q(R))$$

(by 15.3 (ii)).

(iv)+(v) Apply (iii) and 15.3 (iii).

$\square$

DEFINITION 15.12 (i) The *q-connective visible symmetric L-groups* of (R, K) are the cobordism groups

$$VL^n\langle q\rangle(R, K) = NL^n(\Lambda\langle q\rangle(R, K)) \quad (n \in \mathbb{Z})$$

of q-connective n-dimensional normal globally Poincaré complexes (C, ϕ) in $\mathbb{A}(R, K)$.

(ii) The *q-connective visible symmetric $\mathbb{L}$-spectrum* of (R, K) is the algebraic $\mathbb{L}$-spectrum

$$\mathbb{VL}^{\cdot}\langle q\rangle(R, K) = \mathbb{NL}^{\cdot}(\Lambda\langle q\rangle(R, K))$$

with homotopy groups

$$\pi_*(\mathbb{VL}^{\cdot}\langle q\rangle(R, K)) = VL^*\langle q\rangle(R, K) .$$

$\square$

By analogy with 15.7:

PROPOSITION 15.13 *For any commutative ring R and simplicial complex K there is defined a commutative braid of exact sequences of algebraic L-groups*

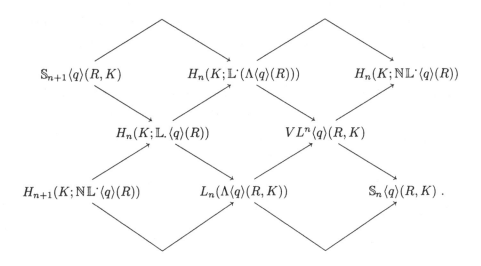

$\square$

In view of the topological applications it is convenient to introduce the following '1/2-connective' hybrids of 0-connective and 1-connective algebraic $\mathbb{L}$-spectra, making use of the following algebraic bordism categories.

DEFINITION 15.14 (i) The 1/2-*connective algebraic bordism categories* of a ring with involution R are

$$\Lambda\langle 1/2 \rangle(R) = (\mathbb{A}(R), \mathbb{B}\langle 0\rangle(R), \mathbb{C}\langle 1\rangle(R)) ,$$

$$\widehat{\Lambda}\langle 1/2 \rangle(R) = (\mathbb{A}(R), \mathbb{B}\langle 0\rangle(R), \mathbb{B}\langle 1\rangle(R)) .$$

(ii) The 1/2-*connective algebraic bordism categories* of a commutative ring R and a simplicial complex K are

$$\Lambda\langle 1/2 \rangle(R, K) = (\mathbb{A}(R, K), \mathbb{B}\langle 0\rangle(R, K), \mathbb{C}\langle 1\rangle(R, K)) ,$$

$$\widehat{\Lambda}\langle 1/2 \rangle(R, K) = (\mathbb{A}(R, K), \mathbb{B}\langle 0\rangle(R, K), \mathbb{B}\langle 1\rangle(R, K)) .$$

(iii) An n-dimensional normal complex (C, ϕ) in $\widehat{\Lambda}(R, K)$ is *1/2-connective* if it is defined in $\widehat{\Lambda}\langle 1/2 \rangle(R, K)$, i.e. if it is 0-connective and locally 1-Poincaré.

(iv) The 1/2-*connective* $\begin{cases} \textit{visible symmetric} \\ \textit{normal} \end{cases}$ *L-groups of* (R, K) are the cobordism groups

$$\begin{cases} VL^*\langle 1/2 \rangle(R, K) = NL^*(\Lambda\langle 1/2\rangle(R, K)) \\ NL^*\langle 1/2 \rangle(R, K) = NL^*(\widehat{\Lambda}\langle 1/2\rangle(R, K)) \end{cases}$$

of n-dimensional 1/2-connective $\begin{cases} \text{globally Poincaré} \\ - \end{cases}$ normal complexes in $\widehat{\Lambda}(R, K)$.

□

DEFINITION 15.15 The 1/2-*connective normal* $\mathbb{L}$-*spectrum* of a ring with involution R is the Ω-spectrum of Kan Δ-sets

$$\mathbb{NL}^{\cdot}\langle 1/2 \rangle(R) = \mathbb{NL}^{\cdot}(\widehat{\Lambda}\langle 1/2 \rangle(R))$$

with

$$NL^n\langle 1/2 \rangle(R)^{(m)}$$

$$= \{ (C, \phi) \in \mathbb{NL}^n(\widehat{\Lambda}\langle 0\rangle(R))^{(m)} \,|\, (\partial C, \psi) \in \mathbb{L}_{n-1}(\Lambda\langle 1\rangle(R))^{(m)} \} ,$$

and homotopy groups

$$\pi_*(\mathbb{NL}^{\cdot}\langle 1/2 \rangle(R)) = NL^*(\widehat{\Lambda}\langle 1/2 \rangle(R)) .$$

□

PROPOSITION 15.16 (i) *The 1/2-connective normal* $\mathbb{L}$*-spectrum* $\mathbb{NL}^{\cdot}\langle 1/2\rangle(R)$ *fits into a commutative braid of fibrations of* Ω*-spectra*

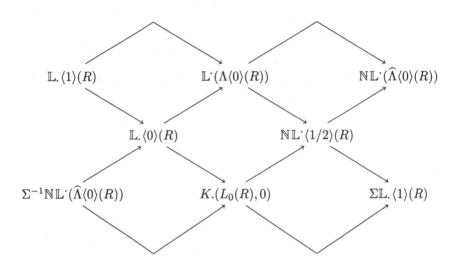

(ii) *The 1/2-connective normal L-groups are such that*

$$NL^n\langle 1/2\rangle(R) = \begin{cases} NL^n(R) & \text{if } n > 1 \\ \text{im}(L^1(R) \longrightarrow NL^1(R)) & \text{if } n = 1 \\ L^0(R) & \text{if } n = 0 \\ 0 & \text{if } n < 0 , \end{cases}$$

with a long exact sequence

$$\cdots \longrightarrow L_n\langle 1\rangle(R) \longrightarrow L^n\langle 0\rangle(R) \longrightarrow NL^n\langle 1/2\rangle(R) \longrightarrow L_{n-1}\langle 1\rangle(R) \longrightarrow \cdots .$$

(iii) *For a commutative ring R and a simplicial complex K there are natural identifications*

$$NL^*\langle 1/2\rangle(R, K) = H_*(K; \mathbb{NL}^{\cdot}\langle 1/2\rangle(R)) .$$

$\square$

DEFINITION 15.17 *The 1/2-connective visible symmetric* $\mathbb{L}$*-spectrum* of a commutative ring R and a simplicial complex K is the Ω-spectrum of Kan Δ-sets

$$\mathbb{VL}^{\cdot}\langle 1/2\rangle(R, K) = \mathbb{NL}^{\cdot}(\Lambda\langle 1/2\rangle(R, K))$$

with

$$\mathbb{VL}^n\langle 1/2\rangle(R, K)^{(m)}$$
$$= \{ (C, \phi) \in \mathbb{NL}^n(\Lambda\langle 0\rangle(R, K))^{(m)} \,|\, (\partial C, \psi) \in \mathbb{L}_{n-1}(\Lambda\langle 1\rangle(R, K))^{(m)} \} ,$$

and homotopy groups

$$\pi_*(\mathbb{VL}^{\cdot}\langle 1/2\rangle(R,K)) = VL^*\langle 1/2\rangle(R,K) .$$

□

PROPOSITION 15.18 (i) *The 1/2-connective visible symmetric L-groups $VL^*\langle 1/2\rangle(R,K)$ fit into a commutative braid of exact sequences of algebraic L-groups*

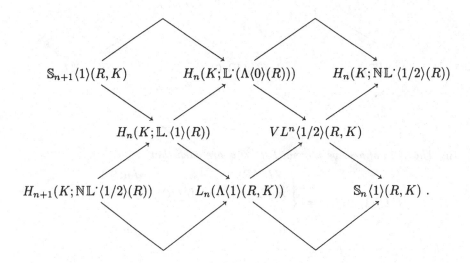

The map

$$L_n(R[\pi_1(K)]) = L_n(\Lambda\langle 1\rangle(R,K)) \longrightarrow VL^n\langle 1/2\rangle(R,K) ;$$

$$(C,\psi) \longrightarrow (C',(1+T)\psi')$$

sends an n-dimensional quadratic complex (C,ψ) in $\Lambda(R,K)$ to the symmetrization of any globally Poincaré cobordant quadratic complex (C',ψ') in $\Lambda\langle 1\rangle(R,K)$.
(ii) *The 1/2-connective visible symmetric L-groups $VL^*\langle 1/2\rangle(R,K)$ are related to the 0-connective visible symmetric L-groups*

$$VL^n\langle 0\rangle(R,K) = NL^n(\Lambda\langle 0\rangle(R,K)) \ (n \geq 0)$$

by a commutative braid of exact sequences

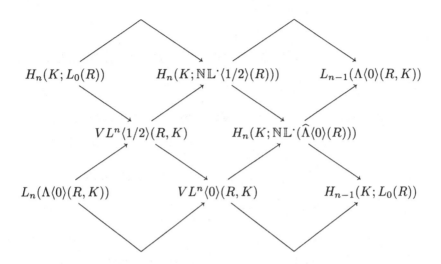

(iii) *The 1/2-connective visible symmetric L-groups* $VL^*\langle 1/2\rangle(R,K)$ *fit into a commutative braid of exact sequences*

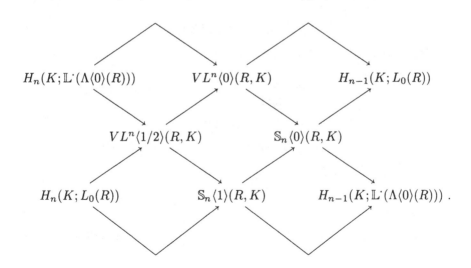

PROOF (i) The 1/2-connective visible symmetric $\mathbb{L}$-spectrum $\mathbb{V}\mathbb{L}^{\cdot}\langle 1/2\rangle(R,K)$ fits into a commutative braid of fibrations of Ω-spectra

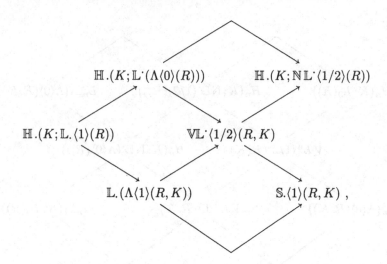

inducing a commutative braid of exact sequences of homotopy groups.
(ii) and (iii) follow from (i).

□

Note that for a ring R with $L^*(R)$ 4-periodic and $L^0(R) \longrightarrow NL^0(R)$ onto
(e.g. $R = \mathbb{Z}$) the 0-connective n-dimensional L-groups of (R, K) are 4-periodic for $n \geq \dim(K)$, with

$$VL^n\langle 0\rangle(R, K) = VL^n(R, K) = VL^{n+4}(R, K)$$

$$\mathbb{S}_n\langle 0\rangle(R, K) = \mathbb{S}_n(R, K) = \mathbb{S}_{n+4}(R, K)$$

$$H_n(K; \mathbb{L}^{\cdot}(\Lambda\langle 0\rangle(R))) = H_n(K; \mathbb{L}^{\cdot}\langle 0\rangle(R))$$
$$= H_n(K; \mathbb{L}^{\cdot}(R)) = H_{n+4}(K; \mathbb{L}^{\cdot}(R))$$

$$H_n(K; \mathbb{L}.(\Lambda\langle 0\rangle(R))) = H_n(K; \mathbb{L}.\langle 0\rangle(R))$$
$$= H_n(K; \mathbb{L}.(R)) = H_{n+4}(K; \mathbb{L}.(R))$$

$$H_n(K; \mathbb{NL}^{\cdot}(\widehat{\Lambda}\langle 0\rangle(R))) = H_n(K; \mathbb{NL}^{\cdot}\langle 0\rangle(R))$$
$$= H_n(K; \mathbb{NL}^{\cdot}(R)) = H_{n+4}(K; \mathbb{NL}^{\cdot}(R)) .$$

Also, for $n \geq 2$

$$L_n(\Lambda(R, K)) = L_n(\Lambda\langle 0\rangle(R, K)) = L_n(\Lambda\langle 1\rangle(R, K)) = L_n(R[\pi_1(K)]) .$$

DEFINITION 15.19 The *algebraic surgery exact sequence* of a simplicial complex K is the exact sequence

$$\ldots \longrightarrow H_n(K; \mathbb{L}.) \xrightarrow{A} L_n(\mathbb{Z}[\pi_1(K)])$$

$$\xrightarrow{\partial} \mathbb{S}_n(K) \longrightarrow H_{n-1}(K; \mathbb{L}.)) \longrightarrow \ldots$$

given by 15.18 in the special case $R = \mathbb{Z}$, with

$$\mathbb{S}_*(K) = \mathbb{S}_*\langle 1 \rangle(K) \; , \; \mathbb{L}. \; = \; \mathbb{L}.\langle 1 \rangle(\mathbb{Z}) \; .$$

□

The algebraic surgery exact sequence will be identified in §18 with the Sullivan–Wall geometric surgery exact sequence for the topological manifold structure set.

Part II

Topology

§16. The L-theory orientation of topology

The algebraic theory of §§1–15 is now applied to construct the L-theory orientations which distinguish topological bundles and manifolds from spherical fibrations and geometric Poincaré complexes. The geometric interpretation of such orientations has already been discussed in Ranicki [140], Levitt and Ranicki [91]: the L-theory orientations are algebraic images of the geometric Poincaré orientations, which are the homotopy theoretic consequences of the transversality properties characteristic of topological bundles and manifolds

Topological bundles and spherical fibrations are already distinguished by the rational homotopy groups of the classifying spaces

$$\pi_*(BG) \otimes \mathbb{Q} = \pi_{*-1}^s \otimes \mathbb{Q} = 0 \ (* > 0) \ ,$$

$$\pi_*(BTOP) \otimes \mathbb{Q} = \pi_*(G/TOP) \otimes \mathbb{Q}$$

$$= L_*(\mathbb{Z}) \otimes \mathbb{Q} = \begin{cases} \mathbb{Q} & \text{if } * \equiv 0(\mathrm{mod}\ 4) \\ 0 & \text{if } * \not\equiv 0(\mathrm{mod}\ 4) \ . \end{cases}$$

The rational cohomology ring $H^*(BSTOP;\mathbb{Q}) = H^*(BSO;\mathbb{Q})$ of the classifying space $BSTOP$ for stable oriented topological bundles is the polynomial algebra over $\mathbb{Q}$ generated by the universal Pontrjagin classes $p_* \in H^{4*}(BSO;\mathbb{Q})$ (Milnor and Stasheff [111], Novikov [120]). The Pontrjagin classes are not defined for spherical fibrations, since $H^*(BSG;\mathbb{Q}) = 0$ for $* > 0$.

Abbreviate

$$\mathbb{L}^{\cdot}\langle 0 \rangle(\mathbb{Z}) = \mathbb{L}^{\cdot} \ , \quad \mathbb{L}.\langle 1 \rangle(\mathbb{Z}) = \mathbb{L}. \ , \quad \mathbb{NL}^{\cdot}\langle 1/2 \rangle(\mathbb{Z}) = \widehat{\mathbb{L}}^{\cdot} \ ,$$

$$VL^*\langle 1/2 \rangle(\mathbb{Z}, X) = VL^*(X) \ , \quad NL^*\langle 1/2 \rangle(\mathbb{Z}, X) = \widehat{L}^*(X) \ .$$

A spherical fibration $\nu: X \longrightarrow BG(k)$ will now be given a canonical $\widehat{\mathbb{L}}^{\cdot}$-cohomology Thom class $\widehat{U}_\nu \in \dot{H}^k(T(\nu); \widehat{\mathbb{L}}^{\cdot})$, with $T(\nu)$ the Thom complex, $\dot{H}^*$ reduced cohomology. Topological reductions $\tilde{\nu}: X \longrightarrow \widetilde{BTOP}(k)$ of ν (if any) are in one–one correspondence with lifts of $\widehat{U}_\nu$ to an $\mathbb{L}^{\cdot}$-cohomology Thom class $U_{\tilde{\nu}} \in \dot{H}^k(T(\nu); \mathbb{L}^{\cdot})$, with any two lifts differing by an element of $\dot{H}^k(T(\nu); \mathbb{L}.)$. Rationally, such lifts correspond to the Pontrjagin classes $p_*(\tilde{\nu}) \in H^{4*}(X;\mathbb{Q})$, or equivalently the $\mathcal{L}$-genus $\mathcal{L}(\tilde{\nu}) \in H^{4*}(X;\mathbb{Q})$.

The normal signature of an n-dimensional geometric Poincaré complex X is a canonical $\widehat{\mathbb{L}}^{\cdot}$-homology fundamental class

$$[X]_{\widehat{\mathbb{L}}} = \hat{\sigma}^*(X) \in H_n(X; \widehat{\mathbb{L}}^{\cdot}) = \widehat{L}^n(X) \ .$$

In §17 it will be proved that for $n \geq 5$ topological manifold structures in the homotopy type of X (if any) are in one–one correspondence with lifts of $[X]_{\widehat{\mathbb{L}}}$ to an $\mathbb{L}^{\cdot}$-homology fundamental class $[X]_{\mathbb{L}} \in H_n(X; \mathbb{L}^{\cdot})$ with assembly the '1/2-connective visible symmetric signature' $A([X]_{\mathbb{L}}) = \sigma^*(X) \in VL^n(X)$.

(Recall that $1/2$-connective $=$ 0-connective and locally 1-Poincaré). In the first instance, only the oriented case is considered: see Appendix A for the modifications required for the nonorientable case. From now on the same terminology is used for a simplicial complex X and its polyhedron $|X|$, and both are denoted by X.

The difference between the stable theories of spherical fibrations and topological bundles can be formulated as a fibration sequence of the classifying spaces

$$G/TOP \longrightarrow BTOP \xrightarrow{J} BG ,$$

and also in terms of the algebraic $\mathbb{L}$-spectra. See Appendix B below for an account of the multiplicative structures on the algebraic $\mathbb{L}$-spectra involved. See Rourke and Sanderson [151] for the theory of topological block bundles. For $k \geq 3$ the classifying space $\widetilde{BTOP}(k)$ for k-dimensional topological block bundles fits into a fibration sequence

$$G/TOP \longrightarrow \widetilde{BTOP}(k) \longrightarrow BG(k)$$

with $BG(k)$ the classifying space for $(k-1)$-spherical fibrations. For $k \leq 2$ there is no difference between spherical fibrations, topological block bundles and vector bundles, so that $BG(k) = \widetilde{BTOP}(k) = BO(k)$.

PROPOSITION 16.1 (Ranicki [140], Levitt and Ranicki [91, 1.12]) *Let $k \geq 3$.*
(i) *A $(k-1)$-spherical fibration $\nu\colon X {\longrightarrow} BG(k)$ has a canonical $\widehat{\mathbb{L}}$-cohomology orientation*

$$\widehat{U}_\nu \in \dot{H}^k(T(\nu);\widehat{\mathbb{L}}\cdot) .$$

(ii) *A topological block bundle $\tilde{\nu}\colon X {\longrightarrow} \widetilde{BTOP}(k)$ has a canonical $\mathbb{L}\cdot$-cohomology orientation*

$$U_{\tilde{\nu}} \in \dot{H}^k(T(\nu);\mathbb{L}\cdot)$$

with image $J(U_{\tilde{\nu}}) = \widehat{U}_\nu \in \dot{H}^k(T(\nu);\widehat{\mathbb{L}}\cdot)$ the canonical $\widehat{\mathbb{L}}\cdot$-cohomology orientation of the associated $(k-1)$-spherical fibration $\nu = J(\tilde{\nu})\colon X {\longrightarrow} BG(k)$.
(iii) *The topological reducibility obstruction of a $(k-1)$-spherical fibration $\nu\colon X {\longrightarrow} BG(k)$*

$$t(\nu) = \delta(\widehat{U}_\nu) \in \dot{H}^{k+1}(T(\nu);\mathbb{L}.)$$

is such that $t(\nu) = 0$ if and only if there exists a topological block bundle reduction $\tilde{\nu}\colon X {\longrightarrow} \widetilde{BTOP}(k)$. Here, δ is the connecting map in the exact sequence

$$\ldots \longrightarrow \dot{H}^k(T(\nu);\mathbb{L}.) \xrightarrow{1+T} \dot{H}^k(T(\nu);\mathbb{L}\cdot) \xrightarrow{J} \dot{H}^k(T(\nu);\widehat{\mathbb{L}}\cdot)$$

$$\xrightarrow{\delta} \dot{H}^{k+1}(T(\nu);\mathbb{L}.) \longrightarrow \ldots .$$

(iv) *The simply connected surgery obstruction defines a homotopy equivalence between the classifying space G/TOP for fibre homotopy trivialized* *topological bundles and the 0th Kan Δ-set $\mathbb{L}_0 = \mathbb{L}\langle 1 \rangle_0(\mathbb{Z})$ of the 1-connective quadratic $\mathbb{L}$-spectrum $\mathbb{L}.$ of $\mathbb{Z}$*

$$G/TOP \xrightarrow{\;\simeq\;} \mathbb{L}_0 \ .$$

(v) *The difference between two topological bundle reductions $\tilde{\nu}, \tilde{\nu}' \colon X \longrightarrow \widetilde{BTOP}(k)$ of the same $(k-1)$-spherical fibration $\nu \colon X \longrightarrow BG(k)$ is classified by* a difference element

$$t(\tilde{\nu}, \tilde{\nu}') \in [X, G/TOP] \ = \ H^0(X; \mathbb{L}.)$$

such that

$$U_{\tilde{\nu}'} - U_{\tilde{\nu}} \ = \ (1+T)(U_{\tilde{\nu}} \cup t(\tilde{\nu}, \tilde{\nu}')) \in \dot{H}^k(T(\nu); \mathbb{L}\cdot) \ ,$$

with $U_{\tilde{\nu}} \cup - \colon H^0(X; \mathbb{L}.) \xrightarrow{\;\simeq\;} \dot{H}^k(T(\nu); \mathbb{L}.)$ the $\mathbb{L}.$-cohomology Thom isomorphism. If $\tilde{\nu}'' \colon X \longrightarrow \widetilde{BTOP}(k)$ is yet another reduction of ν then

$$t(\tilde{\nu}, \tilde{\nu}'') \ = \ t(\tilde{\nu}, \tilde{\nu}') + t(\tilde{\nu}', \tilde{\nu}'') + t(\tilde{\nu}, \tilde{\nu}') \cup t(\tilde{\nu}', \tilde{\nu}'') \in H^0(X; \mathbb{L}.) \ .$$

PROOF The singular complex of the Thom complex $T(\nu)$ of a spherical fibration $\nu \colon X \longrightarrow BG(k)$ contains as a deformation retract the subcomplex of the singular simplexes $\rho \colon \Delta^n \longrightarrow T(\nu)$ which are normal transverse at the zero section $X \subset T(\nu)$, with $M = \rho^{-1}(X)$ an $(n-k)$-dimensional geometric normal complex n-ad. The canonical $\widehat{\mathbb{L}}\cdot$-cohomology orientation $\widehat{U}_\nu \in \dot{H}^k(T(\nu); \widehat{\mathbb{L}}\cdot)$ is represented by the Δ-map $\widehat{U}_\nu \colon T(\nu) \longrightarrow \widehat{\mathbb{L}}^{-k}$ sending ρ to the $(n-k)$-dimensional normal complex $\widehat{\sigma}^*(M) = (C, \phi)$ in $\widehat{\Lambda}(\mathbb{Z})^*(\Delta^n)$ defined in 9.15. A topological block bundle reduction $\tilde{\nu} \colon X \longrightarrow \widetilde{BTOP}(k)$ corresponds to a further deformation retraction of the singular complex of $T(\nu)$ to the subcomplex consisting of the singular simplexes $\rho \colon \Delta^n \longrightarrow T(\nu)$ which are $\mathbb{Z}$-coefficient Poincaré transverse at the zero section $X \subset T(\nu)$, with $M = \rho^{-1}(X)$ a $\mathbb{Z}$-coefficient geometric Poincaré n-ad. The reduction is equivalent to the lift of $\widehat{U}_\nu$ to the $\mathbb{L}\cdot$-cohomology Thom class $U_{\tilde{\nu}} \in \dot{H}^k(T(\nu); \mathbb{L}\cdot)$ represented by the Δ-map $U_{\tilde{\nu}} \colon T(\nu) \longrightarrow \mathbb{L}^{-k}$ sending ρ to the $(n-k)$-dimensional symmetric Poincaré complex $\sigma^*(M) = (C, \phi)$ in $\Lambda(\mathbb{Z})^*(\Delta^n)$ defined in 9.13. For further details see [91] and [140].

$\square$

For $k \geq 3$ the classifying space $\widetilde{BTOP}(k)$ for k-dimensional topological block bundles fits into a fibre square

$$
\begin{array}{ccc}
\widetilde{BTOP}(k) & \longrightarrow & B\mathbb{L}\cdot G(k) \\
\downarrow & & \downarrow \\
BG(k) & \longrightarrow & B\widehat{\mathbb{L}}\cdot G(k)
\end{array}
$$

with $B\mathbb{L}\cdot G(k)$ the classifying space for $(k-1)$-spherical fibrations with a

w_1-twisted $\mathbb{L}^{\cdot}$-orientation, and similarly for $B\widehat{\mathbb{L}^{\cdot}}G(k)$.

REMARK 16.2 The canonical $\mathbb{L}^{\cdot}$-cohomology orientation of an oriented topological bundle $\tilde{\nu}: X \longrightarrow B\widetilde{STOP}(k)$ is given rationally by the inverse $\mathcal{L}$-genus

$$U_{\tilde{\nu}} \otimes \mathbb{Q} = \mathcal{L}^{-1}(\tilde{\nu})$$

$$\in \dot{H}^k(T(\nu); \mathbb{L}^{\cdot}) \otimes \mathbb{Q} = \sum_{j \geq 0} \dot{H}^{4j+k}(T(\tilde{\nu}); \mathbb{Q}) = \sum_{j \geq 0} H^{4j}(X; \mathbb{Q}) \ .$$

Both the $\mathcal{L}$-genus and the symmetric signature determine and (modulo torsion) are determined by the signatures of submanifolds, as used by Thom to characterize the $\mathcal{L}$-genus as a combinatorial invariant (Milnor and Stasheff [111, §20]).

□

REMARK 16.3 The characterization of topological block bundles as $\mathbb{L}^{\cdot}$-oriented spherical fibrations generalizes the characterization due to Sullivan [164] of topological block bundles away from 2 as $KO[1/2]$-oriented spherical fibrations, which is itself a generalization of the Atiyah–Bott–Shapiro KO-orientation of spin bundles. See Madsen and Milgram [99, 5A] for a homotopy-theoretic account of the $KO[1/2]$-orientation of PL-bundles. The characterization of topological block bundles as spherical fibrations with algebraic Poincaré transversality (i.e. an $\mathbb{L}^{\cdot}$-orientation) corresponds to the characterization of topological block bundles as spherical fibrations with geometric Poincaré transversality due to Levitt and Morgan [90], Brumfiel and Morgan [20].

□

If X is an n-dimensional geometric Poincaré complex with Spivak normal structure $(\nu_X: X \longrightarrow BG(k), \rho_X: S^{n+k} \longrightarrow T(\nu_X))$ then $X_+ = X \cup \{pt.\}$ is the S-dual to $T(\nu_X)$, with S-duality isomorphisms

$$\dot{h}^{n+k-*}(T(\nu_X)) \cong h_*(X)$$

for any generalized homology theory h. The topological reducibility obstruction of ν_X

$$t(\nu_X) \in \dot{H}^{k+1}(T(\nu_X); \mathbb{L}_{\cdot}) = H_{n-1}(X; \mathbb{L}_{\cdot})$$

will now be interpreted as the obstruction to lifting the fundamental $\widehat{\mathbb{L}}^{\cdot}$-homology class

$$[X]_{\widehat{\mathbb{L}}} = \widehat{U}_{\nu_X} \in \dot{H}^k(T(\nu_X); \widehat{\mathbb{L}^{\cdot}}) = H_n(X; \widehat{\mathbb{L}^{\cdot}})$$

to a fundamental $\mathbb{L}^{\cdot}$-homology class $[X]_{\mathbb{L}} \in H_n(X; \mathbb{L}^{\cdot})$. In the first instance it is shown that every finite geometric Poincaré complex X is homotopy equivalent to a compact polyhedron with a 1/2-connective symmetric normal structure, allowing the direct construction of $[X]_{\widehat{\mathbb{L}}}$ as the cobordism class of an n-dimensional 1/2-connective symmetric normal complex in

$\mathbb{A}(\mathbb{Z}, X)$. This will also allow the refinement of the visible symmetric signature $\sigma^*(X) \in VL^n(\mathbb{Z}, X)$ defined in §9 to a $1/2$-connective visible symmetric signature $\sigma^*(X) \in VL^n(X)$. The total surgery obstruction $s(X) \in \mathbb{S}_n(X)$ will be defined in §17 as the boundary of $\sigma^*(X) \in VL^n(X)$, such that $s(X) = 0$ if and only if $\sigma^*(X) = A([X]_{\mathbb{L}}) \in VL^n(X)$ for a fundamental $\mathbb{L}^{\textbf{.}}$-homology class $[X]_{\mathbb{L}} \in H_n(X; \mathbb{L}^{\textbf{.}})$.

DEFINITION 16.4 An *n-circuit* is a finite n-dimensional simplicial complex X such that the sum of all the n-simplexes is a cycle

$$[X] = \sum_{\tau \in X^{(n)}} \tau \in \ker(d\colon \Delta(X)_n \longrightarrow \Delta(X)_{n-1}) \, ,$$

possibly using twisted coefficients (in the nonorientable case).

□

By the Poincaré disc theorem of Wall [174, 2.4] every connected finite n-dimensional geometric Poincaré complex X is homotopy equivalent to $Y \cup e^n$ for a finite $(n-1)$-dimensional CW complex Y, and hence to an n-circuit. Thus in dealing with the homotopy theory of finite geometric Poincaré complexes there is no loss of generality in only considering circuits, and for the remainder of §16 only such complexes will be considered.

Let then X be a finite n-dimensional geometric Poincaré complex which is an n-circuit. (It is not assumed that each $(n-1)$-simplex in X is the face of two n-simplexes, cf. 16.8.) As in 9.13 define an n-dimensional globally Poincaré normal complex (C, ϕ) in $\mathbb{A}(\mathbb{Z}, X)$ with

$$C(X) = \Delta(X') \, , \quad C(\tau) = \Delta(D(\tau, X), \partial D(\tau, X)) \ (\tau \in X) \, .$$

The $(\mathbb{Z}, X)$-module duality chain map

$$\phi_0(X) = [X] \cap - \colon C^{n-*}(X) \simeq \Delta(X)^{n-*} \longrightarrow C(X) \simeq \Delta(X)$$

has components

$$\phi_0(\tau) = [D(\tau, X)] \cap - \colon$$

$$C^{n-*}(\tau) = \Delta(D(\tau, X))^{n-|\tau|-*} \longrightarrow C(\tau) = \Delta(D(\tau, X), \partial D(\tau, X)) \, .$$

For every simplex $\tau \in X$ up to chain equivalence

$$\phi_0(\tau) \colon S^{n-|\tau|}\mathbb{Z} \longrightarrow S^{-|\tau|}\Delta(X, X \backslash \{\hat{\tau}\}) \, ; \ 1 \longrightarrow [X][\tau] \, ,$$

and

$$H_*(\phi_0(\tau)) =$$

$$H_*([D(\tau, X)] \cap - \colon \Delta(D(\tau, X))^{n-|\tau|-*} \longrightarrow \Delta(D(\tau, X), \partial D(\tau, X))) \, ,$$

with an exact sequence

$$\ldots \longrightarrow H^{n-|\tau|-r}(\{\hat{\tau}\}) \xrightarrow{[X] \cap -} H_{r+|\tau|}(X, X \backslash \{\hat{\tau}\}) \longrightarrow$$

$$H_r(\phi_0(\tau)) \longrightarrow H^{n-|\tau|-r+1}(\{\hat{\tau}\}) \longrightarrow \ldots \, .$$

The n-dimensional normal complex (C, ϕ) in $\mathbb{A}(\mathbb{Z}, X)$ is 0-connective and globally Poincaré, and the boundary $(n-1)$-dimensional quadratic complex in $\mathbb{A}(\mathbb{Z}, X)$

$$\partial(C, \phi) = (\partial C, \psi)$$

is 0-connective, locally Poincaré and globally contractible, with

$$\partial C(\tau) = S^{-1}C(\phi_0(\tau): C^{n-*}(\tau) \longrightarrow C(\tau)),$$

$$H_*(\partial C(\tau)) = H_{*+1}(\phi_0(\tau)) \quad (\tau \in X).$$

For each n-simplex $\rho \in X^{(n)}$ the (-1)-dimensional quadratic complex $(\partial C(\rho), \psi(\rho))$ in $\mathbb{A}(\mathbb{Z})$ is contractible (since $D(\rho, X) = \{\hat{\rho}\}$ is a 0-dimensional Poincaré complex), so that for each $(n-1)$-simplex $\tau \in X^{(n-1)}$ the 0-dimensional quadratic complex $(\partial C(\tau), \psi(\tau))$ in $\mathbb{A}(\mathbb{Z})$ is Poincaré. In view of the exact sequence given by 15.11 (iii)

$$\ldots \longrightarrow \mathbb{S}_n(X) \longrightarrow \mathbb{S}_n\langle 0\rangle(\mathbb{Z}, X) \longrightarrow H_{n-1}(X; \mathbb{L}_0(\mathbb{Z})) \longrightarrow \mathbb{S}_{n-1}(X) \longrightarrow \ldots$$

the image of $(\partial C, \psi) \in \mathbb{S}_n\langle 0\rangle(\mathbb{Z}, X)$ is the element

$$c(X) = \sum_{\tau \in X^{(n-1)}} \tau(\partial C(\tau), \psi(\tau))$$

$$\in H_{n-1}(X; \mathbb{L}_0(\mathbb{Z})) = H_{n-1}(X; \mathbb{L}.\langle 1\rangle(\mathbb{Z}) \longrightarrow \mathbb{L}.\langle 0\rangle(\mathbb{Z}))$$

which is the obstruction to the existence of a 0-connective locally Poincaré globally contractible quadratic cobordism $(\partial C \oplus \partial C' \longrightarrow D, (\delta\psi, \psi \oplus -\psi'))$ between $(\partial C, \psi)$ and a 1-connective locally Poincaré globally contractible quadratic complex $(\partial C', \psi')$ in $\mathbb{A}(\mathbb{Z}, X)$. Such a complex is the boundary of the union n-dimensional normal complex in $\mathbb{A}(\mathbb{Z}, X)$

$$(C', \phi') = (C, \phi) \cup_\partial (D, (1+T)\delta\psi)$$

which is 1/2-connective and globally Poincaré with

$$(\partial C', \psi') = \partial(C', \phi').$$

Each $(n-1)$-simplex $\tau \in X^{(n-1)}$ is the face of an even number (say $2m_\tau$) of n-simplexes, and $D(\tau, X)$ is the one-vertex union of $2m_\tau$ 1-simplexes. The 1-dimensional normal pair $(D(\tau, X), \partial D(\tau, X))$ may be resolved by a normal degree 1 map

$$(\overline{D}(\tau, X), \partial D(\tau, X)) \longrightarrow (D(\tau, X), \partial D(\tau, X))$$

from a 1-dimensional manifold with boundary $(\overline{D}(\tau, X), \partial D(\tau, X))$, with $\overline{D}(\tau, X)$ the disjoint union of m_τ 1-simplexes. The resolution determines a vanishing of the obstruction $c(X) \in H_{n-1}(X; \mathbb{L}_0(\mathbb{Z}))$ on the chain level, corresponding to a 1/2-connective globally Poincaré n-dimensional normal complex (C', ϕ') in $\mathbb{A}(\mathbb{Z}, X)$.

DEFINITION 16.5 The *1/2-connective visible symmetric signature* of a finite n-dimensional geometric Poincaré complex X is the cobordism class

$$\sigma^*(X) \ = \ (C', \phi') \in VL^n(X) ,$$

with (C', ϕ') as defined above.

□

The visible symmetric signature $\sigma^*(X) = (C, \phi) \in VL^n(\mathbb{Z}, X)$ of 9.13 is the image of the 1/2-connective visible symmetric signature under the natural map $VL^n(X) \longrightarrow VL^n(\mathbb{Z}, X)$ which forgets the 1/2-connective structure.

DEFINITION 16.6 A chain map $f \colon C \longrightarrow D$ in $\mathbb{A}(\mathbb{Z}, X)$ is a *global 1-equivalence* if the algebraic mapping cone $\mathcal{C}(f)$ is 2-connective and globally contractible.

□

The following conditions on an n-dimensional symmetric complex (C, ϕ) in $\mathbb{A}(\mathbb{Z}, X)$ are equivalent:
  (i) (C, ϕ) is locally 1-Poincaré and globally Poincaré,
  (ii) the duality chain map $\phi_0 \colon C^{n-*} \longrightarrow C$ is a global 1-equivalence,
  (iii) the $(n-1)$-dimensional quadratic complex $\partial(C, \phi)$ is 1-connective, locally Poincaré and globally contractible.

PROPOSITION 16.7 *The following conditions on a finite n-dimensional geometric Poincaré complex X are equivalent:*
(i) *the 1/2-connective visible symmetric signature $\sigma^*(X) \in VL^n(X)$ is the assembly of an $\mathbb{L}^{\cdot}$-homology fundamental class $[X]_{\mathbb{L}} \in H_n(X; \mathbb{L}^{\cdot})$*

$$\sigma^*(X) \ = \ A([X]_{\mathbb{L}}) \in VL^n(X) ,$$

(ii) *$\sigma^*(X) \in VL^n(X)$ is represented by a 0-connective n-dimensional globally Poincaré normal complex (C', ϕ') in $\mathbb{A}(\mathbb{Z}, X)$ which is globally 1-equivalent to a 0-connective n-dimensional locally Poincaré normal complex (B, θ) in $\mathbb{A}(\mathbb{Z}, X)$, with*

$$[X]_{\mathbb{L}} \ = \ (B, \theta) \in H_n(X; \mathbb{L}^{\cdot}) \ , \quad \sigma^*(X) \ = \ (C', \phi') \in VL^n(X) .$$

PROOF (ii) $\Longrightarrow$ (i) Globally 1-equivalent globally 1-Poincaré complexes are globally 1-Poincaré cobordant.
(i) $\Longrightarrow$ (ii) Let (C'', ϕ'') be a 0-connective n-dimensional locally Poincaré normal complex in $\mathbb{A}(\mathbb{Z}, X)$ realizing $[X]_{\mathbb{L}} \in H_n(X; \mathbb{L}^{\cdot})$, and let $(C' \oplus C'' \longrightarrow D, (\delta\phi, \phi' \oplus -\phi''))$ be a 0-connective globally Poincaré cobordism in $\mathbb{A}(\mathbb{Z}, X)$ realizing $\sigma^*(X) - A([X]_{\mathbb{L}}) = 0 \in VL^n(X)$. The relative boundary construction gives a 0-connective $(n+1)$-dimensional locally Poincaré normal triad in $\mathbb{A}(\mathbb{Z}, X)$

$$\left(\begin{array}{ccc} \partial(C' \oplus C'') & \longrightarrow & \partial D \\ \downarrow & & \downarrow \\ (C' \oplus C'')^{n-*} & \longrightarrow & (D/(C' \oplus C''))^{n+1-*} \end{array} \quad , \quad \begin{array}{ccc} \partial\phi' \oplus -\partial\phi'' & \longrightarrow & \partial\delta\phi \\ \downarrow & & \downarrow \\ 0 & \longrightarrow & 0 \end{array} \right)$$

with
$$\partial C'' \;=\; S^{-1}C(\phi_0'' \colon C''^{n-*} \longrightarrow C'')$$
locally contractible and
$$\partial C' \;=\; S^{-1}C(\phi_0 \colon C'^{n-*} \longrightarrow C') \,,$$
$$\partial D \;=\; S^{-1}C(\delta\phi_0 \colon (D/(C' \oplus C''))^{n+1-*} \longrightarrow D)$$
globally contractible. The union n-dimensional normal complex
$$(B,\theta) \;=\; (C'^{n-*},0) \cup_{(\partial C', \partial\phi')} (\partial D, \partial\delta\phi)/\partial C''$$
is locally Poincaré, and the projection
$$(B,\theta) \;\longrightarrow\; (B,\theta)/\partial D \;=\; (C',\phi')$$
is a global 1-equivalence.

$\square$

REMARK 16.8 An n-dimensional pseudomanifold X is an n-circuit such that each $(n-1)$-simplex is the face of two n-simplexes (cf. 8.5). An n-dimensional pseudomanifold X is *normal* if the natural maps define isomorphisms
$$H_n(X) \;\xrightarrow{\;\simeq\;}\; H_n(X, X\backslash\{x\}) \quad (x \in X) \,.$$
Normal pseudomanifolds are called normal circuits by McCrory [101]. The following conditions on an n-dimensional pseudomanifold X are equivalent:
(i) X is normal,
(ii) the link of each simplex of dimension $\leq n-2$ is connected,
(iii) the local homology groups $H_n(X, X\backslash\{\hat\tau\})$ $(\tau \in X)$ are infinite cyclic, with generators
$$[D(\tau, X)] \;=\; [X][\tau] \;=\; \sum_{\rho \geq \tau, |\rho|=n} \rho$$
$$\in H_{n-|\tau|}(D(\tau, X), \partial D(\tau, X)) \;=\; H_n(X, X\backslash\{\hat\tau\})$$
the images of the fundamental class of X
$$[X] \;=\; \sum_{\tau \in X, |\tau|=n} \tau \in H_n(X) \,,$$

(iv) the 0-connective n-dimensional normal complex (C, ϕ) in $\mathbb{A}(\mathbb{Z}, X)$ with

$$C(X) = \Delta(X') \ , \ \ C(\tau) = \Delta(D(\tau, X), \partial D(\tau, X)) \ ,$$

$$\phi_0(\tau) = [D(\tau, X)] \cap - : C^{n-*}(\tau) = \Delta(D(\tau, X))^{n-|\tau|-*}$$

$$\longrightarrow C(\tau) = \Delta(D(\tau, X), \partial D(\tau, X)) \ \ (\tau \in X)$$

is locally 1-Poincaré, with

$$H_r(\phi_0(\tau)) = 0 \ \ (r \leq 1, \tau \in X) \ ,$$

(v) the locally Poincaré $(n-1)$-dimensional quadratic complex $(\partial C, \psi)$ in $\mathbb{A}(\mathbb{Z}, X)$ is 1-connective, with

$$H_r(\partial C(\tau)) = 0 \ \ (r \leq 0, \ \tau \in X) \ .$$

The equivalence of (i) and (ii) is due to Goresky and MacPherson [59, p. 151].
The equivalence of (i) and (iv) is the special case $q = 1$ of 15.6.
For an n-dimensional geometric Poincaré complex which is a normal pseudo-manifold the 1/2-connective visible symmetric signature $\sigma^*(X) \in VL^n(X)$ is represented by the 0-connective locally 1-Poincaré globally Poincaré symmetric complex (C, ϕ) in $\mathbb{A}(\mathbb{Z}, X)$

$$\sigma^*(X) = (C, \phi) \in VL^n(X) \ .$$

$\square$

DEFINITION 16.9 (i) The *canonical $\widehat{\mathbb{L}}$-homology fundamental class* of an n-dimensional normal complex X is the cobordism class

$$[X]_{\widehat{\mathbb{L}}} = (C, \phi) \in H_n(X; \widehat{\mathbb{L}}) \ ,$$

with $C(X) = \Delta(X')$.
(ii) An n-dimensional geometric Poincaré complex X is *topologically reducible* if the Spivak normal fibration $\nu_X : X \longrightarrow BG$ admits a topological reduction $\tilde{\nu}_X : X \longrightarrow BTOP$.
(iii) The *topological reducibility obstruction* of an n-dimensional geometric Poincaré complex X is the image

$$t(X) = \partial [X]_{\widehat{\mathbb{L}}} \in H_{n-1}(X; \mathbb{L}.)$$

of $[X]_{\widehat{\mathbb{L}}} \in H_n(X; \widehat{\mathbb{L}})$ under the connecting map ∂ in the exact sequence

$$\ldots \longrightarrow H_n(X; \mathbb{L}.) \overset{1+T}{\longrightarrow} H_n(X; \mathbb{L}^{\cdot})$$

$$\overset{J}{\longrightarrow} H_n(X; \widehat{\mathbb{L}}^{\cdot}) \overset{\partial}{\longrightarrow} H_{n-1}(X; \mathbb{L}.) \longrightarrow \ldots .$$

$\square$

PROPOSITION 16.10 *An n-dimensional geometric Poincaré complex X is topologically reducible if and only if $t(X) = 0 \in H_{n-1}(X; \mathbb{L}.)$.*

PROOF Let $(\nu_X \colon X \longrightarrow BG(k), \rho_X \colon S^{n+k} \longrightarrow T(\nu_X))$ be a Spivak normal structure. The fundamental $\widehat{\mathbb{L}}^{\cdot}$-homology class of X is the S-dual of the canonical $\widehat{\mathbb{L}}^{\cdot}$-orientation of ν_X

$$[X]_{\widehat{\mathbb{L}}^{\cdot}} = \widehat{U}_{\nu_X} \in H_n(X; \widehat{\mathbb{L}}^{\cdot}) = \dot{H}^k(T(\nu_X); \widehat{\mathbb{L}}^{\cdot}),$$

and $t(X)$ is the S-dual of the topological reducibility obstruction of ν_X

$$t(X) = \delta(\widehat{U}_{\nu_X}) = t(\nu_X) \in H_{n-1}(X; \mathbb{L}.) = \dot{H}^{k+1}(T(\nu_X); \mathbb{L}.).$$

□

A polyhedron K is an *n-dimensional combinatorial* $\left\{ \begin{array}{l} \overline{homotopy} \; manifold \\ homology \end{array} \right.$

if the links of i-simplexes are $\left\{ \begin{array}{l} PL \\ homotopy \\ homology \end{array} \right.$ $(n-i-1)$-spheres.

REMARK 16.11 (i) A *triangulation* (K, h) of a topological space M is a polyhedron K with a homeomorphism $h \colon K \longrightarrow M$. If M is an n-dimensional topological manifold then K is an n-dimensional combinatorial homology manifold. Siebenmann [156] showed that for $n \geq 5$ an n-dimensional combinatorial homotopy manifold is an n-dimensional topological manifold.

(ii) A triangulation (K, h) of a topological manifold M is *combinatorial* if K is a combinatorial manifold. A *PL* manifold is a topological manifold with a *PL* equivalence class of combinatorial triangulations. The Hauptvermutung for manifolds was that every homeomorphism of compact *PL* manifolds is homotopic to a *PL* homeomorphism. The Casson–Sullivan invariant for a homeomorphism $f \colon N \longrightarrow M$ of compact n-dimensional *PL* manifolds (Armstrong et al. [5])

$$\kappa(f) = \kappa(M \longrightarrow TOP/PL) \in H^3(M; \mathbb{Z}_2) = H_{n-3}(M; \mathbb{Z}_2)$$

is such that $\kappa(f) = 0$ if (and for $n \geq 5$ only if) f is homotopic to a *PL* homeomorphism (13.1), with $M \longrightarrow TOP/PL = K(\mathbb{Z}_2, 3)$ the classifying map for the topological trivialization determined by f of the difference $\nu_M - (f^{-1})^* \nu_N \colon M \longrightarrow BPL$ of stable *PL* normal bundles. For $n \geq 5$ every element

$$\kappa \in \mathbb{S}^{PL}(T^n) = [T^n, TOP/PL] = H^3(T^n; \mathbb{Z}_2)$$

is realized as $\kappa = \kappa(f)$ for a homeomorphism $f \colon T'^n \longrightarrow T^n$ from a fake *PL* n-dimensional torus T'^n, with a normal map

$$(F, B) \colon (W^{n+1}; T^n, T'^n) \longrightarrow T^n \times ([0, 1]; \{0\}, \{1\})$$

on a *PL* cobordism $(W^{n+1}; T^n, T'^n)$, such that $F|_{T^n} = \mathrm{id}.$, $F|_{T'^n} = f$, providing counterexamples to the Hauptvermutung for manifolds. The rel ∂ surgery obstruction

$$\sigma_*(F, B) = (C, \psi) \in L_{n+1}(\mathbb{Z}[\mathbb{Z}^n]) = H_{n+1}(T^n; \mathbb{L}.)$$

is represented by an $(n+1)$-dimensional quadratic Poincaré complex (C, ψ) in $\mathbb{A}(\mathbb{Z})_*(T^n)$, and

$$\kappa(f) \;=\; \sum_{\sigma \in (T^n)^{(n-3)}} (\text{signature}(C(\sigma), \psi(\sigma))/8) \, \sigma$$

$$\in H^3(T^n; \mathbb{Z}_2) \;=\; H_{n-3}(T^n; \mathbb{Z}_2)$$

is an image of $\sigma_*(F, B) \in L_{n+1}(\mathbb{Z}[\mathbb{Z}^n])$. The surgery-theoretic classification of the PL structures on T^n (by Casson, Hsiang, Shaneson and Wall) is an essential ingredient of the obstruction theory of Kirby and Siebenmann [84] for the existence and uniqueness of combinatorial triangulations on compact topological manifolds in dimensions ≥ 5.

(iii) The Kirby–Siebenmann invariant of a compact n-dimensional topological manifold M

$$\kappa(M) \in H^4(M; \mathbb{Z}_2) \;=\; H_{n-4}(M; \mathbb{Z}_2)$$

is such that $\kappa(M) = 0$ if (and for $n \geq 5$ only if) M admits a combinatorial triangulation. By construction, $\kappa(M)$ is the homotopy class of the composite

$$\kappa(M) : \; M \xrightarrow{\;\nu_M\;} BTOP \longrightarrow B(TOP/PL) \;=\; K(\mathbb{Z}_2, 4) \;,$$

and is such that $\kappa(M) = 0$ if and only if $\nu_M \colon M \longrightarrow BTOP$ lifts to a PL reduction $\tilde{\nu}_M \colon M \longrightarrow BPL$. The invariant is realized by compact topological manifolds in each dimension ≥ 5 which do not admit combinatorial triangulation. For example, if $f \colon T^n \longrightarrow T'^n$, (F, B), W^{n+1} are as in (ii) then the $(n+1)$-dimensional topological manifold

$$N^{n+1} \;=\; W^{n+1} \cup_{f \sqcup \mathrm{id}.} T^n \times [0, 1]$$

is equipped with a normal map $(g, c) : N^{n+1} \longrightarrow T^{n+1}$ such that

$$\sigma_*(g, c) \;=\; (\sigma_*(F, B), 0)$$

$$\in L_{n+1}(\mathbb{Z}[\mathbb{Z}^{n+1}]) \;=\; L_{n+1}(\mathbb{Z}[\mathbb{Z}^n]) \oplus L_n(\mathbb{Z}[\mathbb{Z}^n]) \;,$$

$$g_*\kappa(N) \;=\; (\kappa(f), 0)$$

$$\in H_{n-3}(T^{n+1}; \mathbb{Z}_2) \;=\; H_{n-3}(T^n; \mathbb{Z}_2) \oplus H_{n-4}(T^n; \mathbb{Z}_2) \;.$$

(iv) Let θ_3 be the cobordism group of oriented 3-dimensional combinatorial manifolds which are homotopy spheres, modulo those which bound contractible 4-dimensional combinatorial manifolds. Cohen [37, §4] defined an invariant of a compact n-dimensional combinatorial homotopy manifold K

$$c(K) \;=\; \sum_{\sigma \in K^{(n-4)}} [\mathrm{link}_K(\sigma)] \, \sigma \in H^4(K; \theta_3) = H_{n-4}(K; \theta_3)$$

such that $c(K) = 0$ if and only if K admits a PL resolution, i.e. a transversely cellular PL map $M \longrightarrow K$ from an n-dimensional combinatorial manifold M.

(v) Let θ_3^H be the cobordism group of oriented 3-dimensional combinatorial homology manifolds which are homology spheres, modulo those which bound acyclic 4-dimensional combinatorial manifolds. Let $\alpha : \theta_3^H \longrightarrow \mathbb{Z}_2$ be the Kervaire–Milnor–Rohlin epimorphism, with

$$\alpha(\Sigma^3) \ = \ \text{signature}\,(W)/8 \in \mathbb{Z}_2$$

for any parallelizable 4-dimensional combinatorial manifold W with boundary $\partial W = \Sigma^3$. If $\Delta = (K, h)$ is a triangulation of a compact n-dimensional topological manifold M the element

$$\kappa_\Delta(M) \ = \ \sum_{\sigma \in K^{(n-4)}} [\text{link}_K(\sigma)]\,\sigma \in H^4(M; \theta_3^H) \ = \ H_{n-4}(M; \theta_3^H)$$

is such that $\kappa_\Delta(M) = 0$ if (and for $n \geq 5$ only if) Δ is a combinatorial triangulation of M. The combinatorial triangulation obstruction is an image of the triangulation obstruction

$$\kappa(M) \ = \ \alpha(\kappa_\Delta(M)) \ = \ \sum_{\sigma \in K^{(n-4)}} (\text{signature}\,W(\sigma)/8)\,\sigma$$

$$\in H^4(M; \mathbb{Z}_2) \ = \ H_{n-4}(M; \mathbb{Z}_2)\,,$$

with $W(\sigma)$ a parallelizable 4-dimensional combinatorial manifold with boundary $\partial W(\sigma) = \text{link}_K(\sigma)$.

(vi) A triangulation (K, h) of a topological manifold M is *non-combinatorial* if K is not a combinatorial manifold. Non-simply connected combinatorial homology $(n-2)$-spheres H provided examples of non-combinatorial triangulations $(\Sigma^2 H, h)$ of S^n ($n \geq 5$), with a copy of H as the link of each 1-simplex in the suspension circle of the double suspension $\Sigma^2 H$ (Edwards, see Daverman [41, II.12]).

(vii) Galewski and Stern [55], [56] showed that for $n \geq 5$ a compact n-dimensional combinatorial homology manifold has the homotopy type of a compact n-dimensional topological manifold, and that a compact n-dimensional topological manifold M admits a triangulation if and only if the Kirby–Siebenmann invariant $\kappa(M) \in H^4(M; \mathbb{Z}_2)$ is such that

$$\delta\kappa(M) \ = \ 0 \in H^5(M; \ker(\alpha))\,,$$

with δ the connecting map in the coefficient exact sequence

$$\ldots \longrightarrow H^4(M; \ker(\alpha)) \longrightarrow H^4(M; \theta_3^H) \xrightarrow{\alpha} H^4(M; \mathbb{Z}_2)$$

$$\xrightarrow{\delta} H^5(M; \ker(\alpha)) \longrightarrow \ldots\,.$$

(viii) The Casson invariant of 3-dimensional combinatorial homology spheres shows that certain compact 4-dimensional topological manifolds are not triangulable (Akbulut and McCarthy [1, p. xvi]). In particular, the Freedman

manifold M^4 with
$$\sigma^*(M) = (\mathbb{Z}^8, E_8) = 8 \in L^4(\mathbb{Z}) = \mathbb{Z},$$
$$\kappa(M) = 1 \in H^4(M; \mathbb{Z}_2) = \mathbb{Z}_2$$
is not triangulable (Freedman and Quinn [53, 10.1]).

(ix) Compact n-dimensional topological manifolds with $n \geq 5$ are finite CW complexes, by virtue of the topological handlebody decomposition obtained by Kirby and Siebenmann [84] for $n \geq 6$ and by Quinn for $n = 5$ ([53, 9.1]). At present, it is not known if every compact topological manifold of dimension ≥ 5 is triangulable.

□

The following conditions on a finite n-dimensional geometric Poincaré complex X are equivalent:

(i) X is an n-dimensional combinatorial homology manifold,

(ii) the algebraic normal complex (C, ϕ) in $\mathbb{A}(\mathbb{Z}, X)$ with $C(X) = \Delta(X')$ is locally Poincaré,

(iii) the quadratic boundary $(\partial C, \psi)$ is locally contractible.

Transversality is a generic property of maps on manifolds, but not of maps on geometric Poincaré complexes.

DEFINITION 16.12 Let X, Y be compact polyhedra, with Y an n-dimensional geometric Poincaré complex. A simplicial map $h : Y \longrightarrow X'$ is *Poincaré transverse* if each
$$(Y(\tau), \partial Y(\tau)) = h^{-1}(D(\tau, X), \partial D(\tau, X)) \quad (\tau \in X)$$
is an $(n - |\tau|)$-dimensional $\mathbb{Z}$-coefficient Poincaré pair.

□

EXAMPLE 16.13 If Y is an n-dimensional combinatorial homology manifold then every simplicial map $h : Y \longrightarrow X'$ is Poincaré transverse, since each $(Y(\tau), \partial Y(\tau))$ $(\tau \in Y)$ is an $(n - |\tau|)$-dimensional combinatorial homology manifold with boundary.

□

In dealing with the L-theoretic properties of topological manifolds in §17 use will be made of the following version of an 'intrinsic transversality structure' of Levitt and Ranicki [91].

DEFINITION 16.14 A *transversality structure* $\Pi = (X, Y, g, h)$ on a finite n-dimensional Poincaré space Z consists of compact polyhedra X, Y together with homotopy equivalences $g : Y \longrightarrow Z$, $h : Y \longrightarrow X'$ such that h is simplicial and Poincaré transverse.

□

PROPOSITION 16.15 *A transversality structure* $\Pi = (X, Y, g, h)$ *on a finite n-dimensional Poincaré space* Z *determines a fundamental* $\mathbb{L}^{\cdot}$-*homology class*

$$[Z]_\Pi = (gh^{-1})_*(C, \phi) \in H_n(Z; \mathbb{L}^{\cdot})$$

with (C, ϕ) *the n-dimensional locally Poincaré normal complex in* $\mathbb{A}(\mathbb{Z})_*(X)$ *defined by*

$$C(\tau) = \Delta(Y(\tau), \partial Y(\tau)) \quad (\tau \in X) .$$

The 1/2-connective visible symmetric signature of Z *is the assembly of* $[Z]_\Pi$

$$\sigma^*(Z) = A([Z]_\Pi) \in VL^n(Z)$$

and the total surgery obstruction is $s(Z) = 0 \in \mathbb{S}_n(Z)$.

$\square$

In §17 it will be proved that a finite n-dimensional Poincaré space Z admits a transversality structure Π if (and for $n \geq 5$ only if) Z is homotopy equivalent to a compact n-dimensional topological manifold. In the first instance we have:

PROPOSITION 16.16 *If* M *is a finite n-dimensional Poincaré space which is*
   either (i) *a combinatorial homology manifold*
  or   (ii) *a topological manifold*
then M *has a canonical transversality structure* $\Pi = (X, Y, g, h)$ *and hence a canonical fundamental* $\mathbb{L}^{\cdot}$-*homology class*

$$[M]_\mathbb{L} = [M]_\Pi \in H_n(M; \mathbb{L}^{\cdot})$$

with the following properties:
(a) *The assembly of* $[M]_\mathbb{L}$ *is the 1/2-connective visible symmetric signature*

$$\sigma^*(M) = A([M]_\mathbb{L}) \in VL^n(M) .$$

(b) $[M]_\mathbb{L}$ *has image the canonical* $\widehat{\mathbb{L}}^{\cdot}$-*homology fundamental class*

$$J[M]_\mathbb{L} = [M]_{\widehat{\mathbb{L}}} \in H_n(M; \widehat{\mathbb{L}}^{\cdot}) .$$

(c) *The canonical* $\mathbb{L}^{\cdot}$-*cohomology orientation* $U_{\tilde{\nu}_M} \in \dot{H}^k(T(\nu_M); \mathbb{L}^{\cdot})$ *of the topological normal block bundle* $\tilde{\nu}_M : M \longrightarrow B\widetilde{TOP}(k)$ *of an embedding* $M^n \subset S^{n+k}$ *is the S-dual of* $[M]_\mathbb{L} \in H_n(M; \mathbb{L}^{\cdot})$.
(d) *If* $N^{n-k} \subset M^n$ *is a codimension* k *submanifold with a normal block bundle* $\nu = \nu_{N \subset M} : N \longrightarrow B\widetilde{TOP}(k)$ *then the canonical* $\mathbb{L}^{\cdot}$-*homology fundamental classes* $[M]_\mathbb{L} \in H_n(M; \mathbb{L}^{\cdot})$, $[N]_\mathbb{L} \in H_{n-k}(N; \mathbb{L}^{\cdot})$ *and the canonical* $\mathbb{L}^{\cdot}$-*cohomology orientation* $U_\nu \in \dot{H}^k(T(\nu); \mathbb{L}^{\cdot})$ *are related by*

$$j_*[M]_\mathbb{L} \cap U_\nu = [N]_\mathbb{L} \in H_{n-k}(N; \mathbb{L}^{\cdot}) ,$$

$$i_*[N]_\mathbb{L} = [M]_\mathbb{L} \cap j^* U_\nu \in H_{n-k}(M; \mathbb{L}^{\cdot})$$

with

$i = $ inclusion $: N \longrightarrow M$, $j = $ projection $: M_+ = M \cup \{\text{pt.}\} \longrightarrow T(\nu)$.

PROOF (i) The canonical transversality structure is defined by

$$(X, Y, g, h) = (M, M', \mathrm{id.}, \mathrm{id.}) ,$$

and the corresponding canonical $\mathbb{L}$-homology fundamental class of M is the cobordism class

$$[M]_{\mathbb{L}} = (C, \phi) \in H_n(M; \mathbb{L}^{\cdot})$$

of the n-dimensional symmetric Poincaré complex (C, ϕ) in $\mathbb{A}(\mathbb{Z})_*(M)$ with $C(M) = \Delta(M')$.

(ii) Any map $f : M \longrightarrow X$ to a compact polyhedron X can be made topologically transverse, with the inverse images

$$(M(\sigma), \partial M(\sigma)) = f^{-1}(D(\sigma, X), \partial D(\sigma, X)) \quad (\sigma \in X)$$

$(n-|\sigma|)$-dimensional submanifolds, some of which may be empty. Let (Y, Z) be a closed neighbourhood of M in $\mathbb{R}^{n+k}$ (k large), a compact $(n+k)$-dimensional PL manifold with boundary which is the total space of a topological (D^k, S^{k-1})-bundle $\nu_M : M \longrightarrow BTOP(k)$. By Quinn [130] Y can be taken to be the mapping cylinder of a map $e : Z \longrightarrow M$. Make e PL transverse, and define an X-dissection $\{Y(\sigma) \, | \, \sigma \in X\}$ of Y by

$$Y(\sigma) = \text{mapping cylinder of } e| : e^{-1}M(\sigma) \longrightarrow M(\sigma) \quad (\sigma \in X) .$$

The projection $g : Y \longrightarrow M$ is a hereditary homotopy equivalence, so that each $(Y(\sigma), \partial Y(\sigma))$ is a simplicial $(n-|\sigma|)$-dimensional geometric Poincaré pair homotopy equivalent to $(M(\sigma), \partial M(\sigma))$. The composite

$$h = fg : Y \xrightarrow{\ g\ } M \xrightarrow{\ f\ } X$$

is such that

$$h^{-1}D(\sigma, X) = g^{-1}M(\sigma) = Y(\sigma) \quad (\sigma \in X) .$$

In particular, if $f : M \longrightarrow X$ is a homotopy equivalence in the preferred simple homotopy type of M (e.g. the inclusion $M \subset Y$), then (X, Y, g, h) defines the canonical transversality structure on M.

□

REMARK 16.17 (i) By 16.2 the canonical $\mathbb{L}$-homology fundamental class of an oriented n-dimensional manifold M^n is given rationally by the Poincaré dual of the $\mathcal{L}$-genus $\mathcal{L}(M) \in H^{4*}(M; \mathbb{Q})$

$$[M]_{\mathbb{L}} \otimes \mathbb{Q} = \mathcal{L}(M) \cap [M]_{\mathbb{Q}} \in H_n(M; \mathbb{L}^{\cdot}) \otimes \mathbb{Q} = \sum_{k \geq 0} H_{n-4k}(M; \mathbb{Q}) ,$$

with $[M]_{\mathbb{Q}} \in H_n(M; \mathbb{Q})$ the $\mathbb{Q}$-coefficient fundamental class. See 24.2 (i) for the evaluation of the signatures of submanifolds $N^{4k} \subset M^n$ in terms of $[M]_{\mathbb{L}} \otimes \mathbb{Q} \in H_n(M; \mathbb{L}^{\cdot}) \otimes \mathbb{Q}$.

(ii) The identity $\sigma^*(M) = A([M]_{\mathbb{L}}) \in VL^n(M)$ is a non-simply connected generalization of the Hirzebruch signature formula in the case $n = 4k$

$$\text{signature}\,(M) \; = \; \langle \mathcal{L}(M), [M]_{\mathbb{Q}} \rangle \in L^{4k}(\mathbb{Z}) \; = \; \mathbb{Z} \; .$$

Also, for any free action of a finite group G on M the identity

$$\sigma^*(M/G) \; = \; A([M/G]_{\mathbb{L}}) \in VL^n(M/G)$$

gives the corresponding special case of the Atiyah–Singer index theorem, that the G-signature of such an action is a multiple of the character of the regular representation. See §22 for rational surgery obstruction theory with finite fundamental group.

$\square$

§17. The total surgery obstruction

The total surgery obstruction $s(X) \in \mathbb{S}_n(X)$ of a finite n-dimensional geometric Poincaré complex X is the invariant introduced in Ranicki [140], such that $s(X) = 0$ if (and for $n \geq 5$ only if) X is homotopy equivalent to a compact n-dimensional topological manifold M^n. Moreover, if $s(X) = 0$ and $n \geq 5$ the manifold structure set $\mathbb{S}^{TOP}(X)$ is in unnatural bijective correspondence with $\mathbb{S}_{n+1}(X)$, as will be shown in §18. Provided the fundamental group $\pi_1(X)$ is 'good' in the sense of Freedman and Quinn [53] these results also hold for $n = 4$. In view of the close connections between the obstruction theories for the existence and uniqueness of manifold structures it is convenient to treat the actual invariants simultaneously, as will be done in §20 in the simply connected case, in §22 for finite fundamental groups, and in §23 for generalized free products and HNN extensions.

The total surgery obstruction unifies the two stages of the obstruction provided by the Browder–Novikov–Sullivan–Wall surgery theory for the existence of a manifold structure in the homotopy type of a geometric Poincaré complex. The first stage is the topological K-theory obstruction to the existence of a topological bundle. The second stage is the algebraic L-theory surgery obstruction to the existence of a homotopy equivalence respecting a choice of topological bundle reduction. As in §16 only the oriented case is considered: see Appendix A for the nonorientable case.

The various generalized homology groups, L-groups and structure groups are related by the following commutative braid of exact sequences, the special case of 15.18 (i) for $R = \mathbb{Z}$, $n \geq 2$:

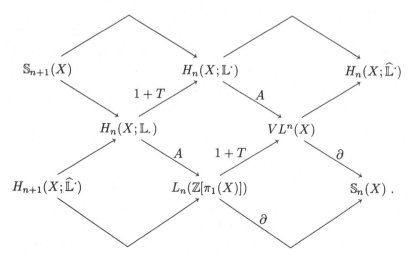

The terminology is as in §16, with

$$\mathbb{L}^{\cdot}\langle 0\rangle(\mathbb{Z}) = \mathbb{L}^{\cdot} , \quad \mathbb{L}^{\cdot}\langle 1\rangle(\mathbb{Z}) = \mathbb{L}_{\cdot} , \quad \mathbb{NL}^{\cdot}\langle 1/2\rangle(\mathbb{Z}) = \widehat{\mathbb{L}}^{\cdot} ,$$

$$\mathbb{S}_*\langle 1\rangle(\mathbb{Z}, X) = \mathbb{S}_*(X) , \quad VL^*\langle 1/2\rangle(\mathbb{Z}, X) = VL^*(X) .$$

Given a finite n-dimensional geometric Poincaré complex X let (C', ϕ') be the 1/2-connective globally Poincaré n-dimensional normal complex in $\mathbb{A}(\mathbb{Z}, X)$ used in 16.5 to define the 1/2-connective visible symmetric signature $\sigma^*(X) = (C', \phi') \in VL^n(X)$.

DEFINITION 17.1 The *total surgery obstruction* of a finite n-dimensional geometric Poincaré complex X is the cobordism class

$$s(X) = \partial\sigma^*(X) \in \mathbb{S}_n(X)$$

represented by the boundary 1-connective locally Poincaré globally contractible $(n-1)$-dimensional quadratic complex $\partial(C', \phi') = (\partial C', \psi')$ in $\mathbb{A}(\mathbb{Z}, X)$.

□

Since $\partial C'(\tau)$ is contractible for n-simplexes $\tau \in X^{(n)}$ the quadratic complex $\partial\sigma^*(X)$ is locally equivalent to a complex in $\mathbb{A}(\mathbb{Z}, X^{[n-1]})$, so that for $n \geq 3$ the total surgery obstruction can be regarded as an element

$$s(X) = \partial\sigma^*(X) \in \mathbb{S}_n(X) = \mathbb{S}_n\langle 0\rangle(\mathbb{Z}, X^{[n-1]}) = \mathbb{S}_n(\mathbb{Z}, X^{[n-1]}) ,$$

using 15.11 (ii) to identify the $\mathbb{S}$-groups.

PROPOSITION 17.2 *The following conditions on a finite n-dimensional geometric Poincaré complex X are equivalent:*

(i) *the total surgery obstruction vanishes*

$$s(X) = 0 \in \mathbb{S}_n(X) ,$$

(ii) *the 1/2-connective visible symmetric signature of X is the assembly $A([X]_{\mathbb{L}})$ of an $\mathbb{L}^{\cdot}$-homology fundamental class $[X]_{\mathbb{L}} \in H_n(X; \mathbb{L}^{\cdot})$*

$$\sigma^*(X) = A([X]_{\mathbb{L}}) \in VL^n(X) .$$

PROOF Immediate from the exact sequence given by 15.18 (i)

$$\cdots \longrightarrow H_n(X; \mathbb{L}^{\cdot}) \overset{A}{\longrightarrow} VL^n(X) \overset{\partial}{\longrightarrow} \mathbb{S}_n(X) \longrightarrow H_{n-1}(X; \mathbb{L}^{\cdot}) \longrightarrow \cdots .$$

□

REMARK 17.3 For a finite n-dimensional geometric Poincaré complex X which is a normal pseudomanifold (16.8) the 1/2-connective visible symmetric signature of X is represented by the 0-connective locally 1-Poincaré globally Poincaré symmetric complex (C, ϕ) in $\mathbb{A}(\mathbb{Z}, X)$ with $C(X) = \Delta(X')$

$$\sigma^*(X) = (C, \phi) \in VL^n(X) ,$$

and the total surgery obstruction is represented by the 1-connective locally Poincaré globally contractible $(n - 1)$-dimensional quadratic complex $\partial(C, \phi) = (\partial C, \psi)$ in $\mathbb{A}(\mathbb{Z}, X)$

$$s(X) = (\partial C, \psi) \in \mathbb{S}_n(X) \ .$$

Note that X is an n-dimensional combinatorial homology manifold if and only if (C, ϕ) is locally Poincaré, in which case the $1/2$-connective visible symmetric signature is the assembly

$$\sigma^*(X) = A([X]_{\mathbb{L}}) \in VL^n(X)$$

of the canonical $\mathbb{L}\cdot$-homology fundamental class

$$[X]_{\mathbb{L}} = (C, \phi) \in H_n(X; \mathbb{L}\cdot) \ ,$$

and the total surgery obstruction is

$$s(X) = \partial \sigma^*(X) = 0 \in \mathbb{S}_n(X) \ .$$

$\square$

The total surgery obstruction $s(X) \in \mathbb{S}_n(X)$ of a finite n-dimensional geometric Poincaré complex X measures the failure of the links of the simplexes $\tau \in X$ to be homology $(n - |\tau| - 1)$-spheres up to chain cobordism: this is the equivalence relation appropriate for deciding if X is homotopy equivalent to a compact topological manifold.

THEOREM 17.4 (Ranicki [140]) *The total surgery obstruction $s(X) \in \mathbb{S}_n(X)$ of a finite n-dimensional geometric Poincaré complex X is such that $s(X) = 0$ if (and for $n \geq 5$ only if) X is homotopy equivalent to a compact n-dimensional topological manifold.*
PROOF Let

$$(\nu \colon X \longrightarrow BG(k) , \ \rho \colon S^{n+k} \longrightarrow T(\nu))$$

be the Spivak normal structure determined by an embedding $X \subset S^{n+k}$ (k large). The topological reducibility obstruction

$$t(X) = [s(X)] = \delta(\widehat{U}_\nu) = t(\nu) \in H_{n-1}(X; \mathbb{L}.) = \dot{H}^{k+1}(T(\nu); \mathbb{L}.)$$

is the primary obstruction both to the vanishing of $s(X)$ and to the existence of a topological manifold in the homotopy type of X. Assume this obstruction vanishes.

Given a choice of reduction $\tilde{\nu} \colon X \longrightarrow \widetilde{BTOP}(k)$ apply the Browder–Novikov transversality construction to obtain a degree 1 normal map

$$(f = \rho|, b) : M = \rho^{-1}(X) \longrightarrow X$$

from an n-dimensional topological manifold M, making $\rho \colon S^{n+k} \longrightarrow T(\nu) = T(\tilde{\nu})$ transverse regular at the zero section $X \subset T(\tilde{\nu})$. Let $\Pi = (Y, Z, g, h)$ be the canonical transversality structure on M given by 16.16. The degree

1 normal map of n-dimensional geometric Poincaré complexes

$$(F, B) : Y \simeq Z \simeq M \xrightarrow{\ (f, b)\ } X$$

has the same surgery obstruction as (f, b)

$$\sigma_*(F, B) \ = \ \sigma_*(f, b) \in L_n(\mathbb{Z}[\pi_1(X)]) \ .$$

Choosing a simplicial approximation $F \colon Y \longrightarrow X'$ there is obtained a degree 1 normal map

$$\{(F(\tau), B(\tau))\} : \{Y(\tau) = F^{-1}D(\tau, X)\} \longrightarrow \{D(\tau, X)\} \quad (\tau \in X)$$

from a cycle of $(n - |\tau|)$-dimensional geometric $\mathbb{Z}$-coefficient Poincaré pairs $(Y(\tau), \partial Y(\tau))$ to a cycle of $(n - |\tau|)$-dimensional geometric normal pairs $(D(\tau, X), \partial D(\tau, X))$ with geometric Poincaré assembly $\bigcup_\tau D(\tau, X) = X'$. The quadratic kernel is an n-dimensional quadratic globally Poincaré complex in $\mathbb{A}(\mathbb{Z}, X)$

$$(C, \psi) \ = \ \{(C(F(\tau)^!), \psi(B(\tau))) \,|\, \tau \in X\} \ ,$$

with quadratic signature the surgery obstruction

$$(C(\widetilde{X}), \psi(\widetilde{X})) \ = \ \sigma_*(F, B) \ = \ \sigma_*(f, b) \in L_n(\mathbb{Z}[\pi_1(X)])$$

and image

$$\partial \sigma_*(F, B) \ = \ -\partial \sigma^*(X) \ = \ -s(X) \in \mathbb{S}_n(X) \ .$$

The surgery obstruction is 0 if (and for $n \geq 5$ only if) (f, b) is normal bordant to a homotopy equivalence.

Now suppose that $\tilde{\nu}, \tilde{\nu}' \colon X \longrightarrow B\widetilde{TOP}(k)$ are two topological block bundle reductions of the Spivak normal fibration ν, giving rise to degree 1 normal maps $(f, b) \colon M \longrightarrow X$, $(f', b') \colon N \longrightarrow X$. The quadratic kernel complexes (C, ψ), (C', ψ') have the same boundary $(n-1)$-dimensional quadratic globally contractible locally Poincaré complex in $\mathbb{A}(\mathbb{Z}, X)$ (up to homotopy equivalence)

$$\partial(C, \psi) \ = \ \partial(C', \psi') \ = \ -\partial \sigma^*(X) \ ,$$

and the union $(C, \psi) \cup (C', -\psi')$ is a 1-connective n-dimensional quadratic locally Poincaré complex in $\mathbb{A}(\mathbb{Z}, X)$. The assembly of the element

$$(C, \psi) \cup (C', -\psi') \in L_n(\Lambda\langle 1 \rangle(\mathbb{Z})_*(X)) \ = \ H_n(X; \mathbb{L}.)$$

is the difference of the surgery obstructions

$$A((C, \psi) \cup (C', -\psi')) \ = \ (C(\widetilde{X}), \psi(\widetilde{X})) - (C'(\widetilde{X}), \psi'(\widetilde{X}))$$

$$= \ \sigma_*(f, b) - \sigma_*(f', b') \in L_n(\mathbb{Z}[\pi_1(X)]) \ .$$

The symmetric $\mathbb{L}$-spectrum $\mathbb{L}^{\cdot}$ is a ring spectrum. (See Appendix B for the multiplicative structure of the $\mathbb{L}$-spectra). The S-dual of the $\mathbb{L}^{\cdot}$-coefficient Thom class $U_{\tilde{\nu}} \in \dot{H}^k(T(\nu); \mathbb{L}^{\cdot})$ of a topological block bundle reduction $\tilde{\nu} \colon X \longrightarrow B\widetilde{TOP}(k)$ of ν is a fundamental $\mathbb{L}^{\cdot}$-coefficient class $[X]_{\tilde{\nu}} \in$

$H_n(X; \mathbb{L}^{\cdot})$. The quadratic $\mathbb{L}$-spectrum $\mathbb{L}_{\cdot}$ is an $\mathbb{L}^{\cdot}$-module spectrum, and there is defined an $\mathbb{L}_{\cdot}$-coefficient Poincaré duality isomorphism

$$[X]_{\tilde{\nu}} \cap - : [X, G/TOP] = H^0(X; \mathbb{L}_{\cdot}) \xrightarrow{U_{\tilde{\nu}} \cup -} \dot{H}^k(T(\nu); \mathbb{L}_{\cdot}) \cong H_n(X; \mathbb{L}_{\cdot}) .$$

The topological block bundle reductions $\tilde{\nu}' : X \longrightarrow B\widetilde{TOP}(k)$ (k large) of ν are classified relative to $\tilde{\nu}$ by the homotopy classes of maps $X \longrightarrow G/TOP$. The difference $t(\tilde{\nu}, \tilde{\nu}') \in [X, G/TOP]$ (16.1 (v)) corresponds to the element $(C, \psi) \cup (C', -\psi') \in H_n(X; \mathbb{L}_{\cdot})$ constructed above, so that

$$\sigma_*(f, b) - \sigma_*(f', b') = A(t(\tilde{\nu}, \tilde{\nu}')) \in \operatorname{im}(A : H_n(X; \mathbb{L}_{\cdot}) \longrightarrow L_n(\mathbb{Z}[\pi_1(X)])) .$$

Thus if $s(X) \in \ker(\mathbb{S}_n(X) \longrightarrow H_{n-1}(X; \mathbb{L}_{\cdot}))$ there exist topological block bundle reductions $\tilde{\nu}$ of ν, and the surgery obstructions $\sigma_*(f, b)$ of the associated degree 1 normal maps $(f, b) : M \longrightarrow X$ define a coset of the image of the assembly map

$$\operatorname{im}(A : H_n(X; \mathbb{L}_{\cdot}) \longrightarrow L_n(\mathbb{Z}[\pi_1(X)])) \subseteq L_n(\mathbb{Z}[\pi_1(X)])$$

(confirming the suggestion of Wall [178, §9]).

The total surgery obstruction $s(X)$ is therefore such that $s(X) = 0 \in \mathbb{S}_n(X)$ if and only if there exists a reduction $\tilde{\nu}$ for which $\sigma_*(f, b) = 0 \in L_n(\mathbb{Z}[\pi_1(X)])$. For $n \geq 5$ this is the necessary and sufficient condition given by the Browder–Novikov–Sullivan–Wall theory for X to be homotopy equivalent to a compact n-dimensional topological manifold.

□

EXAMPLE 17.5 The total surgery obstruction of a compact n-dimensional combinatorial homology manifold X is $s(X) = 0 \in \mathbb{S}_n(X)$, by virtue of the canonical fundamental $\mathbb{L}^{\cdot}$-homology class $[X]_{\mathbb{L}} \in H_n(X; \mathbb{L}^{\cdot})$ (16.16). 17.4 gives an alternative proof of the result of Galewski and Stern [55] that for $n \geq 5$ X is homotopy equivalent to a compact n-dimensional topological manifold.

□

COROLLARY 17.6 *A finite n-dimensional geometric Poincaré complex X admits a transversality structure $\Pi = (Y, Z, g, h)$ if (and for $n \geq 5$ only if) X is homotopy equivalent to a compact n-dimensional topological manifold.*

□

COROLLARY 17.7 *The total surgery obstruction of a topologically reucible finite n-dimensional geometric Poincaré complex X is given by*

$$s(X) = -\partial \sigma_*(f, b)$$

$$\in \operatorname{im}(\partial : L_n(\mathbb{Z}[\pi_1(X)]) \longrightarrow \mathbb{S}_n(X)) = \ker(\mathbb{S}_n(X) \longrightarrow H_{n-1}(X; \mathbb{L}_{\cdot})) ,$$

with $\sigma_*(f, b) \in L_n(\mathbb{Z}[\pi_1(X)])$ the surgery obstruction of any degree 1 normal *map* $(f,b)\colon M \longrightarrow X$ *from a compact n-dimensional manifold* M.

□

See 19.7 below for the generalization of 17.7 to a degree 1 normal map $(f,b)\colon Y \longrightarrow X$ of finite n-dimensional geometric Poincaré complexes, with $b\colon \nu_Y \longrightarrow \nu_X$ a fibre map of the Spivak normal fibrations rather than a bundle map of topological reductions. The formula of 19.7 is

$$s(Y) - s(X) \ = \ \partial \sigma_*(f,b) \in \mathbb{S}_n(X) \ ,$$

expressing the difference of the total surgery obstructions in terms of the quadratic signature $\sigma_*(f,b) \in L_n(\mathbb{Z}[\pi_1(X)])$.

REMARK 17.8 The algebraic surgery exact sequence of a polyhedron X

$$\ldots \longrightarrow H_n(X;\mathbb{L}.) \overset{A}{\longrightarrow} L_n(\mathbb{Z}[\pi_1(X)]) \overset{\partial}{\longrightarrow} \mathbb{S}_n(X) \longrightarrow H_{n-1}(X;\mathbb{L}.) \longrightarrow \ldots$$

can be viewed as the L-theory localization exact sequence for the assembly functor

$$A : \{\,\text{locally Poincaré complexes}\,\} \longrightarrow \{\,\text{globally Poincaré complexes}\,\}$$

inverting all the globally contractible chain complexes. The total surgery obstruction $s(X) \in \mathbb{S}_n(X)$ of an n-dimensional geometric Poincaré complex X is thus an analogue of the boundary construction of quadratic forms on finite abelian groups from integral lattices in rational quadratic forms (cf. 3.13 and Ranicki [143, §§3,4]). The peripheral invariant of Conner and Raymond [38] and Alexander, Hamrick and Vick [2] for actions of cyclic groups on manifolds and the intersection homology peripheral invariant of Goresky and Siegel [61] and Cappell and Shaneson [27] are defined similarly.

□

The connections between the total surgery obstruction and geometric Poincaré transversality are described in §19 below.

§18. The structure set

The relative version of the total surgery obstruction theory of §17 will now be used to identify the Sullivan–Wall surgery exact sequence of a manifold with the algebraic surgery exact sequence of §15. For $n \geq 5$ the structure set $\mathbb{S}^{TOP}(M)$ of an n-dimensional manifold M is identified with the quadratic structure group $\mathbb{S}_{n+1}(M)$.

DEFINITION 18.1 The *structure set* $\mathbb{S}^{TOP}(X)$ of a finite n-dimensional geometric Poincaré complex X is the set of the h-cobordism classes of pairs (compact n-dimensional manifold M, homotopy equivalence $f \colon M \longrightarrow X$) .

□

By 17.4 for $n \geq 5$ the structure set $\mathbb{S}^{TOP}(X)$ is non-empty if and only if $s(X) = 0 \in \mathbb{S}_n(X)$. The structure set $\mathbb{S}^{TOP}(M)$ of a manifold M is pointed, with base point $(M, 1) \in \mathbb{S}^{TOP}(M)$.

More generally, the structure set $\mathbb{S}_{\partial}^{TOP}(X)$ of a finite n-dimensional geometric Poincaré pair $(X, \partial X)$ with compact manifold boundary ∂X is defined to be the set of the rel ∂ h-cobordism classes of homotopy equivalences $f \colon (M, \partial M) \longrightarrow (X, \partial X)$ from compact manifolds with boundary such that $f| \colon \partial M \longrightarrow \partial X$ is a homeomorphism. By the rel ∂ version of 17.4 for $n \geq 5$ $\mathbb{S}_{\partial}^{TOP}(X)$ is non-empty if and only if $s(X) = 0 \in \mathbb{S}_n(X)$. Note that $\mathbb{S}_{\partial}^{TOP}(X) = \mathbb{S}^{TOP}(X)$ in the closed case $\partial X = \emptyset$.

DEFINITION 18.2 Let $(M, \partial M)$ be a compact n-dimensional manifold with boundary, with $n \geq 5$. The *geometric surgery exact sequence* computing the structure sets $\mathbb{S}_{\partial}^{TOP}(M \times D^i)$ $(i \geq 0)$ is the exact sequence of Sullivan [163] and Wall [176, 10.8]

$$\ldots \longrightarrow L_{n+i+1}(\mathbb{Z}[\pi_1(M)]) \longrightarrow \mathbb{S}_{\partial}^{TOP}(M \times D^i)$$

$$\longrightarrow [M \times D^i, \partial(M \times D^i); G/TOP, \{*\}] \xrightarrow{\theta} L_{n+i}(\mathbb{Z}[\pi_1(M)])$$

$$\longrightarrow \ldots \longrightarrow L_{n+1}(\mathbb{Z}[\pi_1(M)]) \longrightarrow \mathbb{S}_{\partial}^{TOP}(M)$$

$$\longrightarrow [M, \partial M; G/TOP, \{*\}] \xrightarrow{\theta} L_n(\mathbb{Z}[\pi_1(M)]) .$$

□

An element $t \in [M, \partial M; G/TOP, \{*\}]$ classifies a topological block bundle reduction $\tilde{\nu} \colon M \longrightarrow B\widetilde{TOP}(k)$ of the Spivak normal fibration $J\nu_M \colon M \longrightarrow BG(k)$ (k large) such that $\tilde{\nu}| = \nu_{\partial M} \colon \partial M \longrightarrow B\widetilde{TOP}(k)$. The surgery obstruction map

$$\theta : [M, \partial M; G/TOP, \{*\}] \longrightarrow L_n(\mathbb{Z}[\pi_1(M)])$$

sends such an element t to the surgery obstruction

$$\theta(t) = \sigma_*(f, b) \in L_n(\mathbb{Z}[\pi_1(M)])$$

of the degree 1 normal map of n-dimensional manifolds with boundary obtained by the Browder–Novikov transversality construction on the degree 1 map $\rho: (D^{n+k}, S^{n+k-1}) \longrightarrow (T(\tilde{\nu}), T(\tilde{\nu}|_{\partial M}))$ determined by an embedding $(M, \partial M) \subset (D^{n+k}, S^{n+k-1})$ (k large)

$$(f, b) = \rho| : (N, \partial N) = \rho^{-1}(M, \partial M) \longrightarrow (M, \partial M)$$

with $\partial f: \partial N \longrightarrow \partial M$ a homeomorphism. The group $L_{n+1}(\mathbb{Z}[\pi_1(M)])$ acts on $\mathbb{S}_{\partial}^{TOP}(M)$ by

$$L_{n+1}(\mathbb{Z}[\pi_1(M)]) \times \mathbb{S}^{TOP}(M) \longrightarrow \mathbb{S}^{TOP}(M) ;$$

$$(x, (N_0, f_0)) \longrightarrow x(N_0, f_0) = (N_1, f_1) ,$$

with $f_0: N_0 \longrightarrow M$, $f_1: N_1 \longrightarrow M$ homotopy equivalences of n-dimensional manifolds with boundary which are related by a degree 1 normal bordism

$$(g, c) : (W^{n+1}; N_0, N_1) \longrightarrow M \times ([0, 1]; \{0\}, \{1\})$$

with rel ∂ surgery obstruction

$$\sigma_*(g, c) = x \in L_{n+1}(\mathbb{Z}[\pi_1(M)]) .$$

Two elements (N_1, f_1), $(N_2, f_2) \in \mathbb{S}_{\partial}^{TOP}(M)$ have the same image in $[M, \partial M; G/TOP, \{*\}]$ if and only if

$$(N_2, f_2) = x(N_1, f_1) \in \mathbb{S}_{\partial}^{TOP}(M)$$

for some $x \in L_{n+1}(\mathbb{Z}[\pi_1(M)])$.

For the remainder of §18 only the closed case $\partial M = \emptyset$ is considered, but there are evident relative versions for the bounded case. In particular, $\mathbb{S}_{\partial}^{TOP}(M)$ is identified with the quadratic structure group $\mathbb{S}_{n+1}(M)$ also in the case $\partial M \neq \emptyset$.

The following invariants are the essential ingredients in the passage from the geometric surgery exact sequence of 18.2 to the algebraic surgery exact sequence of 15.19.

PROPOSITION 18.3 (i) *A normal map of closed n-dimensional manifolds* $(f, b): N \longrightarrow M$ *determines an element, the* normal invariant

$$[f, b]_{\mathbb{L}} \in H_n(M; \mathbb{L}.) ,$$

with assembly the surgery obstruction of (f, b)

$$A([f, b]_{\mathbb{L}}) = \sigma_*(f, b) \in \text{im}(A: H_n(M; \mathbb{L}.) \longrightarrow L_n(\mathbb{Z}[\pi_1(M)]))$$

$$= \ker(L_n(\mathbb{Z}[\pi_1(M)]) \longrightarrow \mathbb{S}_n(M)) ,$$

and symmetrization the difference of the canonical $\mathbb{L}^{\cdot}$-*homology fundamental classes*

$$(1 + T)[f, b]_{\mathbb{L}} = f_*[N]_{\mathbb{L}} - [M]_{\mathbb{L}} \in H_n(M; \mathbb{L}^{\cdot}) .$$

Let $t(b) \in H^0(M; \mathbb{L}.) = [M, G/TOP]$ *be the normal invariant classifying the fibre homotopy trivialized stable bundle* $\tilde{\nu}_M - \nu_M: M \longrightarrow BTOP$, *with*

ν_M the stable normal bundle of M and $\tilde{\nu}_M$ the target of $b\colon \nu_N \longrightarrow \tilde{\nu}_M$. The normal invariant is the image of $t(b) \in [M, G/TOP]$ under the $\mathbb{L}.$-coefficient Poincaré duality isomorphism

$$[M]_{\mathbb{L}} \cap - \colon H^0(M;\mathbb{L}.) \xrightarrow{\simeq} H_n(M;\mathbb{L}.)$$

defined by cap product with the canonical $\mathbb{L}\cdot$-coefficient fundamental class $[M]_{\mathbb{L}} \in H_n(M;\mathbb{L}\cdot)$. The normal invariant is such that $[f,b]_{\mathbb{L}} = 0 \in H_n(M;\mathbb{L}\cdot)$ if and only if (f,b) is normal bordant to the $1\colon M \longrightarrow M$.
(ii) A homotopy equivalence of closed n-dimensional manifolds $f\colon N \longrightarrow M$ determines an element, the structure invariant

$$s(f) \in \mathbb{S}_{n+1}(M) ,$$

with image the normal invariant of the normal map $(f,b)\colon N \longrightarrow M$ with $b\colon \nu_N \longrightarrow (f^{-1})^* \nu_N$ the induced map of stable bundles over f

$$t(f) = [s(f)] = [f,b]_{\mathbb{L}} \in \mathrm{im}(\mathbb{S}_{n+1}(M) \longrightarrow H_n(M;\mathbb{L}.))$$

$$= \ker(A\colon H_n(M;\mathbb{L}.) \longrightarrow L_n(\mathbb{Z}[\pi_1(M)])) .$$

As in (i) the normal invariant is such that $t(f) = 0$ if and only if $(f,b)\colon N \longrightarrow M$ is normal bordant to $1\colon M \longrightarrow M$, in which case the structure invariant $s(f)$ is the image of the rel∂ surgery obstruction of any normal bordism

$$((g;1,f),(c;1,b)) \colon (W^{n+1};M,N) \longrightarrow M \times ([0,1];\{0\},\{1\}) ,$$

that is

$$s(f) = [\sigma_*(g,c)] \in \mathrm{im}(L_{n+1}(\mathbb{Z}[\pi_1(M)]) \longrightarrow \mathbb{S}_{n+1}(M))$$

$$= \ker(\mathbb{S}_{n+1}(M) \longrightarrow H_n(M;\mathbb{L}.)) .$$

PROOF (i) Let X be the polyhedron of an n-dimensional geometric Poincaré complex with a homotopy equivalence $g\colon M \longrightarrow X$, such that both g and $gf\colon N \longrightarrow X$ are topologically transverse across the dual cell decomposition $\{ D(\tau, X) \,|\, \tau \in X \}$ of X. The restrictions of f define a cycle of degree 1 normal maps of $(n - |\tau|)$-dimensional manifolds with boundary

$$\{(f(\tau), b(\tau))\} \colon \{N(\tau)\} \longrightarrow \{M(\tau)\}$$

with

$$M(\tau) = g^{-1} D(\tau, X) , \quad N(\tau) = (gf)^{-1} D(\tau, X) \ (\tau \in X) ,$$

such that $M(\tau) = \{ \mathrm{pt.} \}$ for n-simplexes $\tau \in X^{(n)}$. The kernel cycle

$$\{ (C(f(\tau)^!), \psi(b(\tau))) \,|\, \tau \in X \}$$

of $(n - |\tau|)$-dimensional quadratic Poincaré pairs in $\mathbb{A}(\mathbb{Z})$ is a 1-connective n-dimensional quadratic Poincaré complex in $\mathbb{A}(\mathbb{Z})_*(X)$ allowing the definition

$$[f,b]_{\mathbb{L}} = \{(C(f(\tau)^!), \psi(b(\tau)))\}$$

$$\in L_n(\Lambda\langle 1\rangle(\mathbb{Z})_*(X)) = H_n(X;\mathbb{L}.) = H_n(M;\mathbb{L}.) .$$

(ii) If $f: N \longrightarrow M$ is a homotopy equivalence the quadratic complex of (i) is globally contractible, allowing the definition

$$s(f) = \{(C(f(\tau)^!), \psi(b(\tau)))\} \in \mathbb{S}_{n+1}(X) = \mathbb{S}_{n+1}(M) \ .$$

(Equivalently, define the structure invariant $s(f)$ of a homotopy equivalence $f: N \longrightarrow M$ of compact n-dimensional manifolds to be rel ∂ total surgery obstruction of a finite $(n + 1)$-dimensional geometric Poincaré pair with compact manifold boundary

$$s(f) = s_\partial(W, N \sqcup -M) \in \mathbb{S}_{n+1}(W) = \mathbb{S}_{n+1}(M) \ ,$$

with $W = N \times I \cup_f M$ the mapping cylinder.)

□

EXAMPLE 18.4 The normal invariant of a normal map of closed oriented n-dimensional manifolds $(f, b): N \longrightarrow M$ is given modulo torsion by the difference between the Poincaré duals of the $\mathcal{L}$-genera of M and N

$$[f, b]_{\mathbb{L}} \otimes \mathbb{Q} = f_*(\mathcal{L}(N) \cap [N]_{\mathbb{Q}}) - \mathcal{L}(M) \cap [M]_{\mathbb{Q}}$$

$$\in H_n(M; \mathbb{L}.) \otimes \mathbb{Q} = H_{n-4*}(M; \mathbb{Q}) \ .$$

□

THEOREM 18.5 (Ranicki [140]) *The Sullivan–Wall geometric surgery exact sequence of a compact n-dimensional manifold M with $n \geq 5$ is isomorphic to the algebraic surgery exact sequence, by an isomorphism*

$$\ldots \longrightarrow L_{n+1}(\mathbb{Z}[\pi_1(M)]) \longrightarrow \mathbb{S}^{TOP}(M) \longrightarrow [M, G/TOP] \overset{\theta}{\longrightarrow} L_n(\mathbb{Z}[\pi_1(M)])$$

$$\Big\| \qquad\qquad s\Big\downarrow\simeq \qquad\qquad t\Big\downarrow\simeq \qquad\qquad \Big\|$$

$$\ldots \longrightarrow L_{n+1}(\mathbb{Z}[\pi_1(M)]) \overset{\partial}{\longrightarrow} \mathbb{S}_{n+1}(M) \longrightarrow H_n(M; \mathbb{L}.) \overset{A}{\longrightarrow} L_n(\mathbb{Z}[\pi_1(M)])$$

and for all $i \geq 0$

$$\mathbb{S}_\partial^{TOP}(M \times D^i, M \times S^{i-1}) = \mathbb{S}_{n+i+1}(M) \ ,$$

$$[M \times D^i, M \times S^{i-1}; G/TOP, \{*\}] = H^{-i}(M; \mathbb{L}.) = H_{n+i}(M; \mathbb{L}.) \ .$$

In particular, $H_n(M; \mathbb{L}.) = [M, G/TOP]$ is the bordism group of normal maps $(f, b): N \longrightarrow M$ of closed n-dimensional manifolds.
PROOF An embedding $M \subset S^{n+k}$ (k large) determines a topological normal structure

$$(\tilde{\nu}: M \longrightarrow B\widetilde{TOP}(k), \ \rho: S^{n+k} \longrightarrow T(\tilde{\nu})) \ .$$

By 18.3 (i) the normal invariant defines a bijection

$$t : [M, G/TOP] \overset{\simeq}{\longrightarrow} H_n(M; \mathbb{L}.) \ ; \ c \longrightarrow [f, b]_{\mathbb{L}} \ ,$$

namely the Poincaré duality isomorphism

$$t = [M]_{\mathbb{L}} \cap - : [M, G/TOP] = H^0(M; \mathbb{L}.) \xrightarrow{\simeq} H_n(M; \mathbb{L}.) \ .$$

The surgery obstruction map θ thus factorizes as the composite

$$\theta : [M, G/TOP] = H^0(M; \mathbb{L}.) \xrightarrow{t} H_n(M; \mathbb{L}.) \xrightarrow{A} L_n(\mathbb{Z}[\pi_1(M)]) \ .$$

Use the structure invariant of 18.3 (ii) to define a bijection

$$s : \mathbb{S}^{TOP}(M) \xrightarrow{\simeq} \mathbb{S}_{n+1}(M) \ ; \ (N, f) \longrightarrow s(f) \ .$$

Similarly for the higher structures.

□

In particular, for any closed n-dimensional manifold M and any element $x \in \mathbb{S}_{n+1}(M)$ there exists a closed n-manifold N with a homotopy equivalence $f : N \longrightarrow M$ such that $s(f) = x$.

COROLLARY 18.6 *Let K be a space with finitely presented $\pi_1(K)$.*
(i) *$H_n(K; \mathbb{L}.)$ consists of the images of the normal invariants $[f, b]_{\mathbb{L}}$ of normal maps $(f, b): N \longrightarrow M$ of closed n-dimensional manifolds with a reference map $M \longrightarrow K$.*
(ii) *The image of the assembly map $A: H_n(K; \mathbb{L}.) \longrightarrow L_n(\mathbb{Z}[\pi_1(K)])$ consists of the surgery obstructions $\sigma_*(f, b)$ of the normal maps $(f, b): N \longrightarrow M$ of closed n-dimensional manifolds with a reference map $M \longrightarrow K$.*
(iii) *$\mathbb{S}_{n+1}(K)$ consists of the images of the structure invariants $s(f)$ of homotopy equivalences $(f, b): N \longrightarrow M$ of closed n-dimensional manifolds with a reference map $M \longrightarrow K$.*
(iv) *The image of $\mathbb{S}_{n+1}(K) \longrightarrow H_n(K; \mathbb{L}.)$ consists of the images of the normal invariants $[f, b]_{\mathbb{L}}$ of homotopy equivalences $(f, b): N \longrightarrow M$ of closed n-dimensional manifolds with a reference map $M \longrightarrow K$.*

□

EXAMPLE 18.7 For $n \geq 4$ the manifold structure set of the n-sphere S^n is

$$\mathbb{S}^{TOP}(S^n) = \mathbb{S}_{n+1}(S^n) = 0 \ .$$

This is the TOP version of the n-dimensional Poincaré conjecture (Smale, Stallings, Newman, Freedman), according to any homotopy equivalence $M^n \simeq S^n$ from a compact n-dimensional topological manifold M is homotopic to a homeomorphism.

□

See §20 for $\mathbb{S}_*(M)$ in the simply connected case $\pi_1(M) = \{1\}$.

REMARK 18.8 The simply connected surgery classifying space $\mathbb{L}_0 \simeq G/TOP$

is such that
$$G/TOP \otimes \mathbb{Z}[1/2] \simeq BO \otimes \mathbb{Z}[1/2]$$
$$G/TOP \otimes \mathbb{Z}_{(2)} \simeq \prod_{j \geq 1} K(\mathbb{Z}_{(2)}, 4j) \times \prod_{j \geq 0} K(\mathbb{Z}_2, 4j + 2)$$
with $\mathbb{Z}_{(2)} = \mathbb{Z}[1/odd]$ the localization of $\mathbb{Z}$ at 2, so that for any space X
$$H_*(X; \mathbb{L}.)[1/2] = KO_*(X)[1/2] = \Omega_*(X) \otimes_{\Omega_*(\{pt.\})} L_*(\mathbb{Z})[1/2]$$
$$H_*(X; \mathbb{L}.)_{(2)} = \prod_{j \geq 1} H_{*-4j}(X; \mathbb{Z}_{(2)}) \times \prod_{j \geq 0} H_{*-4j-2}(X; \mathbb{Z}_2) \ .$$
Wall [176, p. 266] used bordism theory and the surgery product formula to define the L-theory assembly map away from 2
$$A : \ H_*(X; \mathbb{L}.)[1/2] \longrightarrow L_*(\mathbb{Z}[\pi_1(X)])[1/2]$$
by sending the bordism class of an n-dimensional manifold M equipped with a reference map $M \longrightarrow X$ to the symmetric signature of Mishchenko [112]
$$A(M) = \sigma^*(M) \in L_n(\mathbb{Z}[\pi_1(X)])[1/2] = L^n(\mathbb{Z}[\pi_1(X)])[1/2] \ .$$
Up to a power of 2 this is a surgery obstruction
$$8\sigma^*(M) = (1 + T)\sigma_*(1 \times (f, b): M \times Q^8 \longrightarrow M \times S^8) \in L^n(\mathbb{Z}[\pi_1(X)]) \ ,$$
with $(f, b): Q^8 \longrightarrow S^8$ the 8-dimensional normal map determined by the framed 3-connected 8-dimensional Milnor PL manifold Q^8 with signature
$$\sigma^*(Q^8) = (\mathbb{Z}^8, E_8) = 8 \in L^8(\mathbb{Z}) = \mathbb{Z} \ .$$
The factorization of the surgery map as
$$\theta : \ [M, G/TOP] \longrightarrow \Omega_n^{TOP}(B\pi \times G/TOP, B\pi \times \{*\})$$
$$\longrightarrow L_n(\mathbb{Z}[\pi]) \ (\pi = \pi_1(M))$$
is due to Sullivan and Wall [176, 13B.3] (originally in the PL category), with
$$[M, G/TOP] \longrightarrow \Omega_n^{TOP}(B\pi \times G/TOP, B\pi \times \{*\}) \ ;$$
$$(g: M \longrightarrow G/TOP) = ((f, b): N \longrightarrow M) \longrightarrow$$
$$(N \xrightarrow{f \times g} (M \times G/TOP, M \times \{*\}) \longrightarrow (B\pi \times G/TOP, B\pi \times \{*\})) \ .$$
See Appendix B for an expression of this factorization using the multiplicative properties of the algebraic $\mathbb{L}$-spectra. The factorization of θ through the assembly map A was first proposed by Quinn [128]: see Mishchenko and Solovev [115], Nicas [118, §3.3], Levitt and Ranicki [91, §3.2] for the geometric construction of A in the case when M is a PL manifold. In Ranicki [140] the factorization of θ through the algebraic assembly map A was obtained by means of the theory of normal complexes and geometric Poincaré complexes due to Quinn [129]: see the Appendix to Hambleton, Milgram, Taylor and Williams [66] for an exposition of this approach. The factorization was used

in [66] and Milgram [106] to compute the surgery obstructions of normal maps of closed manifolds (= the image of $A: H_*(B\pi; \mathbb{L}.) \longrightarrow L_*(\mathbb{Z}[\pi]))$ for finite groups π.

$\square$

REMARK 18.9 (i) A simplicial map $f: J \longrightarrow K'$ is *transversely cellular* if J is an n-dimensional PL manifold and the inverse images of the dual cells of i-simplexes in $\sigma \in K$ are $(n-i)$-dimensional PL balls $f^{-1}D(\sigma, K) \subset J$. Cohen [36], [37] proved that a transversely cellular map of compact PL manifolds is homotopic to a PL homeomorphism, and that for $n \geq 5$ a proper surjective PL map of n-dimensional combinatorial homotopy manifolds with contractible point inverses is homotopic to a homeomorphism.
(ii) A map $f: N \longrightarrow M$ of ANR spaces (e.g. manifolds) is *cell-like* if it is proper, surjective and such that for each $x \in M$ and each neighbourhood U of $f^{-1}(x)$ in N there exists a neighbourhood $V \subseteq U$ of $f^{-1}(x)$ such that the inclusion $V \longrightarrow U$ is null-homotopic. A proper surjective map of finite-dimensional ANR spaces is cell-like if and only if it is a hereditary proper homotopy equivalence, i.e. such that the restriction $f|: f^{-1}(U) \longrightarrow U$ is a proper homotopy equivalence for every open subset $U \subseteq M$. A PL map $f: N \longrightarrow M$ of compact polyhedra is cell-like if and only if the point inverses $f^{-1}(x)$ are contractible, in which case $\tau(f) = 0 \in Wh(\pi_1(M))$ (as is true for any cell-like map of compact ANR spaces). Siebenmann [158] proved that for $n \geq 5$ a proper surjective map $f: N \longrightarrow M$ of n-dimensional manifolds is cell-like if and only if f is a uniform limit of homeomorphisms. More generally, Chapman and Ferry [33] showed that for $n \geq 5$ any sufficiently controlled homotopy equivalence of n-dimensional manifolds can be approximated by a homeomorphism. The structure invariant $s(f) \in \mathbb{S}_{n+1}(M)$ of a homotopy equivalence $f: N \longrightarrow M$ of compact n-dimensional manifolds measures the failure of f to be cell-like on the chain level, i.e. for the point inverses $f^{-1}(x)$ $(x \in M)$ to be acyclic, up to the chain level cobordism relation appropriate for deciding if f is homotopic to a homeomorphism (at least for $n \geq 5$). If f is cell-like then each of the simplicial maps

$$f(\tau) = f| : N(\tau) = (gf)^{-1}D(\tau, X) \longrightarrow M(\tau) = g^{-1}D(\tau, X) \quad (\tau \in X)$$

in the definition of $s(f)$ can be chosen to be a homotopy equivalence, with $g: M \simeq X$ as in 18.3, so that

$$s(f) = 0 \in \mathbb{S}_{n+1}(M) .$$

Thus for $n \geq 5$ a cell-like map $f: N \longrightarrow M$ of compact n-dimensional manifolds is homotopic to a homeomorphism and

$$(N, f) = (M, 1) = 0 \in \mathbb{S}^{TOP}(M) = \mathbb{S}_{n+1}(M) .$$

$\square$

§19. Geometric Poincaré complexes

The total surgery obstruction of §17 and the structure invariant of §18 are now interpreted in terms of geometric Poincaré bordism theory. The total surgery obstruction $s(X) \in \mathbb{S}_n(X)$ of a finite n-dimensional geometric Poincaré complex X is identified with the obstruction to the identity $X \longrightarrow X$ being bordant to a Poincaré transverse map.

The main source of geometric Poincaré complexes is of course:

EXAMPLE 19.1 A compact n-dimensional topological manifold is a finite n-dimensional geometric Poincaré complex.

□

EXAMPLE 19.2 Browder [14] showed that finite H-spaces are geometric Poincaré complexes, providing the first examples of Poincaré spaces other than manifolds or quotients of finite group actions on manifolds (which are $\mathbb{Q}$-coefficient Poincaré complexes). Finite H-spaces are topologically reducible, with trivial Spivak normal fibration, so that simply-connected ones are homotopy equivalent to compact topological manifolds.

□

EXAMPLE 19.3 Gitler and Stasheff [58] used the first exotic class $e_1 \in H^*(BG; \mathbb{Z}_2)$ to show that a certain simply-connected finite 5-dimensional geometric Poincaré complex $X = (S^2 \vee S^3) \cup e^5$ is not topologically reducible, and hence not homotopy equivalent to a compact topological manifold. In fact, X can be chosen to be the total space of a fibration $S^2 \longrightarrow X \longrightarrow S^3$ classified by an element in $\pi_3(BG(3))$ with image $1 \in \pi_3(B(G/TOP)) = \pi_2(G/TOP) = \mathbb{Z}_2$. See Madsen and Milgram [99, pp. 32-34] for the classification of all the 5-dimensional geometric Poincaré complexes of the type $(S^2 \vee S^3) \cup e^5$. See Frank [52] for non-reducible geometric Poincaré complexes detected by the exotic classes $e_1 \in H^*(BG; \mathbb{Z}_p)$ for odd prime p.

□

EXAMPLE 19.4 Wall [174, 5.4.1] constructed for each prime p a reducible finite 4-dimensional geometric Poincaré complex X with $\pi_1(X) = \mathbb{Z}_p$

$$X = e^0 \cup e^1 \cup \bigcup_{10} e^2 \cup e^3 \cup e^4$$

such that X and the universal cover $\widetilde{X}$ are orientable with signature

$$\sigma^*(X) = \sigma^*(\widetilde{X}) = 8 \in L^4(\mathbb{Z}) = \mathbb{Z} \ .$$

Signature is multiplicative for orientable finite covers of orientable compact manifolds, and $\sigma^*(\widetilde{X}) \neq p\,\sigma^*(X)$, so X cannot be homotopy equivalent to a manifold; higher-dimensional examples are obtained by considering the products $X \times (\mathbb{CP}^2)^k$ $(k \geq 1)$. See 22.28 for the systematic construction of such examples, which are detected by the multisignature invariant.

      □

The realization theorem of Wall [176] for the surgery obstruction groups $L_n(\mathbb{Z}[\pi])$ $(n \geq 5)$ provides the following systematic construction of topologically reducible finite geometric Poincaré complexes. Every finitely presented group π is the fundamental group $\pi = \pi_1(M)$ of a compact $(n-1)$-dimensional manifold M^{n-1}. Every element $x \in L_n(\mathbb{Z}[\pi])$ is the rel ∂ surgery obstruction $x = \sigma_*(f, b)$ of a normal map

$$(f, b) : (W^n; M^{n-1}, M'^{n-1}) \longrightarrow M \times ([0, 1]; \{0\}, \{1\})$$

with

$$f|_M = \text{identity} : M \longrightarrow M \times \{0\} \,,$$

$$f|_{M'} = \text{homotopy equivalence} : M' \longrightarrow M \times \{1\} \,.$$

The topologically reducible n-dimensional geometric Poincaré complex

$$X = W \cup_{\partial f} M \times [0, 1]$$

has fundamental group $\pi_1(X) = \pi \times \mathbb{Z}$ but the extraneous $\mathbb{Z}$-factor can be ignored (or removed by Poincaré π_1-surgery as in Browder [17]). The normal map of n-dimensional geometric Poincaré complexes

$$(f, b) \cup 1 : X = W \cup_{\partial f} M \times [0, 1] \longrightarrow M \times S^1 = M \times [0, 1] \cup_{\partial} M \times [0, 1]$$

has quadratic signature $\sigma_*((f, b) \cup 1) = x \in L_n(\mathbb{Z}[\pi])$. Also, if $(g, c) : N \longrightarrow X$ is a normal map from a closed n-dimensional manifold N corresponding to the topological reduction of X then $\sigma_*(g, c) = -x \in L_n(\mathbb{Z}[\pi])$. See Ranicki [142] for the definition and the composition formula for the quadratic signature of a normal map of geometric Poincaré complexes.

PROPOSITION 19.5 *The topologically reducible finite n-dimensional geometric Poincaré complex X with $\pi_1(X) = \pi$ constructed from $x \in L_n(\mathbb{Z}[\pi])$ has total surgery obstruction*

$$s(X) = \partial(x) \in \text{im}(\partial : L_n(\mathbb{Z}[\pi]) \longrightarrow \mathbb{S}_n(X)) = \ker(\mathbb{S}_n(X) \longrightarrow H_{n-1}(X; \mathbb{L}.)),$$

and $s(X) = 0 \in \mathbb{S}_n(X)$ if and only if $x \in \text{im}(A : H_n(X; \mathbb{L}.) \longrightarrow L_n(\mathbb{Z}[\pi]))$.

PROOF The Spivak normal fibration ν_X has a topological reduction such that the corresponding normal map $(g, c) : N^n \longrightarrow X$ has surgery obstruction

$$\sigma_*(g, c) = -\sigma_*(f, b) = -x \in L_n(\mathbb{Z}[\pi]) \,.$$

The total surgery obstruction of X is given by 17.7 to be

$$s(X) = -\partial \sigma_*(g, c) = \partial(x) \in \mathbb{S}_n(X) \,.$$

The equivalence of $s(X) = 0$ and $x \in \text{im}(A)$ is immediate from the exact sequence

$$H_n(X; \mathbb{L}.) \xrightarrow{A} L_n(\mathbb{Z}[\pi_1(X)]) \xrightarrow{\partial} \mathbb{S}_n(X) \,.$$

      □

The construction of geometric Poincaré complexes from surgery obstructions defines a map

$$L_n(\mathbb{Z}[\pi_1(K)]) \longrightarrow \Omega_n^P(K) \; ; \; x \longrightarrow X$$

for any space K with finitely presented $\pi_1(K)$ and $n \geq 5$.

The exact sequence of Levitt [89], Jones [77], Quinn [129], Hausmann and Vogel [72] relating geometric Poincaré and normal cobordism

$$\ldots \longrightarrow \Omega_{n+1}^N(K) \longrightarrow L_n(\mathbb{Z}[\pi_1(K)]) \longrightarrow \Omega_n^P(K) \longrightarrow \Omega_n^N(K) \longrightarrow \ldots$$

has the following generalization:

PROPOSITION 19.6 (Ranicki [140])
(i) *For any polyhedron K with finitely presented $\pi_1(K)$ and $n \geq 5$ there is defined a commutative braid of exact sequences*

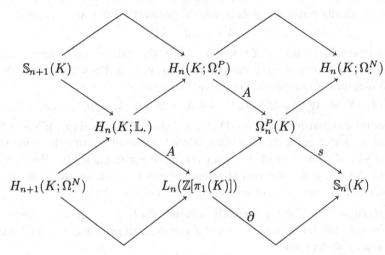

with $\Omega_.^P = \Omega_.^P(\{*\})$ (resp. $\Omega_.^N = \Omega_.^N(\{*\})$) the geometric Poincaré (resp. normal) bordism spectrum of a point and

$$s : \Omega_n^P(K) \longrightarrow \mathbb{S}_n(K) \; ; \; (f \colon X \longrightarrow K) \longrightarrow f_* s(X)$$

the total surgery obstruction map. The quadratic structure group $\mathbb{S}_n(K)$ is the bordism group of maps $(f, \partial f) \colon (X, \partial X) \longrightarrow K$ from finite n-dimensional geometric Poincaré pairs $(X, \partial X)$ such that $\partial f \colon \partial X \longrightarrow K$ is Poincaré transverse.

(ii) *A finite n-dimensional geometric Poincaré complex X has total surgery obstruction $s(X) = 0 \in \mathbb{S}_n(X)$ if (and for $n \geq 5$ only if) there exists an $\Omega_.^P$-homology fundamental class $[X]_P \in H_n(X; \Omega_.^P)$ with assembly the Poincaré bordism class of $1 \colon X \longrightarrow X$*

$$A([X]_P) = (1 \colon X \longrightarrow X) \in \Omega_n^P(X) .$$

PROOF (i) The geometric normal complex bordism spectrum of a point is the Thom spectrum of the universal oriented spherical fibration over the classifying space BSG

$$\Omega^N_. = \{\Omega^N_.(\{*\})_n \mid n \in \mathbb{Z}\} = \underline{MSG} , \ \Omega^N_.(\{*\})_n = \varinjlim_j \Omega^j MSG(j-n) .$$

The normal complex assembly maps are isomorphisms

$$A : H_*(K;\Omega^N_.) \xrightarrow{\sim} \Omega^N_*(K)$$

by normal complex transversality (Quinn [129]). The map s is defined by the total surgery obstruction

$$s : \Omega^P_n(K) \longrightarrow \mathbb{S}_n(K) ; \ (X \longrightarrow K) \longrightarrow s(X) .$$

The geometric Poincaré bordism spectrum of a point

$$\Omega^P_. = \{\Omega^P_.(\{*\})_n \mid n \in \mathbb{Z}\}$$

consists of the Δ-sets with

$\Omega^P_.(\{*\})_n^{(k)} = \{(n+k)$-dimensional oriented finite geometric Poincaré

k-ads $(X;\partial_0 X, \partial_1 X, \ldots, \partial_k X)$ such that $\partial_0 X \cap \partial_1 X \cap \ldots \cap \partial_k X = \emptyset\}$,

with the empty complexes as base simplexes $\emptyset$. As in §12 assume that K is a subcomplex of $\partial \Delta^{m+1}$ for some $m \geq 0$. By 12.6 $H_n(K;\Omega^P_.)$ is the cobordism group of n-dimensional $\Omega^P_.$-cycles in K

$$X = \{X(\tau) \in \Omega^P_.(\{*\})_{n-m}^{(m-|\tau|)} \mid \tau \in K\} ,$$

so that $(X(\tau);\partial_0 X(\tau), \ldots, \partial_{m-|\tau|} X(\tau))$ is an $(n-|\tau|)$-dimensional geometric Poincaré $(m-|\tau|)$-ad with

$$\partial_i X(\tau) = \begin{cases} X(\delta_i \tau) & \text{if } \delta_i \tau \in K \\ \emptyset & \text{if } \delta_i \tau \notin K \end{cases} \ (0 \leq i \leq m - |\tau|) .$$

The assembly of X is the bordism class $(A(X), f) \in \Omega^P_n(K)$ of the union n-dimensional geometric Poincaré complex

$$A(X) = \bigcup_{\tau \in K} X(\tau)$$

with $f : A(X) \longrightarrow K'$ a Poincaré transverse simplicial map such that

$$f^{-1}D(\tau, K) = X(\tau) \ (\tau \in K) .$$

(ii) Immediate from (i).

$\square$

See Levitt and Ranicki [91] for a geometric interpretation of an $\Omega^P_.$-homology fundamental class $[X]_P \in H_n(X;\Omega^P_.)$ such that

$$A([X]_P) = (1 : X \longrightarrow X) \in \Omega^P_n(X)$$

as an 'intrinsic transversality structure'.

COROLLARY 19.7 *If $(f,b): Y \longrightarrow X$ is a normal map of finite n-dimensional geometric Poincaré complexes then the difference of the total surgery obstructions is the image of the quadratic signature $\sigma_*(f,b) \in L_n(\mathbb{Z}[\pi_1(X)])$*

$$s(Y) - s(X) = \partial\sigma_*(f,b) \in \text{im}(\partial: L_n(\mathbb{Z}[\pi_1(X)]) \longrightarrow \mathbb{S}_n(X)) .$$

PROOF The mapping cylinder $W = Y \times I \cup_f X$ of f defines an $(n+1)$-dimensional normal pair $(W, Y \sqcup -X)$ with boundary the n-dimensional geometric Poincaré complex $Y \sqcup -X$, such that

$$\sigma_*(W, Y \sqcup -X) = \sigma_*(f,b) \in \Omega_{n+1}^{N,P}(X) = L_n(\mathbb{Z}[\pi_1(X)]) .$$

$\square$

The symmetric L-groups are not geometrically realizable, in that the symmetric signature map

$$\sigma^* : \Omega_n^P(K) \longrightarrow L^n(\mathbb{Z}[\pi_1(K)]) ; (X \longrightarrow K) \longrightarrow \sigma^*(X)$$

is not onto in general. For example, the $(2k-1)$-connected $4k$-dimensional symmetric Poincaré complex $(S^{2k}\mathbb{Z}[\mathbb{Z}_2], T)$ over $\mathbb{Z}[\mathbb{Z}_2]$ is not in the image of $\sigma^*: \Omega_{4k}^P(B\mathbb{Z}_2) \longrightarrow L^{4k}(\mathbb{Z}[\mathbb{Z}_2])$ for any $k \geq 1$ (Ranicki [143, 7.6.8], see also 9.17).

The fibre of the 1/2-connective visible symmetric signature map

$$\sigma^* : \Omega_.^P(K) \longrightarrow VL^.(K) ; (X \longrightarrow K) \longrightarrow \sigma^*(X)$$

is a homology theory:

COROLLARY 19.8 *For any polyhedron K with finitely presented $\pi_1(K)$ and $n \geq 5$ there is defined a commutative braid of exact sequences*

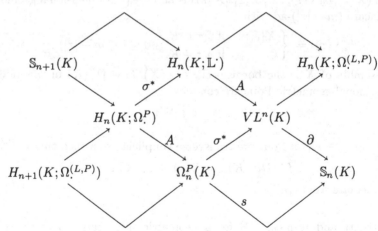

with $\Omega_.^{(L,P)}$ the fibre of the simply connected symmetric signature map $\sigma^*: \Omega_.^P \longrightarrow \mathbb{L}.$.

$\square$

REMARK 19.9 The simply connected normal signature map $\widehat{\sigma}^* : \Omega_.^N \longrightarrow \widehat{\mathbb{L}}^.$ is a rational homotopy equivalence, with both spectra having the rational homotopy type of the Eilenberg-MacLane spectrum $K.(\mathbb{Q}, 0)$ for rational homology:

$$\Omega_.^N \otimes \mathbb{Q} \simeq \underline{MSG} \otimes \mathbb{Q} \simeq K.(\mathbb{Q}, 0)$$

by the finiteness of the stable homotopy groups of spheres $\pi_*^s = \pi_{*+1}(BSG)$ for $* \geq 1$, and

$$\widehat{\mathbb{L}}^. \otimes \mathbb{Q} = \text{cofibre} (1 + T : \mathbb{L}. \longrightarrow \mathbb{L}^.) \otimes \mathbb{Q} \simeq K.(\mathbb{Q}, 0)$$

by virtue of the symmetrization map $1 + T : L_*(\mathbb{Z}) \longrightarrow L^*(\mathbb{Z})$ being an isomorphism modulo 8-torsion. The natural map

$$\Omega_.^{(L,P)} = \text{fibre} (\sigma^* : \Omega_.^P \longrightarrow \mathbb{L}^.) \longrightarrow \text{fibre} (\widehat{\sigma}^* : \Omega_.^N \longrightarrow \widehat{\mathbb{L}}^.)$$

induces isomorphisms of homotopy groups, except possibly in dimensions 4, 5 (in which it at least induces isomorphisms modulo torsion). The 1/2-connective visible symmetric signature map

$$\sigma^* : \Omega_n^P(X) \longrightarrow VL^n(X)$$

is a rational isomorphism for all $n \geq 0$.

$\square$

Given a map $f : Y \longrightarrow X$ there are defined relative $\mathbb{S}$-groups $\mathbb{S}_*(f)$ to fit into a commutative diagram

$$
\begin{array}{ccccccc}
\vdots & & \vdots & & \vdots & & \vdots \\
\downarrow & & \downarrow & & \downarrow & & \downarrow \\
\cdots \longrightarrow H_n(Y; \mathbb{L}.) & \longrightarrow & L_n(\mathbb{Z}[\pi_1(Y)]) & \longrightarrow & \mathbb{S}_n(Y) & \longrightarrow & H_{n-1}(Y; \mathbb{L}.) \longrightarrow \cdots \\
f_* \downarrow & & f_* \downarrow & & f_* \downarrow & & f_* \downarrow \\
\cdots \longrightarrow H_n(X; \mathbb{L}.) & \longrightarrow & L_n(\mathbb{Z}[\pi_1(X)]) & \longrightarrow & \mathbb{S}_n(X) & \longrightarrow & H_{n-1}(X; \mathbb{L}.) \longrightarrow \cdots \\
\downarrow & & \downarrow & & \downarrow & & \downarrow \\
\cdots \longrightarrow H_n(f; \mathbb{L}.) & \longrightarrow & L_n(f) & \longrightarrow & \mathbb{S}_n(f) & \longrightarrow & H_{n-1}(f; \mathbb{L}.) \longrightarrow \cdots \\
\downarrow & & \downarrow & & \downarrow & & \downarrow \\
\cdots \longrightarrow H_{n-1}(Y; \mathbb{L}.) & \longrightarrow & L_{n-1}(\mathbb{Z}[\pi_1(Y)]) & \longrightarrow & \mathbb{S}_{n-1}(Y) & \longrightarrow & H_{n-2}(Y; \mathbb{L}.) \longrightarrow \cdots \\
\downarrow & & \downarrow & & \downarrow & & \downarrow \\
\vdots & & \vdots & & \vdots & & \vdots
\end{array}
$$

with exact rows and columns. The total surgery obstruction of a finite

n-dimensional geometric Poincaré pair (X, Y) is an element

$$s(X, Y) \in \mathbb{S}_n(X, Y) = \mathbb{S}_n(Y \longrightarrow X)$$

of the relative S-group of the inclusion $Y \longrightarrow X$, with image the total surgery obstruction of Y

$$[s(X, Y)] = s(Y) \in \mathbb{S}_{n-1}(Y) .$$

As in the absolute case $(Y = \emptyset)$ the image

$$t(X, Y) = [s(X, Y)] \in H_{n-1}(X, Y; \mathbb{L}.) = \dot{H}^{k+1}(T(\nu_X); \mathbb{L}.)$$

is the obstruction to a topological reduction of the Spivak normal fibration $\nu_X \colon X \longrightarrow BG(k)$. The total surgery obstruction is such that $s(X, Y) = 0$ if (and for $n \geq 6$ only if) (X, Y) is homotopy equivalent to a compact n-dimensional topological manifold with boundary $(M^n, \partial M)$. For $n \geq 6$ the structure set of $(M^n, \partial M)$ is given by

$$\mathbb{S}^{TOP}(M, \partial M) = \mathbb{S}_{n+1}(M, \partial M) .$$

REMARK 19.10 The total surgery obstruction of a finite n-dimensional geometric Poincaré pair (X, Y) such that

$$\pi_1(Y) \cong \pi_1(X)$$

is just the topological reducibility obstruction

$$s(X, Y) = t(X, Y) \in \mathbb{S}_n(X, Y) = H_{n-1}(X, Y; \mathbb{L}.) .$$

Thus $\nu_X \colon X \longrightarrow BG$ is topologically reducible if (and for $n \geq 6$ only if) (X, Y) is homotopy equivalent to a compact n-dimensional topological manifold with boundary – this is the π-π theorem of Wall [176, 3.3] and its trivial converse.

□

§20. The simply connected case

We now turn to the simply connected case $\pi_1(X) = \{1\}$. The total surgery obstruction $s(X) \in \mathbb{S}_n(X)$ of an n-dimensional geometric Poincaré complex X has image the obstruction $t(X) \in H_{n-1}(X; \mathbb{L})$ to a topological reduction of the Spivak normal fibration of X. The simply connected case has the distinctive feature that $\mathbb{S}_n(X) \longrightarrow H_{n-1}(X; \mathbb{L})$ is injective, so that $s(X)$ is determined by $t(X)$. See Browder [16] for a detailed exposition of simply connected surgery obstruction theory in dimensions ≥ 5, and Freedman and Quinn [53] for the extension to the 4-dimensional case.

The simply connected surgery obstruction groups are given by

$$L_n(\mathbb{Z}) = \begin{cases} \mathbb{Z} \\ 0 \\ \mathbb{Z}_2 \\ 0 \end{cases} \text{if } n \equiv \begin{cases} 0 \\ 1 \\ 2 \\ 3 \end{cases} \pmod 4 .$$

The cobordism class of an n-dimensional quadratic Poincaré complex (C, ψ) over $\mathbb{Z}$ is given by

$$(C, \psi) = \begin{cases} (1/8) \text{ signature } (H_{2k}(C)/\text{torsion}, \lambda, \mu) \\ \text{Arf invariant } (H_{2k+1}(C; \mathbb{Z}_2), \lambda, \mu) \end{cases}$$

$$\in L_n(\mathbb{Z}) = \begin{cases} \mathbb{Z} \\ \mathbb{Z}_2 \end{cases} \text{if } n = \begin{cases} 4k \\ 4k+2 \end{cases}$$

with (λ, μ) the $(-)^{n/2}$-quadratic form determined by ψ_0. The surgery obstruction $\sigma_*(f, b) \in L_n(\mathbb{Z})$ of an n-dimensional normal map $(f, b) : M \longrightarrow X$ with $\pi_1(X) = \{1\}$ is the cobordism class of the kernel n-dimensional quadratic Poincaré complex (C, ψ) over $\mathbb{Z}$, with

$$H_*(C) = K_*(M) = \ker(f_* : H_*(M) \longrightarrow H_*(X)) ,$$
$$K_*(M) \oplus H_*(X) = H_*(M) ,$$

so that

$$\sigma_*(f, b) = (C, \psi) = \begin{cases} \text{signature } (K_{2k}(M; \mathbb{R}), \lambda, \mu)/8 \\ \text{Arf invariant } (K_{2k+1}(M; \mathbb{Z}_2), \lambda, \mu) \end{cases}$$

$$\in L_n(\mathbb{Z}) = \begin{cases} \mathbb{Z} \\ \mathbb{Z}_2 \end{cases} \text{if } n = \begin{cases} 4k \\ 4k+2 \end{cases}$$

with (λ, μ) the $(-)^{n/2}$-quadratic form on the kernel module

$$K_{n/2}(M) = \ker(f_* : H_{n/2}(M) \longrightarrow H_{n/2}(X))$$

defined by geometric intersection and self-intersection numbers.

See Kervaire and Milnor [83] and Levine [88] for the original applications of simply connected surgery theory to the classification of differentiable homotopy spheres. For $i \geq 3$ every element $x \in L_{2i}(\mathbb{Z})$ is the surgery obstruction $x = \sigma_*(g, c)$ of a normal map $(g, c) : Q^{2i} \longrightarrow S^{2i}$ with Q a closed framed

$(i-1)$-connected $2i$-dimensional PL manifold constructed by plumbing. For $i = 4$ and $x = 1 \in L_8(\mathbb{Z}) = \mathbb{Z}$ such a manifold Q^8 may be obtained by coning off the boundary of a differentiable 8-dimensional manifold with boundary one of the 7-dimensional exotic spheres of Milnor [108]. For $i = 5$ and $x = 1 \in L_{10}(\mathbb{Z}) = \mathbb{Z}_2$ this gives the PL manifold Q^{10} without differentiable structure of Kervaire [82].

REMARK 20.1 The structure invariant of a homotopy equivalence $f \colon N^n \overset{\simeq}{\longrightarrow} M^n$ of closed simply connected n-dimensional manifolds is given modulo torsion by the difference between the Poincaré duals of the $\mathcal{L}$-genera of M and N (cf. 18.4)

$$s(f) \otimes \mathbb{Q} = \mathcal{L}(M) \cap [M]_{\mathbb{Q}} - f_*(\mathcal{L}(N) \cap [N]_{\mathbb{Q}})$$

$$\in \mathbb{S}_{n+1}(M) \otimes \mathbb{Q} = \ker\left(H_{n-4*}(M;\mathbb{Q}) \longrightarrow H_{n-4*}(\{\text{pt.}\};\mathbb{Q})\right)$$

$$= \sum_{4k \neq n} H_{n-4k}(M;\mathbb{Q}) \ .$$

□

For a simply connected polyhedron K the assembly maps

$$A \colon H_n(K;\mathbb{L}.) \longrightarrow H_n(\{\text{pt.}\};\mathbb{L}.) = L_n(\mathbb{Z}) \quad (n \geq 1)$$

are onto. It follows that the normal invariant maps

$$\mathbb{S}_n(K) \longrightarrow H_{n-1}(K;\mathbb{L}.) \ ; \ r_* s(X) \longrightarrow r_* t(X)$$

are injective, with $r_* s(X) \in \mathbb{S}_n(K)$ the image of the total surgery obstruction $s(X) \in \mathbb{S}_n(X)$ of an n-dimensional geometric Poincaré complex X with a reference map $r \colon X \longrightarrow K$, and $r_* t(X) \in H_{n-1}(K;\mathbb{L}.)$ the image of the topological reducibility obstruction $t(X) = t(\nu_X) \in H_{n-1}(X;\mathbb{L}.)$.

EXAMPLE 20.2 For a simply connected n-dimensional geometric Poincaré complex X the total surgery obstruction $s(X) \in \mathbb{S}_n(X)$ is such that $s(X) = 0$ if and only if $t(X) = 0$. If $t(X) = 0$ there exists a topological reduction $\tilde{\nu} \colon X \longrightarrow BSTOP$ for which the corresponding normal map $(f, b) \colon M^n \longrightarrow X$ has surgery obstruction $\sigma_*(f, b) = 0 \in L_n(\mathbb{Z})$, and if also $n \geq 4$ then (f, b) is normal bordant to a homotopy equivalence $M' \overset{\simeq}{\longrightarrow} X$ for a manifold M'^n.

□

Thus for $n \geq 4$ a simply connected n-dimensional geometric Poincaré complex X is homotopy equivalent to a topological manifold if and only if the Spivak normal fibration $\nu_X \colon X \longrightarrow BSG$ admits a topological reduction $\tilde{\nu} \colon X \longrightarrow BSTOP$. In the even-dimensional case not every such reduction corresponds to a normal map $(f, b) \colon M^n \longrightarrow X$ with zero surgery obstruction. If the corresponding normal map (f, b) has surgery obstruction $\sigma_*(f, b) = x \in L_n(\mathbb{Z})$ and $-x = \sigma_*(g, c)$ for a normal map $(g, c) \colon N^n \longrightarrow S^n$ then the normal map obtained by connected sum

$$(f',b') = (f,b)\#(g,c): M'^n = M^n\# N^n \longrightarrow X = X\# S^n$$

has surgery obstruction

$$\sigma_*(f',b') = \sigma_*(f,b)+\sigma_*(g,c) = x-x = 0 \in L_n(\mathbb{Z})$$

and (f',b') is normal bordant to a homotopy equivalence $M''^n \xrightarrow{\simeq} X$.

PROPOSITION 20.3 *For $n \geq 4$ the structure set of a simply connected n-dimensional topological manifold M is given by*

$$\mathbb{S}^{TOP}(M) = \mathbb{S}_{n+1}(M)$$

$$= \begin{cases} \ker(A\colon H_n(M;\mathbb{L}.)\longrightarrow L_n(\mathbb{Z})) \\ H_n(M;\mathbb{L}.) \end{cases} \text{if } n \equiv \begin{cases} 0 \\ 1 \end{cases} (\text{mod } 2) \;.$$

PROOF This is immediate from $L_{2*+1}(\mathbb{Z}) = 0$ and the exact sequence

$$\ldots \longrightarrow H_n(M;\mathbb{L}.) \xrightarrow{A} L_n(\mathbb{Z}) \longrightarrow \mathbb{S}_n(M) \longrightarrow H_{n-1}(M;\mathbb{L}.) \longrightarrow \ldots .$$

□

EXAMPLE 20.4 The topological manifold structure set of $S^k \times S^{n-k}$ for $n \geq 4$, $k \geq 2$ is

$$\mathbb{S}^{TOP}(S^k \times S^{n-k}) = \mathbb{S}_{n+1}(S^k \times S^{n-k})$$

$$= \ker(\theta\colon [S^k \times S^{n-k}, G/TOP]\longrightarrow L_n(\mathbb{Z}))$$

$$= L_k(\mathbb{Z}) \oplus L_{n-k}(\mathbb{Z}) \;,$$

giving concrete examples of homotopy equivalences of manifolds which are not homotopic to homeomorphisms, as in Novikov [119] (in the smooth case). In particular, in the stable range $2k+1 < n$ a non-zero element

$$x \neq 0 \in L_k(\mathbb{Z}) = \pi_k(G/TOP)$$

$$= \pi_{k+1}(\widetilde{BTOP}(n-k+1)\longrightarrow BG(n-k+1))$$

(such as $x = 1 \in L_2(\mathbb{Z}) = \mathbb{Z}_2$ for $k = 2$, $n = 6$) is realized by a fibre homotopy trivialized topological block bundle $\eta\colon S^k \longrightarrow \widetilde{BTOP}(n-k+1)$. The total space of the sphere bundle

$$S^{n-k} \longrightarrow S(\eta) \longrightarrow S^k$$

is an n-dimensional manifold equipped with a homotopy equivalence $f\colon S(\eta)^n \xrightarrow{\simeq} S^k \times S^{n-k}$ such that the structure invariant is non-zero

$$s(f) = (x,0) \neq 0 \in \mathbb{S}^{TOP}(S^k \times S^{n-k}) = L_k(\mathbb{Z}) \oplus L_{n-k}(\mathbb{Z}) \;,$$

so that f is not homotopic to a homeomorphism.

□

The simply connected symmetric signature of a $4k$-dimensional geometric Poincaré complex X is just the ordinary signature

$$\sigma^*(X) \ = \ \text{signature}\,(H^{2k}(X), \phi) \in L^{4k}(\mathbb{Z}) \ = \ \mathbb{Z} \ .$$

The Hirzebruch formula expresses the signature of an oriented $4k$-dimensional manifold M^{4k} in terms of the $\mathcal{L}$-genus $\mathcal{L}(M) = \mathcal{L}(\tau_M) \in H^{4*}(M; \mathbb{Q})$

$$\sigma^*(M) \ = \ \text{signature}\,(M) \ = \ \langle \mathcal{L}(M), [M]_{\mathbb{Q}} \rangle \in L^{4k}(\mathbb{Z}) \ = \ \mathbb{Z} \ .$$

The defect in the signature formula for Poincaré complexes was used by Browder [15] to detect the failure of a simply connected $4k$-dimensional geometric Poincaré complex X to be homotopy equivalent to a (differentiable) manifold, just as the defect in the signature formula for manifolds with boundary had been previously used by Milnor [108] and Kervaire and Milnor [83] in the detection and classification of exotic spheres. A topological reduction $\tilde{\nu} \colon X \longrightarrow BSTOP$ of the Spivak normal fibration $\nu_X \colon X \longrightarrow BSG$ determines a normal map $(f, b) \colon M^{4k} \longrightarrow X$ with surgery obstruction given by the difference between the evaluation of the $\mathcal{L}$-genus $\mathcal{L}(-\tilde{\nu}) \in H^{4*}(X; \mathbb{Q})$ on $[X]_{\mathbb{Q}} \in H_{4k}(X; \mathbb{Q})$ and the signature of X

$$\sigma_*(f, b) \ = \ \text{signature}\,(K_{2k}(M), \lambda, \mu)/8$$

$$= \ (\text{signature}\,(M) - \text{signature}\,(X))/8$$

$$= \ ((\langle \mathcal{L}(-\tilde{\nu}), [X]_{\mathbb{Q}} \rangle - \sigma^*(X))/8 \in L_{4k}(\mathbb{Z}) \ = \ \mathbb{Z} \ .$$

If $\tilde{\nu}, \tilde{\nu}' \colon X \longrightarrow BSTOP$ are two topological reductions then the surgery obstructions of corresponding normal maps $(f, b) \colon M^{4k} \longrightarrow X$, $(f', b') \colon M'^{4k} \longrightarrow X$ differ by the assembly of the difference element

$$t(\tilde{\nu}, \tilde{\nu}') \in H_{4k}(X; \mathbb{L}.) \ = \ [X, G/TOP] \ ,$$

that is

$$\sigma_*(f, b) - \sigma_*(f', b') \ = \ A(t(\tilde{\nu}, \tilde{\nu}')) \in L_{4k}(\mathbb{Z}) \ .$$

For $k \geq 2$ a topological reduction $\tilde{\nu}$ is realized by a $4k$-dimensional topological manifold M^{4k} with a homotopy equivalence $h \colon M \xrightarrow{\simeq} X$ such that $\nu_M = h^* \tilde{\nu} \colon M \longrightarrow BSTOP$ if and only if the signature satisfies

$$\sigma^*(X) \ = \ \langle \mathcal{L}(-\tilde{\nu}), [X]_{\mathbb{Q}} \rangle \in L^{4k}(\mathbb{Z}) \ = \ \mathbb{Z} \ .$$

For a simply connected $(4k+2)$-dimensional geometric Poincaré complex X with a topological reduction $\tilde{\nu} \colon X \longrightarrow BSTOP$ the surgery obstruction of the corresponding normal map $(f, b) \colon M^{4k+2} \longrightarrow X$ is given by

$$\sigma_*(f, b) \ = \ \text{Arf invariant}\,(K_{2k+1}(M; \mathbb{Z}_2), \lambda, \mu)$$

$$\in L_{4k+2}(\mathbb{Z}) \ = \ L_{4k+2}(\mathbb{Z}_2) \ = \ \mathbb{Z}_2 \ ,$$

with $(K_{2k+1}(M; \mathbb{Z}_2), \lambda, \mu)$ the nonsingular quadratic form defined on the kernel $\mathbb{Z}_2$-module

$$K_{2k+1}(M; \mathbb{Z}_2) \ = \ \ker(f_* \colon H_{2k+1}(M; \mathbb{Z}_2) \longrightarrow H_{2k+1}(X; \mathbb{Z}_2))$$

by geometric intersection and self-intersection numbers, or (equivalently) by functional Steenrod squares. There exists a $(4k+2)$-dimensional topological manifold M^{4k+2} with a homotopy equivalence $h: M \xrightarrow{\simeq} X$ for which $\nu_M = h^*\tilde{\nu}: M \longrightarrow BSTOP$ if and only if this Arf invariant is 0.

For a simply connected $2i$-dimensional geometric Poincaré complex X with a topologically reducible $\nu_X: X \longrightarrow BSG$ there exists a normal map $(f, b): M^{2i} \longrightarrow X$ with surgery obstruction $\sigma_*(f, b) = 0 \in L_{2i}(\mathbb{Z})$, so that $s(X) = 0 \in \mathbb{S}_{2i}(X)$ and X is homotopy equivalent to a manifold. This follows formally from

$$\text{im}(A: H_{2i}(X; \mathbb{L}.) \longrightarrow L_{2i}(\mathbb{Z})) = L_{2i}(\mathbb{Z})$$

and $\pi_*(G/TOP) = L_*(\mathbb{Z})$. For every $i \geq 3$ and every $x \in L_{2i}(\mathbb{Z})$ plumbing can be used to construct a differentiable $2i$-dimensional manifold with boundary $(W^{2i}, \partial W)$ and a normal map

$$(F, B) : (W, \partial W) \longrightarrow (D^{2i}, S^{2i-1})$$

which restricts to a homotopy equivalence $F|: \partial W \xrightarrow{\simeq} S^{2i-1}$ with

$$\sigma_*(F, B) = x \in L_{2i}(\mathbb{Z}) .$$

(See Browder [16, V] for details.) By the $(2i-1)$-dimensional PL Poincaré conjecture the homotopy equivalence $F|: \partial W \xrightarrow{\simeq} S^{2i-1}$ may be taken to be a PL homeomorphism. Thus if X is a simply connected $2i$-dimensional geometric Poincaré complex with a topological reduction $\tilde{\nu}: X \longrightarrow BSTOP$ for which the corresponding normal map $(f, b): M^{2i} \longrightarrow X$ has surgery obstruction $\sigma_*(f, b) = -x \in L_{2i}(\mathbb{Z})$ there exists a normal map

$$(f', b') = (f, b) \cup (F, B) : M'^{2i} = \text{cl}(M \backslash D^{2i}) \cup_\partial W \longrightarrow X$$

with surgery obstruction

$$\sigma_*(f', b') = \sigma_*(f, b) + \sigma_*(F, B) = -x + x = 0 \in L_{2i}(\mathbb{Z}) ,$$

so that (f', b') is normal bordant to a homotopy equivalence $M''^{2i} \xrightarrow{\simeq} X$.

For a simply connected $(2i+1)$-dimensional geometric Poincaré complex X with $i \geq 2$ every topological reduction $\tilde{\nu}: X \longrightarrow BSTOP$ is such that there exists a topological manifold M^{2i+1} with a homotopy equivalence $h: M \xrightarrow{\simeq} X$ and $\nu_M = h^*\tilde{\nu}: M \longrightarrow BSTOP$, since the surgery obstruction takes values in $L_{2i+1}(\mathbb{Z}) = 0$.

EXAMPLE 20.5 A finite H-space X is a geometric Poincaré complex (Browder [14]) with fibre homotopy trivial Spivak normal fibration ν_X, so that in the simply connected case $s(X) = 0$ and (at least for $n \geq 4$) X is homotopy equivalent to a topological manifold. See Cappell and Weinberger [29] for manifold structures on non-simply connected finite H-spaces.

$\square$

§21. Transfer

The L-theory transfer maps associated to fibrations give generalized product formulae for the various signatures, and also the total surgery obstructions. The transfer maps for coverings give the Morita theory isomorphisms in the projective L-groups, which are used in §22 to describe the rational L-theory of finite fundamental groups.

See Ranicki [141, §8], [142, §8] for the L-theory products for any rings with involution R, S

$$L^m(R) \otimes L^n(S) \longrightarrow L^{m+n}(R \otimes S) ,$$

$$L_m(R) \otimes L^n(S) \longrightarrow L_{m+n}(R \otimes S)$$

and for the applications to topology, generalizing the Eilenberg–Zilber theorem

$$\Delta(X \times Y) \simeq \Delta(X) \otimes \Delta(Y) .$$

On the chain level the L-theory products are given by the tensor product pairing

$$\{\,R\text{-module chain complexes}\,\} \times \{\,S\text{-module chain complexes}\,\}$$

$$\longrightarrow \{\,R \otimes S\text{-module chain complexes}\,\} \;;\; (C, D) \longrightarrow C \otimes D .$$

The product of an m-dimensional $\begin{cases} \text{geometric Poincaré complex } X \\ \text{normal map } (f, b) \colon M \longrightarrow X \end{cases}$ and an n-dimensional geometric Poincaré complex Y is an $(m + n)$-dimensional $\begin{cases} \text{geometric Poincaré complex} \\ \text{normal map} \end{cases}$ with $\begin{cases} \text{symmetric} \\ \text{quadratic} \end{cases}$ signature

$$\begin{cases} \sigma^*(X \times Y) \;=\; \sigma^*(X) \otimes \sigma^*(Y) \in L^{m+n}(\mathbb{Z}[\pi_1(X \times Y)]) \\ \sigma_*((f, b) \times 1 \colon M \times Y \longrightarrow X \times Y) \\ \qquad =\; \sigma_*(f, b) \otimes \sigma^*(Y) \in L_{m+n}(\mathbb{Z}[\pi_1(X \times Y)]) . \end{cases}$$

In the simply connected case $\pi_1(X) = \pi_1(Y) = \{1\}$ these are the usual product formulae for the signature and Kervaire–Arf invariant (Browder [16, III.5]). See Appendix B for the corresponding product structures on the algebraic $\mathbb{L}$-spectra. On the cycle level these structures define products in the $1/2$-connective visible symmetric L-groups

$$VL^m(X) \times VL^n(Y) \longrightarrow VL^{m+n}(X \times Y) \;;\; (C, \phi) \otimes (D, \theta) \longrightarrow (C \otimes D, \phi \otimes \theta)$$

for any polyhedra X, Y.

PROPOSITION 21.1 *The product of a finite m-dimensional geometric Poincaré complex X and a finite n-dimensional geometric Poincaré complex Y is a finite $(m + n)$-dimensional geometric Poincaré complex $X \times Y$ with $1/2$-connective visible symmetric signature*

$$\sigma^*(X \times Y) \;=\; \sigma^*(X) \otimes \sigma^*(Y) \in VL^{m+n}(X \times Y)$$

and total surgery obstruction

$$s(X \times Y) = \partial\sigma^*(X \times Y) = \partial(\sigma^*(X) \otimes \sigma^*(Y)) \in \mathbb{S}_{m+n}(X \times Y) .$$

□

A fibration $F \longrightarrow E \overset{p}{\longrightarrow} B$ with the fibre F a finite m-dimensional geometric Poincaré complex induces *transfer* maps in the quadratic L-groups

$$p^! : L_n(\mathbb{Z}[\pi_1(B)]) \longrightarrow L_{m+n}(\mathbb{Z}[\pi_1(E)]) ,$$

which were described geometrically by Quinn [127] and algebraically in Lück and Ranicki [96]. An n-dimensional normal map $(f,b): M \longrightarrow X$ and a reference map $X \longrightarrow B$ lift to an $(m+n)$-dimensional normal map $(f^!,b^!): M^! \longrightarrow X^!$ and a reference map $X^! \longrightarrow E$ such that

$$p^!\sigma_*(f,b) = \sigma_*(f^!,b^!) \in L_{m+n}(\mathbb{Z}[\pi_1(E)]) .$$

From now on, it will be assumed that the fibration is defined by a simplicial map $p: E \longrightarrow B$ of finite simplicial complexes which is a PL fibration in the sense of Hatcher [71], with the fibre $F = p^{-1}(\{*\})$ a finite m-dimensional geometric Poincaré complex. In terms of the cycle theory of §14 the quadratic L-theory transfer maps are given by

$$p^! : L_n(\mathbb{Z}[\pi_1(B)]) = L_n(\Lambda(\mathbb{Z}, B))$$

$$\longrightarrow L_{m+n}(\mathbb{Z}[\pi_1(E)]) = L_{m+n}(\Lambda(\mathbb{Z}, E)) ; (C, \psi) \longrightarrow (C^!, \psi^!)$$

with $(C, \psi) = \{C(\tau), \psi(\tau) \,|\, \tau \in B\}$ a globally Poincaré cycle of $(n - |\tau|)$-dimensional quadratic complexes over $(\mathbb{Z}, B)$ ($=n$-dimensional quadratic Poincaré complex in $\Lambda(\mathbb{Z}, B)$), and

$$(C^!, \psi^!) = \{(C^!(\sigma), \psi^!(\sigma)) \,|\, \sigma \in E\}$$

the lifted globally Poincaré cycle of $(m + n - |\tau|)$-dimensional quadratic complexes over $(\mathbb{Z}, E)$ with

$$C^!(\sigma) = \Delta(D(\sigma, E), \partial D(\sigma, E)) \otimes C(p\sigma) .$$

The cycle approach extends to define compatible transfer maps in the $1/2$-connective visible symmetric L-groups

$$p^! : VL^n(B) \longrightarrow VL^{m+n}(E) ; (C, \phi) \longrightarrow (C^!, \phi^!)$$

and also in the normal L-theory $\widehat{\mathbb{L}}^\cdot$-homology groups

$$p^! : H_n(B; \widehat{\mathbb{L}}^\cdot) \longrightarrow H_{m+n}(E; \widehat{\mathbb{L}}^\cdot) .$$

If F is an m-dimensional homology manifold locally Poincaré cycles lift to locally Poincaré cycles, so in this case the method also gives transfer maps in the $\mathbb{L}.$-homology groups

$$p^! : H_n(B; \mathbb{L}.) \longrightarrow H_{m+n}(E; \mathbb{L}.)$$

and the structure groups

$$p^! : \mathbb{S}_n(B) \longrightarrow \mathbb{S}_{m+n}(E) ; (C, \psi) \longrightarrow (C^!, \psi^!) ,$$

with a map of exact sequences

$$\ldots \longrightarrow H_n(B;\mathbb{L}^{\textrm{.}}) \longrightarrow VL^n(B) \longrightarrow \mathbb{S}_n(B) \longrightarrow H_{n-1}(B;\mathbb{L}^{\textrm{.}}) \longrightarrow \ldots$$

$$\downarrow p^! \qquad\qquad \downarrow p^! \qquad\qquad \downarrow p^! \qquad\qquad \downarrow p^!$$

$$\ldots \rightarrow H_{m+n}(E;\mathbb{L}^{\textrm{.}}) \rightarrow VL^{m+n}(E) \rightarrow \mathbb{S}_{m+n}(E) \rightarrow H_{m+n-1}(E;\mathbb{L}^{\textrm{.}}) \rightarrow \ldots \; .$$

PROPOSITION 21.2 *Let* $F \longrightarrow E \overset{p}{\longrightarrow} B$ *be a PL fibration with the base* B *a finite n-dimensional geometric Poincaré complex and the fibre F a finite m-dimensional geometric Poincaré complex, so that the total space E is a finite $(m+n)$-dimensional geometric Poincaré complex.*

(i) *The 1/2-connective visible symmetric signature of E is the transfer*

$$\sigma^*(E) \; = \; p^! \sigma^*(B) \in VL^{m+n}(E)$$

of the 1/2-connective visible symmetric signature $\sigma^*(B) \in VL^n(B)$, and the total surgery obstruction is

$$s(E) \; = \; \partial\sigma^*(E) \; = \; \partial p^! \sigma^*(B) \in \mathbb{S}_{m+n}(E) \; .$$

(ii) *If F is an m-dimensional homology manifold the total surgery obstruction of E is the transfer*

$$s(E) \; = \; p^! s(B) \in \mathbb{S}_{m+n}(E)$$

of the total surgery obstruction $s(B) \in \mathbb{S}_n(B)$.

$\square$

REMARK 21.3 For any *PL* fibration $F \longrightarrow E \overset{p}{\longrightarrow} B$ with the fibre F a finite m-dimensional geometric Poincaré complex the composite

$$p_! p^! : L_n(\mathbb{Z}[\pi_1(B)]) \longrightarrow L_{m+n}(\mathbb{Z}[\pi_1(E)]) \longrightarrow L_{m+n}(\mathbb{Z}[\pi_1(B)])$$

is shown in Lück and Ranicki [97] to depend only the $\pi_1(B)$-equivariant Witt class $\sigma^*(F,p) \in L^m(\pi_1(B),\mathbb{Z})$ of the symmetric Poincaré complex of F over $\mathbb{Z}$ with the chain homotopy $\pi_1(B)$-action by fibre transport. See [97] for the equivariant L-groups $L^*(\pi,\mathbb{Z})$ and the assembly map $A: H^{-m}(B;\mathbb{L}^{\textrm{.}}) \longrightarrow L^m(\pi_1(B),\mathbb{Z})$. If B is a compact n-dimensional homology manifold and F is a compact m-dimensional homology manifold then E is a compact $(m+n)$-dimensional homology manifold, and the Δ-map $B \longrightarrow \mathbb{L}^{-m}(\mathbb{Z})$ sending each simplex $\tau \in B$ to the symmetric Poincaré fibre $\sigma^*(p^{-1}\tau)$ over $\mathbb{Z}$ represents an element $[F,p]_{\mathbb{L}} \in H^{-m}(B;\mathbb{L}^{\textrm{.}})$ with assembly $A([F,p]_{\mathbb{L}}) = \sigma^*(F,p) \in L^m(\pi_1(B),\mathbb{Z})$. The canonical $\mathbb{L}^{\textrm{.}}$-homology fundamental class $[E]_{\mathbb{L}} \in H_{m+n}(E;\mathbb{L}^{\textrm{.}})$ has image

$$p_![E]_{\mathbb{L}} \; = \; [F,p]_{\mathbb{L}} \cap [B]_{\mathbb{L}} \in H_{m+n}(B;\mathbb{L}^{\textrm{.}}) \; .$$

This is a generalization of the characteristic class formula of Atiyah [6] expressing the signature of the total space E of a differentiable fibre bundle in the case $m = 2i$, $m+n \equiv 0(\bmod 4)$ as a higher signature (cf. 24.3 below)

$$\text{signature}\,(E) \ = \ p_!\sigma^*(E) \ = \ A(p_![E]_{\mathbb{L}})$$

$$= \ \langle \mathcal{L}(B) \cup x, [B]_{\mathbb{Q}} \rangle \in L^{m+n}(\mathbb{Z}) \ = \ \mathbb{Z}$$

with $x = \widetilde{\mathrm{ch}}([\Gamma]_{\mathbb{K}}) \in H^{2*}(B;\mathbb{Q})$ the modified Chern character (involving multiplication by powers of 2) of the $\begin{cases} \text{real} \\ \text{complex} \end{cases}$ K-theory signature $[\Gamma]_{\mathbb{K}} \in$ $\begin{cases} KO(B) \\ KU(B) \end{cases}$ of the flat bundle Γ of nonsingular $(-)^i$-symmetric forms over B with fibres $H^i(F_x;\mathbb{R})$ ($x \in B$) for $i \equiv \begin{cases} 0 \\ 1 \end{cases} (\bmod 2)$

$$[F,p]_{\mathbb{L}} \otimes 1 \ = \ \widetilde{\mathrm{ch}}([\Gamma]_{\mathbb{K}}) \in H^{-2i}(B;\mathbb{L}^{\cdot}) \otimes \mathbb{Q} \subseteq H^{2*}(B;\mathbb{Q}) \ .$$

In the special case when $\pi_1(B)$ acts trivially on $H^*(F;\mathbb{R})$ this gives the product formula of Chern, Hirzebruch and Serre [34]

$$\text{signature}\,(E) \ = \ \text{signature}\,(B)\,\text{signature}\,(F) \in \mathbb{Z} \ .$$

$\square$

REMARK 21.4 A finite d-sheeted covering is a fibration $F \longrightarrow E \overset{p}{\longrightarrow} B$ with the fibre F a 0-dimensional manifold consisting of d points. It is convenient to write $B = X$, $E = \overline{X}$. The covering is classified by the subgroup

$$\overline{\pi} \ = \ \pi_1(\overline{X}) \subset \pi \ = \ \pi_1(X)$$

of finite index d. The transfer maps in the quadratic L-groups

$$p^! \ : \ L_n(\mathbb{Z}[\pi]) \longrightarrow L_n(\mathbb{Z}[\overline{\pi}])$$

are given algebraically by the functor

$$p^! \ : \ \{\,\mathbb{Z}[\pi]\text{-modules}\,\} \longrightarrow \{\,\mathbb{Z}[\overline{\pi}]\text{-modules}\,\} \ ; \ M \longrightarrow M^!$$

sending a $\mathbb{Z}[\pi]$-module M to the $\mathbb{Z}[\overline{\pi}]$-module $M^!$ obtained by restricting the action to $\mathbb{Z}[\overline{\pi}] \subset \mathbb{Z}[\pi]$. The transfer maps define a map of exact sequences

$$\begin{array}{ccccccccc}
\cdots \longrightarrow & H_n(X;\mathbb{L}^{\cdot}) & \longrightarrow & VL^n(X) & \longrightarrow & \mathbb{S}_n(X) & \longrightarrow & H_{n-1}(X;\mathbb{L}^{\cdot}) & \longrightarrow \cdots \\
& \downarrow{\scriptstyle p^!} & & \downarrow{\scriptstyle p^!} & & \downarrow{\scriptstyle p^!} & & \downarrow{\scriptstyle p^!} & \\
\cdots \longrightarrow & H_n(\overline{X};\mathbb{L}^{\cdot}) & \longrightarrow & VL^n(\overline{X}) & \longrightarrow & \mathbb{S}_n(\overline{X}) & \longrightarrow & H_{n-1}(\overline{X};\mathbb{L}^{\cdot}) & \longrightarrow \cdots \ .
\end{array}$$

Also, there are defined commutative diagrams

$$\begin{array}{ccc} H_n(X;\mathbb{L}^{\cdot}) & \xrightarrow{\ \ A\ \ } & L^n(\mathbb{Z}) \\ \ \ \downarrow{\scriptstyle p^{!}} & & \ \ \downarrow{\scriptstyle d\cdot} \\ H_n(\overline{X};\mathbb{L}^{\cdot}) & \xrightarrow{\ \ A\ \ } & L^n(\mathbb{Z}) \end{array} \qquad \begin{array}{ccc} H_n(X;\widehat{\mathbb{L}}^{\cdot}) & \xrightarrow{\ \ A\ \ } & \widehat{L}^n(\mathbb{Z}) \\ \ \ \downarrow{\scriptstyle p^{!}} & & \ \ \downarrow{\scriptstyle d\cdot} \\ H_n(\overline{X};\widehat{\mathbb{L}}^{\cdot}) & \xrightarrow{\ \ A\ \ } & \widehat{L}^n(\mathbb{Z}) \end{array}$$

with A the simply connected assembly, and $d\cdot$ multiplication by

$$\sigma^*(p^{-1}(\text{pt.})) \ = \ d \in L^0(\mathbb{Z}) = \mathbb{Z}\ .$$

If X is a finite n-dimensional geometric Poincaré complex then so is $\overline{X}$, with total surgery obstruction given by 21.2 (ii) to be

$$s(\overline{X}) \ = \ p^{!}s(X) \in \mathbb{S}_n(\overline{X})\ .$$

The normal L-theory fundamental class $[X]_{\widehat{\mathbb{L}}} \in H_n(X;\widehat{\mathbb{L}}^{\cdot})$ of X lifts to the normal L-theory fundamental class of $\overline{X}$

$$p^{!}[X]_{\widehat{\mathbb{L}}} \ = \ [\overline{X}]_{\widehat{\mathbb{L}}} \in H_n(\overline{X};\widehat{\mathbb{L}}^{\cdot})\ ,$$

so that for $n = 4k$ the mod 8 signature is multiplicative

$$\text{signature}(\overline{X}) \ = \ d \cdot \text{signature}(X) \in \widehat{L}^{4k}(\mathbb{Z}) \ = \ \mathbb{Z}_8\ .$$

If $s(X) = 0$ then $s(\overline{X}) = 0$ and there exists a symmetric L-theory fundamental class $[X]_{\mathbb{L}} \in H_n(X;\mathbb{L}^{\cdot})$ such that

$$p^{!}[X]_{\mathbb{L}} \ = \ [\overline{X}]_{\mathbb{L}} \in H_n(\overline{X};\mathbb{L}^{\cdot})$$

is a symmetric L-theory fundamental class for $\overline{X}$. Thus for $n = 4k$ the actual signature is multiplicative for finite geometric Poincaré complexes X with $s(X) = 0$

$$\text{signature}(\overline{X}) \ = \ d \cdot \text{signature}(X) \in L^{4k}(\mathbb{Z}) \ = \ \mathbb{Z}\ .$$

See §22 for further discussion of the multiplicativity of signature for finite coverings.

$$\square$$

Next, we consider the Morita theory for projective K- and L-groups.

Given a ring R and an integer $d \geq 1$ let $M_d(R)$ ring of $d \times d$ matrices with entries in R. Regard $R^d = \sum_d R$ as an $(R, M_d(R))$-bimodule by

$$R \times R^d \times M_d(R) \ \longrightarrow \ R^d \ ; \ (x,(y_i),(z_{jk})) \ \longrightarrow \ (\sum_{j=1}^{d} xy_j z_{jk})\ ,$$

and as an $(M_d(R), R)$-bimodule by

$$M_d(R) \times R^d \times R \ \longrightarrow \ R^d \ ; \ ((x_{ij}),(y_k),z) \ \longrightarrow \ (\sum_{j=1}^{d} x_{ij}y_j z)\ .$$

The Morita equivalence of categories

$$\mu : \{\,\text{f.g. projective } M_d(R)\text{-modules}\,\} \xrightarrow{\;\approx\;} \{\,\text{f.g. projective } R\text{-modules}\,\}\;;$$

$$P \longrightarrow P^! \;=\; R^d \otimes_{M_d(R)} P$$

has inverse

$$\mu^{-1} : \{\,\text{f.g. projective } R\text{-modules}\,\} \xrightarrow{\;\approx\;}$$

$$\{\,\text{f.g. projective } M_d(R)\text{-modules}\,\}\;;\; Q \longrightarrow R^d \otimes_R Q\;.$$

The Morita isomorphism of the projective class groups

$$\mu : K_0(M_d(R)) \xrightarrow{\;\approx\;} K_0(R)\;;\; [P] \longrightarrow [P^!]$$

is such that

$$\mu[M_d(R)] \;=\; d[R]\;,\;\; \mu[R^d] \;=\; [R] \in K_0(R)\;.$$

For any ring with involution R and $\epsilon = \pm 1$ let $\begin{cases} L^*(R,\epsilon) \\ L_*(R,\epsilon) \end{cases}$ be the

$\begin{cases} \epsilon\text{-symmetric} \\ \epsilon\text{-quadratic} \end{cases}$ L-groups of R (Ranicki [141]), such that

$$\begin{cases} L^*(R,1) \;=\; L^*(R) \\ L_*(R,1) \;=\; L_*(R)\;. \end{cases}$$

The 0-dimensional L-group $\begin{cases} L^0(R,\epsilon) \\ L_0(R,\epsilon) \end{cases}$ is the Witt group of nonsingular

$\begin{cases} \epsilon\text{-symmetric} \\ \epsilon\text{-quadratic} \end{cases}$ forms over R. The ϵ-symmetrization maps

$$1 + T_\epsilon : L_*(R,\epsilon) \longrightarrow L^*(R,\epsilon)$$

are isomorphisms modulo 8-torsion, so that

$$L_*(R,\epsilon)[1/2] \;=\; L^*(R,\epsilon)[1/2]\;.$$

The ϵ-quadratic L-groups are 4-periodic

$$L_*(R,\epsilon) \;=\; L_{*+2}(R,-\epsilon) \;=\; L_{*+4}(R,\epsilon)\;.$$

The ϵ-symmetric L-groups are 4-periodic for a Dedekind ring with involution R, and are 4-periodic modulo 2-primary torsion for any R.

DEFINITION 21.5 Given a ring with involution R and a nonsingular ϵ-symmetric form (R^d,ϕ) over R let $M_d(R)^\phi$ denote the $d \times d$ matrix ring $M_d(R)$

$$M_d(R)^\phi \;=\; \operatorname{Hom}_R(R^d, R^d)$$

with the involution

$$M_d(R)^\phi \xrightarrow{\;\approx\;} M_d(R)^\phi\;;\; f \longrightarrow \phi^{-1} f^* \phi\;.$$

$\square$

PROPOSITION 21.6 *The projective L-groups of R and $M_d(R)^\phi$ are related by Morita isomorphisms of projective* $\begin{cases} \eta\text{-symmetric} \\ \eta\text{-quadratic} \end{cases}$ *L-groups*

$$\begin{cases} \mu : L_p^*(R, \eta) \xrightarrow{\simeq} L_p^*(M_d(R)^\phi, \epsilon\eta) \\ \mu : L_*^p(R, \eta) \xrightarrow{\simeq} L_*^p(M_d(R)^\phi, \epsilon\eta) \end{cases}$$

with $\eta = \pm 1$. *The Morita isomorphism* $\mu : L_p^0(M_d(R)^\phi) \xrightarrow{\simeq} L_p^0(R, \epsilon)$ *sends the unit element* $1 = (M_d(R)^\phi, 1) \in L_p^0(M_d(R)^\phi)$ *to*

$$\mu(1) = [(M_d(R)^\phi)^!, 1^!] = (R^d, \phi) \in L_p^0(R, \epsilon) .$$

PROOF The Morita equivalence of additive categories with involution

$$\mu : \{\text{f.g. projective } M_d(R)^\phi\text{-modules}\} \xrightarrow{\simeq}$$

$$\{\text{f.g. projective } R\text{-modules}\} ; P \longrightarrow P^!$$

induces an isomorphism of the projective ± 1-quadratic L-groups

$$\mu : L_*^p(M_d(R)^\phi, \eta) \xrightarrow{\simeq} L_*^p(R, \epsilon\eta) ; [P, \theta] \longrightarrow [P^!, \theta^!] .$$

Similarly for the projective ± 1-symmetric L-groups L_p^*.

$\square$

REMARK 21.7 Let $p : \overline{X} \longrightarrow X$ be a finite d-sheeted covering as in 21.4, so that the fibre $F = p^{-1}(\{*\})$ is the discrete space with d points and

$$p_! : \overline{\pi} = \pi_1(\overline{X}) \longrightarrow \pi = \pi_1(X)$$

is the inclusion of a subgroup of finite index d. The algebraic K-theory transfer maps associated to p are the composites

$$p^! = \mu i_! : K_*(\mathbb{Z}[\pi]) \longrightarrow K_*(M_d(\mathbb{Z}[\overline{\pi}])) \xrightarrow{\simeq} K_*(\mathbb{Z}[\overline{\pi}])$$

with $i_!$ induced by the inclusion of rings

$$i : \mathbb{Z}[\pi] \longrightarrow \text{Hom}_{\mathbb{Z}[\overline{\pi}]}(i^! \mathbb{Z}[\pi], i^! \mathbb{Z}[\pi]) = M_d(\mathbb{Z}[\overline{\pi}])$$

and μ the Morita isomorphisms, such that $p^! \mathbb{Z}[\pi] = \mathbb{Z}[\overline{\pi}]^d$. The projective L-theory transfer maps associated to p are the composites

$$p^! = \mu i_! : L_*^p(\mathbb{Z}[\pi]) \longrightarrow L_*^p(M_d(\mathbb{Z}[\overline{\pi}])^\phi) \xrightarrow{\simeq} L_*^p(\mathbb{Z}[\overline{\pi}])$$

with μ the Morita isomorphisms of 21.6 for the nonsingular symmetric form $\sigma^*(F) = (p^! \mathbb{Z}[\pi], \phi)$ over $\mathbb{Z}[\overline{\pi}]$, with $\phi = 1 \oplus 1 \oplus \ldots \oplus 1$. For the free L-groups actually considered in 21.4 the transfer maps are

$$p^! = \mu i_! : L_*(\mathbb{Z}[\pi]) \longrightarrow L_*^I(M_d(\mathbb{Z}[\overline{\pi}])^\phi) \xrightarrow{\simeq} L_*(\mathbb{Z}[\overline{\pi}])$$

with

$$I = \text{im}(K_0(\mathbb{Z})) = d\mathbb{Z} \subset K_0(M_d(\mathbb{Z}[\overline{\pi}])) = \mathbb{Z} .$$

$\square$

§22. Finite fundamental group

The computation of the structure groups $\mathbb{S}_*(X)$ of a space X requires the calculation of the generalized homology groups $H_*(X;\mathbb{L}.)$, the L-groups $L_*(\mathbb{Z}[\pi])$ $(\pi = \pi_1(X))$ and the assembly map $A: H_*(X;\mathbb{L}.) \longrightarrow L_*(\mathbb{Z}[\pi])$. The classical methods of algebraic topology can deal with $H_*(X;\mathbb{L}.)$, but the more recent methods of algebraic K- and L-theory are required for $L_*(\mathbb{Z}[\pi])$ and A. In fact, it is quite difficult to obtain $\mathbb{S}_*(X)$ in general, but for finite π there is a highly evolved computational technique fulfilling the programme set out by Wall [172, 4.9] for using localization and completion to determine the L-theory of $\mathbb{Z}[\pi]$ from the classification of quadratic forms over algebraic number fields and rings of algebraic integers. Apart from Wall himself, this has involved the work (in alphabetic order) of Bak, Carlsson, Connolly, Hambleton, Kolster, Milgram, Pardon, Taylor, Williams and others.

The topological spherical space form problem is the study of free actions of finite groups on spheres, or equivalently of compact manifolds with finite fundamental group and the sphere S^n as universal cover. A finite group π acts freely on a CW complex X homotopy equivalent to S^n with trivial action on $H_*(X)$ if and only if the cohomology of π is periodic of order q dividing $n + 1$, with q necessarily even and n necessarily odd. The quotient X/π is a finitely dominated n-dimensional geometric Poincaré complex with fundamental group π and universal cover X. There exists such an action of π on X with X/π homotopy equivalent to a compact n-dimensional manifold if (and for $n \geq 5$ only if) π acts freely on S^n. Swan [168] applied algebraic K-theory to the spherical space form problem. The subsequent investigation of the spherical space form problem was one of the motivations for the development of non-simply-connected surgery theory in general, and the computation of $L_*(\mathbb{Z}[\pi])$ for finite π in particular. Madsen, Thomas and Wall [100] used surgery theory to classify the finite groups which act freely on spheres. Madsen and Milgram then classified the actions in dimensions ≥ 5. See Davis and Milgram [42] for a survey.

The computations of $L_*(\mathbb{Z}[\pi])$ have included the determination of the assembly map $A: H_*(B\pi;\mathbb{L}.) \longrightarrow L_*(\mathbb{Z}[\pi])$ for finite π by Hambleton, Milgram, Taylor and Williams [66] and Milgram [106]. The multisignature, Arf invariants, various semi-invariants and Whitehead torsion are used there to detect the surgery obstructions in $\mathrm{im}(A) \subseteq L_*(\mathbb{Z}[\pi])$ of normal maps of closed manifolds with finite fundamental group π. It appears that such invariants also suffice to detect the surgery obstructions in $L_*(\mathbb{Z}[\pi])$ of normal maps of finite geometric Poincaré complexes with finite fundamental group π. Such a detection should allow the total surgery obstruction $s(X) \in \mathbb{S}_n(X)$ of a finite geometric Poincaré complex X with finite $\pi_1(X)$ to be expressed in terms of the underlying homotopy type and these surgery invariants.

See Hambleton and Madsen [64] for the detection of the projective surgery obstructions in $L_*^p(\mathbb{Z}[\pi])$ of normal maps of finitely dominated geometric Poincaré complexes with finite fundamental group π in terms of the multisignature, Arf invariants and various semi-invariants as well as the Wall finiteness obstruction, which together with the underlying homotopy type can be used to at least express the projective total surgery obstruction $s^p(X) \in \mathbb{S}_n^p(X)$ (Appendix C) in terms of computable invariants.

The multisignature is the fundamental invariant of surgery obstruction theory with finite fundamental group π. It is a collection of integers indexed by the irreducible real representations of π, generalizing the signature in the simply connected case. The multisignature suffices for the computation of the projective L-groups $L_*^p(\mathbb{R}[\pi]) = L_p^*(\mathbb{R}[\pi])$, and for the determination of the quadratic L-groups $L_*(\mathbb{Z}[\pi])$ and the quadratic structure groups $\mathbb{S}_*(B\pi)$ modulo torsion. In 22.36 below it is explicitly verified that for an oriented finite n-dimensional geometric Poincaré complex X with a map $\pi_1(X) \longrightarrow \pi$ to a finite group π the multisignature determines the image of the total surgery obstruction $s(X) \in \mathbb{S}_n(X)$ in $\mathbb{S}_n(B\pi)$ modulo torsion. For the sake of brevity only the oriented case is considered in §22.

There are two distinct approaches to the multisignature, both of which were applied to the L-theory of finite groups by Wall [176, 13A,B]:

(i) The K-theoretic G-signature method of Atiyah and Singer [7] and Petrie [125], which depends on the character theory of finite-dimensional F-representations of a compact Lie group G, with $F = \mathbb{R}$ or $\mathbb{C}$. Only the case of a discrete finite group is considered here, with $G = \pi$. The 'K-theory F-multisignature' for $L_p^{2*}(F[\pi])$ consists of the rank invariants of the algebraic K-group $K_0(F[\pi])$ giving a natural isomorphism $L_p^{4*}(F[\pi]) \cong K_0(F[\pi])$, with the complex conjugation involution if $F = \mathbb{C}$. There is a similar (but more complicated) result for $L_p^{4*+2}(F[\pi])$.

(ii) The L-theoretic method of Wall [172], [176], Fröhlich and McEvett [54] and Lewis [92], which depends on the algebraic properties of the ring $F[\pi]$ for a finite group π, with F any field of characteristic 0. The 'L-theory F-multisignature' for $L_p^{2*}(F[\pi])$ consists of the signature invariants of the L-groups of the division rings appearing in the Wedderburn decomposition of $F[\pi]$ as a product of matrix algebras over division rings.

The K- and L-theory F-multisignatures coincide whenever both are defined. The $\mathbb{Q}$-multisignature coincides with the $\mathbb{R}$-multisignature.

DEFINITION 22.1 (i) Given a commutative ring with involution F and a group π let the group ring $F[\pi]$ have the involution

$$\bar{}: F[\pi] \longrightarrow F[\pi] \; ; \; \sum_{g \in \pi} a_g g \longrightarrow \sum_{g \in \pi} \bar{a}_g g^{-1} \; (a_g \in F) \; .$$

The involution on $F[\pi]$ is *real* if it is the identity on F.

The involution on $F[\pi]$ is *hermitian* if it is not the identity on F.

(ii) For $F = \mathbb{C}$ let $\mathbb{C}^+$ (resp. $\mathbb{C}^-$) denote $\mathbb{C}$ with the identity (resp. complex conjugation) involution, so that $\mathbb{C}^+[\pi]$ (resp. $\mathbb{C}^-[\pi]$) is $\mathbb{C}[\pi]$ with the real (resp. hermitian) involution.

□

For a finite group π and any field F of characteristic not divisible by $|\pi|$ the ring $F[\pi]$ is semi-simple, by Maschke's theorem, so every $F[\pi]$-module is projective. For any involution on F the $F[\pi]$-dual of a f.g. $F[\pi]$-module M is a f.g. $F[\pi]$-module $M^* = \text{Hom}_{F[\pi]}(M, F[\pi])$, with $F[\pi]$ acting by

$$F[\pi] \times M^* \longrightarrow M^* \ ; \ (ag, f) \longrightarrow (x \longrightarrow f(x)\bar{a}g^{-1}) \ (a \in F, g \in \pi) \ .$$

The F-module isomorphism

$$\text{Hom}_F(M, F) \xrightarrow{\cong} M^* \ ; \ f \longrightarrow (x \longrightarrow \sum_{g \in \pi} f(gx)g^{-1})$$

is an $F[\pi]$-module isomorphism, with $F[\pi]$ acting by

$$F[\pi] \times \text{Hom}_F(M, F) \longrightarrow \text{Hom}_F(M, F) \ ; \ (ag, f) \longrightarrow (x \longrightarrow f(gx)\bar{a}) \ .$$

For $\epsilon = \pm 1$ the ϵ-symmetric forms (M, ϕ) over $F[\pi]$ are in one–one correspondence with the ϵ-symmetric forms $(M, \phi^!)$ over F which are π-equivariant, that is

$$\phi^!(gx, gy) = \phi^!(x, y) \in \mathbb{R} \ (x, y \in M, g \in \pi) \ .$$

The forms (M, ϕ), $(M, \phi^!)$ correspond if

$$\phi(x, y) = \sum_{g \in \pi} \phi^!(gx, y)g \in F[\pi] \ ,$$

or equivalently

$$\phi^!(x, y) = \text{coefficient of } 1 \text{ in } \phi(x, y) \in F \subset F[\pi] \ .$$

LEMMA 22.2 *Let $F = \mathbb{R}$ or $\mathbb{C}^-$. A f.g. $F[\pi]$-module M supports a nonsingular symmetric form (M, θ) over $F[\pi]$ which is positive definite:*

$$\theta^!(x, x) > 0 \ (x \in M \backslash \{0\}) \ .$$

Any two such forms $\theta(0)$, $\theta(1)$ are homotopic, i.e. related by a continuous map $\theta \colon I \longrightarrow \text{Hom}_{F[\pi]}(M, M^*)$ with each $(M, \theta(t))$ $(t \in I)$ positive definite.

PROOF The underlying F-module of M supports a positive definite symmetric form (M, θ_0) over F, which is unique up to homotopy. The symmetric form $(M, \theta^!)$ over F obtained by averaging

$$\theta^!(x, y) = (1/|\pi|) \sum_{g \in \pi} \theta_0(gx, gy) \in F \ (x, y \in M)$$

is positive definite and π-equivariant, corresponding to a nonsingular symmetric form (M, θ) over $F[\pi]$.

□

Let $F = \mathbb{R}$ or $\mathbb{C}^-$, as before. Given a f.g. $F[\pi]$-module M and an endomorphism $f: M \longrightarrow M$ let

$$f^t = \theta^{-1} f^* \theta : M \longrightarrow M$$

be the endomorphism adjoint with respect to the nonsingular symmetric form (M, θ) over $F[\pi]$ with the form (M, θ') over F positive definite.

The following definition of the multisignature is just a translation into the language of algebraic K-theory of the definition of the G-signature due to Atiyah and Singer [7, pp. 578–579] in the case of a discrete finite group $G = \pi$.

DEFINITION 22.3 Let $F = \mathbb{R}$ or $\mathbb{C}^-$. The K-theory F-multisignature of a projective nonsingular ϵ-symmetric form (M, ϕ) over $F[\pi]$ is the element

$$[M, \phi] \in K_0(F[\pi], \epsilon)$$

defined as follows:

(i) If $\epsilon = +1$ then $K_0(F[\pi], \epsilon) = K_0(F[\pi])$. The $F[\pi]$-module morphism $f = \theta^{-1}\phi: M \longrightarrow M$ is self-adjoint, that is $f^t = f$, and may be diagonalized by the spectral theorem with real eigenvalues. The positive and negative eigenspaces M_+, M_- are π-invariant, so that they are f.g. projective $F[\pi]$-modules, and

$$[M, \phi] = [M_+] - [M_-] \in K_0(F[\pi]) .$$

(ii) If $F = \mathbb{C}^-$ and $\epsilon = -1$ then $K_0(F[\pi], \epsilon) = K_0(\mathbb{C}[\pi])$. The K-theory F-multisignature of (M, ϕ) is defined to be the K-theory F-multisignature (as in (i)) of the nonsingular symmetric form $(M, i\phi)$ over $\mathbb{C}^-[\pi]$

$$[M, \phi] = [M, i\phi] = [M_+] - [M_-] \in K_0(\mathbb{C}[\pi]) .$$

(iii) If $F = \mathbb{R}$ and $\epsilon = -1$ then

$$K_0(F[\pi], \epsilon) = \{ x - x^* \mid x \in K_0(\mathbb{C}[\pi]) \} \subset K_0(\mathbb{C}[\pi]) .$$

The $\mathbb{R}[\pi]$-module morphism $f = \theta^{-1}\phi: M \longrightarrow M$ is skew-adjoint, that is $f^t = -f$. If $(ff^t)^{1/2}$ denotes the positive square root of ff^t the automorphism

$$J = f/(ff^t)^{1/2} : M \longrightarrow M$$

is such that $J^2 = -1$ and commutes with the action of π. Let (M, J), $(M, -J)$ be the f.g. projective $\mathbb{C}[\pi]$-modules defined by the two π-invariant complex structures $J, -J$ on M. The K-theory $\mathbb{R}$-multisignature of (M, ϕ) is given by

$$[M, \phi] = [M, J] - [M, -J] \in K_0(\mathbb{R}[\pi], -1) \subset K_0(\mathbb{C}[\pi]) .$$

This is the K-theory $\mathbb{C}$-multisignature (as in (i)) of the nonsingular symmetric form $(\mathbb{C} \otimes_{\mathbb{R}} M, i \otimes \phi)$ over $\mathbb{C}^-[\pi]$, with

$$(\mathbb{C} \otimes_{\mathbb{R}} M)_{\pm} = \{ 1 \otimes x \mp i \otimes Jx \mid x \in M \} \cong (M, \pm J) .$$

□

PROPOSITION 22.4 *The K-theory F-multisignature defines isomorphisms*

$$L_p^0(F[\pi], \epsilon) \xrightarrow{\sim} K_0(F[\pi], \epsilon) \ ; \ (M, \phi) \longrightarrow [M, \phi] \ \ (F = \mathbb{R} \ or \ \mathbb{C}^-) \ .$$

PROOF For $\epsilon = 1$ the inverse isomorphism is defined by sending a projective class $[M] \in K_0(F[\pi])$ to the Witt class $[M, \theta] \in L_p^0(F[\pi])$ of the positive definite nonsingular symmetric form (M, θ) over $F[\pi]$ given by 22.2. Similarly for $(F, \epsilon) = (\mathbb{C}^-, -1)$, with $[M]$ sent to $(M, i\theta)$. For $(F, \epsilon) = (\mathbb{R}, -1)$ see 22.19 below.

□

Let F be a field of characteristic 0, and let π be a finite group. The L-theory multisignature for $L_p^*(F[\pi])$ is an analogue of 'multirank' for the projective class group $K_0(F[\pi])$. Both the multirank and the multisignature are collections of integer-valued rank invariants indexed by the irreducible F-representations of the finite group π, obtained as follows.

By Wedderburn's theorem $F[\pi]$ is a finite product of simple rings

$$F[\pi] \ = \ S_1(F, \pi) \times S_2(F, \pi) \times \ldots \times S_{\alpha(F, \pi)}(F, \pi) \ ,$$

starting with $S_1(F, \pi) = F$. Each of the factors is a matrix algebra

$$S_j(F, \pi) \ = \ M_{d_j(F, \pi)}(D_j(F, \pi))$$

over a simple finite-dimensional F-algebra

$$D_j(F, \pi) \ = \ \mathrm{End}_{F[\pi]}(P_j) \ ,$$

which is the endomorphism ring of the corresponding simple f.g. projective $F[\pi]$-module $P_j = D_j(F, \pi)^{d_j(F, \pi)}$, with centre F. Let G be the Galois group of the field extension of F obtained by adjoining the $|\pi|$th roots of 1. G is a subgroup of $\mathbb{Z}_{|\pi|}^{\bullet}$, the multiplicative group of units in $\mathbb{Z}_{|\pi|} \backslash \{0\}$. Two elements $x, y \in \pi$ are *F-conjugate* if

$$x^g \ = \ h^{-1}yh \in \pi$$

for some $g \in G$, $h \in \pi$. The number of simple factors in $F[\pi]$ is given by $\alpha(F, \pi)$

= no. of isomorphism classes of irreducible F-representations of π

= no. of F-conjugacy classes in π .

See Serre [154, 12.4] or Curtis and Reiner [40, 21.5] for the details. For each isomorphism class of simple finite-dimensional algebras D over F let $\alpha_D(F, \pi)$ be the number of factors $S_j(F, \pi)$ in $F[\pi]$ with $D_j(F, \pi) = D$, so that

$$\alpha(F, \pi) \ = \ \sum_D \alpha_D(F, \pi) \ .$$

For a division ring R f.g. projective R-modules are f.g. free, and rank defines an isomorphism

$$K_0(R) \xrightarrow{\simeq} \mathbb{Z} \ ; \ [R^m] - [R^n] \longrightarrow m - n \ .$$

The algebraic K-groups of a product of rings $R = R_1 \times R_2$ are given by

$$K_*(R_1 \times R_2) \ = \ K_*(R_1) \oplus K_*(R_2) \ .$$

For any finite group π

$$K_0(F[\pi]) \ = \ \sum_{j=1}^{\alpha(F,\pi)} K_0(S_j(F,\pi)) \ = \ \sum_{j=1}^{\alpha(F,\pi)} K_0(D_j(F,\pi)) \ = \ \sum_{j=1}^{\alpha(F,\pi)} \mathbb{Z} \ .$$

The F-*multirank* of a f.g. projective $F[\pi]$-module P is the collection of $\alpha(F,\pi)$ rank invariants

$$r_j(P) \ = \ [S_j(F,\pi) \otimes_{F[\pi]} P] \in K_0(S_j(F,\pi)) \ = \ K_0(D_j(F,\pi)) \ = \ \mathbb{Z} \ ,$$

one for each simple factor $S_j(F,\pi)$ in $F[\pi]$. The F-multirank defines an isomorphism

$$r_*(P) : K_0(F[\pi]) \xrightarrow{\simeq} \sum_{j=1}^{\alpha(F,\pi)} \mathbb{Z} \ ; \ [P] \longrightarrow (r_1(P), r_2(P), \ldots, r_{\alpha(F,\pi)}(P)) \ ,$$

with $r_*((D_j)^{d_j}) = (0, \ldots, 0, 1, 0, \ldots, 0)$ and $r_*(S_j) = (0, \ldots, 0, d_j, 0, \ldots, 0)$ $(d_j = d_j(F,\pi))$. The inclusion $i: F \longrightarrow F[\pi]$ induces a rudimentary algebraic K-theory assembly map

$$i_! \ = \ \begin{pmatrix} d_1 \\ d_2 \\ \vdots \\ d_{\alpha(F,\pi)} \end{pmatrix} :$$

$$H_0(B\pi; \mathbb{K}(F)) \ = \ K_0(F) \ = \ \mathbb{Z} \longrightarrow K_0(F[\pi]) \ = \ \sum_{j=1}^{\alpha(F,\pi)} \mathbb{Z} \ ;$$

$$[F] \ = \ 1 \longrightarrow r_*(F[\pi]) \ = \ (d_1, d_2, \ldots, d_{\alpha(F,\pi)})$$

with $\mathbb{K}(F)$ the algebraic K-theory spectrum of F. The transfer map is given by

$$i^! \ = \ (c_1 d_1 \ c_2 d_2 \ \ldots \ c_{\alpha(F,\pi)} d_{\alpha(F,\pi)}) : K_0(F[\pi]) \ = \ \sum_{j=1}^{\alpha(F,\pi)} \mathbb{Z} \longrightarrow K_0(F) \ = \ \mathbb{Z}$$

with $c_j = \dim_F(D_j(F,\pi))$, and

$$i^! i_! \ = \ \sum_{j=1}^{\alpha(F,\pi)} c_j(d_j)^2 \ = \ |\pi| : K_0(F) \ = \ \mathbb{Z} \longrightarrow K_0(F) \ = \ \mathbb{Z} \ .$$

The reduced projective class group $\tilde{K}_0(\mathbb{Z}[\pi])$ is finite for a finite group π by a theorem of Swan, and every f.g. projective $\mathbb{Z}[\pi]$-module P induces a f.g. free $\mathbb{Q}[\pi]$-module $\mathbb{Q}[\pi] \otimes_{\mathbb{Z}[\pi]} P$, so that

$$\operatorname{im}(\tilde{K}_0(\mathbb{Z}[\pi]) \longrightarrow \tilde{K}_0(\mathbb{Q}[\pi])) = \{0\}$$

and the $\mathbb{Q}$-multirank is not useful for detecting $\tilde{K}_0(\mathbb{Z}[\pi])$. The F-multitorsion is defined for any field F of characteristic 0 by means of the identification

$$K_1(F[\pi]) = \sum_{j=1}^{\alpha(F,\pi)} K_1(D_j(F,\pi)) .$$

By a theorem of Bass the torsion group $K_1(\mathbb{Z}[\pi])$ and the Whitehead group $Wh(\pi)$ are finitely generated for finite π, with the same rank

$$\dim_{\mathbb{Q}} \mathbb{Q} \otimes K_1(\mathbb{Z}[\pi]) = \dim_{\mathbb{Q}} \mathbb{Q} \otimes Wh(\pi) = \alpha(\mathbb{R}, \pi) - \alpha(\mathbb{Q}, \pi)$$

detected by the $\mathbb{Q}$-multitorsion subject to the restrictions given by the Dirichlet unit theorem: each of the $\alpha(\mathbb{Q}, \pi)$ simple factors $S = M_d(D)$ in $\mathbb{Q}[\pi]$ contributes $\alpha(\mathbb{R}, S) - 1$, with $\alpha(\mathbb{R}, S)$ the number of simple factors in $\mathbb{R} \otimes_{\mathbb{Q}} S$.

The character of an F-representation $\rho: \pi \longrightarrow GL_d(F)$ is the (conjugacy) class function

$$\chi(\rho) : \pi \longrightarrow F ; g \longrightarrow \operatorname{tr}(\rho(g)) .$$

Let $R_F(\pi)$ be the F-coefficient character group of π, the free abelian group of $\mathbb{Z}$-linear combinations of the characters of the irreducible F-representations. The F-multirank also defines an isomorphism

$$K_0(F[\pi]) \xrightarrow{\simeq} R_F(\pi) ; [P] \longrightarrow \sum_{j=1}^{\alpha(F,\pi)} r_j(P)\chi(\rho_j)$$

with ρ_j the irreducible F-representation

$$\rho_j : \pi \longrightarrow \operatorname{Aut}_F(D_j(F,\pi)^{d_j}) = GL_{c_j d_j}(F)$$

of degree $c_j d_j$ defined by the composite

$$\pi \longrightarrow F[\pi] \longrightarrow S_j(F,\pi) = \operatorname{End}_{D_j(F,\pi)}(D_j(F,\pi)^{d_j}) .$$

EXAMPLE 22.5 (i) The element $i_!(F] = [F[\pi]] \in K_0(F[\pi])$ corresponds to the character

$$\chi : \pi \longrightarrow F ; g \longrightarrow \begin{cases} |\pi| \\ 0 \end{cases} \text{ if } \begin{cases} g = 1 \\ g \neq 1 \end{cases}$$

of the regular F-representation $F[\pi]$ of π with degree $|\pi|$.
(ii) Regarded as a character, the K-theory F-multisignature (22.3) of a nonsingular ϵ-symmetric form (M, ϕ) over $F[\pi]$ ($F = \mathbb{R}$ or $\mathbb{C}^-$) is the class

function

$$[M, \phi] : \pi \longrightarrow \begin{cases} F \\ \mathbb{C} \end{cases} ; g \longrightarrow \sigma(g, (M, \phi)) = \begin{cases} \text{tr}(g|_{M^+}) - \text{tr}(g|_{M^-}) \\ \text{tr}(g|_{(M,J)}) - \overline{\text{tr}(g|_{(M,J)})} \end{cases}$$

$$\text{if } (F, \epsilon) = \begin{cases} (\mathbb{R}, 1) \text{ or } (\mathbb{C}^-, \pm 1) \\ (\mathbb{R}, -1) \ . \end{cases}$$

In particular, for $(M, \phi) = (F[\pi], 1)$ this is the character of the regular F-representation, as in (i).

$\square$

If $\{\rho_1, \rho_2, \ldots, \rho_{\alpha(F,\pi)}\}$ is a complete set of irreducible F-representations of π with characters $\{\chi_1, \chi_2, \ldots, \chi_{\alpha(F,\pi)}\}$ then the central idempotent

$$e_j(F, \pi) = e_j(F, \pi)^2 \in F[\pi]$$

with

$$e_j(F, \pi)F[\pi] = S_j(F, \pi) \ , \ e_j(F, \pi)e_k(F, \pi) = 0 \ (j \neq k)$$

is given by

$$e_j(F, \pi) = (f_j/|\pi|) \sum_{g \in \pi} \chi_j(g)g^{-1} \in F[\pi]$$

for some $f_j \in F$.

As a purely algebraic invariant the multisignature is a generalization of the signatures used by Hasse [70] and Landherr [86] to classify quadratic and hermitian forms over algebraic number fields. The *total signature* map on the symmetric Witt group $L^0(F)$ of a field F with the identity involution

$$\sigma = \sum_{j=1}^{\alpha} \sigma_j : L^0(F) \longrightarrow \sum_{j=1}^{\alpha} L^0(\mathbb{R}) = \sum_{j=1}^{\alpha} \mathbb{Z}$$

has one component for each embedding $\sigma_j : F \subset \mathbb{R}$ (Milnor and Husemoller [110, 3.3.10], Scharlau [153, 3.6]). The kernel of σ is the torsion subgroup of $L^0(F)$, with 2-primary torsion only. The image of σ is constrained by the congruences

$$\sigma_j(M, \phi) \equiv \dim_F(M) \, (\text{mod} \, 2) \ (1 \leq j \leq \alpha)$$

for any nonsingular symmetric form (M, ϕ) over F. For an algebraic number field F the image of σ is such that

$$2(\sum_{j=1}^{\alpha} \mathbb{Z}) \subseteq \text{im}(\sigma) \subseteq \sum_{j=1}^{\alpha} \mathbb{Z}$$

and σ is an isomorphism modulo 2-primary torsion [110, p. 65]. For any field F of characteristic $\neq 2$ $L_*(F) = L^*(F)$ and every nonsingular skew-symmetric form over F is hyperbolic, so that $L_2(F) = L^2(F) = 0$.

The product decomposition

$$F[\pi] \;=\; \prod_{j=1}^{\alpha(F,\pi)} M_{d_j(F,\pi)}(D_j(F,\pi))$$

reduces the computation of $L_*(F[\pi])$ for finite π to that of $L_*(D)$ for division rings with involution D which are finite-dimensional algebras over F. By assumption F has characteristic 0, so that $1/2 \in F$ and there is no difference between the quadratic and symmetric L-groups

$$L_*(F[\pi]) \;=\; L^*(F[\pi]) \;.$$

The calculations are particularly easy for projective L-theory L_p^*, since this has better categorical properties than the free L-theory L^*, while differing from it in at most 2-primary torsion:

PROPOSITION 22.6 *For any ring with involution A the forgetful maps $L^*(A)$ $\longrightarrow L_p^*(A)$ from the free to the projective L-groups are isomorphisms modulo 2-primary torsion, so that*

$$L^*(A)[1/2] \;=\; L_p^*(A)[1/2] \;.$$

PROOF Immediate from the exact sequence of Ranicki [136]

$$\ldots \longrightarrow L^n(A) \longrightarrow L_p^n(A) \longrightarrow \widehat{H}^n(\mathbb{Z}_2 \,; \widetilde{K}_0(A)) \longrightarrow L^{n-1}(A) \longrightarrow \ldots \;,$$

since the Tate $\mathbb{Z}_2$-cohomology groups $\widehat{H}^*$ are of exponent 2.

□

PROPOSITION 22.7 (i) *The odd-dimensional projective L-groups of a semi-simple ring A with involution vanish:*

$$L_p^{2*+1}(A) \;=\; 0 \;.$$

(ii) *For a finite group π and any field F with $|\pi| \nmid \mathrm{char}(F)$*

$$L_p^{2*+1}(F[\pi]) \;=\; 0 \;.$$

PROOF (i) The proof of $L_{2*+1}^p(A) = 0$ in Ranicki [138] extends to symmetric L-theory.
(ii) Immediate from (i), since $F[\pi]$ is semi-simple.

□

A division ring D is such that $\widetilde{K}_0(D) = 0$, and so $L^*(D) = L_p^*(D)$ for any involution on D. Also, D is simple, so that $L^{2*+1}(D) = 0$. Let $D^\bullet = D\backslash\{0\}$, and for $\epsilon = \pm 1$ let

$$D_\epsilon^\bullet \;=\; \{x \in D^\bullet \,|\, \bar{x} = \epsilon x\} \;.$$

Every nonsingular ϵ-symmetric form over D is equivalent in the Witt group to a diagonal form $\sum_{m=1}^{n} (D, x_m)$ with $x_m \in D_\epsilon^\bullet$, so that the morphism

$$\mathbb{Z}[D_\epsilon^\bullet] \;\longrightarrow\; L^0(D, \epsilon) \;; \; [x] \longrightarrow (D, x)$$

is onto.

PROPOSITION 22.8 *The ϵ-symmetric Witt group $L^0(D,\epsilon)$ of a division ring with involution D is given in terms of generators and relations by*

$$L^0(D,\epsilon) = \mathbb{Z}[D_\epsilon^\bullet]/N_\epsilon$$

with N_ϵ the subgroup of $\mathbb{Z}[D_\epsilon^\bullet]$ generated by elements of the type

$$[x] - [ax\bar{a}] \ , \ [x] + [-x] \ , \ [x] + [y] - [x+y] - [x(x+y)^{-1}y]$$

for any $a \in D^\bullet$, $x,y \in D_\epsilon^\bullet$ with $x + y \neq 0$.

PROOF See Scharlau [153, 2.9] and Cibils [35]. (For a field F of characteristic $\neq 2$ with the identity involution such a presentation of $L^0(F,1)$ was originally obtained by Witt himself).   □

   The projective L-theory of products is given by:

PROPOSITION 22.9 *Let R be a ring which is a product*

$$R = R_1 \times R_2 \ .$$

For an involution on R which preserves the factors $(\overline{R}_i = R_i)$

$$L_p^*(R) = L_p^*(R_1) \oplus L_p^*(R_2) \ ,$$

while for an involution which interchanges the factors $(\overline{R}_1 = R_2)$

$$L_p^*(R) = 0 \ .$$

Similarly for the quadratic L-groups L_*.

PROOF The central idempotents

$$e_1 = (1,0) \ , \ e_2 = (0,1) \in R = R_1 \times R_2$$

are such that

$$e_i R = R_i \ , \ (e_i)^2 = e_i \ , \ e_1 + e_2 = 1 \ , \ e_1 e_2 = 0 \in R \ (i = 1,2) \ .$$

An involution on R preserves the factors if and only if $\bar{e}_i = e_i$ in which case there are defined isomorphisms

$$L_p^*(R) \xrightarrow{\simeq} L_p^*(R_1) \oplus L_p^*(R_2) \ ; \ (C,\phi) \longrightarrow (e_1 C, e_1 \phi) \oplus (e_2 C, e_2 \phi) \ .$$

An involution on R interchanges the factors if and only if $\bar{e}_1 = e_2$, in which case for every projective symmetric Poincaré complex (C,ϕ) over R there is defined a null-cobordism $(C \longrightarrow e_1 C, (0,\phi))$, and so $L_p^*(R) = 0$.   □

   A simple factor $S_j(F,\pi)$ of $F[\pi]$ is preserved by the involution

$$\overline{S_j(F,\pi)} = S_j(F,\pi)$$

if and only if the idempotent $e_j(F,\pi) \in F[\pi]$ is such that

$$\overline{e_j(F,\pi)} = e_j(F,\pi) \in F[\pi] \ .$$

PROPOSITION 22.10 *The projective L-groups of $F[\pi]$ are such that*

$$L_p^*(F[\pi]) = \sum_{j \in J(F,\pi)} L_p^*(S_j(F,\pi))$$

with

$$J(F,\pi) = \{j \mid \overline{S}_j = S_j\} \subseteq \{1, 2, \ldots, \alpha(F,\pi)\}$$

the indexing set for the simple factors $S_j = S_j(F,\pi)$ preserved by the involution on $F[\pi]$, depending on the choice of involution on the ground field F. (In fact, $L_p^{2*+1}(F[\pi]) = 0$, by 22.7.)
PROOF Immediate from 22.9, since the simple factors $S_j(F,\pi)$ of $F[\pi]$ not preserved by the involution come in pairs $S_j(F,\pi) \times S_j(F,\pi)^{op}$ with the hyperbolic involution $(x,y) \longmapsto (y,x)$.

□

From now on, only the ground fields $F = \mathbb{C}, \mathbb{R}, \mathbb{Q}$ will be considered.

PROPOSITION 22.11 *Let D be a division ring such that $M_d(D)$ is a simple factor of $F[\pi]$ for some finite group π. For any involution on D and $\epsilon = \pm 1$ the ϵ-symmetric Witt group $L^0(D, \epsilon)$ is a countable abelian group of finite rank, with 2-primary torsion only.*
PROOF See Wall [177].

□

The 2-primary torsion in $L^0(D, \epsilon)$ may well be infinitely generated in the case $F = \mathbb{Q}$ (Hasse–Witt invariants), e.g. if $D = \mathbb{Q}$, $\epsilon = +1$

$$L^0(\mathbb{Q}, 1) = L^0(\mathbb{Q}) = L^0(\mathbb{R}) \oplus \bigoplus_{p \text{ prime}} L^0(\mathbb{F}_p) = \mathbb{Z} \oplus (\mathbb{Z}_2)^\infty \oplus (\mathbb{Z}_4)^\infty$$

with $\mathbb{F}_p$ the finite field of p elements (Milnor and Husemoller [110, IV.2]).

TERMINOLOGY 22.12 Given a division ring with involution D as in 22.11 let $r^k(D) \geq 0$ be the rank of the $(-)^k$-symmetric Witt group of D, so that

$$L^0(D, (-)^k)[1/2] = \sum_{r^k(D)} \mathbb{Z}[1/2] .$$

□

The rank of the Witt group $L^{2k}(D) = L^0(D, (-)^k)$ of a division ring with involution D is the number of the signatures given by the embeddings of D in $\mathbb{R}, \mathbb{H}$ and $\mathbb{C}^-$, whose L-theory is tabulated in 22.16 below.

The following definition of the multisignature is just a translation into the language of algebraic L-theory of the definition due to Wall [172, 4.9], [176, p. 164].

DEFINITION 22.13 The *L-theory F-multisignature* of a nonsingular $(-1)^k$-symmetric form (M, ϕ) over $F[\pi]$ for a finite group π is the collection of $\alpha^k(F, \pi)$ signature invariants

$$\sigma_j(M, \phi) = [S_j(F, \pi) \otimes_{F[\pi]} (M, \phi)] \in \operatorname{im}\left(L^{2k}(S_j(F, \pi)) \longrightarrow \sum_{r^k(D_j(F,\pi))} \mathbb{Z}\right)$$

with $\alpha^k(F, \pi) = \sum_{j \in J(F,\pi)} r^k(D_j(F, \pi))$.

□

PROPOSITION 22.14 *The L-theory F-multisignature map*

$$\sigma = \sum_{j \in J(F,\pi)} \sigma_j : L^{2k}(F[\pi]) \longrightarrow \sum_{j \in J(F,\pi)} \sum_{r^k(D_j(F,\pi))} \mathbb{Z} = \sum_{\alpha^k(F,\pi)} \mathbb{Z}$$

is an isomorphism modulo 2-primary torsion, with

$$L^{2k}(F[\pi])[1/2] = \sum_{j \in J(F,\pi)} L^{2k}(D_j(F, \pi))[1/2] \xrightarrow{\simeq} \sum_{\alpha^k(F,\pi)} \mathbb{Z}[1/2] .$$

PROOF Immediate from 21.6 and 22.10.

□

The (α, β)-*quaternion algebra* over a field F is the division F-algebra with centre F defined for any $\alpha, \beta \in F^\bullet$ by

$$\left(\frac{\alpha, \beta}{F}\right) = \{w + xi + yj + zk \mid w, x, y, z \in F\}$$

with

$$i^2 = \alpha , \quad j^2 = \beta , \quad ij = -ji = k , \quad k^2 = -\alpha\beta .$$

Now specialize to the case $F = \mathbb{R}$. The ring $\mathbb{R}[\pi]$ is a product of simple finite-dimensional algebras over $\mathbb{R}$. Such an algebra is a matrix ring $M_d(D)$ with D one of $\mathbb{R}, \mathbb{H}, \mathbb{C}$.

The quaternion ring

$$\mathbb{H} = \left(\frac{-1, -1}{\mathbb{R}}\right) = \{w + xi + yj + zk \mid w, x, y, z \in \mathbb{R}\}$$

is given the quaternion conjugation involution

$$\mathbb{H} \longrightarrow \mathbb{H} ; v = w + xi + yj + zk \longrightarrow \bar{v} = w - xi - yj - zk .$$

DEFINITION 22.15 Let D be one of the rings with involution $\mathbb{R}, \mathbb{H}, \mathbb{C}^-$. The *signature* of a nonsingular symmetric form (M, ϕ) over D is defined by

$$\operatorname{signature}(M, \phi) = \sum_{m=1}^{n} \operatorname{sign} x_m \in \mathbb{Z}$$

using any diagonalization $(M, \phi) \cong \sum_{m=1}^{n} (D, x_m)$, with $x_m \in D_{+1}^{\bullet} = \mathbb{R} \backslash \{0\}$.
Equivalently,

$$\text{signature} \, (M, \phi) \; = \; [M_+] - [M_-] \in K_0(D) \; = \; \mathbb{Z}$$

for any decomposition $(M, \phi) = (M_+, \phi_+) \oplus (M_-, \phi_-)$ into positive definite and negative definite parts.

□

PROPOSITION 22.16 (i) *The L-groups of $\mathbb{R}$ are given by*

$$L^n(\mathbb{R}) \; = \; \begin{cases} \mathbb{Z} \\ 0 \end{cases} \textit{if } n \begin{cases} \equiv 0 \\ \not\equiv 0 \end{cases} (\text{mod } 4)$$

with isomorphisms

$$\text{signature} : \; L^{4*}(\mathbb{R}) \; \xrightarrow{\simeq} \; K_0(\mathbb{R}) \; = \; \mathbb{Z} \, ,$$

so that $r^0(\mathbb{R}) = 1$, $r^1(\mathbb{R}) = 0$.
(ii) *The L-groups of $\mathbb{H}$ are given by*

$$L^n(\mathbb{H}) \; = \; \begin{cases} \mathbb{Z} \\ \mathbb{Z}_2 \\ 0 \end{cases} \textit{if } n \equiv \begin{cases} 0 \\ 2 \\ 1, 3 \end{cases} (\text{mod } 4)$$

with isomorphisms

$$\text{signature} : \; L^{4*}(\mathbb{H}) \; \xrightarrow{\simeq} \; K_0(\mathbb{H}) \; = \; \mathbb{Z} \, ,$$

so that $r^0(\mathbb{H}) = 1$, $r^1(\mathbb{H}) = 0$. The generator $1 \in L^{4*+2}(\mathbb{H}) = \mathbb{Z}_2$ is represented by the nonsingular skew-symmetric form $(\mathbb{H}, i)$.
(iii) *The L-groups of $\mathbb{C}^-$ are given by*

$$L^n(\mathbb{C}^-) \; = \; \begin{cases} \mathbb{Z} \\ 0 \end{cases} \textit{if } n \equiv \begin{cases} 0 \\ 1 \end{cases} (\text{mod } 2)$$

with isomorphisms

$$\text{signature} : \; L^{2*}(\mathbb{C}^-) \; \xrightarrow{\simeq} \; K_0(\mathbb{C}) \; = \; \mathbb{Z} \, ,$$

so that $r^0(\mathbb{C}^-) = r^1(\mathbb{C}^-) = 1$.
(iv) *The L-groups of $\mathbb{C}^+$ are given by*

$$L^n(\mathbb{C}^+) \; = \; \begin{cases} \mathbb{Z}_2 \\ 0 \end{cases} \textit{if } n \begin{cases} \equiv 0 \\ \not\equiv 0 \end{cases} (\text{mod } 4)$$

so that $r^0(\mathbb{C}^+) = r^1(\mathbb{C}^+) = 0$.

□

The number of simple factors $S_j = S_j(\mathbb{R}, \pi) = M_{d_j}(D_j(\mathbb{R}, \pi))$ in $\mathbb{R}[\pi]$ is

$\alpha(\mathbb{R}, \pi) \; = \;$ no. of irreducible $\mathbb{R}$-representations of π

$= \;$ no. of conjugacy classes of unordered pairs $\{g, g^{-1}\}$ in π

$= \; \alpha_{\mathbb{R}}(\mathbb{R}, \pi) + \alpha_{\mathbb{C}}(\mathbb{R}, \pi) + \alpha_{\mathbb{H}}(\mathbb{R}, \pi)$

with $\alpha_D(\mathbb{R}, \pi)$ the number of simple factors S_j such that $D_j(\mathbb{R}, \pi) = D$. The projective class group of $\mathbb{R}[\pi]$ is given by

$$K_0(\mathbb{R}[\pi]) = \sum_{j=1}^{\alpha(\mathbb{R},\pi)} K_0(S_j(\mathbb{R}, \pi)) = \sum_{j=1}^{\alpha(\mathbb{R},\pi)} K_0(D_j(\mathbb{R}, \pi))$$

$$= \sum_{\alpha_\mathbb{R}(\mathbb{R},\pi)} K_0(\mathbb{R}) \oplus \sum_{\alpha_\mathbb{H}(\mathbb{R},\pi)} K_0(\mathbb{H}) \oplus \sum_{\alpha_\mathbb{C}(\mathbb{R},\pi)} K_0(\mathbb{C}) = \sum_{\alpha(\mathbb{R},\pi)} \mathbb{Z} \, .$$

Every simple factor $S_j(\mathbb{R}, \pi)$ in $\mathbb{R}[\pi]$ is preserved by the involution, and the duality involution $*: K_0(\mathbb{R}[\pi]) \longrightarrow K_0(\mathbb{R}[\pi])$ is the identity.

In order to obtain the corresponding computation of $L_p^{2*}(\mathbb{R}[\pi])$ it is necessary to consider the action of the involution on $\mathbb{R}[\pi]$ on the simple factors $S_j(\mathbb{R}, \pi)$.

Let A be a central simple algebra over a field K of characteristic $\neq 2$, with $\dim_K(A) = d^2$. Involutions

$$I : A \xrightarrow{\;\simeq\;} A \; ; \; a \longrightarrow \bar{a}$$

are classified by the dimensions of the I-invariant subspaces

$$A^+ = H^0(\mathbb{Z}_2 \, ; A) = \{a \in A \,|\, \bar{a} = a\} \, ,$$

$$A^- = H^1(\mathbb{Z}_2 \, ; A) = \{a \in A \,|\, \bar{a} = -a\}$$

with $A = A^+ \oplus A^-$, as follows:

(I) (first kind, orthogonal type)

$$\dim_K(A^+) = d(d+1)/2 \, , \; \dim_K(A^-) = d(d-1)/2 \, ,$$

in which case $I|: K \longrightarrow K$ is the identity,

(II) (first kind, symplectic type)

$$\dim_K(A^+) = d(d-1)/2 \, , \; \dim_K(A^-) = d(d+1)/2 \, ,$$

in which case $I|: K \longrightarrow K$ is the identity,

(III) (second kind, unitary type) d is even and

$$\dim_K(A^+) = \dim_K(A^-) = d^1/2 \, ,$$

in which case $I|: K \longrightarrow K$ is not the identity.

See Scharlau [153, §8.7] for further details.

EXAMPLE 22.17 Let (V, ϕ) be a nonsingular ϵ-symmetric form over a field with involution K of characteristic $\neq 2$, and let $\dim_K(V) = d$. Define an involution on the d^2-dimensional central simple K-algebra $A = \mathrm{Hom}_K(V, V)$ by

$$I : A \xrightarrow{\;\simeq\;} A \; ; \; f \longrightarrow \phi^{-1} f^* \phi \, .$$

Let $M_d(K)^\phi$ be the matrix ring with involution defined in 21.5. A choice of basis for V determines an identification $A = M_d(K)^\phi$. Use the isomorphism

$$A = \mathrm{Hom}_K(V,V) \xrightarrow{\simeq} \mathrm{Hom}_K(V,V^*) \; ; \; f \longrightarrow \phi f$$

to identify I with the ϵ-duality involution

$$I : \mathrm{Hom}_K(V,V^*) \xrightarrow{\simeq} \mathrm{Hom}_K(V,V^*) \; ; \; f \longrightarrow \epsilon f^* \; .$$

The I-invariant subspaces

$$A^\pm = \{ f \in \mathrm{Hom}_K(V,V^*) \,|\, \epsilon f^* = \pm f \}$$

are the spaces of $\pm\epsilon$-symmetric forms on V. The involution $I: A \longrightarrow A$ corresponds to the ϵ-transposition involution $x \otimes y \longrightarrow \epsilon y \otimes x$ on $V^* \otimes_K V^*$ under the isomorphism

$$V^* \otimes_K V^* \xrightarrow{\simeq} \mathrm{Hom}_K(V,V^*) \; ; \; f \otimes g \longrightarrow (x \longrightarrow (y \longrightarrow \overline{f(x)}g(y))) \; ,$$

allowing the identifications

$$A^\epsilon = \mathrm{Sym}(V^* \otimes_K V^*) \; , \quad A^{-\epsilon} = \mathrm{Alt}(V^* \otimes_K V^*) \; .$$

For the identity involution on F and $\epsilon = +1$ (resp. -1) the involution on A is of the first kind and the orthogonal (resp. symplectic) type (I) (resp. (II)). If (V,ϕ) admits a complex structure, an automorphism $J : (V,\phi) \longrightarrow (V,\phi)$ such that $J^2 = -1$, there is defined an isomorphism

$$A^+ \xrightarrow{\simeq} A^- \; ; \; \theta \longrightarrow J\theta$$

and the involution on A is of the second kind and unitary type.

□

The *round free quadratic L-groups* $L^r_*(R)$ are the quadratic L-groups of a ring with involution R defined using f.g. free R-modules of even rank, which differ from the projective and free L-groups by the exact sequences

$$\ldots \longrightarrow \widehat{H}^{n+1}(\mathbb{Z}_2 \,;\, K_0(R)) \longrightarrow L^r_n(R) \longrightarrow L^p_n(R)$$

$$\longrightarrow \widehat{H}^n(\mathbb{Z}_2 \,;\, K_0(R)) \longrightarrow \ldots \; ,$$

$$\ldots \longrightarrow \widehat{H}^{n+1}(\mathbb{Z}_2 \,;\, \mathrm{im}(K_0(\mathbb{Z}) \to K_0(R))) \longrightarrow L^r_n(R) \longrightarrow L^h_n(R)$$

$$\longrightarrow \widehat{H}^n(\mathbb{Z}_2 \,;\, \mathrm{im}(K_0(\mathbb{Z}) \to K_0(R))) \longrightarrow \ldots \; .$$

Similarly for the *round free symmetric L-groups* $L^*_r(R)$. See Hambleton, Ranicki and Taylor [67] for further details.

THEOREM 22.18 *Let π be a finite group.*
(i) *The projective L-groups of $\mathbb{R}[\pi]$ are given by*

$$L^n_p(\mathbb{R}[\pi]) = \begin{cases} \displaystyle\sum_{\alpha(\mathbb{R},\pi)} \mathbb{Z} \\ \displaystyle\sum_{\alpha_{\mathrm{H}}(\mathbb{R},\pi)} \mathbb{Z}_2 \oplus \sum_{\alpha_{\mathrm{C}}(\mathbb{R},\pi)} \mathbb{Z} \quad \textit{if } n \equiv \\ 0 \end{cases} \begin{cases} 0 \\ 2 \quad (\mathrm{mod}\ 4) \; , \\ 1,3 \end{cases}$$

with the $\mathbb{Z}$-components detected by the $\mathbb{R}$-multisignature.
(ii) *The round free L-groups of $\mathbb{R}[\pi]$ are given by*

$$
L_r^n(\mathbb{R}[\pi]) = \begin{cases} \displaystyle\sum_{\alpha(\mathbb{R},\pi)} 2\mathbb{Z} \\ \displaystyle\sum_{\alpha_\mathbb{R}(\mathbb{R},\pi)} \mathbb{Z}_2 \\ \displaystyle\sum_{\alpha_\mathbb{C}(\mathbb{R},\pi)} 2\mathbb{Z} \\ 0 \end{cases} \quad if\ n \equiv \begin{cases} 0 \\ 1 \\ 2 \\ 3 \end{cases} (\mathrm{mod}\ 4)\ ,
$$

with $2\mathbb{Z}$ denoting the corresponding subgroup of $\mathbb{Z} \subseteq L_p^n(\mathbb{R}[\pi])$.
(iii) *The free L-groups of $\mathbb{R}[\pi]$ are given by*

$$
L^n(\mathbb{R}[\pi]) = \begin{cases} \mathbb{Z} \oplus \displaystyle\sum_{\alpha(\mathbb{R},\pi)-1} 2\mathbb{Z} \\ \displaystyle\sum_{\alpha_\mathbb{R}(\mathbb{R},\pi)-1} \mathbb{Z}_2 \\ \displaystyle\sum_{\alpha_\mathbb{C}(\mathbb{R},\pi)} 2\mathbb{Z} \\ 0 \end{cases} \quad if\ n \equiv \begin{cases} 0 \\ 1 \\ 2 \\ 3 \end{cases} (\mathrm{mod}\ 4)\ .
$$

PROOF (i) Each of the idempotents $e_j = e_j(\mathbb{R}, \pi) \in \mathbb{R}[\pi]$ is such that $\bar{e}_j = e_j$, so that the involution on $\mathbb{R}[\pi]$ preserves each simple factor

$$
S_j(\mathbb{R}, \pi) = M_{d_j}(D_j(\mathbb{R}, \pi)) \quad (d_j = d_j(\mathbb{R}, \pi))\ ,
$$

and as a ring with involution

$$
\mathbb{R}[\pi] = S_1(\mathbb{R}, \pi) \times S_2(\mathbb{R}, \pi) \times \ldots \times S_{\alpha(\mathbb{R},\pi)}(\mathbb{R}, \pi)\ .
$$

Each $D_j(\mathbb{R}, \pi)$ is one of $\mathbb{R}$, $\mathbb{H}$, $\mathbb{C}$ with the standard involution, respectively the identity, quaternion conjugation, and complex conjugation. The three types are distinguished by the type of the involution on $S_j(\mathbb{R}, \pi)$, or by the ring structure of $\mathbb{C} \otimes_\mathbb{R} S_j(\mathbb{R}, \pi)$, as follows:

(I) (orthogonal) $D_j(\mathbb{R}, \pi) = \mathbb{R}$ if the involution on $S_j(\mathbb{R}, \pi)$ is of the orthogonal type, with

$$
\mathbb{C} \otimes_\mathbb{R} S_j(\mathbb{R}, \pi) = M_{d_j}(\mathbb{C})\ .
$$

(II) (symplectic) $D_j(\mathbb{R}, \pi) = \mathbb{H}$ if the involution on $S_j(\mathbb{R}, \pi)$ is of the symplectic type, with

$$
\mathbb{C} \otimes_\mathbb{R} S_j(\mathbb{R}, \pi) = M_{2d_j}(\mathbb{C})\ .
$$

(III) (unitary) $D_j(\mathbb{R}, \pi) = \mathbb{C}$ if the involution on $S_j(\mathbb{R}, \pi)$ is of the unitary type, with

$$
\mathbb{C} \otimes_\mathbb{R} S_j(\mathbb{R}, \pi) = M_{d_j}(\mathbb{C}) \times M_{d_j}(\mathbb{C})\ .
$$

By the Morita isomorphisms of 21.6

$$L_p^{2k}(\mathbb{R}[\pi]) = \sum_{j=1}^{\alpha^k(\mathbb{R},\pi)} L_p^{2k}(S_j(\mathbb{R},\pi))$$

$$= \sum_{\alpha_\mathbb{R}(\mathbb{R},\pi)} L^{2k}(\mathbb{R}) \oplus \sum_{\alpha_\mathbb{H}(\mathbb{R},\pi)} L^{2k}(\mathbb{H}) \oplus \sum_{\alpha_\mathbb{C}(\mathbb{R},\pi)} L^{2k}(\mathbb{C}^-)$$

and by 22.16

$$r^0(D_j(\mathbb{R},\pi)) = 1 ,$$

$$\alpha^0(\mathbb{R},\pi) = \alpha_\mathbb{R}(\mathbb{R},\pi) + \alpha_\mathbb{H}(\mathbb{R},\pi) + \alpha_\mathbb{C}(\mathbb{R},\pi) = \alpha(\mathbb{R},\pi) ,$$

$$r^1(D_j(\mathbb{R},\pi)) = \begin{cases} 1 \\ 0 \end{cases} \text{ if } D_j(\mathbb{R},\pi) = \begin{cases} \mathbb{C} \\ \mathbb{R} \text{ or } \mathbb{H} , \end{cases}$$

$$\alpha^1(\mathbb{R},\pi) = \alpha_\mathbb{C}(\mathbb{R},\pi) .$$

(ii) Immediate from (i) and the exact sequence

$$\dots \longrightarrow L_r^n(\mathbb{R}[\pi]) \longrightarrow L_p^n(\mathbb{R}[\pi]) \longrightarrow \widehat{H}^n(\mathbb{Z}_2 ; K_0(\mathbb{R}[\pi]))$$
$$\longrightarrow L_r^{n-1}(\mathbb{R}[\pi]) \longrightarrow \dots$$

with

$$\widehat{H}^n(\mathbb{Z}_2 ; K_0(\mathbb{R}[\pi])) = \begin{cases} K_0(\mathbb{R}[\pi])/2K_0(\mathbb{R}[\pi]) \\ 0 \end{cases} \text{ if } n \equiv \begin{cases} 0 \\ 1 \end{cases} (\text{mod } 2) .$$

(iii) Immediate from (i) and the exact sequence

$$\dots \longrightarrow L^n(\mathbb{R}[\pi]) \longrightarrow L_p^n(\mathbb{R}[\pi]) \longrightarrow \widehat{H}^n(\mathbb{Z}_2 ; \widetilde{K}_0(\mathbb{R}[\pi]))$$
$$\longrightarrow L^{n-1}(\mathbb{R}[\pi]) \longrightarrow \dots .$$

$\square$

PROPOSITION 22.19 *Let $F = \mathbb{R}$ or $\mathbb{C}^-$. The K-theory F-multisignature (22.3) coincides with the L-theory F-multisignature σ (22.13), defining isomorphisms*

$$\sigma : L_p^{4*}(\mathbb{R}[\pi]) \xrightarrow{\simeq} K_0(\mathbb{R}[\pi]) = \sum_{\alpha(\mathbb{R},\pi)} \mathbb{Z} ,$$

$$\sigma : L_p^{4*+2}(\mathbb{R}[\pi]) \xrightarrow{\simeq} K_0(\mathbb{R}[\pi],-1) = \sum_{\alpha_\mathbb{C}(\mathbb{R},\pi)} \mathbb{Z}$$

with

$$K_0(\mathbb{R}[\pi],-1) = \{\chi - \bar{\chi} \,|\, \chi \in \sum_{\alpha_\mathbb{C}(\mathbb{R},\pi)} \mathbb{Z}\} \subseteq K_0(\mathbb{C}[\pi]) = \sum_{\alpha(\mathbb{C},\pi)} \mathbb{Z}$$

indexed by the conjugate pairs of irreducible $\mathbb{C}$-representations of π of the unitary type (III).

PROOF Immediate from 22.18.

□

The only simple finite-dimensional algebra over $\mathbb{C}$ is $\mathbb{C}$ itself, so that the simple factors in the Wedderburn decomposition

$$\mathbb{C}[\pi] = \prod_{j=1}^{\alpha(\mathbb{C},\pi)} S_j(\mathbb{C},\pi)$$

are matrix algebras $S_j(\mathbb{C},\pi) = M_{d_j}(\mathbb{C})$, one for each degree d_j irreducible $\mathbb{C}$-representation of π. The type of an irreducible $\mathbb{C}$-representation ρ of degree d is distinguished by the Frobenius–Schur number associated to its character χ

$$c(\rho) = (1/|\pi|) \sum_{g \in \pi} \chi(g^2) \in \mathbb{C}.$$

This is the coefficient of the trivial representation $\mathbb{C}$ in the $\mathbb{C}$-representation of degree $d(d+1)/2$

$$\mathrm{Sym}(V \otimes_{\mathbb{C}} V) = H^0(\mathbb{Z}_2 ; V \otimes_{\mathbb{C}} V)$$

of symmetric forms on $V^* = \mathrm{Hom}_{\mathbb{C}}(V,\mathbb{C})$ over $\mathbb{C}^+$, with $V = \mathbb{C}^d$ the representation space of ρ (cf. 22.17). Equivalently, the type of the representation is determined by the type of form supported by V over $\mathbb{C}^+[\pi]$, as follows:

(I) (orthogonal) $c(\rho) = 1$ if and only if $\chi = \bar{\chi}$ is real and ρ is equivalent to an $\mathbb{R}$-representation of degree d, i.e. if there exists a $\mathbb{C}[\pi]$-module isomorphism $V \cong \mathbb{C}[\pi] \otimes_{\mathbb{R}[\pi]} V_0$ for some f.g. $\mathbb{R}[\pi]$-module V_0 which is a d-dimensional real vector space. This is the case if and only if there exists a nonsingular symmetric form (V,ϕ) over $\mathbb{C}^+[\pi]$. The simple factor $M_d(\mathbb{R})$ of $\mathbb{R}[\pi]$ induces the simple factor $\mathbb{C} \otimes_{\mathbb{R}} M_d(\mathbb{R}) = M_d(\mathbb{C})$ of $\mathbb{C}[\pi]$.

(II) (symplectic) $c(\rho) = -1$ if and only if $\chi = \bar{\chi}$ is real but ρ is not equivalent to an $\mathbb{R}$-representation, in which case d is even and there exists an irreducible $\mathbb{R}$-representation σ of degree $d/2$ of quaternionic type. This is the case if and only if there exists a nonsingular skew-symmetric form (V,ϕ) over $\mathbb{C}^+[\pi]$. The simple factor $M_{d/2}(\mathbb{H})$ of $\mathbb{R}[\pi]$ induces the simple factor $\mathbb{C} \otimes_{\mathbb{R}} M_{d/2}(\mathbb{H}) = M_d(\mathbb{C})$ of $\mathbb{C}[\pi]$.

(III) (unitary) $c(\rho) = 0$ if and only if $\chi \neq \bar{\chi}$ is not real, so that it is purely imaginary and ρ is not isomorphic to the complex conjugate representation $\bar{\rho}$. This is the case if and only if V is not $\mathbb{C}[\pi]$-module isomorphic to its $\mathbb{C}^+[\pi]$-dual V^*. The simple factor $M_d(\mathbb{C})$ of $\mathbb{R}[\pi]$ induces a product of simple factors $\mathbb{C} \otimes_{\mathbb{R}} M_d(\mathbb{C}) = M_d(\mathbb{C}) \times M_d(\mathbb{C})$ in $\mathbb{C}[\pi]$ interchanged by the real involution.

See Serre [154, §13] and Curtis and Reiner [40, §73A] for further details.

EXAMPLE 22.20 (i) For any finite group π the trivial irreducible $\mathbb{C}$-represent-
ation of degree 1

$$\pi \longrightarrow GL_1(\mathbb{C}) \; ; \; g \longrightarrow 1$$

is of the orthogonal type (I).
(ii) Let $Q_8 = \langle x, y \,|\, x^4 = 1, x^2 = y^2, xyx^{-1} = y^{-1} \rangle$ be the quaternion group
of order 8. The irreducible $\mathbb{C}$-representation ρ of degree 2 defined by

$$Q_8 \longrightarrow GL_2(\mathbb{C}) \; ; \; x \longrightarrow \begin{pmatrix} i & 0 \\ 0 & -i \end{pmatrix} , \; y \longrightarrow \begin{pmatrix} 0 & -1 \\ 1 & 0 \end{pmatrix}$$

is of the symplectic type (II).
(iii) Let $\mathbb{Z}_m = \langle T \,|\, T^m = 1 \rangle$ be the cyclic group of order m. The irreducible
$\mathbb{C}$-representations of $\mathbb{Z}_m$ of degree 1 defined by

$$\rho_j : \mathbb{Z}_m \longrightarrow GL_1(\mathbb{C}) \; ; \; T \longrightarrow e^{2\pi i j/m} \quad (0 \le j < m)$$

are of the orthogonal type (I) if $j = 0$ or $m/2$ (m even), and of the unitary
type (III) otherwise, with $\bar{\rho}_j = \rho_{m-j}$.

$\square$

The number of simple factors $S_j(\mathbb{C}, \pi)$ in $\mathbb{C}[\pi]$ is

$$\begin{aligned} \alpha(\mathbb{C}, \pi) &= \text{no. of irreducible } \mathbb{C}\text{-representations of } \pi \\ &= \text{no. of conjugacy classes in } \pi \\ &= \alpha_{\mathbb{R}}(\mathbb{R}, \pi) + \alpha_{\mathbb{H}}(\mathbb{R}, \pi) + 2\alpha_{\mathbb{C}}(\mathbb{R}, \pi) \end{aligned}$$

with

$$\begin{cases} \text{(I) } \alpha_{\mathbb{R}}(\mathbb{R}, \pi) \\ \text{(II) } \alpha_{\mathbb{H}}(\mathbb{R}, \pi) \\ \text{(III)} 2\alpha_{\mathbb{C}}(\mathbb{R}, \pi) \end{cases} = \text{no. of irreducible } \mathbb{C}\text{-representations } \rho \text{ with } c(\rho) = \begin{cases} 1 \\ -1 \\ 0 . \end{cases}$$

PROPOSITION 22.21 (i) *The projective L-groups of* $\mathbb{C}^+[\pi]$ *are given by*

$$L_p^n(\mathbb{C}^+[\pi]) = \begin{cases} \sum\limits_{\alpha_{\mathbb{R}}(\mathbb{R},\pi)} \mathbb{Z}_2 \\ \sum\limits_{\alpha_{\mathbb{H}}(\mathbb{R},\pi)} \mathbb{Z}_2 \quad \text{if } n \equiv \\ 0 \end{cases} \begin{cases} 0 \\ 2 \quad (\text{mod } 4) . \\ 1, 3 \end{cases}$$

The inclusion $i^+ : \mathbb{R}[\pi] \longrightarrow \mathbb{C}^+[\pi]$ *induces*

$$i_!^+ = \sum_{\alpha_{\mathbb{R}}(\mathbb{R},\pi)} 1 \oplus 0 \oplus 0 : L_p^{4*}(\mathbb{R}[\pi]) = \sum_{\alpha_{\mathbb{R}}(\mathbb{R},\pi)} \mathbb{Z} \oplus \sum_{\alpha_{\mathbb{H}}(\mathbb{R},\pi)} \mathbb{Z} \oplus \sum_{\alpha_{\mathbb{C}}(\mathbb{R},\pi)} \mathbb{Z}$$

$$\longrightarrow L_p^{4*}(\mathbb{C}^+[\pi]) = \sum_{\alpha_{\mathbb{R}}(\mathbb{R},\pi)} \mathbb{Z}_2 \ ,$$

$$i_!^+ = 0 : L_p^{4*+2}(\mathbb{R}[\pi]) = \sum_{\alpha_{\mathbb{H}}(\mathbb{R},\pi)} \mathbb{Z}_2 \oplus \sum_{\alpha_{\mathbb{C}}(\mathbb{R},\pi)} \mathbb{Z}$$

$$\longrightarrow L_p^{4*+2}(\mathbb{C}^+[\pi]) = \sum_{\alpha_{\mathbb{H}}(\mathbb{R},\pi)} \mathbb{Z}_2 \ .$$

(ii) *The projective L-groups of* $\mathbb{C}^-[\pi]$ *are given by*

$$L_p^n(\mathbb{C}^-[\pi]) = \begin{cases} \sum_{\alpha(\mathbb{C},\pi)} \mathbb{Z} \\ 0 \end{cases} \text{ if } n \equiv \begin{cases} 0 \\ 1 \end{cases} (\text{mod } 2) \ ,$$

with the $\mathbb{Z}$*-components detected by the* $\mathbb{C}^-$*-multisignature. The inclusion* $i^- : \mathbb{R}[\pi] \longrightarrow \mathbb{C}^-[\pi]$ *induces*

$$i_!^- = \sum_{\alpha_{\mathbb{R}}(\mathbb{R},\pi)} 1 \oplus \sum_{\alpha_{\mathbb{H}}(\mathbb{R},\pi)} 2 \oplus \sum_{\alpha_{\mathbb{C}}(\mathbb{R},\pi)} \begin{pmatrix} 1 \\ 1 \end{pmatrix} :$$

$$L_p^{4*}(\mathbb{R}[\pi]) = \sum_{\alpha_{\mathbb{R}}(\mathbb{R},\pi)} \mathbb{Z} \oplus \sum_{\alpha_{\mathbb{H}}(\mathbb{R},\pi)} \mathbb{Z} \oplus \sum_{\alpha_{\mathbb{C}}(\mathbb{R},\pi)} \mathbb{Z}$$

$$\longrightarrow L_p^{4*}(\mathbb{C}^-[\pi]) = \sum_{\alpha_{\mathbb{R}}(\mathbb{R},\pi)} \mathbb{Z} \oplus \sum_{\alpha_{\mathbb{H}}(\mathbb{R},\pi)} \mathbb{Z} \oplus \sum_{\alpha_{\mathbb{C}}(\mathbb{R},\pi)} (\mathbb{Z} \oplus \mathbb{Z}) \ ,$$

$$i_!^- = 0 \oplus \sum_{\alpha_{\mathbb{C}}(\mathbb{R},\pi)} \begin{pmatrix} 1 \\ -1 \end{pmatrix} :$$

$$L_p^{4*+2}(\mathbb{R}[\pi]) = \sum_{\alpha_{\mathbb{H}}(\mathbb{R},\pi)} \mathbb{Z}_2 \oplus \sum_{\alpha_{\mathbb{C}}(\mathbb{R},\pi)} \mathbb{Z}$$

$$\longrightarrow L_p^{4*+2}(\mathbb{C}^-[\pi]) = \sum_{\alpha_{\mathbb{R}}(\mathbb{R},\pi)} \mathbb{Z} \oplus \sum_{\alpha_{\mathbb{H}}(\mathbb{R},\pi)} \mathbb{Z} \oplus \sum_{\alpha_{\mathbb{C}}(\mathbb{R},\pi)} (\mathbb{Z} \oplus \mathbb{Z}) \ .$$

$\square$

REMARK 22.22 For any ring A and a non-square central unit $a \in A$ let

$$A[\sqrt{a}\,] = A[t]/(t^2 - a)$$

be the quadratic extension ring obtained by adjoining the square roots of a. Given an involution $\bar{} : A \longrightarrow A$ with $\bar{a} = a$ let $A[\sqrt{a}\,]^+$, $A[\sqrt{a}\,]^-$ denote the rings with involution defined by $A[\sqrt{a}\,]$ with the involution on A extended by

$$\bar{} : A[\sqrt{a}\,]^{\pm} \longrightarrow A[\sqrt{a}\,]^{\pm} \ ; \ x + y\sqrt{a} \longrightarrow \bar{x} \pm \bar{y}\sqrt{a} \ .$$

(In the classic case $A = \mathbb{R}$, $a = -1$, $A[\sqrt{a}] = \mathbb{C}$). Jacobson's work on hermitian forms over quadratic field extensions was used by Milnor and Husemoller [110, p. 116] to obtain an exact sequence

$$0 \longrightarrow L^0(A[\sqrt{a}]^-) \longrightarrow L_0(A) \longrightarrow L_0(A[\sqrt{a}]^+)$$

in the case when A is a field with the identity involution. See Wall [176, 12C], Hambleton [62], Harsiladze [69], Hambleton, Taylor and Williams [68], Lewis [93], Ranicki [144] for various generalizations to the L-theory of quadratic extensions of more general rings with involution A. The isomorphisms of relative L-groups of the induction and transfer maps of the inclusions $i^{\pm}: A \longrightarrow A[\sqrt{a}]^{\pm}$ obtained in [144] for any A

$$L_*(i_!^+: A \longrightarrow A[\sqrt{a}]^-) \cong L_{*+1}(i_!^-: A[\sqrt{a}]^+ \longrightarrow A) ,$$
$$L_*((i^-)^!: A \longrightarrow A[\sqrt{a}]^-) \cong L_{*+1}((i^+)^!: A[\sqrt{a}]^+ \longrightarrow A)$$

and the skew-suspension isomorphisms

$$L_n(A[\sqrt{a}]^-) \overset{\simeq}{\longrightarrow} L_{n+2}(A[\sqrt{a}]^-) ; \ (C, \psi) \longrightarrow (SC, \sqrt{a}\,\overline{S}\psi)$$

were combined into a commutative braid of exact sequences

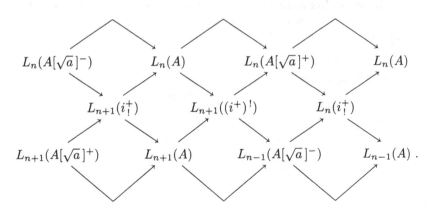

If the rings A and $A[\sqrt{a}]$ are semisimple then

$$L_{2*+1}^p(A) = L_{2*+1}^p(A[\sqrt{a}]^{\pm}) = 0 ,$$

so that in the projective version of the braid with even n the L-groups at the bottom are all 0 and the L-groups at the top fit into an octagon

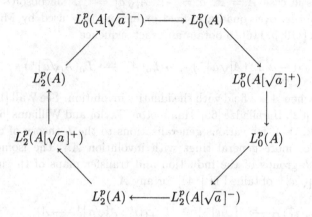

of exact sequences of projective Witt groups. The projective L-groups of $\mathbb{R}[\pi]$, $\mathbb{C}^+[\pi]$, $\mathbb{C}^-[\pi]$ computed in 22.18 and 22.21 fit into this octagon with $A = \mathbb{R}[\pi]$, $a = -1$, $A[\sqrt{a}]^\pm = \mathbb{C}^\pm[\pi]$.

□

The transfer map associated to the inclusion $i\colon \mathbb{R} \longrightarrow \mathbb{R}[\pi]$
$$i^! \colon L^n_p(\mathbb{R}[\pi]) \longrightarrow L^n(\mathbb{R}) \ ; \ (C, \phi) \longrightarrow (C^!, \phi^!)$$
sends a f.g. projective n-dimensional symmetric Poincaré complex (C, ϕ) over $\mathbb{R}[\pi]$ to the f.g. free n-dimensional symmetric Poincaré complex $(C^!, \phi^!)$ over $\mathbb{R}$ with $C^!$ obtained from C by the restriction of the $\mathbb{R}[\pi]$-action to $\mathbb{R} \subset \mathbb{R}[\pi]$ and
$$\phi^!(x)(y) \ = \ \text{coefficient of 1 in } \phi(x)(y) \in \mathbb{R} \subset \mathbb{R}[\pi] \ .$$
For $n = 4k$ the signature
$$i^!(C, \phi) \ = \ (C^!, \phi^!) \in L^{4k}(\mathbb{R}) \ = \ \mathbb{Z}$$
is determined by the L-theory $\mathbb{R}$-multisignature according to
$$i^! \ = \ (c_1 d_1 \ c_2 d_2 \ \ldots \ c_{\alpha(\mathbb{R}, \pi)} d_{\alpha(\mathbb{R}, \pi)}) \ :$$
$$L^{4k}_p(\mathbb{R}[\pi]) \ = \ \sum_{j=1}^{\alpha(\mathbb{R}, \pi)} \mathbb{Z} \longrightarrow L^{4k}(\mathbb{R}) \ = \ \mathbb{Z} \ ,$$
with
$$c_j \ = \ c_j(\mathbb{R}, \pi) \ = \ \dim_{\mathbb{R}}(D_j(\mathbb{R}, \pi)) \ = \ \begin{cases} 1 \\ 2 \ \text{if} \ D_j(\mathbb{R}, \pi) \ = \ \begin{cases} \mathbb{R} \\ \mathbb{C} \ , \\ \mathbb{H} \end{cases} \\ 4 \end{cases}$$
$$d_j \ = \ d_j(\mathbb{R}, \pi) \ .$$
In terms of the character of the K-theory $\mathbb{R}$-multisignature (22.5 (ii))
$$i^!(M, \phi) \ = \ \sigma(1, (M, \phi)) \in \mathbb{Z} \subset \mathbb{R} \ .$$

PROPOSITION 22.23 (i) *The $\mathbb{R}$-coefficient algebraic L-theory assembly map of a finite group π*

$$A : H_*(B\pi; \mathbf{L}^{\bullet}(\mathbb{R})) \longrightarrow L_p^*(\mathbb{R}[\pi])$$

is given by the composite

$$A : H_*(B\pi; \mathbf{L}^{\bullet}(\mathbb{R})) \longrightarrow L^*(\mathbb{R}) \xrightarrow{i_!} L_p^*(\mathbb{R}[\pi])$$

with $i: \mathbb{R} \longrightarrow \mathbb{R}[\pi]$ the inclusion. In the non-zero case $* = 4k$

$$i_! = \left(\sum_{D_j=\mathbb{R}} d_j, \sum_{D_j=\mathbb{H}} d_j, \sum_{D_j=\mathbb{C}} d_j \right) :$$

$$L^{4k}(\mathbb{R}) = \mathbb{Z} \longrightarrow L_p^{4k}(\mathbb{R}[\pi]) = \sum_{\alpha_{\mathbb{R}}(\mathbb{R},\pi)} \mathbb{Z} \oplus \sum_{\alpha_{\mathbb{H}}(\mathbb{R},\pi)} \mathbb{Z} \oplus \sum_{\alpha_{\mathbb{C}}(\mathbb{R},\pi)} \mathbb{Z} ,$$

with $D_j = D_j(\mathbb{R}, \pi)$, $d_j = d_j(\mathbb{R}, \pi)$.

(ii) *The transfer map is given by*

$$i^! = \left(\sum_{D_j=\mathbb{R}} d_j, \sum_{D_j=\mathbb{H}} 4d_j, \sum_{D_j=\mathbb{C}} 2d_j \right) :$$

$$L_p^{4k}(\mathbb{R}[\pi]) = \sum_{\alpha_{\mathbb{R}}(\mathbb{R},\pi)} \mathbb{Z} \oplus \sum_{\alpha_{\mathbb{H}}(\mathbb{R},\pi)} \mathbb{Z} \oplus \sum_{\alpha_{\mathbb{C}}(\mathbb{R},\pi)} \mathbb{Z} \longrightarrow L^{4k}(\mathbb{R}) = \mathbb{Z}$$

with

$$i^! i_! = \sum_{D_j=\mathbb{R}} (d_j)^2 + \sum_{D_j=\mathbb{H}} 4(d_j)^2 + \sum_{D_j=\mathbb{C}} 2(d_j)^2$$

$$= |\pi| : L^{4k}(\mathbb{R}) = \mathbb{Z} \longrightarrow L^{4k}(\mathbb{R}) = \mathbb{Z} .$$

$\square$

EXAMPLE 22.24 The irreducible $\mathbb{C}$-representations of the cyclic group $\mathbb{Z}_m$ are the representations

$$\rho_j : \mathbb{Z}_m \longrightarrow \mathbb{C} ; T \longrightarrow e^{2\pi ij/m} \quad (0 \le j < m)$$

classified in 22.20 (iii), so that

$$\alpha(\mathbb{R}, \mathbb{Z}_m) = \alpha_{\mathbb{R}}(\mathbb{R}, \mathbb{Z}_m) + \alpha_{\mathbb{C}}(\mathbb{R}, \mathbb{Z}_m) , \quad \alpha_{\mathbb{H}}(\mathbb{R}, \mathbb{Z}_m) = 0 ,$$

$$\alpha_{\mathbb{R}}(\mathbb{R}, \mathbb{Z}_m) = \begin{cases} 1 \\ 2 \end{cases} , \quad \alpha_{\mathbb{C}}(\mathbb{R}, \mathbb{Z}_m) = \begin{cases} (m-1)/2 \\ (m-2)/2 , \end{cases}$$

$$\mathbb{R}[\mathbb{Z}_m] = \begin{cases} \mathbb{R} \oplus \underset{(m-1)/2}{\oplus} \mathbb{C} \\ \mathbb{R}^2 \oplus \underset{(m-2)/2}{\oplus} \mathbb{C} \end{cases} \text{ if } m \text{ is } \begin{cases} \text{odd} \\ \text{even} . \end{cases}$$

The projective L-groups of $\mathbb{R}[\mathbb{Z}_m]$ are given by

$$L_p^{4k}(\mathbb{R}[\mathbb{Z}_m]) = \begin{cases} \mathbb{Z} \oplus \displaystyle\bigoplus_{(m-1)/2} \mathbb{Z} \\ \mathbb{Z}^2 \oplus \displaystyle\bigoplus_{(m-2)/2} \mathbb{Z} \end{cases},$$

$$L_p^{4k+2}(\mathbb{R}[\mathbb{Z}_m]) = \begin{cases} \displaystyle\bigoplus_{(m-1)/2} \mathbb{Z} \\ \displaystyle\bigoplus_{(m-2)/2} \mathbb{Z} \end{cases} \text{ if } m \text{ is } \begin{cases} \text{odd} \\ \text{even} \end{cases},$$

$$L_p^{2*+1}(\mathbb{R}[\mathbb{Z}_m]) = 0 .$$

The $4k$-dimensional assembly map

$$A : H_{4k}(B\mathbb{Z}_m; \mathbb{L}^{\boldsymbol{\cdot}}(\mathbb{R})) = \sum_{j=0}^{\infty} H_{4j}(B\mathbb{Z}_m; L^{4k-4j}(\mathbb{R}))$$

$$\longrightarrow H_0(B\mathbb{Z}_m; L^{4k}(\mathbb{R})) = L^{4k}(\mathbb{R}) = \mathbb{Z} \xrightarrow{i_!} L_p^{4k}(\mathbb{R}[\mathbb{Z}_m])$$

has image the cyclic subgroup generated by $A(1) = (1, 1, \ldots, 1)$, and the $(4k+2)$-dimensional assembly map has image 0. The projective Witt class of a nonsingular symmetric form over $\mathbb{R}[\mathbb{Z}_m]$ is in the image of the assembly map if and only if the $\mathbb{R}$-multisignature components are equal. The $4k$-dimensional transfer map is given by

$$i^! = \begin{cases} (1\ 2\ \ldots\ 2) \\ (1\ 1\ 2\ \ldots\ 2) \end{cases}$$

$$: L_p^{4k}(\mathbb{R}[\mathbb{Z}_m]) = \begin{cases} \mathbb{Z} \oplus \displaystyle\bigoplus_{(m-1)/2} \mathbb{Z} \\ \mathbb{Z}^2 \oplus \displaystyle\bigoplus_{(m-2)/2} \mathbb{Z} \end{cases} \longrightarrow L_p^{4k}(\mathbb{R}) = \mathbb{Z} \text{ if } m \text{ is } \begin{cases} \text{odd} \\ \text{even} \end{cases}.$$

$\square$

REMARK 22.25 The $\mathbb{C}^-$-coefficient algebraic L-theory assembly map

$$A : H_*(B\pi; \mathbb{L}^{\boldsymbol{\cdot}}(\mathbb{C}^-)) \longrightarrow L_p^*(\mathbb{C}^-[\pi])$$

is given by the composite

$$A : H_*(B\pi; \mathbb{L}^{\boldsymbol{\cdot}}(\mathbb{C}^-)) \longrightarrow L^*(\mathbb{C}^-) \xrightarrow{i_!} L_p^*(\mathbb{C}^-[\pi]) .$$

The inclusion $i : \mathbb{C}^- \longrightarrow \mathbb{C}^-[\pi]$ induces

$$i_! = \left(\sum_{D_j=\mathbb{R}} d_j \quad \sum_{D_j=\mathbb{H}} 2d_j \quad \sum_{D_j=\mathbb{C}} d_j \begin{pmatrix} 1 \\ 1 \end{pmatrix} \right) :$$

$$L^{2k}(\mathbb{C}^-) = \mathbb{Z} \longrightarrow L_p^{2k}(\mathbb{C}^-[\pi]) = \sum_{\alpha_{\mathbb{R}}(\mathbb{R},\pi)} \mathbb{Z} \oplus \sum_{\alpha_{\mathbb{H}}(\mathbb{R},\pi)} \mathbb{Z} \oplus \sum_{\alpha_{\mathbb{C}}(\mathbb{R},\pi)} (\mathbb{Z} \oplus \mathbb{Z}) ,$$

with $D_j = D_j(\mathbb{R}, \pi)$, $d_j = d_j(\mathbb{R}, \pi)$ as in 22.23. The transfer map is given by

$$i^! = \left(\sum_{D_j=\mathbb{R}} d_j \quad \sum_{D_j=\mathbb{H}} 2d_j \quad \sum_{D_j=\mathbb{C}} d_j (1\ 1) \right) :$$

$$L_p^{2k}(\mathbb{C}^-[\pi]) = \sum_{\alpha_\mathbb{R}(\mathbb{R},\pi)} \mathbb{Z} \oplus \sum_{\alpha_\mathbb{H}(\mathbb{R},\pi)} \mathbb{Z} \oplus \sum_{\alpha_\mathbb{C}(\mathbb{R},\pi)} (\mathbb{Z} \oplus \mathbb{Z}) \longrightarrow L^{2k}(\mathbb{C}^-) = \mathbb{Z}$$

with

$$i^! i_! = \sum_{D_j=\mathbb{R}} (d_j)^2 + \sum_{D_j=\mathbb{H}} 4(d_j)^2 + \sum_{D_j=\mathbb{C}} 2(d_j)^2 = |\pi| :$$

$$L^{2k}(\mathbb{C}^-) = \mathbb{Z} \longrightarrow L^{2k}(\mathbb{C}^-) = \mathbb{Z} .$$

□

A nonsingular symmetric form (M, ϕ) over $\mathbb{R}[\pi]$ is such that

$$(M, \phi) \in \mathrm{im}(A: H_{4k}(B\pi; \mathbb{L}^{\cdot}(\mathbb{R})) \longrightarrow L_p^{4k}(\mathbb{R}[\pi]))$$

if and only if the character of the K-theory $\mathbb{R}$-multisignature (22.5 (ii)) is a multiple of the character of the regular $\mathbb{R}$-representation $\mathbb{R}[\pi]$, as originally proved by Wall [176, 13B.1].

Let $j: \mathbb{R}[\pi] \longrightarrow S_1(\mathbb{R}, \pi) = \mathbb{R}$ be the projection, with kernel

$$\ker(j) = \prod_{m=2}^{\alpha(\mathbb{R},\pi)} S_m(\mathbb{R}, \pi) .$$

The induced map

$$j_! : L_p^{4k}(\mathbb{R}[\pi]) \longrightarrow L^{4k}(\mathbb{R})$$

sends a $4k$-dimensional symmetric Poincaré complex (C, ϕ) over $\mathbb{R}[\pi]$ to the signature of the $4k$-dimensional symmetric Poincaré complex $\mathbb{R} \otimes_{\mathbb{R}[\pi]} (C, \phi)$ over $\mathbb{R}$, with components

$$j_! = (1\ 0\ \dots\ 0) : L_p^{4k}(\mathbb{R}[\pi]) = \sum_{m=1}^{\alpha(\mathbb{R},\pi)} \mathbb{Z} \longrightarrow L^{4k}(\mathbb{R}) = \mathbb{Z} .$$

In terms of the character of the K-theory $\mathbb{R}$-multisignature (22.5 (ii))

$$j_!(M, \phi) = (1/|\pi|) \sum_{g \in \pi} \sigma(g, (M, \phi)) \in \mathbb{Z} \subset \mathbb{R} ,$$

the coefficient of the trivial representation $\mathbb{R}$ in the virtual $\mathbb{R}$-representation $[M, \phi] = [M_+] - [M_-] \in K_0(\mathbb{R}[\pi])$ (cf. Hirzebruch and Zagier [75, p. 31]).

PROPOSITION 22.26 (i) *For any element* $x \in \mathrm{im}(A) \subseteq L_p^{4k}(\mathbb{R}[\pi])$ *the signature of the transfer* $i^!(x) \in L^{4k}(\mathbb{R})$ *is* $|\pi|$ *times the signature of the projection* $j_!(x) \in L^{4k}(\mathbb{R})$, *that is*

$$i^!(x) = |\pi| j_!(x) \in L^{4k}(\mathbb{R}) = \mathbb{Z} .$$

(ii) *For a regular covering $\overline{M}$ of a compact $4k$-dimensional manifold M with finite group of covering translations π*

$$\text{signature}\,(\overline{M})\;=\;|\pi|\,\text{signature}\,(M)\in L^{4k}(\mathbb{R})\;=\;\mathbb{Z}\;.$$

PROOF (i) This is immediate from 22.23.

(ii) Apply (i) to the symmetric signature $\sigma^*(M) = (\Delta(\overline{M}),\phi) \in L_p^{4k}(\mathbb{R}[\pi])$.

$\square$

The multiplicativity of the signature for finite coverings of manifolds (21.4, 22.26) is traditionally proved by the Hirzebruch formula

$$\text{signature}\,(M)\;=\;\langle \mathcal{L}(M),[M]_{\mathbb{Q}}\rangle \in L^{4k}(\mathbb{Z})\;=\;\mathbb{Z}\;.$$

PROPOSITION 22.27 *Let X be a finite $4k$-dimensional geometric Poincaré complex, and let $\overline{X}$ be a regular cover of X with finite group of covering translations π. The symmetric signature*

$$\sigma^*(X)\;=\;(\Delta(\overline{X}),\phi)\in L^{4k}(\mathbb{R}[\pi])$$

is such that

$$i^!\sigma^*(X)\;=\;\text{signature}\,(\overline{X})\;,$$

$$j_!\,\sigma^*(X)\;=\;\text{signature}\,(X)\in L^{4k}(\mathbb{R})\;=\;\mathbb{Z}\;.$$

If X is homotopy equivalent to a compact topological manifold then

$$\sigma^*(X)\;=\;A([X]_{\mathbb{L}})\in \text{im}(A\colon H_{4k}(X;\mathbb{L}^{\cdot}(\mathbb{R}))\!\longrightarrow\!L^{4k}(\mathbb{R}[\pi]))$$

and by 22.26

$$i^!\sigma^*(X)\;=\;|\pi|j_!\,\sigma^*(X)\in L^{4k}(\mathbb{R})\;,$$

$$\text{signature}\,(\overline{X})\;=\;|\pi|\,\text{signature}\,(X)\in\mathbb{Z}\;.$$

$\square$

The examples of geometric Poincaré complexes with non-multiplicative signature constructed by Wall [174] will now be related to elements $x \in L_{4k}(\mathbb{Z}[\mathbb{Z}_q])$ which are not in the image of the assembly map $A\colon H_{4k}(B\mathbb{Z}_q;\mathbb{L}.)\longrightarrow L_{4k}(\mathbb{Z}[\mathbb{Z}_q])$.

EXAMPLE 22.28 The quadratic L-groups $L_*(\mathbb{Z}[\mathbb{Z}_q])$ (q prime) can be computed using the Rim-Milnor cartesian square of rings with involution

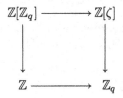

where

$$\mathbb{Z}[\zeta]\;=\;\mathbb{Z}[z]/(1+z+z^2+\ldots+z^{q-1})\;\;(\bar{z}=z^{-1})$$

is the extension of $\mathbb{Z}$ by the primitive qth root of unity $\zeta = e^{2\pi i/q}$ with the complex conjugation involution. As in Ranicki [143, 6.3] there is defined an L-theory Mayer–Vietoris exact sequence

$$0 \longrightarrow L_{4k}(\mathbb{Z}[\mathbb{Z}_q]) \longrightarrow L_{4k}(\mathbb{Z}) \oplus L_{4k}(\mathbb{Z}[\zeta])$$
$$\longrightarrow L_{4k}(\mathbb{Z}_q) \longrightarrow L_{4k-1}(\mathbb{Z}[\mathbb{Z}_q]) \longrightarrow 0 .$$

As in Wall [174] the pullback construction can be used to obtain a nonsingular quadratic form (K, λ, μ) over $\mathbb{Z}[\mathbb{Z}_q]$ with $K = \mathbb{Z}[\mathbb{Z}_q]^8$ a f.g. free $\mathbb{Z}[\mathbb{Z}_q]$-module of rank 8, such that

$$\mathbb{Z} \otimes_{\mathbb{Z}[\mathbb{Z}_q]} (K, \lambda, \mu) = (\mathbb{Z}^8, E_8) = 1 \in L_{4k}(\mathbb{Z}) = \mathbb{Z} ,$$

$$\mathbb{Z}[\zeta] \otimes_{\mathbb{Z}[\mathbb{Z}_q]} (K, \lambda, \mu) = H_+(\mathbb{Z}[\zeta]^4) = 0 \in L_{4k}(\mathbb{Z}[\zeta])$$

with $H_+(\mathbb{Z}[\zeta]^4)$ the hyperbolic form of rank 8 over $\mathbb{Z}[\zeta]$. The Witt class $(K, \lambda, \mu) \in L_{4k}(\mathbb{Z}[\mathbb{Z}_q])$ does not belong to the image of the assembly map

$$(K, \lambda, \mu) \notin \mathrm{im}(A \colon H_{4k}(B\mathbb{Z}_q; \mathbb{L}.) \longrightarrow L_{4k}(\mathbb{Z}[\mathbb{Z}_q]))$$

since the $\mathbb{R}$-multisignature is such that

$$\mathbb{R} \otimes (1 + T)(K, \lambda, \mu) = (8, 0) \notin \mathrm{im}(A) = \{ (s, s, \ldots, s) \mid s \in \mathbb{Z} \}$$

$$\subset L_p^{4k}(\mathbb{R}[\mathbb{Z}_q]) = L^{4k}(\mathbb{R}) \oplus L_p^{4k}(\mathbb{R}[\zeta]) = \begin{cases} \mathbb{Z} \oplus \mathbb{Z} & \text{if } q = 2 \\ \mathbb{Z} \oplus \displaystyle\sum_{(q-1)/2} \mathbb{Z} & \text{if } q \neq 2 . \end{cases}$$

As in 19.5 the element $x = (K, \lambda, \mu) \in L_{4k}(\mathbb{Z}[\mathbb{Z}_q])$ is realized by the surgery obstruction $x = \sigma_*(f, b)$ of a normal map $(f, b) \colon M^{4k} \longrightarrow X$ to a finite $4k$-dimensional geometric Poincaré complex X with $\pi_1(X) = \mathbb{Z}_q$, and

$$s(X) = -[\sigma_*(f, b)] = -[x] \neq 0$$

$$\in \mathrm{im}(L_{4k}(\mathbb{Z}[\mathbb{Z}_q]) \longrightarrow \mathbb{S}_{4k}(B\mathbb{Z}_q)) = \mathrm{coker}(A \colon H_{4k}(B\mathbb{Z}_q; \mathbb{L}.) \longrightarrow L_{4k}(\mathbb{Z}[\mathbb{Z}_q])) .$$

The signature of the universal cover $\widetilde{X}$ of X is not multiplicative, with

$$\sigma^*(X) = \sigma^*(\widetilde{X}) = 8 , \quad \sigma^*(\widetilde{X}) \neq q\sigma^*(X) \in L^{4k}(\mathbb{Z}) = \mathbb{Z} .$$

Thus $s(X) \neq 0 \in \mathbb{S}_{4k}(X)$ and X is not homotopy equivalent to a compact manifold (cf. 19.4).

$\square$

Next, consider the L-theory of the rational group ring $\mathbb{Q}[\pi]$ for a finite group π, which is built up from the Witt groups of quadratic and hermitian forms over algebraic number fields and quaternion algebras.

DEFINITION 22.29 (i) For any field F let $n_{\mathbb{R}}(F)$ be the number of embeddings $F \subset \mathbb{R}$, one for each ordering of F, and let $n_{\mathbb{C}}(F)$ be the number of conjugate pairs of embeddings $F \subset \mathbb{C}$.
(ii) For any field with involution F let $n_{\mathbb{C}}(F, \mathbb{Z}_2)$ be the number of conjugate pairs of embeddings $F \subset \mathbb{C}^-$.

(iii) A field F is *totally real* if $n_{\mathbb{C}}(F) = 0$.

(iv) A field F is *totally imaginary* if $n_{\mathbb{R}}(F) = 0$.

□

An algebraic number field F is a finite extension of $\mathbb{Q}$ with degree
$$\dim_{\mathbb{Q}}(F) = n_{\mathbb{R}}(F) + 2n_{\mathbb{C}}(F).$$

PROPOSITION 22.30 (Milnor and Husemoller [110], Scharlau [153], Wall [177])

(i) *If F is a field of characteristic $\neq 2$ with the identity involution then*
$$r^0(F) = n_{\mathbb{R}}(F) \ , \quad r^1(F) = 0 \ .$$

(ii) *If D is a division ring with an involution with the centre an algebraic number field F with a non-trivial involution then*
$$r^0(D) = r^1(D) = n_{\mathbb{C}}(F, \mathbb{Z}_2) \ .$$

(iii) *If $D = \left(\dfrac{\alpha, \beta}{F} \right)$ is a 4-dimensional quaternion algebra over an algebraic number field F such that either (a) the involution is by $\bar{i} = -i$, $\bar{j} = -j$, $\bar{k} = -k$ and the identity on F, and $\alpha, \beta \in F^{\bullet}$ are totally negative (= have negative valuation for each embedding $F \subset \mathbb{R}$) or (b) the involution is by $\bar{i} = i$, $\bar{j} = j$, $\bar{k} = -k$ and the identity on F, and $\alpha, \beta \in F^{\bullet}$ are not both totally negative, then*
$$r^0(D) = n_{\mathbb{R}}(F) \ , \quad r^1(D) = 0 \ .$$

PROOF (i) The total signature map
$$\sum_{n_{\mathbb{R}}(F)} \text{signature} : L^{4*}(F) \longrightarrow \sum_{n_{\mathbb{R}}(F)} L^{4*}(\mathbb{R}) = \sum_{n_{\mathbb{R}}(F)} \mathbb{Z}$$
is an isomorphism modulo 2-primary torsion.

(ii) Consider first the special case $D = F$. Let $F_0 = \{z \in F \mid \bar{z} = z\}$ be the fixed field of the involution, so that $F = F_0(\sqrt{a})$ is a quadratic extension of F_0 for some $a \in F \backslash F_0$ and $\overline{x + y\sqrt{a}} = x - y\sqrt{a}$ $(x, y \in F_0)$. Let $n_{\mathbb{R}}^+(F_0, a)$ (resp. $n_{\mathbb{R}}^-(F_0, a)$) be the number of embeddings $e : F_0 \subset \mathbb{R}$ such that $e(a) > 0$ (resp. $e(a) < 0$), so that
$$n_{\mathbb{C}}(F, \mathbb{Z}_2) = n_{\mathbb{R}}^-(F_0, a) \ , \quad n_{\mathbb{R}}(F_0) = n_{\mathbb{R}}^+(F_0, a) + n_{\mathbb{R}}^-(F_0, a) \ ,$$
$$n_{\mathbb{R}}(F) = 2n_{\mathbb{R}}^+(F_0, a) \ , \quad n_{\mathbb{C}}(F) = 2n_{\mathbb{C}}(F_0) + n_{\mathbb{R}}^-(F_0, a) \ ,$$
in agreement with the exact octagon of 22.22. The total signature map
$$\sum_{n_{\mathbb{C}}(F, \mathbb{Z}_2)} \text{signature} : L^{2*}(F) \longrightarrow \sum_{n_{\mathbb{C}}(F, \mathbb{Z}_2)} L^{2*}(\mathbb{C}^-) = \sum_{n_{\mathbb{C}}(F, \mathbb{Z}_2)} \mathbb{Z}$$
is an isomorphism modulo 2-primary torsion.

For arbitrary D each complex embedding $F \subset \mathbb{C}$ gives a map
$$L^{2*}(D) \longrightarrow L^{2*}(\mathbb{C} \otimes_F D)[1/2] = L^{2*}(M_d(\mathbb{C}^-))[1/2] = \mathbb{Z}[1/2]$$

and the total signature map

$$\sum_{n_{\mathbb{C}}(F,\mathbb{Z}_2)} \text{signature} : L^{2*}(D) \longrightarrow \sum_{n_{\mathbb{C}}(F,\mathbb{Z}_2)} \mathbb{Z}[1/2]$$

is again an isomorphism modulo 2-primary torsion.

(iii) (a) Each real embedding $F \subset \mathbb{R}$ gives a map

$$L^{2k}(D) \longrightarrow L^{2k}\left(\left(\frac{\alpha, \beta}{\mathbb{R}}\right)\right) = L^{2k}(\mathbb{H}) = \begin{cases} \mathbb{Z} & \text{if } k = 0 \\ \mathbb{Z}_2 & \text{if } k = 1 . \end{cases}$$

The total signature map

$$\sum_{n_{\mathbb{R}}(F)} \text{signature} : L^{4*}\left(\left(\frac{\alpha, \beta}{F}\right)\right) \longrightarrow \sum_{n_{\mathbb{R}}(F)} L^{4*}(\mathbb{H}) = \sum_{n_{\mathbb{R}}(F)} \mathbb{Z}$$

is an isomorphism modulo 2-primary torsion.

(iii) (b) Each real embedding $F \subset \mathbb{R}$ gives a map

$$L^{2k}(D) \longrightarrow L^{2k}\left(\left(\frac{\alpha, \beta}{\mathbb{R}}\right)\right) = L^{2k}(M_2(\mathbb{R}))[1/2]$$

$$= L^{2k}(\mathbb{R})[1/2] = \begin{cases} \mathbb{Z}[1/2] & \text{if } k = 0 \\ 0 & \text{if } k = 1 . \end{cases}$$

The total signature map

$$\sum_{n_{\mathbb{R}}(F)} \text{signature} : L^{4*}\left(\left(\frac{\alpha, \beta}{F}\right)\right) \longrightarrow \sum_{n_{\mathbb{R}}(F)} L^{4*}(\mathbb{R})[1/2] = \sum_{n_{\mathbb{R}}(F)} \mathbb{Z}[1/2]$$

is an isomorphism modulo 2-primary torsion.

□

Given an irreducible $\mathbb{C}$-representation ρ of π let $\mathbb{Q}(\chi)$ be the field extension of $\mathbb{Q}$ obtained by adjoining all the characters $\chi(g) = \text{tr}(\rho(g)) \in \mathbb{C}$ $(g \in \pi)$. Two such representations ρ, ρ' are *Galois conjugate* if $\mathbb{Q}(\chi) = \mathbb{Q}(\chi')$ and $\chi'(g) = \chi(\gamma(g))$ $(g \in \pi)$ for some Galois automorphism $\gamma \in \text{Gal}(\mathbb{Q}(\chi)/\mathbb{Q})$. The $\mathbb{C}$-representation of π induced from an irreducible $\mathbb{Q}$-representation of π is the sum of Galois conjugacy classes of an irreducible $\mathbb{C}$-representation.

The number of simple factors $M_d(D)$ in $\mathbb{Q}[\pi]$ is

$\alpha(\mathbb{Q}, \pi)$

= no. of irreducible $\mathbb{Q}$-representations of π

= no. of conjugacy classes of cyclic subgroups of π

= no. of Galois conjugacy classes of irreducible $\mathbb{C}$-representations of π .

The involution on $\mathbb{Q}[\pi]$ preserves each of the simple factors $S = M_d(D)$. As a ring with involution

$$S = M_d(D)^{\phi}$$

in the terminology of 21.5, for some nonsingular ϵ-symmetric form (D^d, ϕ) over a central simple $\mathbb{Q}$-algebra with involution D with $\epsilon = \pm 1$, so that by 21.6

$$L_*^p(S) = L_*(D, \epsilon) = \begin{cases} L_*(D) \\ L_{*+2}(D) \end{cases} \text{if } \epsilon = \begin{cases} +1 \\ -1 . \end{cases}$$

As before, D is one of three types:

(I) (orthogonal) $\chi = \bar{\chi}$ is real and ρ is equivalent to an $\mathbb{R}$-representation. In this case the centre $\mathbb{Q}(\chi)$ of D is totally real with the identity involution, ϵ can be chosen to be $+1$ and

$$\mathbb{R} \otimes_{\mathbb{Q}} M_d(D) = \prod_{n_{\mathbb{R}}(\mathbb{Q}(\chi))} M_d(\mathbb{R}) .$$

Either $D = \mathbb{Q}(\chi)$ with the identity involution, or $D = \left(\dfrac{\alpha, \beta}{\mathbb{Q}(\chi)} \right)$ with $\alpha, \beta \in \mathbb{Q}(\chi)^\bullet$ not both totally negative and with the involution $\bar{i} = i$, $\bar{j} = j$, $\bar{k} = -k$.

(II) (symplectic) $\chi = \bar{\chi}$ is real but ρ is not equivalent to an $\mathbb{R}$-representation. In this case the centre $\mathbb{Q}(\chi)$ of D is totally real with the identity involution, ϵ can be chosen to be $+1$ and

$$\mathbb{R} \otimes_{\mathbb{Q}} M_d(D) = \prod_{n_{\mathbb{R}}(\mathbb{Q}(\chi))} M_d(\mathbb{H}) .$$

$D = \left(\dfrac{\alpha, \beta}{\mathbb{Q}(\chi)} \right)$ is a quaternion algebra over $\mathbb{Q}(\chi)$, with α, β totally negative and with the involution $\bar{i} = -i, \bar{j} = -j, \bar{k} = -k$.

(III) (unitary) $\chi \neq \bar{\chi}$ is not real. In this case the centre $\mathbb{Q}(\chi)$ of D is totally imaginary with non-trivial involution such that $n_{\mathbb{C}}(\mathbb{Q}(\chi), \mathbb{Z}_2) = n_{\mathbb{C}}(\mathbb{Q}(\chi))$, ϵ can be chosen to be $+1$ and

$$\mathbb{R} \otimes_{\mathbb{Q}} M_d(D) = \prod_{n_{\mathbb{C}}(\mathbb{Q}(\chi))} M_d(\mathbb{C}) .$$

PROPOSITION 22.31 (Wall [177])
(i) Let $S = M_d(D)^\phi$ be a simple factor of the ring with involution $\mathbb{Q}[\pi]$ for a finite group π, with (D^d, ϕ) a nonsingular symmetric form over a division ring with involution D with centre $\mathbb{Q}(\chi)$. The L-groups of S and $\mathbb{R} \otimes_{\mathbb{Q}} D$ coincide modulo 2-primary torsion, with

$$L^n(S)[1/2] = L^n(\mathbb{R} \otimes_{\mathbb{Q}} D)[1/2] = \begin{cases} \displaystyle\sum_{r^k(D)} \mathbb{Z}[1/2] \\ 0 \end{cases} \text{if } n = \begin{cases} 2k \\ 2k+1 \end{cases}$$

given by the multisignature. The rank $\begin{cases} r^0(D) \\ r^1(D) \end{cases}$ of $\begin{cases} L^{4*}(D) \\ L^{4*+2}(D) \end{cases}$ is the number of simple factors in $\mathbb{R} \otimes_{\mathbb{Q}} D$ of $\left\{ \begin{array}{l} any \\ unitary \end{array} \right.$ type, that is

$$r^0(D) = \left\{ \begin{array}{l} n_{\mathbb{R}}(\mathbb{Q}(\chi)) \\ n_{\mathbb{R}}(\mathbb{Q}(\chi)) \\ n_{\mathbb{C}}(\mathbb{Q}(\chi)) \end{array} \right. , \quad r^1(D) = \left\{ \begin{array}{l} 0 \\ 0 \\ n_{\mathbb{C}}(\mathbb{Q}(\chi)) \end{array} \right. \quad in\ the\ case \left\{ \begin{array}{l} (I) \\ (II) \\ (III) \end{array} \right. .$$

(ii) The L-groups of $\mathbb{Q}[\pi] = \prod S$ are such that

$$L^*(\mathbb{Q}[\pi])[1/2] = \sum_S L^*(S)[1/2]$$

$$= \sum_S L^*(\mathbb{R} \otimes_{\mathbb{Q}} D)[1/2] = L^*(\mathbb{R}[\pi])[1/2] ,$$

with $S = M_d(D)^\phi$ as in (i).

□

EXAMPLE 22.32 The Wedderburn decomposition of the rational group ring of the cyclic group $\mathbb{Z}_m$ is

$$\mathbb{Q}[\mathbb{Z}_m] = \prod_{d|m} \mathbb{Q}(d) ,$$

with $\mathbb{Q}(d) = \mathbb{Q}(e^{2\pi i/d})$ the cyclotomic number field obtained from $\mathbb{Q}$ by adjoining the dth roots of unity. Now $\mathbb{Q}(d)$ is totally real for $d = 1, 2$ and totally imaginary for $d \geq 3$, with one embedding

$$\mathbb{Q}(d) \longrightarrow \mathbb{C} ; \ e^{2\pi i/d} \longrightarrow e^{2\pi i u/d}$$

for each unit $u \in \mathbb{Z}_d^\bullet \subset \mathbb{Z}_d$. Thus

$$n_{\mathbb{R}}(\mathbb{Q}(d)) = \left\{ \begin{array}{l} 1 \\ 0 \end{array} \right. if \left\{ \begin{array}{l} d = 1, 2 \\ d \geq 3 , \end{array} \right.$$

$$n_{\mathbb{C}}(\mathbb{Q}(d)) = n_{\mathbb{C}}(\mathbb{Q}(d), \mathbb{Z}_2) = \left\{ \begin{array}{l} 0 \\ \phi(d)/2 \end{array} \right. if \left\{ \begin{array}{l} d = 1, 2 \\ d \geq 3 \end{array} \right.$$

with $\phi(d) = |\mathbb{Z}_d^\bullet|$ the Euler function, the number of positive integers $< d$ which are coprime to d. By 22.30 the symmetric Witt group of $\mathbb{Q}(d)$ is such that

$$L^0(\mathbb{Q}(d))[1/2] = \left\{ \begin{array}{l} \mathbb{Z}[1/2] \\ \mathbb{Z}[1/2]^{\phi(d)/2} \end{array} \right. if \left\{ \begin{array}{l} d = 1, 2 \\ d \geq 3 . \end{array} \right.$$

By 22.31 the symmetric Witt group of $\mathbb{Q}[\mathbb{Z}_m]$ is such that

$$L^0(\mathbb{Q}[\mathbb{Z}_m])[1/2] = \sum_{d|m} L^0(\mathbb{Q}(d))[1/2] = \left\{ \begin{array}{l} \mathbb{Z}[1/2]^{(m+1)/2} \\ \mathbb{Z}[1/2]^{(m+2)/2} \end{array} \right. if\ m\ is \left\{ \begin{array}{l} odd \\ even , \end{array} \right.$$

using $\sum_{d|m} \phi(d) = m$. This agrees with the computation of $L_p^0(\mathbb{R}[\mathbb{Z}_m])$ in 22.24.

□

The $\mathbb{Q}$-multisignature gives as much information in L-theory as the $\mathbb{R}$-multisignature:

PROPOSITION 22.33 *The L-groups of $\mathbb{Q}[\pi]$ and $\mathbb{R}[\pi]$ for a finite group π agree modulo 2-primary torsion*

$$L^n(\mathbb{Q}[\pi])[1/2] = L^n(\mathbb{R}[\pi])[1/2] = \begin{cases} \displaystyle\sum_{\alpha(\mathbb{R},\pi)} \mathbb{Z}[1/2] \\ \displaystyle\sum_{\alpha_\mathbb{C}(\mathbb{R},\pi)} \mathbb{Z}[1/2] \\ 0 \end{cases} \text{ if } n \equiv \begin{cases} 0 \\ 2 \\ 1,3 \end{cases} \pmod 4$$

detected by the $\mathbb{R}$-multisignature. (In fact, $L_p^{2*+1}(\mathbb{Q}[\pi]) = 0$).
PROOF Write the simple factors $M_{d_j(\mathbb{Q},\pi)}(D_j(\mathbb{Q},\pi))$ of $\mathbb{Q}[\pi]$ as

$$S_j = M_{d_j}(D_j) \quad (1 \le j \le \alpha(\mathbb{Q},\pi)) \ .$$

The involution on $\mathbb{Q}[\pi]$ preserves each S_j, so that by 22.10

$$L_p^{2k}(\mathbb{Q}[\pi]) = \sum_{j=1}^{\alpha(\mathbb{Q},\pi)} L_p^{2k}(S_j) \ .$$

Let $(D_j^{d_j}, \phi_j)$ be a nonsingular ϵ_j-symmetric form over D_j such that

$$S_j = M_{d_j}(D_j)^{\phi_j} \ , \quad L_p^{2k}(S_j) = L^{2k}(D_j, \epsilon_j) \ .$$

The projective L-groups $L_p^{2*}(\mathbb{Q}[\pi])$ are given by

$$L_p^{2k}(\mathbb{Q}[\pi]) = \sum_{j=1}^{\alpha(\mathbb{Q},\pi)} L^{2k}(D_j, \epsilon_j) \ .$$

The contributions to the $\mathbb{Q}$-multisignature of all the simple factors S_j of $\mathbb{Q}[\pi]$ are thus just the $\mathbb{R}$-multisignatures of the induced products of simple factors $\mathbb{R} \otimes_\mathbb{Q} S_j$ of $\mathbb{R}[\pi]$, with

$$\alpha^0(\mathbb{Q},\pi) = \sum_{\chi=\bar\chi} n_\mathbb{R}(\mathbb{Q}(\chi)) + \sum_{\chi\neq\bar\chi} n_\mathbb{C}(\mathbb{Q}(\chi)) = \alpha(\mathbb{R},\pi) = \alpha^0(\mathbb{R},\pi) \ ,$$

$$\alpha^1(\mathbb{Q},\pi) = \sum_{\chi\neq\bar\chi} n_\mathbb{C}(\mathbb{Q}(\chi)) = \alpha_\mathbb{C}(\mathbb{R},\pi) = \alpha^1(\mathbb{R},\pi) \ .$$

$\square$

The computation of the L-theory of $\mathbb{Q}[\pi]$ is now applied to the computation of the L-theory of $\mathbb{Z}[\pi]$ modulo 2-primary torsion, and hence the determination of the image of the total surgery obstruction in $\mathbb{S}_*(B\pi)$ modulo torsion.

PROPOSITION 22.34 (i) *The symmetrization and localization maps*
$$L_*(\mathbb{Z}[\pi]) \longrightarrow L^*(\mathbb{Z}[\pi]) \ , \quad L_*(\mathbb{Z}[\pi]) \longrightarrow L_*(\mathbb{Q}[\pi]) \ , \quad L^*(\mathbb{Z}[\pi]) \longrightarrow L^*(\mathbb{Q}[\pi])$$
are isomorphisms modulo 2-primary torsion for any group π, so that
$$L_*(\mathbb{Z}[\pi])[1/2] = L_*(\mathbb{Q}[\pi])[1/2] = L^*(\mathbb{Z}[\pi])[1/2] = L^*(\mathbb{Q}[\pi])[1/2] \ .$$

(ii) *For a finite group π*

$$L_n(\mathbb{Z}[\pi])[1/2] = L_n(\mathbb{Q}[\pi])[1/2] = L_n(\mathbb{R}[\pi])[1/2]$$

$$= \begin{cases} \sum\limits_{\alpha^k(\mathbb{R},\pi)} \mathbb{Z}[1/2] \\ 0 \end{cases} \quad if \ n = \begin{cases} 2k \\ 2k+1 \end{cases}$$

with $\alpha^k(\mathbb{R},\pi) = \alpha^k(\mathbb{Q},\pi) = \begin{cases} \alpha(\mathbb{R},\pi) \\ \alpha_\mathbb{C}(\mathbb{R},\pi) \end{cases}$ *for* $k \equiv \begin{cases} 0 \\ 1 \end{cases}$ (mod 2).

(iii) *The* reduced *quadratic L-groups*

$$\widetilde{L}_*(\mathbb{Z}[\pi]) = L_*(\mathbb{Z}\longrightarrow\mathbb{Z}[\pi])$$

are such that

$$L_*(\mathbb{Z}[\pi]) = L_*(\mathbb{Z}) \oplus \widetilde{L}_*(\mathbb{Z}[\pi]) \ .$$

For a finite group π the reduced L-groups are detected modulo 2-primary torsion by the reduced $\mathbb{R}$-multisignature

$$\widetilde{L}_n(\mathbb{Z}[\pi])[1/2] = \begin{cases} \operatorname{coker}\left(A = \sum\limits_{j=1}^{\alpha(\mathbb{R},\pi)} d_j : \mathbb{Z}\longrightarrow \sum\limits_{\alpha(\mathbb{R},\pi)} \mathbb{Z}\right)[1/2] \\ \sum\limits_{\alpha_\mathbb{C}(\mathbb{R},\pi)} \mathbb{Z}[1/2] \\ 0 \end{cases}$$

$$if \ n \equiv \begin{cases} 0 \\ 2 \\ 1,3 \end{cases} \quad (\text{mod } 4) \ (d_j = d_j(\mathbb{R},\pi)) \ .$$

PROOF (i) The profinite completion of $\mathbb{Z}$ and its fraction field (the finite adeles) are given by

$$\widehat{\mathbb{Z}} = \varprojlim_m \mathbb{Z}/m\mathbb{Z} = \prod\limits_{q \text{ prime}} \widehat{\mathbb{Z}}_q \ , \quad \widehat{\mathbb{Q}} = (\widehat{\mathbb{Z}}\backslash\{0\})^{-1}\widehat{\mathbb{Z}} = \prod\limits_q (\widehat{\mathbb{Q}}_q, \widehat{\mathbb{Z}}_q) \ ,$$

using the q-adic completions of $\mathbb{Z}$ and $\mathbb{Q}$

$$\widehat{\mathbb{Z}}_q = \varprojlim_k \mathbb{Z}/q^k\mathbb{Z} \ , \quad \widehat{\mathbb{Q}}_q = (\widehat{\mathbb{Z}}_q\backslash\{0\})^{-1}\widehat{\mathbb{Z}}_q \ .$$

The L-groups of the inclusions

$$i : \mathbb{Z}[\pi] \longrightarrow \mathbb{Q}[\pi] \ , \quad \widehat{i} : \widehat{\mathbb{Z}}[\pi] \longrightarrow \widehat{\mathbb{Q}}[\pi]$$

are related by a natural transformation of localization exact sequences

$$\begin{array}{ccccccccc}
\cdots \to & L_n^p(\mathbb{Z}[\pi]) & \longrightarrow & L_n^X(\mathbb{Q}[\pi]) & \longrightarrow & L_n^X(i) & \longrightarrow & L_{n-1}^p(\mathbb{Z}[\pi]) & \to \cdots \\
& \downarrow & & \downarrow & & \downarrow{\scriptstyle\cong} & & \downarrow & \\
\cdots \to & L_n^p(\widehat{\mathbb{Z}}[\pi]) & \longrightarrow & L_n^{\widehat{X}}(\widehat{\mathbb{Q}}[\pi]) & \longrightarrow & L_n^{\widehat{X}}(\widehat{i}) & \longrightarrow & L_{n-1}^p(\widehat{\mathbb{Z}}[\pi]) & \to \cdots
\end{array}$$

254 ALGEBRAIC L-THEORY AND TOPOLOGICAL MANIFOLDS

with excision isomorphisms $L_*^X(i) \cong L_*^{\widehat{X}}(\widehat{i})$ of the relative L-groups, where
$$X = \operatorname{im}(\widetilde{K}_0(\mathbb{Z}[\pi]) \longrightarrow \widetilde{K}_0(\mathbb{Q}[\pi])) \ , \ \widehat{X} = \operatorname{im}(\widetilde{K}_0(\widehat{\mathbb{Z}}[\pi]) \longrightarrow \widetilde{K}_0(\widehat{\mathbb{Q}}[\pi])) \ .$$
(If π is finite then $X = \{0\}$ by a result of Swan). The projective symmetric Witt group of $\widehat{\mathbb{Z}}$
$$L_p^0(\widehat{\mathbb{Z}}) = \prod_q L^0(\widehat{\mathbb{Z}}_q) = L^0(\mathbb{Z}_8) \oplus \prod_{q \neq 2} L^0(\mathbb{F}_q)$$
is a ring with 1 of exponent 8, which acts on $L_*^{\widehat{X}}(\widehat{i})$. See Ranicki [139, 4.4], [141, §8], [143, §3.6] for further details.
(ii) Immediate from (i) and 22.33.
(iii) Immediate from (ii).

$\square$

REMARK 22.35 The computation of $L_*(\mathbb{Z}[\pi])$ (π finite) modulo 2-primary torsion was originally obtained by Wall [176, pp. 167–168], [177] using the work of Kneser on Galois cohomology to formulate the L-theory Hasse principle
$$L_*(\mathbb{Z}[\pi])[1/2] = L_*(\widehat{\mathbb{Q}}[\pi])[1/2] \oplus L_*(\mathbb{R}[\pi])[1/2] \ .$$
Ian Hambleton has pointed out that the action of $L^0(\widehat{\mathbb{Q}})$ on $L_*(\widehat{\mathbb{Q}}[\pi])$ gives a direct derivation of
$$L_*(\mathbb{Z}[\pi])[1/2] = L_*(\mathbb{R}[\pi])[1/2] \ ,$$
which avoids the detailed analysis in 22.33 of $L_*(\mathbb{Q}[\pi])[1/2]$, as follows. The symmetric Witt ring of $\widehat{\mathbb{Q}}$
$$L^0(\widehat{\mathbb{Q}}) = \left(\prod_{q \text{ prime}} L^0(\widehat{\mathbb{Z}}_q) \right) \oplus \left(\sum_q L^0(\mathbb{F}_q) \right)$$
$$= \left(L^0(\mathbb{Z}_8) \oplus \prod_{q \neq 2} L^0(\mathbb{F}_q) \right) \oplus \left(\sum_q L^0(\mathbb{F}_q) \right)$$
has exponent 8, so that $L_*(\widehat{\mathbb{Q}}[\pi])[1/2] = 0$. In fact
$$L_*^X(R[\pi])[1/2] = L_*(\mathbb{R}[\pi])[1/2]$$
for any ring R with $\mathbb{Z} \subseteq R \subseteq \mathbb{Q}$, with any decoration subgroup $X \subseteq \widetilde{K}_i(R[\pi])$ $(i = 0, 1)$. See Bak and Kolster [8], Carlsson and Milgram [32], Kolster [85], Hambleton and Madsen [64] for the computation of the torsion in the projective L-groups $L_*^p(\mathbb{Z}[\pi])$, which is all 2-primary.

$\square$

The classifying space $B\pi$ of a finite group π has the rational homotopy type of a point: the transfer map $p^!$ associated to the universal covering projection $p \colon E\pi \longrightarrow B\pi$ is such that
$$p_! p^! = |\pi| : h_*(B\pi) \longrightarrow h_*(E\pi) \longrightarrow h_*(B\pi)$$

for any generalized homology theory h_*, with $E\pi \simeq \{\mathrm{pt.}\}$ and $h_*(E\pi) = h_*(\{\mathrm{pt.}\})$. It follows that the maps

$$H_n(\{\mathrm{pt.}\};\mathbb{L}.) = L_n(\mathbb{Z}) \longrightarrow H_n(B\pi;\mathbb{L}.) \ (n > 0)$$

are isomorphisms modulo torsion. The natural transformation of exact sequences

$$
\begin{array}{ccccccccc}
\cdots \longrightarrow & L_n(\mathbb{Z}) & \longrightarrow & L_n(\mathbb{Z}[\pi]) & \longrightarrow & \widetilde{L}_n(\mathbb{Z}[\pi]) & \xrightarrow{\ 0\ } & L_{n-1}(\mathbb{Z}) & \longrightarrow \cdots \\
& \downarrow & & \| & & \downarrow & & \downarrow & \\
\cdots \longrightarrow & H_n(B\pi;\mathbb{L}.) & \xrightarrow{\ A\ } & L_n(\mathbb{Z}[\pi]) & \longrightarrow & \mathbb{S}_n(B\pi) & \longrightarrow & H_{n-1}(B\pi;\mathbb{L}.) & \longrightarrow \cdots
\end{array}
$$

is an isomorphism modulo torsion, with

$$\mathbb{S}_{2*+1}(B\pi) \otimes \mathbb{Q} = \widetilde{L}_{2*+1}(\mathbb{Z}[\pi]) \otimes \mathbb{Q} = 0 .$$

THEOREM 22.36 *Let X be a finite $2k$-dimensional geometric Poincaré complex, with a regular finite cover $\overline{X}$ classified by a morphism $\pi_1(X) \longrightarrow \pi$ to a finite group π.*
(i) *The symmetric signature $\sigma^*(X) = (C(\overline{X}), \phi) \in L^{2k}(\mathbb{Z}[\pi])$ is determined modulo 2-primary torsion by the $\mathbb{R}$-multisignature of the nonsingular $(-)^k$-symmetric form $(H^k(\overline{X};\mathbb{R}), \phi_0)$ over $\mathbb{R}[\pi]$*

$$\sigma^*(X) = \mathbb{R}\text{-multisignature}\,(H^k(\overline{X};\mathbb{R}), \phi_0)$$

$$\in L_p^{2k}(\mathbb{R}[\pi]) = \begin{cases} \displaystyle\sum_{\alpha(\mathbb{R},\pi)} \mathbb{Z} \\ \displaystyle\sum_{\alpha_{\mathbb{C}}(\mathbb{R},\pi)} \mathbb{Z} \end{cases} if\ k \equiv \begin{cases} 0 \\ 1 \end{cases} (\mathrm{mod}\ 2) ,$$

with $L^{2k}(\mathbb{Z}[\pi])[1/2] = L_p^{2k}(\mathbb{R}[\pi])[1/2]$.
(ii) *The image in $\mathbb{S}_{2k}(B\pi)$ of the total surgery obstruction $s(X) \in \mathbb{S}_{2k}(X)$ is determined up to torsion by the reduced $\mathbb{R}$-multisignature*

$$[s(X)] \otimes \mathbb{Q} = \sigma^*(X) \otimes \mathbb{Q} \in \mathbb{S}_{2k}(B\pi) \otimes \mathbb{Q} = \widetilde{L}_{2k}(\mathbb{Z}[\pi]) \otimes \mathbb{Q}$$

$$= \begin{cases} \mathrm{coker}\left(A = \displaystyle\sum_{j=1}^{\alpha(\mathbb{R},\pi)} d_j(\mathbb{R},\pi) : \mathbb{Q} \longrightarrow \displaystyle\sum_{\alpha(\mathbb{R},\pi)} \mathbb{Q} \right) \\ \displaystyle\sum_{\alpha_{\mathbb{C}}(\mathbb{R},\pi)} \mathbb{Q} \end{cases} if\ k \equiv \begin{cases} 0 \\ 1 \end{cases} (\mathrm{mod}\ 2) .$$

PROOF The symmetrization maps

$$1 + T : \mathbb{S}_*(X) = \mathbb{S}\langle 1\rangle_*(X) \longrightarrow \mathbb{S}\langle 1\rangle^*(X)$$

are isomorphisms modulo 2-primary torsion for any space X, and

$$(1+T)s(X) = [\sigma^*(X)] \in \mathrm{im}(L^n(\mathbb{Z}[\pi_1(X)]) \longrightarrow \mathbb{S}^n(X))$$

for any finite n-dimensional geometric Poincaré complex X. $\qquad\square$

EXAMPLE 22.37 Let X be a finite $2k$-dimensional geometric Poincaré complex, with a regular m-fold cyclic cover $\overline{X}$ classified by a morphism $\pi_1(X) \longrightarrow \mathbb{Z}_m$. The multisignature of X with respect to $\overline{X}$ is an element

$$\sigma^*(X) = (s_1, s_2, \ldots, s_{\alpha^k(\mathbb{R},\mathbb{Z}_m)}) \in L_p^{2k}(\mathbb{R}[\mathbb{Z}_m]) = \sum_{j=1}^{\alpha^k(\mathbb{R},\mathbb{Z}_m)} \mathbb{Z}$$

with

$$\alpha^0(\mathbb{R}, \mathbb{Z}_m) = \begin{cases} (m+1)/2 \\ (m+2)/2 \end{cases}, \quad \alpha^1(\mathbb{R}, \mathbb{Z}_m) = \begin{cases} (m-1)/2 \\ (m-2)/2 \end{cases} \text{ if } m \text{ is } \begin{cases} \text{odd} \\ \text{even} \end{cases}$$

(cf. 22.24, 22.32). The total surgery obstruction $s(X) \in \mathbb{S}_{2k}(X)$ has image $[s(X)] \otimes \mathbb{Q} = (s_1, s_2, \ldots, s_{\alpha^k(\mathbb{R},\mathbb{Z}_m)})$

$$\in \mathbb{S}_{2k}(B\mathbb{Z}_m) \otimes \mathbb{Q} = \begin{cases} \operatorname{coker}\left((1\ 1\ \ldots\ 1) : \mathbb{Q} \longrightarrow \sum_{\alpha^0(\mathbb{R},\mathbb{Z}_m)} \mathbb{Q} \right) \\ \displaystyle\sum_{\alpha^1(\mathbb{R},\mathbb{Z}_m)} \mathbb{Q} \end{cases}$$

$$\text{if } k \equiv \begin{cases} 0 \\ 1 \end{cases} (\bmod\ 2) .$$

For $k \equiv 0 \,(\bmod\, 2)$ there is one multisignature component s_j for each irreducible $\mathbb{R}$-representation of $\mathbb{Z}_m = \langle T \,|\, T^m = 1 \rangle$

$$\rho_j : \mathbb{Z}_m \longrightarrow D_j = \begin{cases} \mathbb{R} & \text{if } j = 1 \text{ or } (m+2)/2 \ (m \text{ even}) \\ \mathbb{C} & \text{otherwise} ; \end{cases}$$

$$T \longrightarrow e^{2\pi i j/m} \quad (0 \le j < \alpha^0(\mathbb{R}, \mathbb{Z}_m)) ,$$

with

$$\text{signature}\,(X) = s_1 \in L^{4k}(\mathbb{R}) = \mathbb{Z} ,$$

$$\text{signature}\,(\overline{X}) = \sum_{j=1}^{\alpha^0(\mathbb{R},\mathbb{Z}_m)} c_j s_j \in L^{4k}(\mathbb{R}) = \mathbb{Z} \ (c_j = \dim_{\mathbb{R}}(D_j)) .$$

The total surgery obstruction is such that

$$[s(X)] \otimes \mathbb{Q} = 0 \in \mathbb{S}_{2k}(B\mathbb{Z}_m) \otimes \mathbb{Q}$$

if and only if the multisignature components are equal

$$\text{signature}\,(X) = s_1 = s_2 = \ldots = s_{\alpha^0(\mathbb{R},\mathbb{Z}_m)} \in \mathbb{Z} ,$$

in which case

$$\text{signature}\,(\overline{X}) = \left(\sum_{j=1}^{\alpha^0(\mathbb{R},\mathbb{Z}_m)} c_j \right) s_1 = m \,\text{signature}\,(X) \in \mathbb{Z}$$

confirming the multiplicativity of the signature for finite covers of manifolds (21.4, 22.26) in the cyclic case.

□

§23. Splitting

The algebraic methods appropriate to the computation of $L_*(\mathbb{Z}[\pi])$ and $\mathbb{S}_*(B\pi)$ for finite groups π do not in general extend to infinite groups π. At present, systematic computations are possible only for infinite groups π which are geometric in some sense, such as the following.

(i) π is an n-dimensional Poincaré duality group, i.e. such that the classifying space $B\pi$ is an n-dimensional geometric Poincaré complex. Differential and hyperbolic geometry provide many examples of Poincaré duality groups π acting freely on an open contractible n-dimensional manifolds with compact quotient, such as the torsion-free crystallographic groups acting on $\mathbb{R}^n$. The generic result expected in this case is that the assembly map $A: H_*(B\pi; \mathbb{L}.) \longrightarrow L_*(\mathbb{Z}[\pi])$ is an isomorphism for $* > n$, with $\mathbb{S}_*(B\pi) = 0$ for $* > n$ and $s(B\pi) = 0 \in \mathbb{S}_n(B\pi) = \mathbb{Z}$, so that $B\pi$ is homotopy equivalent to an aspherical compact n-dimensional topological manifold with topological rigidity. This is the strongest form of the Novikov and Borel conjectures, which will be discussed (but alas not proved) in §24 below.

(ii) π acts on a tree with compact quotient, so that by the Bass–Serre theory π is either an amalgamated free product or an HNN extension. The generic result available in this case is that if π is obtained from the trivial group $\{1\}$ by a sequence of amalgamated free products and HNN extensions then $\mathbb{S}_*(B\pi)$ can be expressed in terms of the Tate $\mathbb{Z}_2$-cohomology groups of the duality involution on the algebraic K-theory of $\mathbb{Z}[\pi]$, the UNil-groups of Cappell [22] and the generalized Browder–Livesay LN-groups of Wall [176, §12C], which arise from the codimension 1 splitting obstruction theory. It is this splitting theory which will be considered now.

DEFINITION 23.1 A homotopy equivalence $f: M' \longrightarrow M$ of compact n-dimensional manifolds *splits* along a compact submanifold $N^{n-q} \subset M^n$ if f is h-cobordant a homotopy equivalence (also denoted by f) transverse regular at $N \subset M$, such that the restriction $f|: N' = f^{-1}(N) \longrightarrow N$ is a homotopy equivalence of compact $(n - q)$-dimensional manifolds.

$\square$

If a homotopy equivalence of compact manifolds $f: M' \longrightarrow M$ is h-cobordant to a homeomorphism then f splits along every submanifold $N \subset M$. Conversely, if $f: M' \longrightarrow M$ does not split along some submanifold $N \subset M$ then f cannot be h-cobordant (let alone homotopic) to a homeomorphism.

In general, homotopy equivalences do not split along submanifolds. Surgery theory provides various K- and L-theory obstructions to splitting, whose vanishing is both necessary and sufficient for splitting if $n - q \geq 5$, and which are also the obstructions to transversality for geometric Poincaré complexes. There is also an obstruction theory for the more delicate problem of splitting up to homotopy, i.e. replacing h-cobordisms by s-cobordisms, which involves Whitehead torsion. See Ranicki [143, §7] for a preliminary account of the splitting obstruction theory from the chain complex point of view.

The geometric codimension q splitting obstruction LS-groups $LS_*(\Phi)$ of Wall [176, §11] are defined using normal maps with reference maps to a space X which is expressed as a union

$$X = E(\xi) \cup_{S(\xi)} Z$$

with $(E(\xi), S(\xi))$ the (D^q, S^{q-1})-bundle associated to a topological block bundle $\xi : Y \longrightarrow B\widetilde{TOP}(q)$ over a subspace $Y \subset X$, for some $q \geq 1$. By the Seifert–Van Kampen theorem the fundamental group(oid)s fit into a pushout square

$$
\begin{array}{ccc}
\pi_1(S(\xi)) & \longrightarrow & \pi_1(Z) \\
\downarrow & \Phi & \downarrow \\
\pi_1(Y) & \longrightarrow & \pi_1(X) \ .
\end{array}
$$

The LS-groups are designed to fit into an exact sequence

$$\ldots \longrightarrow LS_{n-q}(\Phi) \longrightarrow L_n(\xi^! : Y \to Z)$$
$$\longrightarrow L_n(X) \longrightarrow LS_{n-q-1}(\Phi) \longrightarrow \ldots$$

with $L_*(X) = L_*(\mathbb{Z}[\pi_1(X)])$. In the original setting of [176] these were the obstruction groups appropriate to simple homotopy equivalences. Here, only ordinary homotopy equivalences are being considered, with free L-groups and the corresponding modification in the definition of $LS_*(\Phi)$. The free and simple LS-groups differ in 2-primary torsion only, being related by the appropriate Rothenberg-type exact sequence.

A map from a compact n-dimensional manifold $r : M^n \longrightarrow X = E(\xi) \cup_{S(\xi)} Z$ can be made transverse regular at the zero section $Y \subset E(\xi) \subset X$, with

$$r^{-1}(Y) = N^{n-q} \subset M^n$$

a codimension q compact submanifold and the restriction

$$s = r| : N^{n-q} = r^{-1}(Y) \longrightarrow Y$$

such that

$$\nu_{N \subset M} = s^* \xi : N \longrightarrow B\widetilde{TOP}(q) \ , \quad M = E(\nu_{N \subset M}) \cup f^{-1}(Z) \ .$$

PROPOSITION 23.2 *Let M^n be a closed n-dimensional manifold with a π_1-isomorphism reference map $r : M^n \longrightarrow X = E(\xi) \cup_{S(\xi)} Z$ transverse regular at $Y \subset X$, such that the restriction $r| : N^{n-q} = r^{-1}(Y) \longrightarrow Y$ is also a π_1-isomorphism.*

(i) *The codimension q splitting obstruction of a homotopy equivalence* $f\colon M'$
$\longrightarrow M$ *of compact n-dimensional manifolds is the image of the structure
invariant* $s(f) \in \mathbb{S}_{n+1}(X)$
$$s_Y(f) \;=\; [s(f)] \in \mathrm{im}(\mathbb{S}_{n+1}(X)\longrightarrow LS_{n-q}(\Phi)) \;,$$
such that $s_Y(f) = 0$ *if (and for* $n - q \geq 6$ *only if)* f *splits along* $N \subset M$.
The splitting obstruction has image
$$[s_Y(f)] \;=\; \sigma_*(g,c) \in \mathrm{im}(A\colon H_{n-q}(Y;\mathbb{L}.)\longrightarrow L_{n-q}(Y))$$
is the surgery obstruction of the normal map of compact $(n-q)$-*dimensional
manifolds obtained by codimension q transversality*
$$(g,c) \;=\; f| \colon (N')^{n-q} \;=\; f^{-1}(N) \longrightarrow N \;.$$
(ii) *For* $n - q \geq 6$ *every element* $x \in LS_{n-q+1}(\Phi)$ *is realized as the rel* ∂
codimension q splitting obstruction $x = s_Y(F)$ *of a homotopy equivalence
of compact* $(n + 1)$-*dimensional manifolds with boundary*
$$F \colon (W^{n+1}; M^n, M'^n) \xrightarrow{\;\simeq\;} M^n \times ([0,1]; \{0\}, \{1\})$$
such that
$$F|_M \;=\; identity \colon M \longrightarrow M \times \{0\} \;,$$
$$F|_{M'} \;=\; split\ homotopy\ equivalence \colon M' \longrightarrow M \times \{1\} \;.$$
PROOF See Wall [176, §11].

$\square$

PROPOSITION 23.3 *The exact sequence of Ranicki [143, 7.2.6] relating the
codimension q splitting obstruction groups* $LS_*(\Phi)$ *and the quadratic structure groups* $\mathbb{S}_*$ *for* $X = E(\xi) \cup_{S(\xi)} Z$
$$\dots \longrightarrow LS_{n-q}(\Phi) \longrightarrow \mathbb{S}_n(\xi^!\colon Y \to Z) \longrightarrow \mathbb{S}_n(X)$$
$$\longrightarrow LS_{n-q-1}(\Phi) \longrightarrow \dots$$
extends to a commutative braid of exact sequences

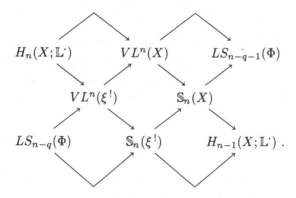

$\square$

The codimension q Poincaré transversality obstruction theory is a delooping of the codimension q splitting obstruction theory for homotopy equivalences of compact manifolds.

PROPOSITION 23.4 (i) *If P is an n-dimensional geometric Poincaré complex with a map $f\colon P \longrightarrow X = E(\xi) \cup_{S(\xi)} Z$ then*

$$s_Y(P) = [\partial \sigma^*(P)] = [s(P)] \in LS_{n-q-1}(\Phi)$$

is the codimension q Poincaré transversality obstruction, such that $s_Y(P) = 0$ if (and for $n - q \geq 6$ only if) there exists a geometric Poincaré bordism

$$(g; f, f') : (Q; P, P') \longrightarrow X$$

such that $(f')^{-1}(Y) \subset P'$ is a codimension q Poincaré subcomplex.
(ii) *The geometric Poincaré bordism groups fit into the commutative braid of exact sequences analogous to the braids of 19.6 (i) and 23.3*

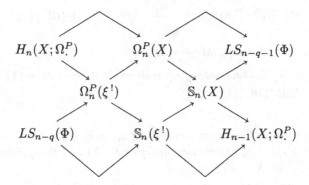

with $\Omega^P_{\boldsymbol{\cdot}} = \Omega^P_{\boldsymbol{\cdot}}(\{*\})$ the Poincaré bordism spectrum of a point.

□

For $q \geq 3$ the fundamental groups are all the same

$$\pi_1(X) = \pi_1(Y) = \pi_1(Z) = \pi_1(S(\xi)) \ (= \pi \text{ , say})$$

and $LS_*(\Phi) = L_*(\mathbb{Z}[\pi])$. For $LS_*(\Phi)$ in the case $q = 2$ see Ranicki [143, 7.8].

For $q = 1$ with X, Y connected, there are the usual three cases:
(A) the normal bundle ξ is trivial, and the complement $Z = X \backslash Y$ is disconnected, with components Z_1, Z_2, so that the fundamental group of X is the amalgamated free product

$$\pi_1(X) = \pi_1(Z_1) *_{\pi_1(Y)} \pi_1(Z_2)$$

determined by the two group morphisms

$$(i_1)_* : \pi_1(Y) \longrightarrow \pi_1(Z_1) \ , \ (i_2)_* : \pi_1(Y) \longrightarrow \pi_1(Z_2)$$

induced by the inclusions $i_1\colon Y \longrightarrow Z_1$, $i_2\colon Y \longrightarrow Z_2$,

(B) the normal bundle ξ is trivial, and the complement Z is connected, so that the fundamental group of X is the HNN extension

$$\pi_1(X) \;=\; \pi_1(Z) *_{\pi_1(Y)} \mathbb{Z}$$

determined by the two group morphisms

$$(i_1)_* \, , \; (i_2)_* : \; \pi_1(Y) \longrightarrow \pi_1(Z)$$

induced by the two inclusions $i_1, i_2 : Y \longrightarrow Z$.

(C) the normal bundle ξ is non-trivial.

If the group morphisms $(i_1)_*$, $(i_2)_*$ in cases (A) and (B) are injections then the LS-groups are direct sums

$$LS_{n-1}(\Phi) \;=\; \mathrm{UNil}_{n+1}(\Phi) \oplus \widehat{H}^n(\mathbb{Z}_2 \,; I)$$

of the UNil-groups of Cappell [22] and the Tate $\mathbb{Z}_2$-cohomology groups with respect to the duality $\mathbb{Z}_2$-action on the algebraic K-group

$$
\begin{aligned}
I \;&=\; \mathrm{im}(\partial\colon Wh(\pi_1(X)) \longrightarrow \widetilde{K}_0(\mathbb{Z}[\pi_1(Y)])) \\
&= \begin{cases} \ker((i_1)_* \oplus (i_2)_* \colon \widetilde{K}_0(\mathbb{Z}[\pi_1(Y)]) \to \widetilde{K}_0(\mathbb{Z}[\pi_1(Z_1)]) \oplus \widetilde{K}_0(\mathbb{Z}[\pi_1(Z_2)])) \\ \ker((i_1)_* - (i_2)_* \colon \widetilde{K}_0(\mathbb{Z}[\pi_1(Y)]) \to \widetilde{K}_0(\mathbb{Z}[\pi_1(Z)])) \end{cases}
\end{aligned}
$$

$$\text{for } \begin{cases} \text{(A)} \\ \text{(B)} \,. \end{cases}$$

Here, $\partial\colon Wh(\pi_1(X)) \longrightarrow \widetilde{K}_0(\mathbb{Z}[\pi_1(Y)])$ is a component of the connecting map in the algebraic K-theory exact sequence of Waldhausen [171]

$$
\begin{cases}
\ldots \longrightarrow Wh(\pi_1(Y)) \oplus \widetilde{\mathrm{Nil}}_1(\Phi) \longrightarrow Wh(\pi_1(Z_1)) \oplus Wh(\pi_1(Z_2)) \\
\qquad\quad \longrightarrow Wh(\pi_1(X)) \xrightarrow{\;\partial\;} \widetilde{K}_0(\mathbb{Z}[\pi_1(Y)]) \oplus \widetilde{\mathrm{Nil}}_0(\Phi) \longrightarrow \ldots \\
\ldots \longrightarrow Wh(\pi_1(Y)) \oplus \widetilde{\mathrm{Nil}}_1(\Phi) \longrightarrow Wh(\pi_1(Z)) \\
\qquad\quad \longrightarrow Wh(\pi_1(X)) \xrightarrow{\;\partial\;} \widetilde{K}_0(\mathbb{Z}[\pi_1(Y)]) \oplus \widetilde{\mathrm{Nil}}_0(\Phi) \longrightarrow \ldots \,.
\end{cases}
$$

The split surjection $LS_{n-1}(\Phi) \longrightarrow \widehat{H}^n(\mathbb{Z}_2 \,; I)$ fits into a commutative square

$$
\begin{array}{ccc}
\mathbb{S}_{n+1}(X) & \longrightarrow & \widehat{H}^{n+1}(\mathbb{Z}_2 \,; Wh(\pi_1(X))) \\
\downarrow & & \downarrow{\scriptstyle \partial} \\
LS_{n-1}(\Phi) & \longrightarrow & \widehat{H}^n(\mathbb{Z}_2 \,; I)
\end{array}
$$

with

$$\mathbb{S}_{n+1}(X) \longrightarrow \widehat{H}^{n+1}(\mathbb{Z}_2 \,; Wh(\pi_1(X))) \,; \; (C, \psi) \longrightarrow \tau(C(\widetilde{X}))$$

the map which sends the cobordism class of an n-dimensional locally Poincaré globally contractible complex (C, ψ) in $\mathbb{A}(\mathbb{Z}, X)$ to the Tate $\mathbb{Z}_2$-cohomology class of the Whitehead torsion of the assembly contractible $\mathbb{Z}[\pi_1(X)]$-module chain complex $C(\widetilde{X})$

$$\tau(C(\widetilde{X})) = (-)^{n+1}\tau(C(\widetilde{X}))^* \in Wh(\pi_1(X)) .$$

For $n \geq 5$ every element $x \in L_{n+1}(X)$ is realized as the rel ∂ surgery obstruction $\sigma_*(F, B)$ of a normal map of compact $(n+1)$-dimensional manifolds with boundary

$$(F, B) : (W^{n+1}; M^n, M'^n) \longrightarrow M \times ([0, 1]; \{0\}, \{1\})$$

such that

$$F|_M = \text{identity} : M \longrightarrow M \times \{0\} ,$$

$$F|_{M'} = f = \text{homotopy equivalence} : M' \longrightarrow M \times \{1\}$$

with a π_1-isomorphism reference map $r : M \longrightarrow X = E(\xi) \cup_{S(\xi)} Z$ transverse regular at $Y \subset X$, such that $N^{n-1} = r^{-1}(Y) \subset M^n$ is a codimension 1 submanifold with $\pi_1(N) \cong \pi_1(Y)$, and such that F is transverse regular at $N \times [0, 1] \subset M \times [0, 1]$ with

$$(V^n; N^{n-1}, N'^{n-1}) = F^{-1}(N \times ([0, 1]; \{0\}, \{1\})) \subset (W^{n+1}; M^n, M'^n)$$

a codimension 1 cobordism. If $x \in \text{UNil}_{n+1}(\Phi) \subseteq L_{n+1}(X)$ the surgery obstruction may be identified with the structure invariant of h and also with the codimension 1 splitting obstruction

$$x = \sigma_*(F, B) = s(f) = s_Y(f)$$

$$\in \text{im}(\text{UNil}_{n+1}(\Phi) \subseteq L_{n+1}(X)) = \text{im}(\text{UNil}_{n+1}(\Phi) \subseteq \mathbb{S}_{n+1}(X))$$

$$= \text{im}(\text{UNil}_{n+1}(\Phi) \subseteq LS_{n-1}(\Phi)) .$$

The identification space

$$P = W \cup_{1 \sqcup f} M \times [0, 1]$$

is an $(n+1)$-dimensional geometric Poincaré complex with a reference map $e : P \longrightarrow X$ such that

$$Q = e^{-1}(Y) = V \cup_{1 \sqcup g} N \times [0, 1] \subset P$$

is a codimension 1 normal subcomplex, with

$$(g, c) = f| : N'^{n-1} = f^{-1}(N) \longrightarrow N^{n-1}$$

the normal map of compact $(n-1)$-dimensional manifolds defined by restriction. The element $x \in \text{UNil}_{n+1}(\Phi)$ may also be identified with the image of the total surgery obstruction of P, and with the codimension 1 Poincaré transversality obstruction to making $Q \subset P$ a codimension 1 Poincaré subcomplex

$$x = [s(P)] = s_Y(P) \in \text{UNil}_{n+1}(\Phi) .$$

EXAMPLE 23.5A Let $Y = \{\text{pt.}\} \subset X = B(\mathbb{Z}_2 * \mathbb{Z}_2) = \mathbb{RP}^\infty \vee \mathbb{RP}^\infty$. Cappell [22], [23] constructed non-trivial elements

$$x \neq 0 \in \mathrm{UNil}_{4k+2}(\Phi) \subset L_{4k+2}(\mathbb{Z}[\mathbb{Z}_2 * \mathbb{Z}_2]),$$

and used them to obtain homotopy equivalences of compact $(4k+1)$-dimensional manifolds

$$f : M^{4k+1} \longrightarrow \mathbb{RP}^{4k+1} \# \mathbb{RP}^{4k+1} \quad (k \geq 1)$$

which do not split along the separating codimension 1 $4k$-sphere $S^{4k} \subset \mathbb{RP}^{4k+1} \# \mathbb{RP}^{4k+1}$, with

$$s(f) = x \neq 0 \in \mathrm{UNil}_{4k+2}(\Phi) = LS_{4k}(\Phi) \subset \mathbb{S}_{4k+2}(B(\mathbb{Z}_2 * \mathbb{Z}_2)).$$

□

EXAMPLE 23.5B Let $X = Y \times S^1$, $\pi_1(Y) = \pi$, so that

$$\pi_1(X) = \pi \times \mathbb{Z}, \quad \mathbb{Z}[\pi_1(X)] = \mathbb{Z}[\pi][z, z^{-1}] \quad (\bar{z} = z^{-1}).$$

The algebraic splitting theorem of Ranicki [137]

$$L_n^h(\mathbb{Z}[\pi][z, z^{-1}]) = L_n^h(\mathbb{Z}[\pi]) \oplus L_{n-1}^p(\mathbb{Z}[\pi])$$

extends to an algebraic splitting theorem

$$\mathbb{S}_n(X) = \mathbb{S}_n(Y) \oplus \mathbb{S}_{n-1}^p(Y)$$

with $\mathbb{S}_*^p(Y)$ the projective $\mathbb{S}$-groups defined to fit into the exact sequence

$$\cdots \longrightarrow H_n(Y; \mathbb{L}.) \longrightarrow L_n^p(\mathbb{Z}[\pi_1(Y)]) \longrightarrow \mathbb{S}_n^p(Y) \longrightarrow H_{n-1}(Y; \mathbb{L}.) \longrightarrow \cdots.$$

(See Appendix C for more on $\mathbb{S}_*^p$). The UNil-groups vanish in this case, and the codimension 1 splitting obstruction groups are given by

$$LS_*(\Phi) = \widehat{H}^{*+1}(\mathbb{Z}_2; \widetilde{K}_0(\mathbb{Z}[\pi])),$$

with an exact sequence

$$\cdots \longrightarrow \mathbb{S}_n(Y) \oplus \mathbb{S}_{n-1}(Y) \longrightarrow \mathbb{S}_n(X) \longrightarrow LS_{n-2}(\Phi)$$
$$\longrightarrow \mathbb{S}_{n-1}(Y) \oplus \mathbb{S}_{n-2}(Y) \longrightarrow \cdots.$$

The codimension 1 splitting obstruction along $Y \times \{*\} \subset X = Y \times S^1$ of a homotopy equivalence of compact $(n-1)$-dimensional manifolds $f : M' \longrightarrow M$ with respect to a map $M \longrightarrow X$ is the image of the structure invariant $s(f) \in \mathbb{S}_n(X)$

$$s_Y(f) = [s(f)] = [B\tau(f)] \in LS_{n-2}(\Phi) = \widehat{H}^{n-1}(\mathbb{Z}_2; \widetilde{K}_0(\mathbb{Z}[\pi]))$$

with $B: Wh(\pi \times \mathbb{Z}) \longrightarrow \widetilde{K}_0(\mathbb{Z}[\pi])$ the Bass–Heller–Swan projection, as in the splitting theorem of Farrell and Hsiang [45]. See Milgram and Ranicki [107], Ranicki [146] for a chain complex treatment of this codimension 1 splitting obstruction, and the extension to lower K- and L-theory.

□

In case (C) if X, Y are connected and $\pi_1(X) \cong \pi_1(Y)$ then Z is connected and $\pi_1(Z) \cong \pi_1(S(\xi))$ is the fundamental group of a nontrivial double cover $S(\xi)$ of Y, so that $\pi_1(X) = \pi$ is an extension of $\pi_1(Z) = \pi'$ by $\mathbb{Z}_2$

$$\{1\} \longrightarrow \pi' \xrightarrow{\ i\ } \pi \xrightarrow{\ \xi\ } \mathbb{Z}_2 \longrightarrow \{1\} \ .$$

It is necessary to use the nonorientable version of the theory here (cf. Appendix A). Given a choice of orientation map $w\colon \pi \longrightarrow \mathbb{Z}_2$ let

$$w' \ = \ wi \ : \ \pi' \longrightarrow \mathbb{Z}_2 \ .$$

The LS-groups are the generalized Browder–Livesay LN-groups of Wall [176, 12C], with

$$LS_*(\Phi) \ = \ LN_*(i^w\colon \mathbb{Z}[\pi']^{w'} \to \mathbb{Z}[\pi]^w) \ = \ L_{*+2}(\mathbb{Z}[\pi'], \alpha)$$

for an appropriate 'antistructure' α on $\mathbb{Z}[\pi']$ depending on the choice of w. (The isomorphism $L_*(i_!^+) \cong L_{*+1}(i_!^-)$ was obtained by Wall [176, 12.9.2] in the split case $\pi' = \pi \times \mathbb{Z}_2$, and by Hambleton [62] in general. See 22.22 for a discussion of this phenomenon in the split case.) See Ranicki [143, §7.6] for the chain complex treatment. The map

$$i^! t \ : \ L_n(\mathbb{Z}[\pi]^w) \longrightarrow LN_{n-2}(i^w) \ = \ L_n(\mathbb{Z}[\pi'], \alpha) \ ; \ (C, \psi) \longrightarrow (i^! C, i^! t\,\psi)$$

in the exact sequence

$$\ldots \longrightarrow L_n((i^w)^!\colon \mathbb{Z}[\pi]^{w\xi} \to \mathbb{Z}[\pi']^{w'}) \longrightarrow L_n(\mathbb{Z}[\pi]^w)$$
$$\xrightarrow{\ i^! t\ } LN_{n-2}(i^w) \longrightarrow L_{n-1}((i^w)^!) \longrightarrow \ldots$$

sends an n-dimensional quadratic Poincaré complex (C, ψ) over $\mathbb{Z}[\pi]^w$ to the n-dimensional quadratic Poincaré complex $(i^! C, i^! t\,\psi)$ over $(\mathbb{Z}[\pi'], \alpha)$, for some fixed choice of $t \in \pi \backslash \pi'$. Similarly for the map

$$i^! t \ : \ VL^n(B\pi^w) \ = \ VL^n(\mathbb{Z}[\pi]^w) \longrightarrow LN_{n-2}(i^w) \ ; \ (C, \phi) \longrightarrow (i^! C, i^! t\,\phi)$$

in the exact sequence

$$\ldots \longrightarrow VL^n((i^w)^!) \longrightarrow VL^n(\mathbb{Z}[\pi]^w) \xrightarrow{\ i^! t\ } LN_{n-2}(i^w)$$
$$\longrightarrow VL^{n-1}((i^w)^!) \longrightarrow \ldots \ .$$

The visible symmetric structure $\phi \in VQ^n(C)$ determines the α-twisted quadratic structure $i^! t\,\phi \in Q_n(i^! C, \alpha)$ by the algebraic analogue of the 'antiquadratic construction' of [143, pp. 687–735].

EXAMPLE 23.5C Let $Y = \mathbb{RP}^{\infty-1} \subset X = \mathbb{RP}^\infty$, with the oriented involution $w = +$ on

$$\mathbb{Z}[\pi_1(X)] \ = \ \mathbb{Z}[\mathbb{Z}_2] \ = \ \mathbb{Z}[T]/(T^2 - 1) \ ,$$

so that $\pi' = \{1\}$, $\pi = \mathbb{Z}_2$, $t = T$, $w(T) = +1$. The codimension 1 splitting obstruction groups in this case are given by Wall [176, 13A.10] to be

$$LS_n(\Phi) \ = \ LN_n(i^+\colon \mathbb{Z} \to \mathbb{Z}[\mathbb{Z}_2]^+) \ = \ L_{n+2}(\mathbb{Z}) \ .$$

From the tabulation of $A \colon H_*(B\mathbb{Z}_2^+; L.(\mathbb{Z})) \longrightarrow L_*(\mathbb{Z}[\mathbb{Z}_2]^+)$ in 9.17 the quadratic $\mathbb{S}$-groups of $B\mathbb{Z}_2 = \mathbb{RP}^\infty$ in the oriented case are given by

$$\mathbb{S}_n(B\mathbb{Z}_2^+) \ = \ LN_{n-2}(i^+) \oplus \begin{cases} \displaystyle\sum_{k \neq -1} H_{n-k}(B\mathbb{Z}_2^+; L_{k-1}(\mathbb{Z})) \\ \displaystyle\sum_{k \in \mathbb{Z}} H_{n-k}(B\mathbb{Z}_2^+; L_{k-1}(\mathbb{Z})) \end{cases}$$

$$\text{if } n \ \equiv \ \begin{cases} 0 \\ 1 \end{cases} \pmod 2 \ .$$

The map

$$i^{\,!}t \colon VL^{4k}(B\mathbb{Z}_2^+) \ \longrightarrow \ \mathbb{S}_{4k}(B\mathbb{Z}_2^+) \ \longrightarrow \ LN_{4k-2}(i^+) \ = \ L_{4k}(\mathbb{Z})$$

sends a $4k$-dimensional visible symmetric Poincaré complex (C, ϕ) over $\mathbb{Z}[\mathbb{Z}_2]^+$ to

$$i^{\,!}t\,(C, \phi) \ = \ (1/8)\,\text{signature}\,(C, i^{\,!}t\,\phi) \in L_{4k}(\mathbb{Z}) \ = \ \mathbb{Z} \ .$$

As in 9.17 let

$$s_\pm(C, \phi) \ = \ \text{signature}\,j_\pm(C, \phi) \in L^{4k}(\mathbb{Z}) \ = \ \mathbb{Z} \ ,$$

with

$$j_\pm \colon \mathbb{Z}[\mathbb{Z}_2] \ \longrightarrow \ \mathbb{Z} \ ; \ a + bT \ \longrightarrow \ a \pm b \ .$$

For any $a + bT \in \mathbb{Z}[\mathbb{Z}_2]$ the eigenvalues of

$$i^{\,!}t\,(a + bT) \ = \ \begin{pmatrix} b & a \\ a & b \end{pmatrix} \ : \ i^{\,!}\mathbb{Z}[\mathbb{Z}_2] \ = \ \mathbb{Z} \oplus \mathbb{Z} \ \longrightarrow \ \mathbb{Z} \oplus \mathbb{Z}$$

are $j_\pm(b + aT) = b \pm a$, so that

$$i^{\,!}t\,(C, \phi) \ = \ (s_+(C, \phi) - s_-(C, \phi))/8 \in L_{4k}(\mathbb{Z}) \ = \ \mathbb{Z} \ .$$

If $f \colon M'^{4k-1} \longrightarrow M^{4k-1}$ is a homotopy equivalence of oriented compact $(4k-1)$-dimensional manifolds and $e \colon M \longrightarrow \mathbb{RP}^\infty$ classifies an oriented double cover $\overline{M} = e^* S^\infty$ then the codimension 1 splitting obstruction of the structure invariant $s(f) \in \mathbb{S}_{4k}(B\mathbb{Z}_2^+)$ is just the desuspension invariant of Browder and Livesay [18]

$$[s(f)] \ = \ i^{\,!}t\,\sigma_*(g, c) \ = \ (1/8)\,\text{signature}\,(i^{\,!}C, i^{\,!}t\,\psi)$$

$$\in LN_{4k}(i^-) \ = \ LN_{4k-2}(i^+) \ = \ L_{4k}(\mathbb{Z}) \ = \ \mathbb{Z} \ ,$$

with $\sigma_*(g, c) = (C, \psi)$ the kernel $(4k-2)$-dimensional quadratic Poincaré complex over $\mathbb{Z}[\mathbb{Z}_2]^-$ of the normal map of nonorientable compact $(4k-2)$-dimensional manifolds

$$(g, c) \ = \ f| \colon N'^{4k-2} \ = \ (ef)^{-1}(\mathbb{RP}^{\infty-1}) \ \longrightarrow \ N^{4k-2} \ = \ e^{-1}(\mathbb{RP}^{\infty-1})$$

obtained by codimension 1 transversality at $\mathbb{RP}^{\infty-1} \subset \mathbb{RP}^\infty$. See Lopez de Medrano [95] for the surgery classification of involutions on simply connected high-dimensional compact manifolds. The splitting obstruction groups $LN_*(i^\pm \colon \mathbb{Z} \to \mathbb{Z}[\mathbb{Z}_2]^\pm)$ are denoted by $BL_{*+1}(\pm)$ in [95].

For the remainder of 23.5C let $(W, \partial W)$ be an oriented $4k$-dimensional geometric Poincaré pair with an oriented double cover $(\overline{W}, \partial \overline{W})$.

The multisignature components of the $4k$-dimensional visible symmetric complex $(\Delta(\overline{W}, \partial \overline{W}), \phi)$ over $\mathbb{Z}[\mathbb{Z}_2]^+$

$$s_{\pm}(W) \;=\; s_{\pm}(\Delta(\overline{W}, \partial \overline{W}), \phi) \;=\; \text{signature}\, j_{\pm}(\Delta(\overline{W}, \partial \overline{W}), \phi) \in \mathbb{Z}$$

are such that

$$\text{signature}\,(W) \;=\; s_+(W) \;, \quad \text{signature}\,(\overline{W}) \;=\; s_+(W) + s_-(W) \in \mathbb{Z} \;.$$

The $\mathbb{Z}_2$-signature of $\overline{W}$ (22.1) is the signature of the $4k$-dimensional quadratic complex $(\Delta(\overline{W}, \partial \overline{W}), i^{\,!} t\, \phi)$ over $\mathbb{Z}$

$$\text{signature}\,(\overline{W}, T) \;=\; \text{signature}\,(\Delta(\overline{W}, \partial \overline{W}), i^{\,!} t\, \phi)$$

$$=\; s_+(W) - s_-(W) \in 8\mathbb{Z} \subset \mathbb{Z} \;.$$

The signature of the cover fails to be multiplicative by

$$2\,\text{signature}\,(W) - \text{signature}\,(\overline{W}) \;=\; s_+(W) - s_-(W)$$

$$=\; \text{signature}\,(\overline{W}, T) \in 8\mathbb{Z} \subset \mathbb{Z} \;.$$

The signature defect for finite covers of compact $4k$-dimensional manifolds with boundary has been studied by Hirzebruch [74] and his school (Jänich, Knapp, Kreck, Neumann, Ossa, Zagier) using the methods of the Atiyah-Singer index theorem, which also apply in the case $k = 1$.

Let $\partial W = \emptyset$, so that W is an oriented finite $4k$-dimensional geometric Poincaré complex with an oriented double cover $\overline{W}$. The total surgery obstruction $s(W) \in \mathbb{S}_{4k}(W)$ has image the codimension 1 Poincaré transversality obstruction

$$s_Y(W) \;=\; [s(W)] \;=\; \text{signature}\,(\overline{W}, T)/8 \;=\; (s_+(W) - s_-(W))/8$$

$$=\; (2\,\text{signature}\,(W) - \text{signature}\,(\overline{W}))/8$$

$$\in LN_{4k-2}(i^+) \;=\; L_{4k}(\mathbb{Z}) \;=\; \mathbb{Z} \;,$$

which has been studied by Hambleton and Milgram [65].

If $(W^{4k}, \partial W)$ is an oriented compact $4k$-dimensional manifold with boundary and $(f, b) \colon (W'^{4k}, \partial W') \longrightarrow (W, \partial W)$ is a normal map which restricts to a homotopy equivalence on the boundaries

$$h \;=\; \partial f \colon \partial W' \overset{\simeq}{\longrightarrow} \partial W$$

then the rel ∂ surgery obstruction is given by

$$\sigma_*(f, b) \;=\; (\, s_+(\overline{W}') - s_+(\overline{W}), \, s_-(\overline{W}') - s_-(\overline{W}) \,)$$

$$\in L_{4k}(\mathbb{Z}[\mathbb{Z}_2]^+) \;=\; \mathbb{Z} \oplus \mathbb{Z} \;.$$

The identification space

$$P \;=\; W' \cup_h -W$$

is an oriented 4k-dimensional geometric Poincaré complex with an oriented double cover $\overline{P}$ classified by a map $P \longrightarrow X = \mathbb{RP}^\infty$. The structure invariant $s(h) \in \mathbb{S}_{4k}(\partial W)$ and the total surgery obstruction $s(P) \in \mathbb{S}_{4k}(P)$ have the same image

$$[s(h)] = [s(P)] = [\sigma_*(f,b)] \in \operatorname{im}(L_{4k}(\mathbb{Z}[\mathbb{Z}_2]^+) \longrightarrow \mathbb{S}_{4k}(B\mathbb{Z}_2^+)),$$

and the codimension 1 splitting obstruction along $Y = \mathbb{RP}^{\infty-1} \subset X$ is given by the Browder–Livesay invariant

$$
\begin{aligned}
s_Y(h) = s_Y(P) &= \operatorname{signature}(\overline{P},T)/8 \\
&= (2\operatorname{signature}(P) - \operatorname{signature}(\overline{P}))/8 \\
&= (\operatorname{signature}(\overline{W}',T) - \operatorname{signature}(\overline{W},T))/8 \\
&\in LN_{4k-2}(i^+) = L_{4k}(\mathbb{Z}) = \mathbb{Z}.
\end{aligned}
$$

If $(W^{4k}, \partial W)$ is an oriented compact 4k-dimensional manifold with boundary then the classifying map

$$e : (W, \partial W) \longrightarrow \mathbb{RP}^\infty$$

for the double cover $(\overline{W}^{4k}, \partial\overline{W})$ can be made transverse regular at $\mathbb{RP}^{\infty-1} \subset \mathbb{RP}^\infty$ with

$$(V^{4k-1}, \partial V) = e^{-1}(\mathbb{RP}^{\infty-1}) \subset (W^{4k}, \partial W)$$

a codimension 1 nonorientable submanifold. The double cover $\overline{V}$ of V is oriented, and separates $\overline{W}$ as

$$\overline{W} = \overline{W}^+ \cup_{\overline{V}} \overline{W}^-$$

with $T(\overline{W}^\pm) = \overline{W}^\mp$. The singular symmetric forms on $H_{2k}(\overline{W}^\pm)$ and $H_{2k}(W)$ have the same radical quotients, so that

$$\operatorname{signature}(\overline{W}^\pm) = \operatorname{signature}(W) \in \mathbb{Z}.$$

The signature defect is

$$
\begin{aligned}
\operatorname{signature}(\overline{W},T) &= 2\operatorname{signature}(W) - \operatorname{signature}(\overline{W}) \\
&= \operatorname{signature}(\overline{W}^+) + \operatorname{signature}(\overline{W}^-) - \operatorname{signature}(\overline{W}) \\
&= \text{the signature non-additivity invariant of Wall [173]} \\
&= \text{the } \rho\text{-invariant of Wall [176, 13B.2]} \in \mathbb{Z}.
\end{aligned}
$$

This invariant depends only on the $(4k-1)$-dimensional boundary manifold ∂W, for if $(W', \partial W')$ is another oriented manifold with the same boundary $\partial W' = \partial W$ the union $P^{4k} = W' \cup_\partial -W$ is a closed oriented 4k-dimensional manifold such that

$$\text{signature}\,(\overline{W}',T) - \text{signature}\,(\overline{W},T)$$
$$= \text{signature}\,(\overline{P},T)$$
$$= 2\,\text{signature}\,(P) - \text{signature}\,(\overline{P})$$
$$= 0 \in \mathbb{Z}$$

by Novikov additivity and the multiplicativity of the signature for finite covers of manifolds. See Hirzebruch and Zagier [75, 4.2] and Neumann [117] for the connections with the Atiyah–Patodi–Singer α-, γ- and η-invariants of odd-dimensional manifolds.

□

§24. Higher signatures

The higher signatures are non-simply connected generalizations of the $\mathcal{L}$-genus, corresponding to the rational part of the canonical $\mathbb{L}^{\cdot}$-orientation of compact topological manifolds. A general discussion of the connections between the algebraic L-theory assembly map and the Novikov conjecture on the homotopy invariance of the higher signatures is followed by the particular discussion of the homotopy types of the classifying spaces $B\pi$ of Poincaré duality groups satisfying the conjecture. The total surgery obstruction of such geometric Poincaré complexes is detected by codimension n signatures.

DEFINITION 24.1 (i) The *higher signature* of an oriented compact n-dimensional manifold M^n with respect to a cohomology class $x \in H^{n-4*}(M;\mathbb{Q})$ is

$$\sigma_x(M) = \langle \mathcal{L}(M) \cup x, [M]_{\mathbb{Q}} \rangle \in \mathbb{Q} ,$$

with $\mathcal{L}(M) = \mathcal{L}(\tau_M) = \mathcal{L}^{-1}(\nu_M) \in H^{4*}(M;\mathbb{Q})$ the $\mathcal{L}$-genus and $[M]_{\mathbb{Q}} \in H_n(M;\mathbb{Q})$ the rational fundamental class.
(ii) A higher signature $\sigma_x(M)$ is *universal* if $x = f^*y$ is the pullback of a class $y \in H^{n-4*}(B\pi;\mathbb{Q})$ $(\pi = \pi_1(M))$ along a classifying map $f: M \longrightarrow B\pi$ for the universal cover of M.

□

A universal higher signature $\sigma_{f^*y}(M)$ is usually written as $\sigma_y(M)$.

EXAMPLE 24.2 (i) For $x = 1 \in H^0(M;\mathbb{Q})$ the universal higher signature $\sigma_x(M) \in \mathbb{Q}$ of an oriented compact n-dimensional manifold M with $n \equiv 0 \pmod 4$ is just the ordinary signature, since by the Hirzebruch formula

$$\sigma_1(M) = \langle \mathcal{L}(M), [M]_{\mathbb{Q}} \rangle = \text{signature}\,(M) \in \mathbb{Z} \subset \mathbb{Q} .$$

If $n \not\equiv 0 \pmod 4$ then $\sigma_1(M) = 0$.
(ii) Given an oriented compact n-dimensional manifold M^n and an oriented compact $4k$-dimensional submanifold $N^{4k} \subseteq M^n$ write the inverse $\mathcal{L}$-genus of the normal block bundle $\nu_{N \subseteq M}: N \longrightarrow B\widetilde{STOP}(n-4k)$ as

$$\mathcal{L}(N,M) = \mathcal{L}^{-1}(\nu_{N \subseteq M}) \in H^{4*}(N;\mathbb{Q}) .$$

Let $i: N \longrightarrow M$ be the inclusion, and let

$$x = i^! \mathcal{L}(N,M) \in H^{n-4k+4*}(M;\mathbb{Q})$$

be the image of $\mathcal{L}(N,M)$ under the Umkehr map

$$i^!: H^{4*}(N;\mathbb{Q}) \cong H_{4k-4*}(N;\mathbb{Q}) \xrightarrow{i_*} H_{4k-4*}(M;\mathbb{Q}) \cong H^{n-4k+4*}(M;\mathbb{Q}) .$$

It follows from the identity $\nu_N = \nu_{N \subseteq M} \oplus i^*\nu_M: N \longrightarrow BSTOP$ that

$$\mathcal{L}(N) = \mathcal{L}(N,M) \cup i^* \mathcal{L}(M) \in H^{4*}(N;\mathbb{Q}) .$$

The corresponding higher signature of M is the ordinary signature of N

$$\sigma_x(M) = \langle \mathcal{L}(M) \cup x, [M]_{\mathbb{Q}} \rangle = \langle \mathcal{L}(N), [N]_{\mathbb{Q}} \rangle = \text{signature}\,(N) \in \mathbb{Z} .$$

The special case $N^{4k} = M^n$ is (i), with $x = 1 \in H^0(M;\mathbb{Q}) = \mathbb{Q}$. In the special case $N^0 = \{\text{pt.}\} \subset M^n$ the element $x = i^!(1) = 1 \in H^n(M;\mathbb{Q}) = \mathbb{Q}$ is such that $\sigma_x(M) = \text{signature}(N) = 1 \in \mathbb{Z} \subset \mathbb{Q}$. As in Thom's combinatorial construction the $\mathcal{L}$-genus $\mathcal{L}(M) \in H^{4*}(M;\mathbb{Q})$ is characterized by the signatures of compact submanifolds $N^{4k} \subseteq M^n$ with trivial normal bundle $\nu_{N\subseteq M} = \epsilon^{n-4k} : N \longrightarrow B\widetilde{STOP}(n-4k)$

$$\mathcal{L}(M)^* : H_{4*}(M;\mathbb{Q}) \cong H^{n-4*}(M;\mathbb{Q}) \longrightarrow \mathbb{Q} ; \ x \longrightarrow \sigma_x(M) = \text{signature}(N)$$

with $i_*[N]_\mathbb{Q} \in H_{4k}(M;\mathbb{Q})$ the Poincaré dual of $x = i^!(1) \in H^{n-4k}(M;\mathbb{Q})$ and $\mathcal{L}(N,M) = 1$. In general, these higher signatures are not universal.

(iii) The cap product of the canonical $\mathbb{L}$-homology class $[B]_\mathbb{L} \in H_n(B;\mathbb{L}^{\cdot})$ of a compact n-dimensional manifold B and an $\mathbb{L}$-cohomology class $\Gamma \in H^{-m}(B;\mathbb{L}^{\cdot})$ is an $\mathbb{L}$-homology class $[B]_\mathbb{L} \cap \Gamma \in H_{m+n}(B;\mathbb{L}^{\cdot})$ (Appendix B). If $m = 2i$, $n = 2j$ with $i + j \equiv 0 \pmod 2$ the product $[B]_\mathbb{L} \cap \Gamma$ determines a nonsingular symmetric form ϕ on the jth cohomology $H^j(B;\{H^i(\Gamma)\})$ of B with coefficients in the flat bundle $H^i(\Gamma)$ of nonsingular $(-)^i$-symmetric forms over $\mathbb{Z}$. The signature of this form is given by the simply connected assembly

$$\text{signature}\,(H^j(B;\{H^i(\Gamma)\}),\phi) = A([B]_\mathbb{L} \cap \Gamma) \in L^{2(i+j)}(\mathbb{Z}) = \mathbb{Z}\,,$$

and hence as a universal higher signature

$$\text{signature}\,(H^j(B;\{H^i(\Gamma)\}),\phi) = \sigma_x(B) \in \mathbb{Z} \subset \mathbb{Q}$$

with $x = \widetilde{\text{ch}}([\Gamma]_\mathbb{K}) \in H^{2*}(B\pi;\mathbb{Q})$ the modified Chern character of the topological K-theory signature $[\Gamma]_\mathbb{K} \in \begin{cases} KO(B) \\ KU(B) \end{cases}$ (for $i \equiv \begin{cases} 0 \\ 1 \end{cases} \pmod 2$) determined by the action of $\pi_1(B) = \pi$ on the local system of $(-)^j$-symmetric forms on $H^j(\Gamma)$, as in the work of Atiyah [6], Lusztig [98] and Meyer [104] on the non-multiplicativity of the signature of a fibre bundle (cf. 21.3). The signature of a compact $2(i+j)$-dimensional manifold E which is the total space of a fibre bundle $F \longrightarrow E \longrightarrow B$ with the base B a compact $2j$-dimensional manifold and the fibre F a compact $2i$-dimensional manifold is given by the higher signature

$$\text{signature}\,(E) = \text{signature}\,(H^j(B;\{H^i(\Gamma)\}),\phi) = \sigma_x(B) \in L^{2(i+j)}(\mathbb{Z}) = \mathbb{Z}$$

with $\Gamma \in H^{-2i}(B;\mathbb{L}^{\cdot})$ such that $H^*(\Gamma) = H^*(F)$, as in Lück and Ranicki [97].

$\square$

PROPOSITION 24.3 (i) *The canonical $\mathbb{L}^{\cdot}$-orientation $[M]_\mathbb{L} \in H_n(M;\mathbb{L}^{\cdot})$ of an oriented compact n-dimensional manifold M determines and is determined modulo torsion by the higher signature map*

$$H^{n-4*}(M;\mathbb{Q}) \longrightarrow \mathbb{Q} ; \ x \longrightarrow \sigma_x(M)\,.$$

(ii) *The normal invariant $[f,b]_\mathbb{L} \in H_n(M;\mathbb{L}_{\cdot})$ of a normal map $(f,b): N \longrightarrow M$ of closed oriented n-dimensional manifolds determines and is determined* *modulo torsion by the differences of the higher signatures*

$$H^{n-4*}(M;\mathbb{Q}) \longrightarrow \mathbb{Q} \; ; \; x \longrightarrow \sigma_{f*x}(N) - \sigma_x(M) \; .$$

PROOF (i) Both the higher signatures and $[M]_\mathbb{L} \otimes \mathbb{Q}$ determine and are determined by the signatures of compact submanifolds of M, with

$$\sigma_x(M) \; = \; (\mathcal{L}(M) \cap [M]_\mathbb{Q}) \cap x \; = \; ([M]_\mathbb{L} \otimes \mathbb{Q}) \cap x \in H_0(B\pi;\mathbb{Q}) \; = \; \mathbb{Q}$$

for any $x \in H^{n-4*}(M;\mathbb{Q})$. The universal coefficient isomorphism

$$\mathrm{Hom}_\mathbb{Q}(H^{n-4*}(M;\mathbb{Q}),\mathbb{Q}) \; \cong \; H_{n-4*}(M;\mathbb{Q})$$

sends the higher signature map $x \longrightarrow \sigma_x(M)$ to the element

$$[M]_\mathbb{L} \otimes \mathbb{Q} \; = \; \mathcal{L}(M) \cap [M]_\mathbb{Q} \in H_{n-4*}(M;\mathbb{Q}) \; .$$

(ii) This follows from (i), since $1 + T: H_*(M;\mathbb{L}.) \longrightarrow H_*(M;\mathbb{L}\cdot\langle 1 \rangle(\mathbb{Z}))$ is an isomorphism modulo 8-torsion, and

$$(1 + T)[f,b]_\mathbb{L} \; = \; f_*[N]_\mathbb{L} - [M]_\mathbb{L} \in H_n(M;\mathbb{L}\cdot\langle 1 \rangle(\mathbb{Z})) \; .$$

□

CONJECTURE 24.4 (Novikov) *The universal higher signatures are homotopy invariant for any group π.*

□

Write the quadratic L-theory assembly map for the classifying space $B\pi$ of a group π as

$$A_\pi \; : \; H_*(B\pi;\mathbb{L}.(\mathbb{Z})) \longrightarrow L_*(\mathbb{Z}[\pi]) \; .$$

PROPOSITION 24.5 *The following versions of the Novikov conjecture are equivalent for any group π:*

(i) *the universal higher signatures are homotopy invariant, i.e. for any homotopy equivalence $h: N \longrightarrow M$ of oriented compact n-dimensional manifolds with $\pi_1(M) = \pi_1(N) = \pi$ and every $x \in H^{n-4*}(B\pi;\mathbb{Q})$*

$$\sigma_x(M) \; = \; \sigma_x(N) \in \mathbb{Q} \; ,$$

(ii) *the rational canonical $\mathbb{L}\cdot$-homology classes are homotopy invariant, i.e. for any homotopy equivalence $h: N \longrightarrow M$ of oriented compact n-dimensional manifolds with $\pi_1(M) = \pi_1(N) = \pi$*

$$[M]_\mathbb{L} \otimes \mathbb{Q} \; = \; h_*[N]_\mathbb{L} \otimes \mathbb{Q} \in H_n(M;\mathbb{L}\cdot) \otimes \mathbb{Q} \; ,$$

(iii) *the rational assembly map*

$$A_\pi \otimes \mathbb{Q} \; : \; H_*(B\pi;\mathbb{L}.(\mathbb{Z})) \otimes \mathbb{Q} \; = \; \sum_{j \in \mathbb{Z}} H_{*-4j}(B\pi;\mathbb{Q}) \longrightarrow L_*(\mathbb{Z}[\pi]) \otimes \mathbb{Q}$$

is injective,

(iv) *the dual of the rational assembly map*

$$(A_\pi \otimes \mathbb{Q})^* \; : \; \mathrm{Hom}_\mathbb{Q}(L_*(\mathbb{Z}[\pi]) \otimes \mathbb{Q}, \mathbb{Q}) \longrightarrow \sum_{j \in \mathbb{Z}} H^{*-4j}(B\pi;\mathbb{Q})$$

is surjective.

PROOF (i) $\iff$ (ii) Working as in the proof of 24.3 (i) the image $f * [M]_{\mathbb{L}} \in H_n(B\pi; \mathbb{L}^{\cdot})$ determines and is determined modulo torsion by the universal higher signatures $\sigma_x(M) \in \mathbb{Q}$ ($x \in H^{n-4*}(B\pi; \mathbb{Q})$).

(iii) $\implies$ (ii) Symmetric and quadratic L-theory only differ in 2-primary torsion, so (iii) is equivalent to the injectivity of the rational assembly map in symmetric L-theory

$$A_\pi \otimes \mathbb{Q} : H_*(B\pi; \mathbb{L}^{\cdot}) \otimes \mathbb{Q} = \sum_{j \in \mathbb{Z}} H_{*-4j}(B\pi; \mathbb{Q}) \longrightarrow L^*(\mathbb{Z}[\pi]) \otimes \mathbb{Q} .$$

For any compact n-dimensional manifold M with $\pi_1(M) = \pi$ and classifying map $f: M \longrightarrow B\pi$ the assembly of $f_*[M]_{\mathbb{L}} \in H_n(B\pi; \mathbb{L}^{\cdot})$ is the homotopy invariant symmetric signature

$$A_\pi f_*[M]_{\mathbb{L}} = A[M]_{\mathbb{L}} = \sigma^*(M) \in L^n(\mathbb{Z}[\pi]) .$$

(ii) $\implies$ (iii) Every element in $\mathbb{S}_{*+1}(B\pi)$ is the image of the structure invariant $s(h)$ of a homotopy equivalence $h: N \longrightarrow M$ of closed manifolds with fundamental group π. The kernel of the assembly map

$$\ker(A_\pi: H_*(B\pi; \mathbb{L}.(\mathbb{Z})) \longrightarrow L_*(\mathbb{Z}[\pi])) = \operatorname{im}(\mathbb{S}_{*+1}(B\pi) \longrightarrow H_*(B\pi; \mathbb{L}.(\mathbb{Z})))$$

consists of the images of the normal invariants $[h]_{\mathbb{L}}$ of such homotopy equivalences h, which are given modulo 2-primary torsion (and a fortiori rationally) by the differences of the canonical $\mathbb{L}^{\cdot}$-homology classes

$$[h]_{\mathbb{L}} \otimes \mathbb{Q} = h_*[N]_{\mathbb{L}} \otimes \mathbb{Q} - [M]_{\mathbb{L}} \otimes \mathbb{Q}$$

$$\in H_*(B\pi; \mathbb{L}.(\mathbb{Z})) \otimes \mathbb{Q} = H_*(B\pi; \mathbb{L}^{\cdot}(\mathbb{Z})) \otimes \mathbb{Q} .$$

A cohomology class $x \in H^*(B\pi; \mathbb{Q})$ is such that the function $M \longrightarrow \sigma_x(M)$ is a homotopy invariant if and only if

$$\ker(A_\pi \otimes \mathbb{Q}) \cap x = 0 \in H_0(B\pi; \mathbb{Q}) = \mathbb{Q} .$$

(This is the case if and only if $x \in H^*(B\pi; \mathbb{Q})$ is in the image of the $\mathbb{Q}$-dual assembly map $(A_\pi \otimes \mathbb{Q})^*$.) If $\ker(A_\pi \otimes \mathbb{Q}) = 0$ then every class $x \in H^*(B\pi; \mathbb{Q})$ satisfies this condition.

(iii) $\iff$ (iv) Trivial.

$\square$

REMARK 24.6 The equivalence of the Novikov conjecture for π and the rational injectivity of the assembly map A_π was first established by Wall [176, §17H], Mishchenko and Solovev [115] and Kaminker and Miller [78].

$\square$

Only infinite groups π need be considered for the Novikov conjecture, since for finite π

$$H_*(B\pi; \mathbb{L}.(\mathbb{Z})) \otimes \mathbb{Q} = H_*(\{ \mathrm{pt.} \}; \mathbb{L}.(\mathbb{Z})) \otimes \mathbb{Q} = L_*(\mathbb{Z}) \otimes \mathbb{Q} ,$$

and $A_\pi \otimes \mathbb{Q}: L_*(\mathbb{Z}) \otimes \mathbb{Q} \longrightarrow L_*(\mathbb{Z}[\pi]) \otimes \mathbb{Q}$ is the injection induced by the inclusion $\mathbb{Z} \longrightarrow \mathbb{Z}[\pi]$.

REMARK 24.7 The Novikov conjecture for the free abelian groups $\mathbb{Z}^n$ ($n \geq$ 1) was proved (more or less explicitly) by Novikov [120], [121], Rohlin [149], Farrell and Hsiang [45], Kasparov [80], Lusztig [98], Shaneson [155], Ranicki [137], Cappell [24] using a variety of topological, analytic and algebraic methods. This case is especially significant, on account of the related properties of the n-torus $B\mathbb{Z}^n = T^n$ used in the work of Novikov [120] on the topological invariance of the rational Pontrjagin classes, and in the work of Kirby and Siebenmann [84] on topological manifolds.

$\square$

REMARK 24.8 Cappell [24] used codimension 1 splitting methods (§23) to construct a class of groups π satisfying the Novikov conjecture. The class is closed under free products with amalgamation and HNN extensions which are 'square root closed', and includes the trivial group $\{1\}$ and the free abelian groups $\mathbb{Z}^n$ ($n \geq 1$). See Stark [162] for an extension of the class.

$\square$

REMARK 24.9 The Novikov conjecture is related to Atiyah–Singer index theory, C^*-algebras, hyperbolic geometry, differential geometry, cyclic homology, equivariant and controlled topology. The following list of references is only a small sample of the literature: Connes and Moscovici [39], Farrell and Hsiang [46], Kasparov [81], Mishchenko and Fomenko [114], Rosenberg [150]. See Mishchenko [113] and Weinberger [179] for surveys.

$\square$

REMARK 24.10 In the analytic approaches to the Novikov conjecture the group ring $\mathbb{Z}[\pi]$ is embedded in the reduced C^*-algebra $C^*_r(\pi)$. The analytic index in $K_0(C^*_r(\pi))$ is identified with the image of the symmetric signature in $L^{2k}(\mathbb{Z}[\pi])$, using an isomorphism $L^{2k}(C^*_r(\pi)) \cong K_0(C^*_r(\pi))$ generalizing the multisignature (see Kaminker and Miller [79], for example). The algebraic L-theory assembly map A_π corresponds to a topological K-theory assembly map $\beta \colon K_*(B\pi) \longrightarrow K_*(C^*_r(\pi))$, and it is β which is proved to be a rational split injection in various cases.

$\square$

The simply connected L-groups are detected by the signatures of nonsingular symmetric and quadratic forms over $\mathbb{Z}$, with isomorphisms

$$L^0(\mathbb{Z}) \xrightarrow{\ \simeq\ } \mathbb{Z} \ ; \ (C, \phi) \longrightarrow \text{signature}\,(C, \phi) \ ,$$

$$L_0(\mathbb{Z}) \xrightarrow{\ \simeq\ } \mathbb{Z} \ ; \ (C, \psi) \longrightarrow (\text{signature}\,(C, \psi))/8 \ .$$

If $K = B\pi$ is an aspherical n-dimensional geometric Poincaré complex satisfying a strong form of the Novikov conjecture the total surgery obstruction

$$s(K) \in \mathbb{S}_n(K) = L_0(\mathbb{Z})$$

is now interpreted as the difference between local and global codimension n signatures.

In the first instance an equivalence is established between three formulations of the algebraic L-theory assembly maps being isomorphisms in the 4-periodic range.

LEMMA 24.11 *For any n-dimensional simplicial complex K the following three conditions are equivalent:*

$N(K)$: *the 0-connective quadratic $\mathbb{S}$-groups of K are such that*

$$\mathbb{S}_m\langle 0\rangle(\mathbb{Z}, K) \;=\; 0 \;\;\text{ for } m \geq n \;,$$

$N(K)_*$: *the 0-connective quadratic L-theory assembly maps*

$$A : H_m(K; \mathbb{L}.\langle 0\rangle(\mathbb{Z})) \;\longrightarrow\; L_m(\mathbb{Z}[\pi]) \;\;(\pi = \pi_1(K))$$

are isomorphisms for $m \geq n$,

$N(K)^*$: *the 0-connective visible symmetric L-theory assembly maps*

$$A : H_m(K; \mathbb{L}^{\cdot}\langle 0\rangle(\mathbb{Z})) \;\longrightarrow\; VL^m\langle 0\rangle(\mathbb{Z}, K)$$

are isomorphisms for $m \geq n$.

PROOF All the groups and maps involved are 4-periodic for dimension reasons and by the 4-periodicity of quadratic L-theory, except that the map $\mathbb{S}_n\langle 0\rangle(\mathbb{Z}, K) \longrightarrow \mathbb{S}_{n+4}\langle 0\rangle(\mathbb{Z}, K)$ may possibly fail to be onto. The cokernel of this map is isomorphic to the cokernel of the first map in the exact sequence

$$H_{n-1}(K; \mathbb{L}.\langle 0\rangle(\mathbb{Z})) \;\longrightarrow\; H_{n+3}(K; \mathbb{L}.\langle 0\rangle(\mathbb{Z})) \;\longrightarrow\; H_n(K; L_3(\mathbb{Z})) \;,$$

which is onto since $L_3(\mathbb{Z}) = 0$. Thus each of the conditions $N(K)$, $N(K)_*$, $N(K)^*$ is 4-periodic. The implications $N(K) \Longleftrightarrow N(K)_* \Longleftrightarrow N(K)^*$ now follow from the commutative braid of exact sequences

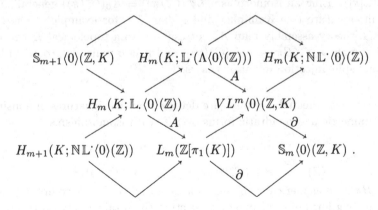

□

DEFINITION 24.12 An *n-dimensional Poincaré duality group* π is a group such that the classifying space $B\pi$ is an n-dimensional Poincaré space.

□

Poincaré duality groups are finitely presented, infinite and torsion-free.

DEFINITION 24.13 An *n-dimensional Novikov group* π is an n-dimensional Poincaré duality group such that the classifying space $K = B\pi$ satisfies any one of the three equivalent conditions $N(K)$, $N(K)_*$, $N(K)^*$ of 24.11.

□

The strong form of the Novikov conjecture (24.4) is that condition $N(K)$ holds for any n-dimensional Poincaré duality group π, and that $s(K) = 0 \in$ $\mathbb{S}_n(K) = L_0(\mathbb{Z})$, so that K is homotopy equivalent to an aspherical compact n-dimensional topological manifold. This includes the Borel conjecture concerning the rigidity of aspherical manifolds, since it implies that for any aspherical compact n-dimensional manifold M with $\pi_1(M) = \pi$, $M \simeq K$

$$\mathbb{S}^{TOP}(M) = \mathbb{S}_{n+1}(M) = \mathbb{S}_{n+1}(K) = \{0\} \ ,$$

so that any homotopy equivalence $f: N \longrightarrow M$ of compact aspherical manifolds is homotopic to a homeomorphism (at least for $n \geq 5$).

REMARK 24.14 Many examples of Novikov groups arise geometrically as the fundamental groups $\pi = \pi_1(M)$ of aspherical compact manifolds $M = B\pi$ with topological rigidity, such that $\mathbb{S}_\partial^{TOP}(M \times D^i, M \times S^{i-1}) = 0$ for $i \geq 0$. See Farrell and Hsiang [46], Farrell and Jones [47], [48], Ferry and Weinberger [51], Yamasaki [187]. The methods of controlled topology are particularly relevant here (see Appendix C).

□

For any n-dimensional simplicial complex K define also the 4-periodic condition:

$N(K)_\mathbb{Q}$: the rational 0-connective quadratic L-theory assembly maps
$$A \otimes \mathbb{Q} : H_m(K; \mathbb{L}.\langle 0 \rangle(\mathbb{Z})) \otimes \mathbb{Q} \longrightarrow L_m(\mathbb{Z}[\pi]) \otimes \mathbb{Q} \ (\pi = \pi_1(K))$$
are monomorphisms for $m \geq n$.

For $K = B\pi$ this is just condition 24.5 (iv), so that $N(B\pi)_\mathbb{Q}$ is equivalent to the Novikov conjecture (24.4). Davis [43, §11] has shown that $N(B\pi)_\mathbb{Q}$ is true for all the groups π with $B\pi$ the homotopy type of a finite complex if and only if $N(B\pi)_\mathbb{Q}$ is true for all the groups π with $B\pi$ the homotopy type of an aspherical compact topological manifold.

PROPOSITION 24.15 *If π is an n-dimensional Novikov group with classifying space $K = B\pi$ then*

$$\mathbb{S}_m(K) = H_m(K; L_0(\mathbb{Z})) = \begin{cases} L_0(\mathbb{Z}) & \text{if } m = n \\ 0 & \text{if } m \geq n+1 \end{cases} .$$

If $s(K) = 0 \in \mathbb{S}_n(K) = L_0(\mathbb{Z})$ then (at least for $n \geq 5$) the homotopy type of K contains an aspherical compact topological n-manifold M, and

$$\mathbb{S}_\partial^{TOP}(M \times D^i, M \times S^{i-1}) = \mathbb{S}_{n+i+1}(M) = \mathbb{S}_{n+i+1}(K) = 0 \ (i \geq 0) .$$

PROOF Immediate from the exact sequence given by 15.11 (iii)

$$\cdots \longrightarrow \mathbb{S}_{m+1}\langle 0 \rangle(\mathbb{Z}, K) \longrightarrow H_m(K; L_0(\mathbb{Z})) \longrightarrow \mathbb{S}_m(K)$$
$$\longrightarrow \mathbb{S}_m\langle 0 \rangle(\mathbb{Z}, K) \longrightarrow \cdots ,$$

and the identification in 18.5 of the Sullivan–Wall geometric surgery exact sequence with the algebraic surgery sequence. □

EXAMPLE 24.16 The free abelian group $\mathbb{Z}^n$ of rank n is an n-dimensional Novikov group, with classifying space $K(\mathbb{Z}^n, 1) = T^n$ the n-torus. The assembly maps are isomorphisms

$$A : H_m(T^n; \mathbb{L}^{\textstyle{\cdot}}\langle 0 \rangle(\mathbb{Z})) \xrightarrow{\simeq} VL^m\langle 0 \rangle(\mathbb{Z}, T^n) = L^m(\mathbb{Z}[\mathbb{Z}^n]) ,$$

$$A : H_m(T^n; \mathbb{L}_{\textstyle{\cdot}}\langle 0 \rangle(\mathbb{Z})) \xrightarrow{\simeq} L_m(\mathbb{Z}[\mathbb{Z}^n])$$

for $m \geq n$ by the Laurent polynomial extension splitting theorems of Shaneson [155], Wall [176, 13A.8], Novikov [121], Ranicki [137], Milgram and Ranicki [107] (and $Wh(\mathbb{Z}^n) = 0$), so that

$$\mathbb{S}_m\langle 0 \rangle(T^n) = 0 \text{ for } m \geq n .$$

T^n is a manifold, and

$$s(T^n) = 0 \in \mathbb{S}_n(T^n) = L_0(\mathbb{Z}),$$

$$\mathbb{S}_\partial^{TOP}(T^n \times D^i, T^n \times S^{i-1}) = \mathbb{S}_{n+i+1}(T^n) = \mathbb{S}_{n+i+1}\langle 0 \rangle(T^n) = 0 \ (i \geq 0).$$

□

REMARK 24.17 Let $\pi, \bar{\pi}$ be n-dimensional Novikov groups such that $\bar{\pi} \subset \pi$ is a subgroup of finite index $[\pi : \bar{\pi}] = d$. As in 21.4 there is defined a d-sheeted covering

$$p : \overline{K} = B\bar{\pi} \longrightarrow K = B\pi$$

with the total surgery obstruction of $\overline{K}$ the transfer of the total surgery obstruction of K

$$s(\overline{K}) = p^!s(K) \in \mathbb{S}_n(\overline{K}) .$$

The transfer map $p^! : H_n(K; L_0(\mathbb{Z})) \longrightarrow H_n(\overline{K}; L_0(\mathbb{Z}))$ is an isomorphism, being the Poincaré dual of

$$p^* = 1 : H^0(K; L_0(\mathbb{Z})) = \mathbb{Z} \xrightarrow{\simeq} H^0(\overline{K}; L_0(\mathbb{Z})) = \mathbb{Z} .$$

It follows from the commutative square

$$
\begin{array}{ccc}
H_n(K; L_0(\mathbb{Z})) & \xrightarrow{\ \simeq\ } & \mathbb{S}_n(K) \\
{\scriptstyle p^!}\Big\downarrow{\scriptstyle\simeq} & & \Big\downarrow{\scriptstyle p^!} \\
H_n(\overline{K}; L_0(\mathbb{Z})) & \xrightarrow{\ \simeq\ } & \mathbb{S}_n(\overline{K})
\end{array}
$$

that $p^!: \mathbb{S}_n(K)\longrightarrow\mathbb{S}_n(\overline{K})$ is also an isomorphism, so that $s(K) = 0$ if and only if $s(\overline{K}) = 0$. If K is homotopy equivalent to a compact topological manifold ($s(K) = 0$) then so is the finite cover $\overline{K}$, i.e. only the converse statement is of interest.

$\square$

DEFINITION 24.18 Let π be an n-dimensional Novikov group, with classifying space $K = B\pi$.

(i) The *codimension n* $\begin{cases} symmetric \\ quadratic \end{cases}$ *signature map* is the composite

$$
\begin{cases}
B : VL^n\langle 0\rangle(\mathbb{Z}, K) \xrightarrow{A^{-1}} H_n(K; \mathbb{L}^{\cdot}\langle 0\rangle(\mathbb{Z})) \longrightarrow H_n(K; L^0(\mathbb{Z})) = L^0(\mathbb{Z}) \\
B : L_n(\mathbb{Z}[\pi]) \xrightarrow{A^{-1}} H_n(K; \mathbb{L}_{\cdot}\langle 0\rangle(\mathbb{Z})) \longrightarrow H_n(K; L_0(\mathbb{Z})) = L_0(\mathbb{Z}) ,
\end{cases}
$$

with

$$
\begin{cases}
H_n(K; \mathbb{L}^{\cdot}\langle 0\rangle(\mathbb{Z})) \longrightarrow H_n(K; L^0(\mathbb{Z})) ; (C, \phi) \longrightarrow \sum_{\tau \in K^{(n)}} (C(\tau), \phi(\tau))\tau \\
H_n(K; \mathbb{L}_{\cdot}\langle 0\rangle(\mathbb{Z})) \longrightarrow H_n(K; L_0(\mathbb{Z})) ; (C, \psi) \longrightarrow \sum_{\tau \in K^{(n)}} (C(\tau), \psi(\tau))\tau .
\end{cases}
$$

(ii) The *global codimension n signature* of an n-dimensional 0-connective globally Poincaré $\begin{cases} normal \\ quadratic \end{cases}$ complex $\begin{cases} (C, \phi) \\ (C, \psi) \end{cases}$ in $\mathbb{A}(\mathbb{Z}, K)$ is

$$
\begin{cases}
B^{global}(C, \phi) = B(C(\widetilde{K}), \phi(\widetilde{K})) \in L^0(\mathbb{Z}) , \\
B^{global}(C, \psi) = B(C(\widetilde{K}), \psi(\widetilde{K})) \in L_0(\mathbb{Z})
\end{cases}
$$

with $\begin{cases} (C(\widetilde{K}), \phi(\widetilde{K})) \in VL^n\langle 0\rangle(\mathbb{Z}, K) \\ (C(\widetilde{K}), \psi(\widetilde{K})) \in L_n(\mathbb{Z}[\pi]) \end{cases}$ the assembly over the universal cover $\widetilde{K}$ of K.

(iii) The *local codimension n signature* of an n-dimensional 1/2-connective $\begin{cases} normal \\ quadratic \end{cases}$ complex $\begin{cases} (C, \phi) \\ (C, \psi) \end{cases}$ in $\mathbb{A}(\mathbb{Z}, K)$ is

$$
\begin{cases}
B^{local}(C, \phi) = \sum_{\tau \in K^{(n)}} (C(\tau), \phi(\tau))\tau \in H_n(K; L^0(\mathbb{Z})) = L^0(\mathbb{Z}) , \\
B^{local}(C, \psi) = \sum_{\tau \in K^{(n)}} (C(\tau), \psi(\tau))\tau \in H_n(K; L_0(\mathbb{Z})) = L_0(\mathbb{Z}) .
\end{cases}
$$

$\square$

Note that for any n-dimensional 0-connective locally Poincaré $\begin{cases} \text{normal} \\ \text{quadratic} \end{cases}$
complex $\begin{cases} (C, \phi) \\ (C, \psi) \end{cases}$ in $\mathbb{A}(\mathbb{Z}, K)$ and any n-simplex $\tau \in K^{(n)}$

$$\begin{cases} B^{global}(C, \phi) = B^{local}(C, \phi) = (C(\tau), \phi(\tau)) \in L^0(\mathbb{Z}) , \\ B^{global}(C, \psi) = B^{local}(C, \psi) = (C(\tau), \psi(\tau)) \in L_0(\mathbb{Z}) . \end{cases}$$

EXAMPLE 24.19 If X is a compact n-dimensional topological manifold with a map $X \longrightarrow K = B\pi$ to the classifying space of an n-dimensional Novikov group π, and (C, ϕ) is the 0-connective locally Poincaré n-dimensional symmetric complex in $\mathbb{A}(\mathbb{Z}, K)$ representing the image in $H_n(K; \mathbb{L}^{\cdot})$ of the fundamental $\mathbb{L}^{\cdot}$-homology class $[X]_{\mathbb{L}} \in H_n(X; \mathbb{L}^{\cdot})$ then

$$B\sigma^*(X) = B^{global}(C, \phi) = B^{local}(C, \phi) \in L^0(\mathbb{Z}) = \mathbb{Z} .$$

In 24.22 below this codimension n symmetric signature will be identified with the degree of the map $X \longrightarrow K$.

□

The difference between local and global codimension n signatures detects the total surgery obstruction for the classifying spaces of Novikov groups:

PROPOSITION 24.20 *Let π be an n-dimensional Novikov group, with classifying space $K = B\pi$.*
(i) *The difference between local and global codimension n quadratic signatures defines an isomorphism*

$$\mathbb{S}_n(K) \xrightarrow{\simeq} L_0(\mathbb{Z}) ;$$

$$(C, \psi) \longrightarrow B^{global}(D/C, \delta\psi/\psi) - B^{local}(D/C, \delta\psi/\psi) .$$

Here, (C, ψ) is an $(n-1)$-dimensional 1-connective locally Poincaré globally contractible quadratic complex in $\mathbb{A}(\mathbb{Z}, K)$, and $(D/C, \delta\psi/\psi)$ is the n-dimensional 0-connective globally Poincaré quadratic complex in $\mathbb{A}(\mathbb{Z}, K)$ defined by the algebraic Thom complex of any 0-connective locally Poincaré null-cobordism $(C \longrightarrow D, (\delta\psi, \psi))$.
(ii) *The n-dimensional 1/2-connective visible symmetric L-group of K is such that*

$$VL^n(K) = H_n(K; \mathbb{L}^{\cdot}\langle 0\rangle(\mathbb{Z})) \oplus L_0(\mathbb{Z})$$

with an isomorphism

$$VL^n(K) \xrightarrow{\simeq} H_n(K; \mathbb{L}^{\cdot}\langle 0\rangle(\mathbb{Z})) \oplus L_0(\mathbb{Z}) ;$$

$$(C, \phi) \longrightarrow (A^{-1}(C(\widetilde{K}), \phi(\widetilde{K})), (B^{global}(C, \phi) - B^{local}(C, \phi))/8) ,$$

and

$$\partial : VL^n(K) \longrightarrow \mathbb{S}_n(K) = L_0(\mathbb{Z}) ;$$

$$(C, \phi) \longrightarrow (B^{global}(C, \phi) - B^{local}(C, \phi))/8 .$$

(iii) *The total surgery obstruction of K is*

$$s(K) \;=\; \partial \sigma^* \langle 1/2 \rangle (K)$$

$$= (B^{global} \sigma^* \langle 1/2 \rangle (K) - B^{local} \sigma^* \langle 1/2 \rangle (K))/8$$

$$= (B \sigma^* \langle 0 \rangle (K) - 1)/8 \in \mathbb{S}_n(K) \;=\; L_0(\mathbb{Z}) \;=\; \mathbb{Z} \,,$$

with $\sigma^* \langle q \rangle (K) \in VL^n \langle q \rangle (\mathbb{Z}, K)$ the q-connective visible symmetric signature of K for $q = 0, 1/2$.

PROOF (i) Note first that for any simplicial complex K there is defined a commutative braid of exact sequences

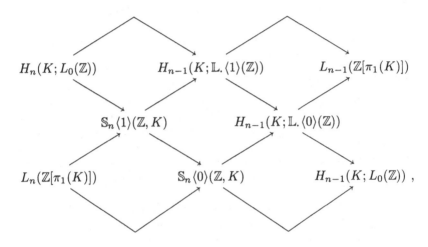

and that there is defined an isomorphism

$$H_n(K; \mathbb{L}. \langle 1 \rangle (\mathbb{Z}) \to \mathbb{L}. \langle 0 \rangle (\mathbb{Z})) \xrightarrow{\;\simeq\;} H_n(K; L_0(\mathbb{Z})) \;;$$

$$(C \longrightarrow D, (\delta\psi, \psi)) \longrightarrow \sum_{\tau \in K^{(n)}} (D(\tau)/C(\tau), \delta\psi(\tau)/\psi(\tau))\tau$$

with $(C \longrightarrow D, (\delta\psi, \psi))$ an n-dimensional locally Poincaré quadratic pair in $\mathbb{A}(\mathbb{Z}, K)$ such that C is 1-connective and D is 0-connective. For $K = B\pi$ it is also the case that $\mathbb{S}_{n+1} \langle 0 \rangle (\mathbb{Z}, K) = \mathbb{S}_n \langle 0 \rangle (\mathbb{Z}, K) = 0$, so there is defined an isomorphism

$$H_n(K; L_0(\mathbb{Z})) \xrightarrow{\;\simeq\;} \mathbb{S}_n \langle 1 \rangle (\mathbb{Z}, K) \;=\; \mathbb{S}_n(K)$$

with the inverse specified in the statement.

(ii) Since $\mathbb{S}_{n+1} \langle 0 \rangle (\mathbb{Z}, K) = \mathbb{S}_n \langle 0 \rangle (\mathbb{Z}, K) = 0$ the diagram of 15.18 (iii) includes a commutative braid of exact sequences

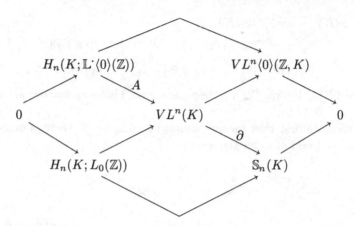

defining a direct sum system

$$VL^n\langle 0\rangle(\mathbb{Z}, K) \;=\; H_n(K; \mathbb{L}^{\cdot}\langle 0\rangle(\mathbb{Z})) \;\overrightarrow{\underset{\longleftarrow}{}}$$

$$VL^n(K) \;\overrightarrow{\underset{\longleftarrow}{}}\; \mathbb{S}_n(K) \;=\; H_n(K; L_0(\mathbb{Z})) \;.$$

In particular, the quadratic boundary map

$$\partial : VL^n(K) \longrightarrow \mathbb{S}_n(K) \;;\; (C,\phi) \longrightarrow \partial(C,\phi) \;=\; (\partial C, \psi)$$

is a split surjection, with kernel isomorphic to $VL^n\langle 0\rangle(\mathbb{Z}, K)$. For any n-dimensional 1/2-connective globally Poincaré normal complex (C,ϕ) in $\mathbb{A}(\mathbb{Z}, K)$ the image of $(C,\phi) \in VL^n(K)$ in the algebraic normal complex cobordism group

$$H_n(K; \widehat{\mathbb{L}^{\cdot}}) \;=\; H_n(K; \mathbb{L}.\langle 1\rangle(\mathbb{Z}) \to \mathbb{L}^{\cdot}\langle 0\rangle(\mathbb{Z}))$$

is represented by the n-dimensional 0-connective locally Poincaré normal pair $(\partial C \longrightarrow D, (0, (1+T)\psi))$ with $D = C^{n+1-*}$, $(D/\partial C, 0/(1+T)\psi) = (C,\phi)$. The symmetric version of (i) now allows the identification

$$(1+T)(\partial C, \psi)$$
$$= B^{global}(D/\partial C, 0/(1+T)\psi) - B^{local}(D/\partial C, 0/(1+T)\psi)$$
$$= B^{global}(C,\phi) - B^{local}(C,\phi) \;\in\; H_n(K; L^0(\mathbb{Z})) \;=\; L^0(\mathbb{Z}) \;,$$

with

$$1+T \;=\; 8 : L_0(\mathbb{Z}) \;=\; \mathbb{Z} \longrightarrow L^0(\mathbb{Z}) \;=\; \mathbb{Z} \;.$$

(iii) The n-dimensional 1/2-connective globally Poincaré normal complex (C,ϕ) in $\mathbb{A}(\mathbb{Z}, K)$ with $C = \Delta(K')$ representing $\sigma^*\langle 1/2\rangle(\mathbb{Z}, K) = (C,\phi) \in$

$VL^n(K)$ has codimension n symmetric signatures

$$B^{global}(C, \phi) = B\sigma^*\langle 0 \rangle (K) \,,$$

$$B^{local}(C, \phi) = \sum_{\tau \in K^{(n)}} \tau = [K] = 1 \in H_n(K; L^0(\mathbb{Z})) = L^0(\mathbb{Z}) = \mathbb{Z} \,.$$

Now apply (ii).

□

EXAMPLE 24.21 The 0-connective visible symmetric signature of T^n is

$$\sigma^*(T^n) = A([T^n]_{\mathbb{L}}) = (0, \dots, 0, 1)$$

$$\in VL^n \langle 0 \rangle (\mathbb{Z}, T^n) = L^n(\mathbb{Z}[\mathbb{Z}^n])$$

$$= L^n(\mathbb{Z}) \oplus \binom{n}{1} L^{n-1}(\mathbb{Z}) \oplus \dots \oplus \binom{n}{k} L^{n-k}(\mathbb{Z}) \oplus \dots \oplus L^0(\mathbb{Z}) \,.$$

The codimension n symmetric signature of T^n is the generator

$$B\sigma^*(T^n) = B^{global}\sigma^*(T^n) = B^{local}\sigma^*(T^n) = (\mathbb{Z}, 1) = 1 \in L^0(\mathbb{Z}) = \mathbb{Z} \,.$$

The 1/2-connective visible symmetric signature of T^n is

$$\sigma^*(T^n) = ((0, \dots, 0, 1), 0) \in VL^n(T^n) = L^n(\mathbb{Z}[\mathbb{Z}^n]) \oplus L_0(\mathbb{Z}) \,,$$

and the total surgery obstruction is $s(T^n) = 0 \in \mathbb{S}_n(T^n) = L_0(\mathbb{Z})$.

□

PROPOSITION 24.22 *Let X be an n-dimensional geometric Poincaré complex with a morphism $e\colon \pi_1(X) \longrightarrow \pi$ to an n-dimensional Novikov group π.*
(i) The total surgery obstruction $s(X) \in \mathbb{S}_n(X)$ has image

$$s(X) = \partial(C, \phi) = (B^{global}(C, \phi) - B^{local}(C, \phi))/8$$

$$\in \mathbb{S}_n(K) = L_0(\mathbb{Z}) = \mathbb{Z} \,,$$

with $(C, \phi) = \sigma^*(X)$ the n-dimensional 1/2-connective globally Poincaré normal complex in $\mathbb{A}(\mathbb{Z}, K)$ associated to a map $e\colon X \longrightarrow K = B\pi$ inducing $e\colon \pi_1(X) \longrightarrow \pi$, and

$$B^{global}(C, \phi) = B\sigma^*(X) \,,$$

$$B^{local}(C, \phi) = (\text{degree of } e\colon X \longrightarrow K) \in L^0(\mathbb{Z}) = \mathbb{Z} \,.$$

(ii) If $(f, b)\colon Y \longrightarrow X$ is a normal map of n-dimensional geometric Poincaré complexes the difference of the images in $\mathbb{S}_n(K)$ of the total surgery obstructions of X, Y is the codimension n quadratic signature of the surgery obstruction $\sigma_*(f, b) \in L_n(\mathbb{Z}[\pi])$

$$s(Y) - s(X) = B\sigma_*(f, b) \in \mathbb{S}_n(K) = L_0(\mathbb{Z}) \,.$$

(iii) If $\nu_X\colon X \longrightarrow BG$ admits a topological reduction $\tilde{\nu}_X\colon X \longrightarrow BTOP$ then the image in $\mathbb{S}_n(K)$ of the total surgery obstruction $s(X) \in \mathbb{S}_n(X)$ is given

up to sign by the codimension n quadratic signature of the surgery obstruction $\sigma_*(f,b) \in L_n(\mathbb{Z}[\pi])$ of the corresponding topological normal map $(f,b): M \longrightarrow X$

$$s(X) = -B\sigma_*(f,b) \in \mathbb{S}_n(K) = L_0(\mathbb{Z}) .$$

PROOF (i) The 1/2-connective visible symmetric signature $\sigma^*(X) \in VL^n(K)$ is represented by the n-dimensional 1/2-connective globally Poincaré normal complex (C, ϕ) in $\mathbb{A}(\mathbb{Z}, K)$ of the n-dimensional normal complex cycle $\{X(\tau)|\tau \in K\}$ defined by the inverse images of the dual cells

$$X(\tau) = e^{-1}D(\tau, K) ,$$

with $(X(\tau), \partial X(\tau))$ an $(n - |\tau|)$-dimensional normal pair and

$$C(\tau) = \Delta(X(\tau), \partial X(\tau)) .$$

As in §16 assume that K is an n-dimensional simplicial complex with fundamental class

$$[K] = \sum_{\tau \in K^{(n)}} \tau = 1 \in H_n(K) = \mathbb{Z}$$

and similarly for X. The degree of $e: X \longrightarrow K$ is the number $d \in \mathbb{Z}$ such that

$$e_*[X] = d[K] \in H_n(K) = \mathbb{Z} ,$$

which on the chain level can be expressed as

$$e_*[X] = \sum_{\rho \in X^{(n)}} e(\rho) = \sum_{\tau \in K^{(n)}} \left(\sum_{\rho \in X^{(n)}, e(\rho) = \tau} 1 \right) \tau$$

$$= d \left(\sum_{\tau \in K^{(n)}} \tau \right) = d[K] \in \Delta_n(K) .$$

The degree d is thus the algebraic number of n-simplexes $\rho \in X^{(n)}$ in the inverse image $e^{-1}(\tau)$ of any n-simplex $\tau \in K^{(n)}$, which is the algebraic number of vertices $\hat{\rho} \in X'^{(0)}$ in the 0-dimensional geometric Poincaré complex

$$X(\tau) = e^{-1}(\hat{\tau}) = \bigcup_{\rho \in X^{(n)}, e(\rho) = \tau} \hat{\rho} ,$$

and also the symmetric signature of $X(\tau)$

$$\sigma^*(X(\tau)) = \sum_{\rho \in X^{(n)}, e(\rho) = \tau} (\mathbb{Z}, 1) = d \in L^0(\mathbb{Z}) = \mathbb{Z} .$$

The local codimension n symmetric signature of $\sigma^*(X)$ is thus

$$B^{local}(C, \phi) = \sum_{\tau \in K^{(n)}} \tau(C(\tau), \phi(\tau)) = d \in H_n(K; L^0(\mathbb{Z})) = L^0(\mathbb{Z}) = \mathbb{Z} .$$

(ii) Apply $Y \longrightarrow K$ to the identification $s(Y) - s(X) = \partial\sigma_*(f,b) \in \mathbb{S}_n(Y)$ given by 19.7.

(iii) Apply $e: X \longrightarrow K$ to the identification $s(X) = -\partial \sigma_*(f, b) \in \mathbb{S}_n(X)$. Alternatively, substitute $s(M) = 0 \in \mathbb{S}_n(M)$ in the formula $s(M) - s(X) = B\sigma_*(f, b)$ given by (ii).

□

REMARK 24.23 A *resolution* (M, f) of a space X is a topological manifold M together with a proper cell-like surjection $f: M \longrightarrow X$. Quinn [132], [133] investigated the resolution of compact ANR homology manifolds by compact topological manifolds, using controlled surgery theory and algebraic Poincaré complexes to formulate the following obstruction. (It is now known that this obstruction is realized, see 25.13). A compact ANR is homotopy equivalent to a finite CW complex by the result of West, so that a compact n-dimensional ANR homology manifold X is a finite n-dimensional Poincaré space. Let $X_1 \subset X$ be a neighbourhood of a point $x_0 \in X$. As in [132, 4.1] there is defined a finite n-dimensional geometric Poincaré complex Y with a normal map $(f, b): Y \longrightarrow T^n$ such that the proper normal map $(\tilde{f}, \tilde{b}): \tilde{Y} \longrightarrow \tilde{T}^n = \mathbb{R}^n$ is bordant to a proper normal map $X_1 \longrightarrow \mathbb{R}^n$. The codimension n symmetric signatures of the associated 0-connective n-dimensional globally Poincaré normal complex $(C, \phi) = \sigma^*(Y)$ in $\mathbb{A}(\mathbb{Z}, T^n)$ are

$$B^{global}(C, \phi) = B\sigma^*(Y) ,$$

$$B^{local}(C, \phi) = (\text{degree of } f: Y \longrightarrow T^n) = 1 \in L^0(\mathbb{Z}) = \mathbb{Z} .$$

The local signature obstruction of [133] to a resolution of X by a compact topological manifold is defined by

$$i(X) = B^{global}(C, \phi) - B^{local}(C, \phi)(B\sigma^*(Y) - 1)/8 \in L_0(\mathbb{Z}) = \mathbb{Z} .$$

(Unfortunately, the local signature $B\sigma^*(Y) \in L^0(\mathbb{Z})$ of [133] arises here as a global codimension n signature.) The total surgery obstruction $s(X) \in \mathbb{S}_n(X) = \mathbb{S}_n\langle 1\rangle(\mathbb{Z}, X)$ is the image of

$$i(X) = \partial \sigma_*(f, b) = s(Y) \in \mathbb{S}_n(T^n) = H_n(X; L_0(\mathbb{Z})) = L_0(\mathbb{Z})$$

under the map in the exact sequence

$$\ldots \longrightarrow \mathbb{S}_{n+1}\langle 0\rangle(\mathbb{Z}, X) \longrightarrow H_n(X; L_0(\mathbb{Z})) \longrightarrow \mathbb{S}_n\langle 1\rangle(\mathbb{Z}, X)$$
$$\longrightarrow \mathbb{S}_n\langle 0\rangle(\mathbb{Z}, X) \longrightarrow \ldots .$$

For $n \geq 5$ X is homotopy equivalent to a topological manifold (not necessarily a resolution) if and only if $i(X) \in \text{im}(\mathbb{S}_{n+1}\langle 0\rangle(\mathbb{Z}, X) \longrightarrow L_0(\mathbb{Z}))$. The resolution obstruction of a homology manifold X is an invariant of the controlled chain equivalence inducing the Poincaré duality $H^{n-*}(X) \cong H_*(X)$. See §25 and Appendix C for some further discussion of the surgery classification of compact ANR homology manifolds and controlled topology.

□

§25. The 4-periodic theory

The 4-periodic theory is the version of surgery in which the 1-connective L-spectrum $\mathbb{L}. = \mathbb{L}.\langle 1\rangle(\mathbb{Z})$ is replaced by the 4-periodic spectrum $\mathbb{L}.(\mathbb{Z})$, corresponding to the difference in the codimension n transversality properties of n-dimensional topological manifolds and n-dimensional ANR homology manifolds. The algebraic and topological properties of the 4-periodic theory will now be investigated, including an interpretation of the difference between the 4-periodic and 1-connective theories in terms of the local and global signatures of §24.

The 4-periodicity of surgery was first observed experimentally by Kervaire and Milnor [83], in the simply connected high-dimensional case arising in the classification of compact $(n-1)$-dimensional differentiable manifolds which are homotopy spheres and bound framed n-dimensional manifolds W^n, with $n \geq 5$. After framed surgery below the middle dimension W can be taken to be $[(n-2)/2]$-connected. The obstruction to making W contractible by surgery in the middle dimension is an element of the simply-connected surgery obstruction group $L_n(\mathbb{Z})$. For $n = 2i$ this is the Witt class of the $(-)^i$-quadratic intersection form on $H_i(W)$. In particular, for $n = 4k \geq 8$ the E_8-plumbing of 8 copies of $\tau_{S^{2k}}: S^{2k} \longrightarrow BSO(2k)$ is a framed $(2k-1)$-connected 4k-dimensional differentiable manifold W^{4k} with boundary an exotic $(4k-1)$-dimensional sphere Σ^{4k-1} and symmetric intersection form

$$(H_{2k}(W), \lambda) = (\mathbb{Z}^8, E_8),$$

such that the corresponding surgery obstruction is

$$\text{signature}(W)/8 = 1 \in L_{4k}(\mathbb{Z}) = \mathbb{Z}.$$

There is no obstruction for $n = 2i + 1$, since $L_{2i+1}(\mathbb{Z}) = 0$. The simply-connected surgery obstruction is 4-periodic since $[(n-2)/2]$-connected n-dimensional manifolds have the same homological intersection properties as $[(n+2)/2]$-connected $(n+4)$-dimensional manifolds. The simply-connected surgery obstruction groups $L_n(\mathbb{Z}) = \pi_n(G/TOP)$ are 4-periodic, but the groups of h-cobordism classes of exotic spheres $\theta_n = \pi_n(TOP/O)$ are not 4-periodic.

The 4-periodicity persists in surgery on n-dimensional normal maps (f, b): $M^n \longrightarrow X$, which can be made $[n/2]$-connected by surgery below the middle dimension. The non-simply connected obstruction to surgery on (f, b) depends only on the middle-dimensional chain level intersection properties of the $\mathbb{Z}[\pi_1(X)]$-module homology kernels

$$K_*(M) = \ker(\tilde{f}_*: H_*(\widetilde{M}) \longrightarrow H_*(\widetilde{X})),$$

which are the same for $[n/2]$-connected n-dimensional normal maps and $[(n+4)/2]$-connected $(n+4)$-dimensional normal maps. The surgery obstruction groups $L_*(\pi) = L_*(\mathbb{Z}[\pi])$ of Wall [176] were defined algebraically to be such that

$$L_*(\mathbb{Z}[\pi]) = L_{*+4}(\mathbb{Z}[\pi]) \ ,$$

with the 4-periodicity isomorphisms realized geometrically as products with the complex projective plane $\mathbb{C}\mathbb{P}^2$

$$- \times \mathbb{C}\mathbb{P}^2 : L_n(\mathbb{Z}[\pi]) \xrightarrow{\simeq} L_{n+4}(\mathbb{Z}[\pi]) \ ;$$

$$\sigma_*((f,b): M \longrightarrow X) \longrightarrow \sigma_*((f,b) \times 1: M \times \mathbb{C}\mathbb{P}^2 \longrightarrow X \times \mathbb{C}\mathbb{P}^2) \ .$$

The expression of $L_*(\mathbb{Z}[\pi])$ as cobordism groups of quadratic Poincaré complexes in Ranicki [140] allowed the 4-periodicity isomorphisms to be realized algebraically as products with the symmetric signature of $\mathbb{C}\mathbb{P}^2$

$$\sigma^*(\mathbb{C}\mathbb{P}^2) = \text{signature}(\mathbb{C}\mathbb{P}^2) = 1 \in L^4(\mathbb{Z}) = \mathbb{Z} \ ,$$

and also as the double skew-suspension maps

$$\bar{S}^2 = - \otimes \sigma^*(\mathbb{C}\mathbb{P}^2) : L_n(\mathbb{Z}[\pi]) \xrightarrow{\simeq} L_{n+4}(\mathbb{Z}[\pi]) \ ;$$

$$(C,\psi) \longrightarrow (S^2 C, \bar{S}^2 \psi) = (C,\psi) \otimes \sigma^*(\mathbb{C}\mathbb{P}^2) \ .$$

The classifying space G/O for differentiable surgery is 4-periodic modulo torsion, since it has the rational homotopy type

$$G/O \otimes \mathbb{Q} \simeq BO \otimes \mathbb{Q} \simeq \prod_{j=1}^{\infty} K(\mathbb{Q}, 4j) \ .$$

The classifying space G/TOP for topological surgery is 4-periodic, with a homotopy equivalence

$$\Omega^4 G/TOP \simeq L_0(\mathbb{Z}) \times G/TOP \ .$$

The geometric surgery spectra of Quinn [127] and the quadratic L-theory spectra of Ranicki [135] with homotopy groups L_* realize the 4-periodicity on the spectrum level, with

$$\mathbb{L}_0(\mathbb{Z}) \simeq L_0(\mathbb{Z}) \times G/TOP \ , \ \mathbb{L}_0 \simeq G/TOP$$

in the simply-connected case.

In order to obtain an algebraic formulation of the surgery exact sequence and the total surgery obstruction for topological manifolds it was necessary to kill the 0th homotopy group $\pi_0(\mathbb{L}\langle 0 \rangle.(\mathbb{Z})) = L_0(\mathbb{Z})$ in $\mathbb{L}\langle 0 \rangle.(\mathbb{Z})$ and work with the 1-connective quadratic L-theory spectrum $\mathbb{L}. = \mathbb{L}\langle 1 \rangle.(\mathbb{Z})$, as in §15. The controlled and bounded surgery of Quinn [132], [133] and Ferry and Pedersen [50] have shown that the original 0-connective 4-periodic surgery spectra are related to the surgery exact sequence and total surgery obstruction for compact ANR homology manifolds.

Products with the $\mathbb{L}^{\cdot}$-coefficient fundamental class $[\mathbb{C}\mathbb{P}^2]_{\mathbb{L}} \in H_4(\mathbb{C}\mathbb{P}^2; \mathbb{L}^{\cdot})$

$$- \otimes [\mathbb{C}\mathbb{P}^2]_{\mathbb{L}} : \mathbb{S}_n(X) \longrightarrow \mathbb{S}_{n+4}(X \times \mathbb{C}\mathbb{P}^2)$$

are not in general isomorphisms, fitting into an exact sequence
$$\ldots \longrightarrow H_{n+4}(X \times S^2; \mathbb{L}.) \longrightarrow \mathbb{S}_n(X) \longrightarrow \mathbb{S}_{n+4}(X \times \mathbb{CP}^2)$$
$$\longrightarrow H_{n+3}(X \times S^2; \mathbb{L}.) \longrightarrow \ldots .$$
However, for an n-dimensional polyhedron X the 1-connective quadratic
$\mathbb{S}$-groups $\mathbb{S}_*(X) = \mathbb{S}_*\langle 1 \rangle(\mathbb{Z}, X)$ are themselves 4-periodic in dimensions $\geq$
$n + 2$, with the double skew-suspension maps
$$\bar{S}^2 : \mathbb{S}_m(X) \longrightarrow \mathbb{S}_{m+4}(X) ; \; (C, \psi) \longrightarrow (S^2C, \bar{S}^2\psi)$$
isomorphisms for $m \geq n + 2$. In this 4-periodicity range the 1-connective
$\mathbb{S}$-groups coincide with the $\mathbb{S}$-groups $\mathbb{S}_*(\mathbb{Z}, X) = \mathbb{S}_*\langle 0 \rangle(\mathbb{Z}, X)$ appearing in
the 4-periodic algebraic surgery exact sequence of §14
$$\ldots \longrightarrow H_m(X; \mathbb{L}.(\mathbb{Z})) \xrightarrow{A} L_m(\mathbb{Z}[\pi_1(X)])$$
$$\longrightarrow \mathbb{S}_m(\mathbb{Z}, X) \longrightarrow H_{m-1}(X; \mathbb{L}.(\mathbb{Z})) \longrightarrow \ldots .$$
Abbreviate
$$\mathbb{L}.\langle 0 \rangle(\mathbb{Z}) = \overline{\mathbb{L}}. \; , \quad VL^{\cdot}(\{*\}) = VL^{\cdot}\langle 0 \rangle(\mathbb{Z}, \{*\}) = \overline{\mathbb{L}}^{\cdot} \; ,$$
$$VL^*\langle 0 \rangle(\mathbb{Z}, X) = \overline{VL}^*(X) \; , \; \mathbb{S}_*\langle 0 \rangle(\mathbb{Z}, X) = \overline{\mathbb{S}}_*(X) \; ,$$
writing the corresponding assembly maps A as $\overline{A}$.

PROPOSITION 25.1 (i) *Up to homotopy equivalence*
$$\overline{\mathbb{L}}^{\cdot} = K.(L_0(\mathbb{Z}), 0) \vee \mathbb{L}^{\cdot}$$
with $K.(L_0(\mathbb{Z}), 0)$ *the Eilenberg–MacLane spectrum of* $L_0(\mathbb{Z})$*-coefficient homology, so that for any space* X
$$H_*(X; \overline{\mathbb{L}}^{\cdot}) = H_*(X; L_0(\mathbb{Z})) \oplus H_*(X; \mathbb{L}^{\cdot}) \; .$$
(ii) *For any space* X *there are defined commutative braids of exact sequences*

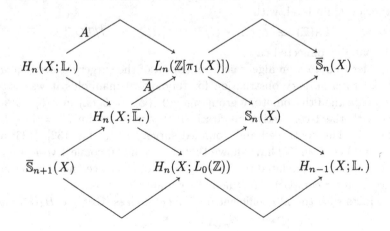

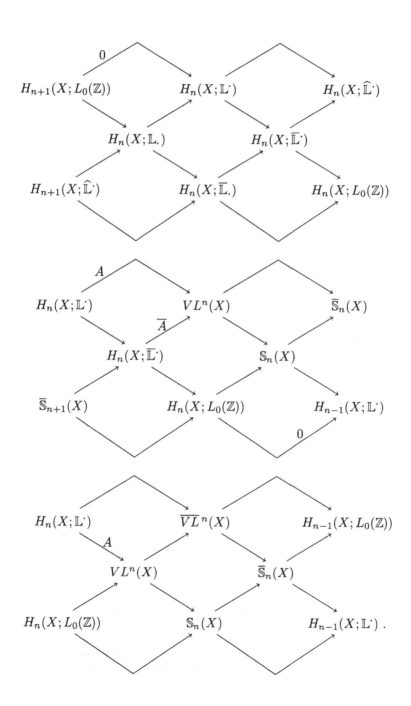

PROOF (i) The inclusion $\mathbb{L}^{\cdot} \longrightarrow \overline{\mathbb{L}}^{\cdot}$ is split by the forgetful map $\overline{\mathbb{L}}^{\cdot} \longrightarrow \mathbb{L}^{\cdot}$.
(ii) The braids of exact sequences are induced from the braids of fibrations of spectra given by 15.18.

$\square$

REMARK 25.2 The fibration

$$\mathbb{L}. \longrightarrow \overline{\mathbb{L}}. \longrightarrow K.(L_0(\mathbb{Z}), 0)$$

splits when localized at 2, but not away from 2. Taylor and Williams [169, Thm. A] show that the 0-connective quadratic $\mathbb{L}$-spectrum of any ring with involution A is such that

$$\mathbb{L}.\langle 0 \rangle (A)[1/2] = bo\Lambda_0 \vee \Sigma bo\Lambda_1 \vee \Sigma^2 bo\Lambda_2 \vee \Sigma^3 bo\Lambda_3$$

$$\mathbb{L}.\langle 0 \rangle (A)_{(2)} = \bigvee_{j=0}^{\infty} K.(L_j(A)_{(2)}, j)$$

where $bo\Lambda_i$ denotes connective KO theory with coefficients in the group $\Lambda_i = L_i(A)[1/2]$. For $A = \mathbb{Z}$ this gives

$$\overline{\mathbb{L}}.[1/2] = \mathbb{L}.\langle 0 \rangle (\mathbb{Z})[1/2] = bo[1/2] \ ,$$

$$\mathbb{L}.[1/2] = \mathbb{L}.\langle 1 \rangle (\mathbb{Z})[1/2] = bo\langle 1 \rangle [1/2] \ .$$

$\square$

PROPOSITION 25.3 *Let X be an n-dimensional polyhedron.*
(i) *The $\overline{\mathbb{S}}_*$-groups of X are such that*

$$\overline{\mathbb{S}}_m(X) = \mathbb{S}_m(\mathbb{Z}, X) = \overline{\mathbb{S}}_{m+4}(X) \ \textit{for } m \geq n \ ,$$

$$\overline{\mathbb{S}}_m(X) = \mathbb{S}_m\langle q \rangle (\mathbb{Z}, X) = \mathbb{S}_m(\mathbb{Z}, X) \ \textit{for } m \geq n+1, q \leq 0,$$

$$\overline{\mathbb{S}}_m(X) = \mathbb{S}_m(X) \ \textit{for } m \geq n+2 \ ,$$

with an exact sequence

$$0 \longrightarrow \mathbb{S}_{n+1}(X) \longrightarrow \overline{\mathbb{S}}_{n+1}(X) \longrightarrow H_n(X; L_0(\mathbb{Z}))$$
$$\longrightarrow \mathbb{S}_n(X) \longrightarrow \overline{\mathbb{S}}_n(X) \ .$$

(ii) *The $\overline{VL}^*$-groups of X are such that*

$$\overline{VL}^m(X) = VL^m(\mathbb{Z}, X) = \overline{VL}^{m+4}(X) \ \textit{for } m \geq n \ ,$$

$$\overline{VL}^m(X) = VL^m\langle q \rangle (\mathbb{Z}, X) = VL^m(\mathbb{Z}, X) \ \textit{for } m \geq n+1, q \leq 0,$$

$$\overline{VL}^m(X) = VL^m(X) \ \textit{for } m \geq n+2 \ ,$$

with an exact sequence

$$0 \longrightarrow VL^{n+1}(X) \longrightarrow \overline{VL}^{n+1}(X) \longrightarrow H_n(X; L_0(\mathbb{Z}))$$
$$\longrightarrow VL^n(X) \longrightarrow \overline{VL}^n(X) \ .$$

(iii) *If $(C \longrightarrow D, (\delta\psi, \psi))$ is an n-dimensional locally Poincaré globally contractible quadratic pair in $\mathbb{A}(\mathbb{Z}, X)$ with C 1-connective and D 0-connective* *then the image in* $\mathbb{S}_n(X)$ *of the homology class*

$$x = \sum_{\tau \in X^{(n)}} ((D/C)(\tau), (\delta\psi/\psi)(\tau))\tau \in H_n(X; L_0(\mathbb{Z}))$$

is the cobordism class

$$[x] = (C, \psi) \in \ker(\mathbb{S}_n(X) \longrightarrow \overline{\mathbb{S}}_n(X)) = \mathrm{im}(H_n(X; L_0(\mathbb{Z})) \longrightarrow \mathbb{S}_n(X)) .$$

(iv) *If* $(E \longrightarrow F, (\delta\phi, \phi))$ *is an* $(n+1)$*-dimensional globally Poincaré normal pair in* $\mathbb{A}(\mathbb{Z}, X)$ *with* (E, ϕ) *1/2-connective and* F *0-connective then the homology class* $x \in H_n(X; L_0(\mathbb{Z}))$ *determined in* (ii) *by the* n*-dimensional locally Poincaré globally contractible quadratic pair* $(\partial E \longrightarrow \partial F, \partial(\delta\phi, \phi))$

$$x = \sum_{\tau \in X^{(n)}} ((\partial F/\partial E)(\tau), \partial(\delta\phi/\phi)(\tau))\tau \in H_n(X; L_0(\mathbb{Z}))$$

is such that

$$[x] = \partial(E, \phi) \in \mathrm{im}(H_n(X; L_0(\mathbb{Z})) \longrightarrow \mathbb{S}_n(X)) ,$$

$$(1 + T)(x) = \sum_{\tau \in X^{(n)}} (E(\tau), \phi(\tau))\tau \in H_n(X; L^0(\mathbb{Z})) .$$

(v) *The diagram*

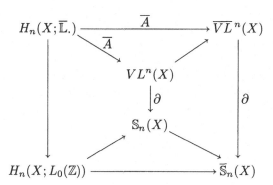

commutes.

PROOF (i) The double skew-suspension maps define an isomorphism of exact sequences

$$\ldots \longrightarrow H_m(X; \mathbb{L}.\langle q+1 \rangle) \longrightarrow H_m(X; \mathbb{L}.\langle q \rangle) \longrightarrow H_{m-q}(X; L_q(\mathbb{Z})) \to \ldots$$

$$\downarrow \qquad\qquad\qquad \downarrow \qquad\qquad\qquad \downarrow$$

$$\ldots \to H_{m+4}(X; \mathbb{L}.\langle q+5 \rangle) \to H_{m+4}(X; \mathbb{L}.\langle q+4 \rangle) \to H_{m-q}(X; L_{q+4}(\mathbb{Z})) \to \ldots$$

for any $m, q \in \mathbb{Z}$, with $\mathbb{L}.\langle q \rangle = \mathbb{L}.\langle q \rangle(\mathbb{Z})$. The natural map

$$H_m(X; \mathbb{L}.\langle 0 \rangle) = H_{m+4}(X; \mathbb{L}.\langle 4 \rangle) \longrightarrow H_{m+4}(X; \mathbb{L}.\langle 0 \rangle)$$

is an isomorphism for $m \geq n - 1$, being the composite of the isomorphisms in the middle of each of the exact sequences

$$H_{m-k+5}(X; L_k(\mathbb{Z})) = 0 \longrightarrow H_{m+4}(X; \mathbb{L}.\langle k + 1 \rangle) \longrightarrow H_{m+4}(X; \mathbb{L}.\langle k \rangle)$$
$$\longrightarrow H_{m-k+4}(X; L_k(\mathbb{Z})) = 0$$

for $k = 0, 1, 2, 3$. A 5-lemma argument applied to these and the 4-periodicity isomorphisms $L_m(\mathbb{Z}[\pi_1(X)]) \cong L_{m+4}(\mathbb{Z}[\pi_1(X)])$ gives that the double skew-suspension maps $\overline{S}_m(X) \longrightarrow \overline{S}_{m+4}(X)$ are isomorphisms for $m \geq n$. The relationship between the S_*- and $\overline{S}_*$-groups is given by the exact sequence of 15.11 (iii)

$$\ldots \longrightarrow H_m(X; L_0(\mathbb{Z})) \longrightarrow S_m(X) \longrightarrow \overline{S}_m(X)$$
$$\longrightarrow H_{m-1}(X; L_0(\mathbb{Z})) \longrightarrow \ldots ,$$

noting that $H_m(X; L_0(\mathbb{Z})) = 0$ for $m \geq n + 1$. Also, 15.18 (iii) gives an exact sequence

$$H_{n+1}(X; L_{-1}(\mathbb{Z})) = 0 \longrightarrow \overline{S}_n(X) \longrightarrow S_n\langle -1 \rangle(\mathbb{Z}, X)$$
$$\longrightarrow H_n(X; L_{-1}(\mathbb{Z})) = 0$$

and by 15.11 (v)

$$\overline{S}_n(X) = S_n\langle -1 \rangle(\mathbb{Z}, X) = S_n(\mathbb{Z}, X) .$$

(ii) This follows from (i) and the commutative braids of exact sequences given by 25.1.

(iii) and (iv). These identities are formal consequences of the identifications in §15 of the 0- and 1-connective quadratic $\mathbb{L}$-spectra with the appropriately connective quadratic Poincaré complexes.

(v) Let (C, ψ) be an n-dimensional quadratic complex in $\Lambda\langle 0 \rangle(\mathbb{Z})_*(X)$, representing an element

$$(C, \psi) \in L_n(\Lambda\langle 0 \rangle(\mathbb{Z})_*(X)) = H_n(X; \overline{\mathbb{L}}.)$$

with images

$$(1 + T)(C, \psi) = (C, (1 + T)\psi) \in \overline{VL}^n(X) ,$$

$$[C, \psi] = \sum_{\tau \in X^{(n)}} (C(\tau), \psi(\tau))\tau \in H_n(X; L_0(\mathbb{Z})) .$$

Since $L_n(\Lambda\langle 0 \rangle(\mathbb{Z}, X)) = L_n(\Lambda\langle 1 \rangle(\mathbb{Z}, X))$ (by 15.11 (i)) there exists an $(n + 1)$-dimensional quadratic pair in $\Lambda\langle 0 \rangle(\mathbb{Z}, X)$

$$P = (C' \oplus C \longrightarrow D, (\delta\psi, \psi' \oplus -\psi))$$

with C' 1-connective and D 0-connective. The assembly of (C, ψ) is represented by (C', ψ')

$$\overline{A}(C, \psi) = (C', \psi') \in L_n(\mathbb{Z}[\pi_1(X)]) = L_n(\Lambda\langle 1 \rangle(\mathbb{Z}, X)) ,$$

so the composite

$$H_n(X;\overline{\mathbb{L}.}) \xrightarrow{\overline{A}} L_n(\mathbb{Z}[\pi_1(X)]) \xrightarrow{1+T} VL^n(X)$$

sends $(C,\psi) \in H_n(X;\overline{\mathbb{L}.})$ to $(C',(1+T)\psi') \in VL^n(X)$. The boundary of P is an n-dimensional locally Poincaré globally contractible quadratic pair in $\Lambda\langle 0\rangle(\mathbb{Z}, X)$

$$\partial P = (\partial C' \oplus \partial C \longrightarrow \partial D, \partial(\delta\psi, \psi' \oplus -\psi))$$

with

$$\partial C_r = C_{r+1} \oplus C^{n-r}, \quad \partial C'_r = C'_{r+1} \oplus C'^{n-r},$$

$$\partial D_r = D_{r+1} \oplus D^{n-r+1} \oplus C'^{n-r} \oplus C^{n-r} \quad (r \in \mathbb{Z})$$

such that ∂C is locally contractible, $\partial C'$ is 1-connective and ∂D is 0-connective. The composite

$$H_n(X;\overline{\mathbb{L}.}) \xrightarrow{\overline{A}} L_n(\mathbb{Z}[\pi_1(X)]) \longrightarrow VL^n(X) \xrightarrow{\partial} \mathbb{S}_n(X)$$

sends $(C,\psi) \in H_n(X;\overline{\mathbb{L}.})$ to $\partial(C',\psi') \in \mathbb{S}_n(X)$. For each n-simplex $\tau \in X^{(n)}$ the 0-dimensional quadratic Poincaré complex in $\Lambda\langle 0\rangle(\mathbb{Z})$

$$\partial P(\tau) = ((\partial D/(\partial C' \oplus \partial C))(\tau), (\delta\psi/(\psi' \oplus -\psi))(\tau))$$

is cobordant to $(C(\tau), \psi(\tau))$, so that

$$[\partial P] = \sum_{\tau \in X^{(n)}} \partial P(\tau)\tau = \sum_{\tau \in X^{(n)}} (C(\tau), \psi(\tau))\tau = [C,\psi] \in H_n(X; L_0(\mathbb{Z})).$$

An application of (ii) gives

$$\partial(C',\psi') = [\partial P] = [C,\psi]$$

$$\in \ker(\mathbb{S}_n(X) \longrightarrow \overline{\mathbb{S}}_n(X)) = \operatorname{im}(H_n(X; L_0(\mathbb{Z})) \longrightarrow \mathbb{S}_n(X)),$$

verifying the commutativity of the diagram.

$\square$

REMARK 25.4 For a compact n-dimensional topological manifold M^n with $n \geq 5$ 18.5 gives that for $i \geq 1$

$$\mathbb{S}^{TOP}_\partial(M \times D^{i+4}, M \times S^{i+3}) = \mathbb{S}_{n+i+5}(M) = \mathbb{S}_{n+i+1}(M)$$

$$= \mathbb{S}^{TOP}_\partial(M \times D^i, M \times S^{i-1}) = \overline{\mathbb{S}}_{n+i+1}(M).$$

Also, the initial part of the exact sequence

$$0 \longrightarrow \mathbb{S}_{n+1}(M) \longrightarrow \overline{\mathbb{S}}_{n+1}(M) \longrightarrow H_n(M; L_0(\mathbb{Z})) \longrightarrow \mathbb{S}_n(M) \longrightarrow \overline{\mathbb{S}}_n(M)$$

can be expressed as

$$0 \longrightarrow \mathbb{S}^{TOP}(M) \longrightarrow \mathbb{S}^{TOP}_\partial(M \times D^4, M \times S^3) \longrightarrow L_0(\mathbb{Z}).$$

See Kirby and Siebenmann [84, Appendix C to Essay V], Nicas [118] and Cappell and Weinberger [28] for geometric interpretations of this almost

4-periodicity of the topological manifold structure sets.

$\square$

DEFINITION 25.5 The *4-periodic visible symmetric signature* of a finite n-dimensional geometric Poincaré complex X

$$\bar{\sigma}^*(X) \in \overline{VL}^n(X)$$

is the 0-connective visible symmetric signature defined in §9, which is an image of the 1/2-connective visible symmetric signature $\sigma^*(X) \in VL^n(X)$ defined in §15.

$\square$

DEFINITION 25.6 The *4-periodic total surgery obstruction* of a finite n-dimensional geometric Poincaré complex X is the image of the total surgery obstruction $s(X) \in \mathbb{S}_n(X)$ in the 4-periodic quadratic structure group

$$\bar{s}(X) \; = \; [s(X)] \in \overline{\mathbb{S}}_n(X) \;,$$

or equivalently as the boundary of the 4-periodic visible symmetric signature

$$\bar{s}(X) \; = \; \partial\bar{\sigma}^*(X) \in \overline{\mathbb{S}}_n(X) \;.$$

$\square$

PROPOSITION 25.7 *Let X be a finite n-dimensional geometric Poincaré complex.*
(i) *The following conditions on X are equivalent:*
  (a) *the 4-periodic total surgery obstruction is*
$$\bar{s}(X) = 0 \in \overline{\mathbb{S}}_n(X) \;,$$
  (b) *there exists an $\mathbb{L}^{\cdot}$-homology fundamental class $[X]_{\mathbb{L}} \in H_n(X; \mathbb{L}^{\cdot})$ with assembly*
$$A([X]_{\mathbb{L}}) \; = \; \bar{\sigma}^*(X) \in \overline{VL}^n(X) \;,$$
  (c) *there exists an $\overline{\mathbb{L}}^{\cdot}$-homology fundamental class $[X]_{\overline{\mathbb{L}}} \in H_n(X; \overline{\mathbb{L}}^{\cdot})$ with assembly*
$$\overline{A}([X]_{\overline{\mathbb{L}}}) \; = \; \sigma^*(X) \in VL^n(X) \;.$$

(ii) *If the Spivak normal fibration $\nu_X : X \longrightarrow BG$ admits a topological reduction $\tilde{\nu} : X \longrightarrow BTOP$ and there exists an element $x \in H_n(X; \overline{\mathbb{L}}.)$ such that the surgery obstruction of a corresponding normal map $(f, b) : M \longrightarrow X$ is*
$$\sigma_*(f, b) \; = \; \overline{A}(x) \in \mathrm{im}(\overline{A} : H_n(X; \overline{\mathbb{L}}.) \longrightarrow L_n(\mathbb{Z}[\pi_1(X)]))$$
then
$$\bar{s}(X) \; = \; 0 \in \overline{\mathbb{S}}_n(X) \;,$$
$$s(X) \; = \; [i(x)] \in \ker(\mathbb{S}_n(X) \longrightarrow \overline{\mathbb{S}}_n(X)) \; = \; \mathrm{im}(H_n(X; L_0(\mathbb{Z})) \longrightarrow \mathbb{S}_n(X)) \;,$$

with $i(x) \in H_n(X; L_0(\mathbb{Z}))$ the image of x under the natural map
$$H_n(X; \overline{\mathbb{L}}\boldsymbol{\cdot}) = H_n(X; \mathbb{L}\boldsymbol{\cdot}\langle 0 \rangle(\mathbb{Z})) \longrightarrow H_n(X; \pi_0(\mathbb{L}\boldsymbol{\cdot}\langle 0 \rangle(\mathbb{Z}))) = H_n(X; L_0(\mathbb{Z})).$$
PROOF (i) Immediate from the exact sequences
$$\ldots \longrightarrow H_n(X; \mathbb{L}\boldsymbol{\cdot}) \xrightarrow{A} \overline{VL}^n(X) \xrightarrow{\partial} \mathbb{S}_n(X) \longrightarrow H_{n-1}(X; \mathbb{L}\boldsymbol{\cdot}) \longrightarrow \ldots,$$
$$\ldots \longrightarrow H_n(X; \overline{\mathbb{L}}\boldsymbol{\cdot}) \xrightarrow{\overline{A}} VL^n(X) \xrightarrow{\partial} \overline{\mathbb{S}}_n(X) \longrightarrow H_{n-1}(X; \overline{\mathbb{L}}\boldsymbol{\cdot}) \longrightarrow \ldots.$$
(ii) The natural map $H_n(X; \overline{\mathbb{L}}\boldsymbol{\cdot}) \longrightarrow H_n(X; L_0(\mathbb{Z}))$ coincides with the composite
$$H_n(X; \overline{\mathbb{L}}\boldsymbol{\cdot}) \xrightarrow{([X]_{\mathbb{L}} \cap -)^{-1}} H^0(X; \overline{\mathbb{L}}\boldsymbol{\cdot}) = [X, L_0(\mathbb{Z}) \times G/TOP]$$
$$\xrightarrow{\text{projection}} [X, L_0(\mathbb{Z})] = H^0(X; L_0(\mathbb{Z})) \xrightarrow{[X] \cap -} H_n(X; L_0(\mathbb{Z})),$$
with $[X]_{\mathbb{L}} = f_*[M]_{\mathbb{L}} \in H_n(X; \mathbb{L}\boldsymbol{\cdot})$ the $\mathbb{L}\boldsymbol{\cdot}$-coefficient fundamental class of X determined by (f, b), and $[X] \in H_n(X)$ the ordinary ($\mathbb{Z}$-coefficient) fundamental class. The identities $\bar{s}(X) = 0$, $s(X) = [i(x)]$ follow from 25.3 (iv).
$\square$

The resolution obstruction of a compact n-dimensional ANR homology manifold M
$$i(M) \in H_n(M; L_0(\mathbb{Z})) = L_0(\mathbb{Z})$$
was defined by Quinn [133] as the difference of local and global codimension n signatures (24.23).

PROPOSITION 25.8 Let X be a finite n-dimensional geometric Poincaré complex which is homotopy equivalent to a compact n-dimensional ANR homology manifold M. The total surgery obstruction of X is the image of the resolution obstruction of M
$$s(X) = [i(M)]$$
$$\in \operatorname{im}(H_n(X; L_0(\mathbb{Z})) \longrightarrow \mathbb{S}_n(X)) = \ker(\mathbb{S}_n(X) \longrightarrow \overline{\mathbb{S}}_n(X)),$$
and the 4-periodic total surgery obstruction of X is
$$\bar{s}(X) = 0 \in \overline{\mathbb{S}}_n(X).$$
Moreover, a choice of homotopy equivalence $M \simeq X$ determines an $\mathbb{L}\boldsymbol{\cdot}$-homology fundamental class $[X]_{\mathbb{L}} \in H_n(X; \mathbb{L}\boldsymbol{\cdot})$ with assembly
$$A([X]_{\mathbb{L}}) = \bar{\sigma}^*(X) \in \overline{VL}^n(X),$$
and an $\overline{\mathbb{L}}\boldsymbol{\cdot}$-homology fundamental class $[X]_{\overline{\mathbb{L}}} \in H_n(X; \overline{\mathbb{L}}\boldsymbol{\cdot})$ with assembly
$$\overline{A}([X]_{\overline{\mathbb{L}}}) = \sigma^*(X) \in VL^n(X).$$
PROOF The total surgery obstruction of M is determined by a normal map $(f, b) \colon N \longrightarrow M$ from a compact topological manifold N associated to the

canonical topological reduction $\tilde{\nu}_M \colon M \longrightarrow BTOP$ of the Spivak normal fibration $\nu_M \colon M \longrightarrow BG$ (Ferry and Pedersen [50])

$$s(M) = -[\sigma_*(f,b)]$$

$$\in \operatorname{im}(L_n(\mathbb{Z}[\pi_1(M)]) \longrightarrow \mathbb{S}_n(M)) = \ker(\mathbb{S}_n(M) \longrightarrow H_{n-1}(M;\mathbb{L}_\bullet)) \ .$$

The canonical $\mathbb{L}_\bullet$-homology fundamental class of M is defined by

$$[M]_{\mathbb{L}} = f_*[N]_{\mathbb{L}} \in H_n(M;\mathbb{L}_\bullet)$$

with assembly the 4-periodic visible symmetric signature of M

$$A([M]_{\mathbb{L}}) = \bar{\sigma}^*(M) \in \overline{VL}^n(M) \ .$$

The canonical $\overline{\mathbb{L}}_\bullet$-homology fundamental class of M is defined by

$$[M]_{\overline{\mathbb{L}}} = (i(M),[M]_{\mathbb{L}}) \in H_n(M;\overline{\mathbb{L}}_\bullet) = H_n(M;L_0(\mathbb{Z})) \oplus H_n(M;\mathbb{L}_\bullet)$$

with assembly the 1/2-connective visible symmetric signature of M

$$\overline{A}([M]_{\overline{\mathbb{L}}}) = \sigma^*(M) \in VL^n(M) \ ,$$

and such that

$$s(M) = \partial \sigma^*(M) = [i(M)] = -[\sigma_*(f,b)] \in \mathbb{S}_n(M) \ .$$

The surgery obstruction of (f,b) is the assembly of

$$(-i(M),0) \in H_n(M;\overline{\mathbb{L}}_\bullet) = H_n(M;L_0(\mathbb{Z})) \oplus H_n(M;\mathbb{L}_\bullet) \ ,$$

that is

$$\sigma_*(f,b) = \overline{A}(-i(M),0) \in \operatorname{im}(\overline{A} \colon H_n(M;\overline{\mathbb{L}}_\bullet) \longrightarrow L_n(\mathbb{Z}[\pi_1(M)]))$$

$$= \ker(\partial \colon L_n(\mathbb{Z}[\pi_1(M)]) \longrightarrow \overline{\mathbb{S}}_n(M)) \ .$$

The 0-connective visible symmetric signature of M is

$$\sigma^*(M) = \sigma^*(N) - (1+T)\sigma_*(f,b) = A([M]_{\mathbb{L}}) + (1+T)\overline{A}(i(M))$$

$$= \overline{A}([M]_{\overline{\mathbb{L}}}) \in \operatorname{im}(\overline{A} \colon H_n(M;\overline{\mathbb{L}}_\bullet) \longrightarrow VL^n(M)) \ .$$

$\square$

Normal maps $(f,b) \colon (N,\nu_N) \longrightarrow (M,\tilde{\nu}_M)$ of closed n-dimensional manifolds are classified by the normal invariant (18.3 (i))

$$[f,b]_{\mathbb{L}} \in [M,G/TOP] = H^0(M;\mathbb{L}_\bullet) = H_n(M;\mathbb{L}_\bullet)$$

represented by the fibre homotopy trivialized difference $\tilde{\nu}_M - \nu_M \colon M \longrightarrow BTOP$, with ν_M the stable normal bundle.

DEFINITION 25.9 The *4-periodic normal invariant* of a normal map $(f,b) \colon (N,\nu_N) \longrightarrow (M,\tilde{\nu}_M)$ of compact n-dimensional ANR homology manifolds is

$$[f,b]_{\overline{\mathbb{L}}} = (i(N) - i(M),[f,b]_{\mathbb{L}})$$

$$\in [M,L_0(\mathbb{Z}) \times G/TOP] = H^0(M;\overline{\mathbb{L}}_\bullet) = H_n(M;\overline{\mathbb{L}}_\bullet)$$

$$= H_n(M;L_0(\mathbb{Z})) \oplus H_n(M;\mathbb{L}_\bullet)$$

with $[f,b]_L = t(b) \in [M, G/TOP]$ represented by the fibre homotopy trivialized difference $\tilde{\nu}_M - \nu_M : M \longrightarrow BTOP$, with ν_M the canonical topological reduction of the Spivak normal fibration.

$\square$

The 4-periodic normal invariant will also be written as

$$\bar{t}(i,b) = [f,b]_{\bar{L}} \in H_n(M; \bar{L}.)$$

in terms of

$$(i,b) = (i(N) - i(M), t(b)) \in H_n(M; L_0(\mathbb{Z})) \oplus H_n(M; L.) .$$

The surgery obstruction of a normal map $(f,b): N \longrightarrow M$ of closed n-dimensional ANR homology manifolds is the assembly of the 4-periodic normal invariant

$$\sigma_*(f,b) = \overline{A}([f,b]_{\bar{L}})$$

$$\in \text{im}(\overline{A}: H_n(M; \bar{L}.) \longrightarrow L_n(\mathbb{Z}[\pi_1(M)])) = \ker(L_n(\mathbb{Z}[\pi_1(M)]) \longrightarrow \bar{\mathbb{S}}_n(M)) .$$

EXAMPLE 25.10 Given a compact n-dimensional ANR homology manifold M let $(f,b): N \longrightarrow M$ be the normal map associated to the canonical topological reduction of M, with N a compact n-dimensional ANR topological manifold. The 4-periodic normal invariant of (f,b) is

$$[f,b]_{\bar{L}} = (-i(M), 0) \in H_n(M; \bar{L}.) = H_n(M; L_0(\mathbb{Z})) \oplus H_n(M; L.) .$$

$\square$

The structure invariant (18.3) of a homotopy equivalence $f: N \longrightarrow M$ of compact n-dimensional topological manifolds is the rel ∂ total surgery obstruction

$$s(f) = \bar{s}_\partial(N \times I \cup_f M, M \sqcup N) \in \mathbb{S}_{n+1}(M)$$

with image the normal invariant

$$[s(f)] = [f,b]_L$$

$$\in \text{im}(\mathbb{S}_{n+1}(M) \longrightarrow H_n(M; L.)) = \ker(A: H_n(M; L.) \longrightarrow L_n(\mathbb{Z}[\pi_1(M)])) .$$

of the normal map $(f,b): (N, \nu_N) \longrightarrow (M, (f^{-1})^* \nu_N)$, with ν_N the stable normal bundle.

DEFINITION 25.11 The *4-periodic structure invariant* of a homotopy equivalence $f: N \longrightarrow M$ of compact n-dimensional ANR homology manifolds is the rel ∂ 4-periodic total surgery obstruction

$$\bar{s}(f) = \bar{s}_\partial(N \times I \cup_f M, M \sqcup N) \in \bar{\mathbb{S}}_{n+1}(M)$$

with image the 4-periodic normal invariant

$$[\bar{s}(f)] = [f,b]_{\bar{L}} = \bar{t}(i(N) - i(M), [f,b]_L)$$

$$\in \text{im}(\bar{\mathbb{S}}_{n+1}(M) \longrightarrow H_n(M; \bar{L}.)) = \ker(\overline{A}: H_n(M; \bar{L}.) \longrightarrow L_n(\mathbb{Z}[\pi_1(M)]))$$

of the normal map $(f, b): (N, \nu_N) \longrightarrow (M, (f^{-1})^* \nu_N)$, with $\nu_N: N \longrightarrow BTOP$ the canonical topological reduction.   □

The resolution obstruction $i(M)$ is not a homotopy invariant, with $i(M) = i(N) \in L_0(\mathbb{Z})$ for a homotopy equivalence $f: N \longrightarrow M$ if and only if

$$\bar{s}(f) \in \ker(\overline{\mathbb{S}}_{n+1}(M) \longrightarrow H_n(M; L_0(\mathbb{Z}))) = \mathrm{im}(\mathbb{S}_{n+1}(M) \longrightarrow \overline{\mathbb{S}}_{n+1}(M)) .$$

As in §19 write the geometric Poincaré and mormal bordism spectra of a point as

$$\Omega_{\cdot}^P = \Omega_{\cdot}^P(\{*\}) \ , \ \Omega_{\cdot}^N = \Omega_{\cdot}^N(\{*\}) \ .$$

Define

$$\overline{\Omega}_{\cdot}^P = \mathrm{cofibre}(\mathbb{L}_{\cdot} \longrightarrow \overline{\mathbb{L}}_{\cdot} \vee \Omega_{\cdot}^P) = \mathrm{fibre}(\Omega_{\cdot}^N \longrightarrow \Sigma\overline{\mathbb{L}}_{\cdot}) ,$$

so that for any space K there is defined a commutative braid of exact sequences

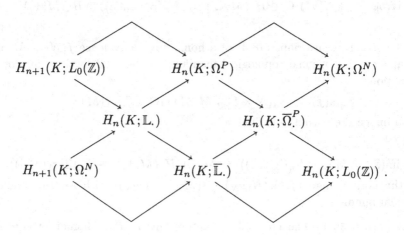

The relation between the 4-periodic theory and geometric Poincaré bordism is given by the following generalization of 19.6:

PROPOSITION 25.12 (i) *For any polyhedron K with finitely presented $\pi_1(K)$*

and $n \geq 5$ there is defined a commutative braids of exact sequences

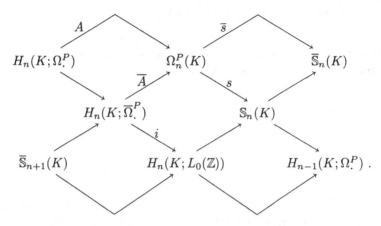

(ii) *A finite n-dimensional geometric Poincaré complex X has 4-periodic total surgery obstruction $\bar{s}(X) = 0 \in \bar{\mathbb{S}}_n(X)$ if (and for $n \geq 5$ only if) there exists an $\overline{\Omega}^{P}_{.}$-homology fundamental class $[X]_{\overline{P}} \in H_n(X; \overline{\Omega}^{P}_{.})$ with assembly the geometric Poincaré bordism class of $1: X \longrightarrow X$,*

$$\overline{A}([X]_{\overline{P}}) = (1: X \longrightarrow X) \in \Omega^{P}_n(X) ,$$

in which case the total surgery obstruction of X is given by

$$s(X) = [i[X]_{\overline{P}}] \in \operatorname{im}(H_n(X; L_0(\mathbb{Z})) \longrightarrow \mathbb{S}_n(X)) = \ker(\mathbb{S}_n(X) \longrightarrow \bar{\mathbb{S}}_n(X)) .$$

□

The structure set $\mathbb{S}^{HTOP}(M)$ of a compact n-dimensional ANR homology manifold M is defined to be the set of h-cobordism classes of pairs

(compact n-dimensional ANR homology manifold N,

homotopy equivalence $f: N \longrightarrow M$) .

REMARK 25.13 In March 1992 Bryant–Mio and Ferry–Weinberger announced the existence of nonresolvable compact n-dimensional ANR homology manifolds realizing the Quinn resolution obstruction in each dimension $n \geq 5$. It follows that the 4-periodic total surgery obstruction of a finite n-dimensional geometric Poincaré complex X is such that $\bar{s}(X) = 0 \in \bar{\mathbb{S}}_n(X)$ if (and for $n \geq 5$ only if) X is homotopy equivalent to a compact n-dimensional ANR homology manifold M, in which case the total surgery obstruction of X is the image of the resolution obstruction of M

$$s(X) = [i(M)] \in \operatorname{im}(H_n(X; L_0(\mathbb{Z})) \longrightarrow \mathbb{S}_n(X)) = \ker(\mathbb{S}_n(X) \longrightarrow \bar{\mathbb{S}}_n(X)) .$$

It also follows that the structure set $\mathbb{S}^{HTOP}(M)$ of a compact n-dimensional

ANR homology manifold M fits into an exact sequence of pointed sets

$$\cdots \longrightarrow L_{n+1}(\mathbb{Z}[\pi_1(M)]) \longrightarrow \mathbb{S}^{HTOP}(M)$$
$$\longrightarrow [M, L_0(\mathbb{Z}) \times G/TOP] \longrightarrow L_n(\mathbb{Z}[\pi_1(M)])$$

with the 4-periodic structure and normal invariants defining a bijection with the 4-periodic algebraic surgery exact sequence

$$\cdots \longrightarrow L_{n+1}(\mathbb{Z}[\pi]) \longrightarrow \mathbb{S}^{HTOP}(M) \longrightarrow [M, L_0(\mathbb{Z}) \times G/TOP] \longrightarrow L_n(\mathbb{Z}[\pi])$$

$$\bar{s}\Big\downarrow \simeq \qquad\qquad \bar{t}\Big\downarrow \simeq$$

$$\cdots \longrightarrow L_{n+1}(\mathbb{Z}[\pi]) \longrightarrow \overline{\mathbb{S}}_{n+1}(M) \longrightarrow H_n(M; \overline{\mathbb{L}}.) \overset{\overline{A}}{\longrightarrow} L_n(\mathbb{Z}[\pi])$$

with $\pi = \pi_1(M)$. The generator $1 \in L_0(\mathbb{Z})$ is realized by a nonresolvable compact n-dimensional ANR homology manifold Σ^n with a homotopy equivalence $f \colon \Sigma^n \longrightarrow S^n$ such that

$$i(\Sigma^n) = \bar{s}(f) = 1 \in \mathbb{S}^{HTOP}(S^n) = \overline{\mathbb{S}}_{n+1}(S^n) = L_0(\mathbb{Z}) = \mathbb{Z} .$$

        □

REMARK 25.14 The 4-periodic total surgery obstruction $\bar{s}(B\pi)$ of the classifying space $B\pi$ of an n-dimensional Novikov group π takes value in $\overline{\mathbb{S}}_n(B\pi) = \{0\}$, so that by 25.13 $B\pi$ is homotopy equivalent to a compact n-dimensional ANR homology manifold M (at least for $n \geq 5$) such that

$$s(M) = i(M) \in \mathbb{S}_n(M) = \mathbb{S}_n(B\pi) = L_0(\mathbb{Z}) ,$$
$$\mathbb{S}^{HTOP}(M) = \overline{\mathbb{S}}_{n+1}(M) = \overline{\mathbb{S}}_{n+1}(B\pi) = \{0\} .$$

In particular, M is resolvable if and only if M is homotopy equivalent to a manifold. Ferry and Pedersen [50] used bounded surgery to show that any compact ANR homology manifold in the homotopy type of a compact aspherical manifold with a Novikov fundamental group π is resolvable.

        □

EXAMPLE 25.15 The 4-periodic quadratic L-theory assembly maps for T^n are isomorphisms (24.16)

$$\overline{A} \colon H_*(T^n; \overline{\mathbb{L}}.) \overset{\simeq}{\longrightarrow} L_*(\mathbb{Z}[\mathbb{Z}^n]) ,$$

so that

$$\overline{\mathbb{S}}_*(T^n) = 0 , \quad \mathbb{S}_*(T^n) = H_*(T^n; L_0(\mathbb{Z})) \quad (* \geq n)$$

and the '4-periodic' geometric Poincaré bordism assembly maps of 25.12 (i) are isomorphisms for T^n

$$\overline{A} \colon H_*(T^n; \overline{\Omega}^P.) \overset{\simeq}{\longrightarrow} \Omega^P_*(T^n) .$$

In each dimension $n \geq 4$ the element

$$i(1) = (1,0) \in \Omega_n^P(T^n) = H_n(T^n; \overline{\Omega}^P_\bullet) = L_0(\mathbb{Z}) \oplus H_n(T^n; \Omega^P_\bullet)$$

is represented by a normal map $(f,b): Y \longrightarrow T^n$ from a topologically reducible finite n-dimensional geometric Poincaré complex Y, with surgery obstruction

$$\sigma_*(f,b) = (1,0) \in L_n(\mathbb{Z}[\mathbb{Z}^n]) = L_0(\mathbb{Z}) \oplus \left(\sum_{k=1}^n \binom{n}{k} L_k(\mathbb{Z}) \right)$$

and codimension n quadratic signature

$$B\sigma_*(f,b) = (\mathbb{Z}^8, E_8) = 1 \in L_0(\mathbb{Z}) = \mathbb{Z} .$$

The visible symmetric signature of Y is

$$\sigma^*(Y) = \sigma^*(T^n) + (1+T)\sigma_*(f,b) = (1,(9,0))$$

$$\in VL^n(T^n) = L_0(\mathbb{Z}) \oplus L^n(\mathbb{Z}[\mathbb{Z}^n]) ,$$

with components the 4-periodic visible symmetric signature

$$\overline{\sigma}^*(Y) = (9,0)$$

$$\in \overline{VL}^n(T^n) = L^n(\mathbb{Z}[\mathbb{Z}^n]) = L^0(\mathbb{Z}) \oplus \left(\sum_{k=1}^n \binom{n}{k} L^k(\mathbb{Z}) \right) ,$$

and the image of the total surgery obstruction

$$s(Y) = 1 \in \mathbb{S}_n(T^n) = L_0(\mathbb{Z}) .$$

The actual total surgery obstruction $s(Y) \in \mathbb{S}_n(Y)$ is the image of $1 \in H_n(Y; L_0(\mathbb{Z})) = L_0(\mathbb{Z})$ under the map in the exact sequence

$$\cdots \longrightarrow \overline{\mathbb{S}}_{n+1}(Y) \longrightarrow H_n(Y; L_0(\mathbb{Z})) \longrightarrow \mathbb{S}_n(Y) \longrightarrow \overline{\mathbb{S}}_n(Y) \longrightarrow \cdots .$$

For $n = 4$ there is an explicit construction of Y^4 in Quinn [133, §2]

$$Y^4 = \left((T^4)^{(3)} \vee \bigvee_{48} S^2 \right) \cup_\alpha e^4 ,$$

attaching a 4-cell to a 3-complex by a Whitehead product α realizing the nonsingular quadratic form over $\mathbb{Z}[\mathbb{Z}^4]$ of rank 48 representing the image of $1 \in L_0(\mathbb{Z})$ under the geometrically significant split injection of Ranicki [137]

$$\sigma^*(T^4) \otimes - : L_0(\mathbb{Z}) \longrightarrow L_4(\mathbb{Z}[\mathbb{Z}^4]) .$$

For $n \geq 5$ the product

$$(f,b) \times \mathrm{id.} : Y^n = Y^4 \times T^{n-4} \longrightarrow T^4 \times T^{n-4} = T^n$$

has codimension n quadratic signature 1 by the surgery product formula of Ranicki [142]. In each case the 4-periodic total surgery obstruction is $\overline{s}(Y) = 0 \in \overline{\mathbb{S}}_n(Y)$, and the total surgery obstruction $s(Y) \in \mathbb{S}_n(Y)$ is the image of $1 \in H_n(Y; L_0(\mathbb{Z})) = L_0(\mathbb{Z})$. By 25.13 each Y^n is homotopy equivalent to a nonresolvable compact ANR homology manifold.

□

See Appendix C for some further discussion of assembly and controlled topology.

REMARK 25.16 See Hambleton and Hausmann [63, p. 234] for the construction in each dimension $n \geq 4$ of a finite n-dimensional geometric Poincaré complex Y such that
  (i) the fundamental group $\pi_1(Y) = \pi$ is an n-dimensional Novikov group (in the class of Cappell [24]) with the homology of S^n, such that
$$L_*(\mathbb{Z}[\pi]) \ = \ H_*(B\pi; \overline{\mathbb{L}}.) \ = \ H_*(S^n; \overline{\mathbb{L}}.) \ = \ L_{*-n}(\mathbb{Z}) \oplus L_*(\mathbb{Z}) \ ,$$
 (ii) the classifying map $Y \longrightarrow B\pi$ induces an isomorphism of integral homology, with
$$H_*(Y) \ = \ H_*(B\pi) \ = \ H_*(S^n) \ ,$$
(iii) the 4-periodic visible symmetric signature of Y is
$$\overline{\sigma}^*(Y) \ = \ (9,0)$$
$$\in \overline{VL}^n(Y) \ = \ H_n(Y; \overline{\mathbb{L}}^{\cdot}) \ = \ H_n(S^n; \overline{\mathbb{L}}^{\cdot}) \ = \ L^0(\mathbb{Z}) \oplus L^n(\mathbb{Z}) \ ,$$
and the 4-periodic total surgery obstruction of Y is
$$\overline{s}(Y) \ = \ \partial\overline{\sigma}^*(Y) \ = \ 0 \in \overline{\mathbb{S}}_n(Y) \ = \ \{0\} \ ,$$
(iv) the visible symmetric signature of Y is
$$\sigma^*(Y) \ = \ (1, \overline{\sigma}^*(Y)) \in VL^n(Y) \ = \ L_0(\mathbb{Z}) \oplus \overline{VL}^n(Y) \ ,$$
and the total surgery obstruction of Y is
$$s(Y) \ = \ \partial\sigma^*(Y) \ = \ 1$$
$$\in \mathbb{S}_n(Y) \ = \ \text{coker}(H_n(Y; \mathbb{L}.) \longrightarrow L_n(\mathbb{Z}[\pi]))$$
$$= \ \text{coker}(H_n(S^n; \mathbb{L}.) \longrightarrow H_n(S^n; \overline{\mathbb{L}}.)) \ = \ L_0(\mathbb{Z}) \ = \ \mathbb{Z} \ .$$
By 25.13 each Y is homotopy equivalent to a nonresolvable compact ANR homology manifold.

□

REMARK 25.17 Let M be a compact n-dimensional ANR homology manifold, and let $(f,b): N \longrightarrow M$ be a normal map associated to the canonical topological reduction ν_M of the Spivak normal fibration, with N a genuine manifold (as in the proof of 25.8). The Poincaré dual of the $\mathcal{L}$-genus
$$\mathcal{L}(M) \ = \ \mathcal{L}(-\nu_M) \in H^{4*}(M; \mathbb{Q})$$
is the rational part of the canonical $\mathbb{L}^{\cdot}$-homology fundamental class $[M]_{\mathbb{L}}$
$$[M]_{\mathbb{Q}} \cap \mathcal{L}(M) \ = \ [M]_{\mathbb{L}} \otimes 1 \in H_n(M; \mathbb{L}^{\cdot}) \otimes \mathbb{Q} \ = \ H_{n-4*}(M; \mathbb{Q}) \ ,$$
with $[M]_{\mathbb{Q}} \in H_n(M; \mathbb{Q})$ the $\mathbb{Q}$-coefficient fundamental class. Every map $g: M \longrightarrow S^{n-i}$ can be made symmetric Poincaré transverse at a point in

S^{n-i}, with $[M]_{\mathbb{Z}}\cap g^*(1) \in H_i(M)$ represented by an i-dimensional symmetric Poincaré complex 'g^{-1}(pt.)' over $\mathbb{Z}$ such that

$$g_*[M]_{\mathbb{L}} = \sigma^*(\,'g^{-1}(\text{pt.})') \in \dot{H}_n(S^{n-i};\mathbb{L}^{\textbf{.}}) = L^i(\mathbb{Z})\ ,$$

with $1 \in H^{n-i}(S^{n-i}) = \mathbb{Z}$. The composite $gf : N \longrightarrow S^{n-i}$ can be made topologically transverse, and '$(gf)^{-1}$(pt.)' is the symmetric complex of a framed i-dimensional submanifold $(gf)^{-1}$(pt.) $\subset N$. For $i = 4j$ the Hirzebruch signature formula gives

$$\begin{aligned}
\sigma^*(\,'(gf)^{-1}(\text{pt.})') &= \sigma^*((gf)^{-1}(\text{pt.})) \\
&= \langle \mathcal{L}_j(-\nu_N), [N]_{\mathbb{Q}} \cap (gf)^*(1)\rangle \\
&= \langle \mathcal{L}_j(-\tilde{\nu}_N), [N]_{\mathbb{Q}} \cap g^*(1)\rangle \in L^{4j}(\mathbb{Z}) = \mathbb{Z}\ .
\end{aligned}$$

The algebraic normal map (2.16 (i)) of $4j$-dimensional symmetric Poincaré complexes over $\mathbb{Z}$

$$'(f,b)|'\ :\ '(gf)^{-1}(\text{pt.})' \longrightarrow\ 'g^{-1}(\text{pt.})'$$

has quadratic signature the assembly of a $4j$-dimensional component of $[i(M)] \in H_n(M;\overline{\mathbb{L}}.)$

$$\sigma_*(\,'(f,b)|') = \left\{ \begin{array}{ll} 0 & \text{if } j > 0 \\ i(M)g_*[M] & \text{if } j = 0 \end{array} \right\} \in L_{4j}(\mathbb{Z}) = \mathbb{Z}\ ,$$

with symmetrization

$$(1+T)\sigma_*(\,'(f,b)|') = \sigma^*(\,'(gf)^{-1}(\text{pt.})') - \sigma^*(\,'g^{-1}(\text{pt.})') \in L^{4j}(\mathbb{Z}) = \mathbb{Z}\ .$$

Every element $x \in H_{4j}(M;\mathbb{Q})$ is of the form $x = [M]_{\mathbb{Q}} \cap g^*(1)/m$ for some $g : M \longrightarrow S^{n-4j}$, $m \in \mathbb{Z}\backslash\{0\}$. The $\mathcal{L}$-genus of a compact ANR homology manifold is thus characterized by the signatures of symmetric Poincaré subcomplexes

$$\langle \mathcal{L}_j(M), - \rangle\ :\ H_{4j}(M;\mathbb{Q}) \longrightarrow L^{4j}(\mathbb{Z}) \otimes \mathbb{Q} = \mathbb{Q}\ ;$$

$$x = [M]_{\mathbb{Q}} \cap g^*(1)/m \longrightarrow$$

$$\langle \mathcal{L}_j(M), x \rangle = \left\{ \begin{array}{ll} \sigma^*(\,'g^{-1}(\text{pt.})')/m & \text{if } j > 0 \\ \sigma^*(\,'g^{-1}(\text{pt.})')/m + 8i(M)x & \text{if } j = 0\ , \end{array} \right.$$

generalizing the combinatorial definition due to Thom of the $\mathcal{L}$-genus of a PL manifold (and hence the rational Pontrjagin classes) using the signatures of submanifolds – see Milnor and Stasheff [111, §20], and also Appendix C.16 below.

□

§26. Surgery with coefficients

There is also a version of the total surgery obstruction theory for the Λ-homology coefficient surgery theory of Cappell and Shaneson [25], which arises in the surgery classification of codimension 2 submanifolds (cf. Ranicki [143, §§7.8,7.9]).

For Λ-homology surgery the Wall L-groups $L_*(\mathbb{Z}[\pi])$ of a group ring $\mathbb{Z}[\pi]$ are replaced by the Γ-groups $\Gamma_*(\mathcal{F})$ of a 'locally epic' morphism $\mathcal{F}\colon \mathbb{Z}[\pi]\longrightarrow\Lambda$ of rings with involution. (A ring morphism is *locally epic* if every finite subset $\Lambda_0 \subset \Lambda$ there exists a unit $u \in \Lambda$ such that $u\Lambda_0 \subseteq \operatorname{im}(\mathcal{F})$.) By definition, $\Gamma_{2i}(\mathcal{F})$ is the Witt group of Λ-nonsingular $(-)^i$-quadratic forms over $\mathbb{Z}[\pi]$, and $\Gamma_{2i+1}(\mathcal{F})$ is the Witt group of Λ-nonsingular $(-)^i$-quadratic formations over $\mathbb{Z}[\pi]$. The forgetful map $\Gamma_n(\mathcal{F})\longrightarrow L_n(\Lambda)$ is $\left\{ \begin{array}{l} \text{onto} \\ \text{one–one} \end{array} \right.$ for $n \left\{ \begin{array}{l} \text{even} \\ \text{odd} \end{array} \right.$, with $\Gamma_*(1\colon \mathbb{Z}[\pi]\longrightarrow\mathbb{Z}[\pi]) = L_*(\mathbb{Z}[\pi])$. In the terminology of §3 the Γ-groups are given by

$$\Gamma_*(\mathcal{F}) = L_*(\mathbb{A}(\mathbb{Z}[\pi]), \mathbb{B}(\mathbb{Z}[\pi]), \mathbb{C}(\mathcal{F}))$$

with $\mathbb{A}(\mathbb{Z}[\pi])$ the additive category of f.g. free $\mathbb{Z}[\pi]$-modules, $\mathbb{B}(\mathbb{Z}[\pi])$ the category of finite chain complexes in $\mathbb{A}(\mathbb{Z}[\pi])$ and $\mathbb{C}(\mathcal{F}) \subseteq \mathbb{B}(\mathbb{Z}[\pi])$ the subcategory of the chain complexes C which are Λ-contractible, i.e. such that $\Lambda\otimes_{\mathbb{Z}[\pi]}C$ is a contractible chain complex in $\mathbb{A}(\Lambda)$. For the fundamental group $\pi = \pi_1(X)$ of a simplicial complex X the Λ-coefficient version of the algebraic π-π theorem of §10 gives the identification

$$\Gamma_n(\mathcal{F}) = L_n(\mathbb{A}(\mathbb{Z}, X), \mathbb{B}(\mathbb{Z}, X), \mathbb{C}(\mathbb{Z}, X, \Lambda))$$

with $\mathbb{C}(\mathbb{Z}, X, \Lambda) \subset \mathbb{C}(\mathbb{Z}, X)$ the subcategory of Λ-contractible complexes, so that $\Gamma_n(\mathcal{F})$ is the cobordism group of n-dimensional quadratic cycles in X which are globally Λ-Poincaré. Define

$$\mathbb{S}_n(X; \Lambda) = L_n(\mathbb{A}(\mathbb{Z}, X), \mathbb{C}\langle 1\rangle(\mathbb{Z}, X, \Lambda), \mathbb{C}\langle 1\rangle(\mathbb{Z})_*(X)) ,$$

the cobordism group of 1-connective $(n-1)$-dimensional quadratic cycles in X which are locally Poincaré and globally Λ-contractible. The groups $\mathbb{S}_*(X; \Lambda)$ are the Λ-coefficient total surgery obstruction groups of Ranicki [143, p. 774], which fit into a Γ-theory assembly exact sequence

$$\ldots \longrightarrow H_n(X; \mathbb{L}.) \xrightarrow{\ A\ } \Gamma_n(\mathcal{F}) \longrightarrow \mathbb{S}_n(X; \Lambda) \longrightarrow H_{n-1}(X; \mathbb{L}.) \longrightarrow \ldots$$

with

$$A\colon H_n(X; \mathbb{L}.) \xrightarrow{\ A\ } L_n(\mathbb{Z}[\pi]) \longrightarrow \Gamma_n(\mathcal{F}) .$$

There is a Λ-coefficient version of the visible symmetric L-theory, with a commutative braid of exact sequences

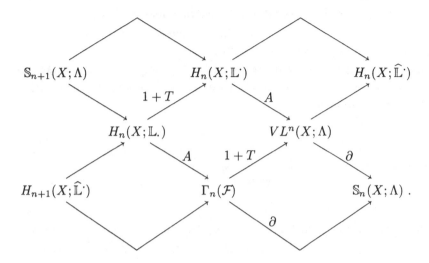

The Λ-*coefficient visible symmetric L-group* $VL^n(X;\Lambda)$ of a simplicial complex X is the cobordism group of 1/2-connective globally Λ-Poincaré n-dimensional algebraic normal complexes in $\mathbb{A}(\mathbb{Z},X)$.

An n-*dimensional geometric* Λ-*Poincaré complex* X is an n-dimensional normal complex with Λ-Poincaré duality isomorphisms

$$[X]\cap : H^{n-*}(X;\Lambda) \longrightarrow H_*(X;\Lambda) ,$$

with respect to a locally epic morphism $\mathcal{F}:\mathbb{Z}[\pi_1(X)]\longrightarrow\Lambda$ of rings with involution. The Λ-*coefficient visible symmetric signature* of X is defined by

$$\sigma^*(X;\Lambda) = (\Delta(X),\phi) \in VL^n(X;\Lambda) ,$$

working as in 16.5 to make $(\Delta(X),\phi)$ 1/2-connective. The Λ-*coefficient total surgery obstruction* of X is defined by

$$s(X;\Lambda) = \partial\sigma^*(X;\Lambda) \in \mathbb{S}_n(X;\Lambda) .$$

As in the absolute case $\mathcal{F} = 1:\mathbb{Z}[\pi]\longrightarrow\Lambda = \mathbb{Z}[\pi]$ (17.4):

PROPOSITION 26.1 *The* Λ-*coefficient total surgery obstruction of a finite n-dimensional geometric* Λ-*Poincaré complex* X *is such that* $s(X;\Lambda) = 0$ *if (and for* $n \geq 5$ *only if)* X *is* Λ-*homology equivalent to a compact n-dimensional topological manifold.*

□

The Λ-*coefficient structure set* $\mathbb{S}^{\Lambda TOP}(M)$ of a compact n-dimensional topological manifold M^n is the pointed set of Λ-coefficient h-cobordism classes of pairs

> (compact n-dimensional topological manifold N^n,
>
> Λ-homology equivalence $h\colon N^n \longrightarrow M^n$)

with base point $(M, \mathrm{id}) = 0 \in \mathbb{S}^{\Lambda TOP}(M)$.

As in the absolute case (15.19, 18.2) for any simplicial complex M and any locally epic $\mathcal{F}\colon \mathbb{Z}[\pi_1(M)] \longrightarrow \Lambda$ there is defined a Λ-*coefficient algebraic surgery exact sequence*

$$\dots \longrightarrow \Gamma_{n+1}(\mathcal{F}) \longrightarrow \mathbb{S}_{n+1}(M;\Lambda) \longrightarrow H_n(M;\mathbb{L}.) \xrightarrow{\ A\ } \Gamma_n(\mathcal{F}) \longrightarrow \dots ,$$

and for any n-dimensional manifold M^n with $n \geq 5$ there is defined a Λ-*coefficient geometric surgery exact sequence*

$$\dots \longrightarrow \Gamma_{n+1}(\mathcal{F}) \longrightarrow \mathbb{S}^{\Lambda TOP}(M) \longrightarrow [M, G/TOP] \longrightarrow \Gamma_n(\mathcal{F}) .$$

As in the absolute case (18.5):

PROPOSITION 26.2 *The Λ-coefficient algebraic and geometric surgery exact sequences of a compact n-dimensional topological manifold M^n with $n \geq 5$ are related by an isomorphism*

$$
\begin{array}{ccccccc}
\dots \longrightarrow \Gamma_{n+1}(\mathcal{F}) & \longrightarrow & \mathbb{S}^{\Lambda TOP}(M) & \longrightarrow & [M, G/TOP] & \longrightarrow & \Gamma_n(\mathcal{F}) \\
\Big\| & & \Big\downarrow s \cong & & \Big\downarrow t \cong & & \Big\| \\
\dots \longrightarrow \Gamma_{n+1}(\mathcal{F}) & \xrightarrow{\ \partial\ } & \mathbb{S}_{n+1}(M;\Lambda) & \longrightarrow & H_n(M;\mathbb{L}.) & \xrightarrow{\ A\ } & \Gamma_n(\mathcal{F}) .
\end{array}
$$

$\square$

The relative Γ-groups $\Gamma_*(\Phi)$ of a commutative square of locally epic morphisms of rings with involution

$$
\begin{array}{ccc}
\mathbb{Z}[\pi] & \xrightarrow{\ 1\ } & \mathbb{Z}[\pi] \\
\Big\downarrow \mathcal{F}' & \Phi & \Big\downarrow \mathcal{F} \\
\Lambda' & \longrightarrow & \Lambda
\end{array}
$$

fit into an exact sequence

$$\dots \longrightarrow \Gamma_n(\mathcal{F}') \longrightarrow \Gamma_n(\mathcal{F}) \longrightarrow \Gamma_n(\Phi) \longrightarrow \Gamma_{n-1}(\mathcal{F}') \longrightarrow \dots .$$

By [143, 2.4.6] (a special case of 3.9)

$$\Gamma_n(\Phi) = L_n(\mathbb{A}(\mathbb{Z}[\pi]), \mathbb{C}(\mathcal{F}), \mathbb{C}(\mathcal{F}'))$$

is the cobordism group of $(n-1)$-dimensional quadratic complexes in $\mathbb{A}(\mathbb{Z}[\pi])$ which are Λ-contractible and Λ'-Poincaré. The various groups are related by a commutative braid of exact sequences

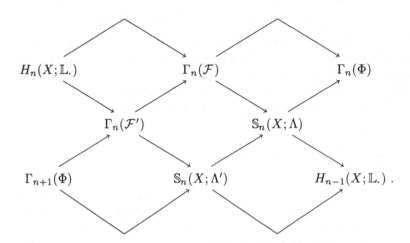

PROPOSITION 26.3 *The Λ-coefficient total surgery obstruction $s(X;\Lambda) \in$ $\mathbb{S}_n(X;\Lambda)$ of a finite n-dimensional geometric Λ-Poincaré complex X has image $[s(X;\Lambda)] = 0 \in \Gamma_n(\Phi)$ if (and for $n \geq 5$ only if) X is Λ-homology equivalent to a finite n-dimensional geometric Λ'-Poincaré complex.*

□

EXAMPLE 26.4 For the augmentation

$$\mathcal{F} : \mathbb{Z}[\mathbb{Z}] = \mathbb{Z}[z,z^{-1}] \longrightarrow \mathbb{Z} \;;\; z \longrightarrow 1 \;\; (\bar{z} = z^{-1})$$

the $\mathbb{Z}$-coefficient rel ∂ total surgery obstruction defines a bijection

$$\mathbb{S}_{\partial}^{\mathbb{Z}\,TOP}(D^{n+1} \times S^1, S^n \times S^1) \xrightarrow{\;\approx\;} \mathbb{S}_{n+3}(S^1;\mathbb{Z}) \;;\; f \longrightarrow s^{\partial}(f;\mathbb{Z}) \;.$$

Since $\mathbb{S}_*(S^1) = 0$ (for $* \geq 2$) the $\mathbb{Z}$-homology structure group $\mathbb{S}_{n+3}(S^1;\mathbb{Z})$ is isomorphic to the relative Γ-group $\Gamma_{n+3}(\Phi)$ of the commutative square

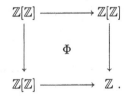

Cappell and Shaneson [25] identified $\Gamma_{n+3}(\Phi)$ for $n \geq 4$ with the cobordism group C_n of locally flat knots $k: S^n \subset S^{n+2}$. The following natural (iso)morphism $C_n \longrightarrow \mathbb{S}_{n+3}(S^1;\mathbb{Z})$ was defined in Ranicki [143, 7.9.4]. The complement of a knot k is the $(n+2)$-dimensional manifold with boundary

$$(X,\partial X) = (\mathrm{cl}(S^{n+2} \backslash U), S^n \times S^1)$$

with $U = k(S^n) \times D^2 \subset S^{n+2}$ a closed regular neighbourhood of $k(S^n)$ in

S^{n+2}. The knot complement is equipped with a normal map

$$(f,b) : (X, \partial X) \longrightarrow (D^{n+1} \times S^1, S^n \times S^1)$$

which is a $\mathbb{Z}$-homology equivalence, and the identity on the boundary. The Blanchfield complex of k ([143, p. 822]) is the $\mathbb{Z}$-contractible $(n+2)$-dimensional quadratic Poincaré complex $\sigma_*(f,b) = (C, \psi)$ over $\mathbb{Z}[\mathbb{Z}]$, with $H_*(C) = \dot{H}_*(\overline{X})$ the reduced homology of the canonical infinite cyclic cover $\overline{X}$ of X. For $n \geq 4$ the cobordism class of the knot k is given by

$$(k \colon S^n \subset S^{n+2}) \;=\; s^\partial(f;\mathbb{Z}) \;=\; (C, \psi)$$

$$\in \mathbb{S}_\partial^{\mathbb{Z}\,TOP}(D^{n+1} \times S^1, S^n \times S^1) \;=\; \mathbb{S}_{n+3}(S^1; \mathbb{Z}) \;=\; \Gamma_{n+3}(\Phi) \;=\; C_n \;.$$

$\square$

REMARK 26.5 Let $f \colon Y \longrightarrow X$ be a map of compact polyhedra, with X an n-dimensional geometric Poincaré complex. If f is a homotopy equivalence the induced maps $f_* \colon \mathbb{S}_*(Y) \longrightarrow \mathbb{S}_*(X)$ are isomorphisms, and Y is an n-dimensional geometric Poincaré complex with total surgery obstruction

$$s(Y) \;=\; (f_*)^{-1} s(X) \in \mathbb{S}_n(Y) \;.$$

If f is a $\mathbb{Z}$-homology equivalence (= stable homotopy equivalence for finite CW complexes) the induced maps $f_* \colon H_*(Y; \mathbb{L}.) \longrightarrow H_*(X; \mathbb{L}.)$ are isomorphisms, and Y is an n-dimensional $\mathbb{Z}$-coefficient geometric Poincaré complex with Spivak normal fibration $\nu_Y = f^* \nu_X \colon Y \longrightarrow BG$ and topological reducibility obstruction

$$t(Y) \;=\; (f_*)^{-1} t(X) \in H_{n-1}(Y; \mathbb{L}.) \;.$$

The image of the total surgery obstruction $s(X) \in \mathbb{S}_n(X)$

$$[s(X)] \in \mathbb{S}_n(f) \;=\; L_n(f \colon \mathbb{Z}[\pi_1(Y)] \longrightarrow \mathbb{Z}[\pi_1(X)])$$

is an obstruction to Y being a geometric Poincaré complex with $f_* s(Y) = s(X)$. See Hambleton and Hausmann [63] for a study of this 'minus' problem for geometric Poincaré complexes, in the context of the Quillen plus construction.

$\square$

REMARK 26.6 The results of §§16–25 for n-dimensional manifolds and geometric Poincaré complexes with $n \geq 5$ also apply to the case $n = 4$, provided the fundamental group π_1 is not too large – see Freedman and Quinn [53]. However, as explained in [53, 11.8] there is a failure of 4-dimensional homology surgery already in the case $\pi_1(X) = \mathbb{Z}$, which is detected by the Casson–Gordon invariants of the cobordism group C_1 of classical knots $k \colon S^1 \subset S^3$. Thus the results of §26 do not in general apply to $n = 4$.

$\square$

Appendix A. The nonorientable case

This appendix deals with the modifications required for the twisted case, in which the simplicial complex K is equipped with a nontrivial double cover K^w. In particular, the universal assembly functors of §9 are generalized to w-twisted universal assembly functors of algebraic bordism categories

$$A : \Lambda(R)_*(K, w) \longrightarrow \Lambda(R, K, w) ,$$

$$A : \Lambda(R, K, w) \longrightarrow \Lambda(R[\pi]^w) ,$$

$$A : \widehat{\Lambda}(R, K, w) \longrightarrow \widehat{\Lambda}(R[\pi]^w) .$$

The nonorientable version of L-theory appears in the codimension 1 splitting obstruction theory of type (C) (as described in §23), and has been used to determine the image of the assembly map $A\colon H_*(B\pi; \mathbb{L}.) \longrightarrow L_*(\mathbb{Z}[\pi])$ for finite groups π in the orientable case, using appropriate index 2 subgroups – see Wall [176, 12C], Hambleton [62], Cappell and Shaneson [26], Harsiladze [69], Hambleton, Taylor and Williams [68] and Hambleton, Milgram, Taylor and Williams [66].

The fundamental group of the double cover K^w

$$\pi_1(K^w) = \pi^w$$

is a subgroup of $\pi_1(K) = \pi$ of index 2, and the orientation character is given by

$$w : \pi \longrightarrow \{\pm 1\} ; \ g \longrightarrow \begin{cases} +1 & \text{if } g \in \pi^w \\ -1 & \text{otherwise .} \end{cases}$$

Let R be a commutative ring, as before, and let $R[\pi]^w$ be the group ring $R[\pi]$ with the w-twisted involution as in 1.4. The tensor product over $R[\pi]^w$ of f.g. free $R[\pi]$-modules M, N is the abelian group

$$M \otimes_{R[\pi]^w} N = M \otimes_{\Lambda(R[\pi]^w)} N$$

$$= M \otimes_R N / \{x \otimes gy - w(g)g^{-1}x \otimes y \mid x \in M, \, y \in N, \, g \in \pi\} .$$

Regard M, N as $R[\pi^w]$-modules via the inclusion $R[\pi^w] \longrightarrow R[\pi]$, and let $\mathbb{Z}[\mathbb{Z}_2]$ act on the abelian group $M \otimes_{R[\pi^w]} N$ by

$$T : M \otimes_{R[\pi^w]} N \longrightarrow M \otimes_{R[\pi^w]} N ; \ x \otimes y \longrightarrow tx \otimes ty ,$$

using any element $t \in \pi \backslash \pi^w$, and the oriented involution on $R[\pi^w]$. Let $\mathbb{Z}^-$ denote the $\mathbb{Z}[\mathbb{Z}_2]$-module defined by $\mathbb{Z}$ with $T \in \mathbb{Z}_2$ acting by $T(1) = -1$. The natural isomorphism of abelian groups

$$\mathbb{Z}^- \otimes_{\mathbb{Z}[\mathbb{Z}_2]} (M \otimes_{R[\pi^w]} N) \overset{\simeq}{\longrightarrow} M \otimes_{R[\pi]^w} N ; \ 1 \otimes (x \otimes y) \longrightarrow x \otimes y$$

will be used to identify

$$\mathbb{Z}^- \otimes_{\mathbb{Z}[\mathbb{Z}_2]} (M \otimes_{R[\pi^w]} N) = M \otimes_{R[\pi]^w} N .$$

Let

$$T : K^w \overset{\simeq}{\longrightarrow} K^w ; \ \sigma \longrightarrow T\sigma$$

be the covering translation, a free involution.

DEFINITION A1 An (R, K, w)-*module* M is a f.g. free (R, K^w)-module such that
$$M(T\sigma) = M(\sigma) \quad (\sigma \in K^w) .$$
A morphism of (R, K, w)-modules $f: M \longrightarrow N$ is an (R, K^w)-module morphism such that
$f(T\tau, T\sigma) = f(\tau, \sigma)$:
$$M(T\sigma) = M(\sigma) \longrightarrow N(T\tau) = N(\tau) \quad (\sigma \leq \tau \in K^w) .$$

□

Given (R, K, w)-modules M,N let $\mathbb{Z}[\mathbb{Z}_2]$ act on the abelian group $M \otimes_{\mathbb{A}(R,K^w)} N$ by
$$T : M \otimes_{\mathbb{A}(R,K^w)} N \longrightarrow M \otimes_{\mathbb{A}(R,K^w)} N ;$$
$$x(\sigma) \otimes y(\sigma) \longrightarrow x(T\sigma) \otimes y(T\sigma) \quad (\sigma \in K^w) .$$

DEFINITION A2 Let $\mathbb{A}(R, K, w)$ be the additive category of (R, K, w)-modules and morphisms, and let
$$\mathbb{B}(R, K, w) = \mathbb{B}(\mathbb{A}(R, K, w))$$
be the additive category of finite chain complexes in $\mathbb{A}(R, K, w)$. $\mathbb{A}(R, K, w)$ has a chain duality
$$T : \mathbb{A}(R, K, w) \longrightarrow \mathbb{B}(R, K, w) ; \quad M \longrightarrow TM$$
characterized by the identities
$$\mathrm{Hom}_{\mathbb{A}(R,K,w)}(TM, N) = M \otimes_{\mathbb{A}(R,K,w)} N = \mathbb{Z}^- \otimes_{\mathbb{Z}[\mathbb{Z}_2]} (M \otimes_{\mathbb{A}(R,K^w)} N) .$$

□

The universal cover $\widetilde{K}$ of K (assumed connected) is also the universal cover of K^w. The universal assembly of an (R, K, w)-module M is a f.g. free $R[\pi]$-module
$$M(\widetilde{K}) = \sum_{\tilde{\sigma} \in \widetilde{K}} M(p^w\tilde{\sigma}) ,$$
with $p^w: \widetilde{K} \longrightarrow K^w$ the covering projection.

DEFINITION A3 Given R, K, K^w, π, π^w there are defined algebraic bordism categories:

(i) the *local* f.g. free (R, K, w)-module bordism category
$$\Lambda(R)_*(K, w) = (\mathbb{A}(R, K, w), \mathbb{B}(R, K, w), \mathbb{C}(R)_*(K, w)) ,$$
with $\mathbb{C}(R)_*(K, w) \subseteq \mathbb{B}(R, K, w)$ the full subcategory of (R, K, w)-module chain complexes C such that each $C(\sigma)$ $(\sigma \in K^w)$ is a contractible R-module chain complex;

(ii) the *global* f.g. free (R, K, w)-module bordism category
$$\Lambda(R, K, w) = (\mathbb{A}(R, K, w), \mathbb{B}(R, K, w), \mathbb{C}(R, K, w)) \,,$$
with $\mathbb{C}(R, K, w) \subseteq \mathbb{B}(R, K, w)$ the full subcategory of (R, K, w)-module chain complexes C such that the assembly $R[\pi]$-module chain complex $C(\widetilde{K})$ is contractible.

□

PROPOSITION A4 (i) *Inclusion defines a twisted universal assembly functor of algebraic bordism categories*
$$A : \Lambda(R)_*(K, w) \longrightarrow \Lambda(R, K, w) \,.$$
(ii) *The universal assembly functor* $A\colon \mathbb{A}(R, K, w) \longrightarrow \mathbb{A}(R[\pi]^w)$ *extends to twisted universal assembly functors of algebraic bordism categories*
$$A : \Lambda(R, K, w) \longrightarrow \Lambda(R[\pi]^w) \,, \quad A : \widehat{\Lambda}(R, K, w) \longrightarrow \widehat{\Lambda}(R[\pi]^w) \,,$$
with
$$\widehat{\Lambda}(R, K, w) = (\mathbb{A}(R, K, w), \mathbb{B}(R, K, w), \mathbb{B}(R, K, w)) \,.$$

□

DEFINITION A5 The assembly chain map for (R, K, w)-module chain complexes C, D
$$\tilde{\alpha}_{C,D} : C \otimes_{\mathbb{A}(R, K^w)} D \longrightarrow C(\widetilde{K}) \otimes_{R[\pi]^w} D(\widetilde{K})$$
is a $\mathbb{Z}[\mathbb{Z}_2]$-module chain map, and so induces a *twisted universal assembly* chain map
$$\alpha_{C,D} = 1 \otimes \tilde{\alpha}_{C,D} : C \otimes_{(R,K,w)} D = \mathbb{Z}^- \otimes_{\mathbb{Z}[\mathbb{Z}_2]} (C \otimes_{(R,K^w)} D)$$
$$\longrightarrow C(\widetilde{K}) \otimes_{R[\pi]^w} D(\widetilde{K}) = \mathbb{Z}^- \otimes_{\mathbb{Z}[\mathbb{Z}_2]} (C(\widetilde{K}) \otimes_{R[\pi^w]} D(\widetilde{K})) ; \phi \longrightarrow \phi(\widetilde{K}) \,,$$
with $C \otimes_{(R,K,w)} D$ short for $C \otimes_{\mathbb{A}(R,K,w)} D$. Here, the $\mathbb{Z}[\mathbb{Z}_2]$-actions are those induced from $T\colon K^w \longrightarrow K^w$.

□

In the special case $C = D$ the assembly of A5 is a $\mathbb{Z}[\mathbb{Z}_2]$-module chain map
$$\alpha = \alpha_{C,C} : C \otimes_{(R,K,w)} C \longrightarrow C(\widetilde{K}) \otimes_{R[\pi]^w} C(\widetilde{K}) ; \phi \longrightarrow \phi(\widetilde{K})$$
inducing abelian group morphisms
$$\alpha^\% : Q^n(C) = H_n(\mathrm{Hom}_{\mathbb{Z}[\mathbb{Z}_2]}(W, (C \otimes_{(R,K,w)} C))) \longrightarrow$$
$$Q^n(C(\widetilde{K})) = H_n(\mathrm{Hom}_{\mathbb{Z}[\mathbb{Z}_2]}(W, (C(\widetilde{K}) \otimes_{R[\pi]^w} C(\widetilde{K})))) \,,$$
$$\alpha_\% : Q_n(C) = H_n(W \otimes_{\mathbb{Z}[\mathbb{Z}_2]} (C \otimes_{(R,K,w)} C)) \longrightarrow$$
$$Q_n(C(\widetilde{K})) = H_n(W \otimes_{\mathbb{Z}[\mathbb{Z}_2]} (C(\widetilde{K}) \otimes_{R[\pi]^w} C(\widetilde{K}))) \ (n \in \mathbb{Z}) \,.$$
Here, the $\mathbb{Z}[\mathbb{Z}_2]$-actions are given by the duality involutions.

The twisted version of 9.11 is given by:

PROPOSITION A6 *Twisted universal assembly defines functors of algebraic bordism categories*

$$A : \Lambda(R, K, w) \longrightarrow \Lambda(R[\pi]^w) \quad , \quad A : \widehat{\Lambda}(R, K, w) \longrightarrow \widehat{\Lambda}(R[\pi]^w)$$

inducing twisted universal assembly maps in the $\begin{cases} symmetric \\ visible\ symmetric \\ quadratic \\ normal \end{cases}$ *L-*

groups

$$\begin{cases} A : L^n(R, K, w) = L^n(\Lambda(R, K, w)) \longrightarrow L^n(R[\pi]^w) \\ A : VL^n(R, K, w) = NL^n(\Lambda(R, K, w)) \longrightarrow VL^n(R[\pi]^w) \\ A : L_n(R, K, w) = L_n(\Lambda(R, K, w)) \longrightarrow L_n(R[\pi]^w) \\ A : NL^n(R, K, w) = NL^n(\widehat{\Lambda}(R, K, w)) \longrightarrow NL^n(R[\pi]^w) \,. \end{cases} \quad (n \in \mathbb{Z})$$

□

The twisted version of 9.16 is given by:

EXAMPLE A7 An n-dimensional geometric $\begin{cases} \text{Poincaré complex } K \\ \text{normal map } (f, b) \colon M \longrightarrow K' \\ \text{normal complex } K \end{cases}$

with orientation map

$$w = w_1(\nu_K) : \pi = \pi_1(K) \longrightarrow \{\pm 1\}$$

has a twisted $\begin{cases} \text{visible symmetric} \\ \text{quadratic} \\ \text{normal} \end{cases}$ signature $\begin{cases} \sigma^*(K) \in VL^n(\mathbb{Z}, K, w) \\ \sigma_*(f, b) \in L_n(\mathbb{Z}, K, w) \\ \widehat{\sigma}^*(K) \in NL^n(\mathbb{Z}, K, w) \end{cases}$

with assembly the twisted $\begin{cases} \text{symmetric} \\ \text{quadratic} \\ \text{normal} \end{cases}$ signature $\begin{cases} \sigma^*(K) \in L^n(\mathbb{Z}[\pi]^w) \\ \sigma_*(f, b) \in L_n(\mathbb{Z}[\pi]^w) \\ \widehat{\sigma}^*(K) \in NL^n(\mathbb{Z}[\pi]^w). \end{cases}$

□

The algebraic π-π theorem of §10 has an evident twisted version for a double cover K^w of K, with the twisted assembly maps defining isomorphisms

$$L_*(R, K, w) \xrightarrow{\cong} L_*(R[\pi_1(K)]^w) \,.$$

DEFINITION A8 The *twisted universal assembly map* on the twisted generalized homology groups with quadratic L-theory coefficients is the composite

$$A : H_n(K, w; \mathbb{L}.(R)) = L_n(\Lambda(R)_*(K, w)) \longrightarrow L_n(R, K, w) \longrightarrow L_n(R[\pi]^w)$$

of the morphisms given by A4 and A6, with $L_n(R, K, w) \cong L_n(R[\pi]^w)$.

□

The cycle approach to generalized homology of §12 can be extended to twisted coefficients, as follows.

Given pointed Δ-sets J, K with involutions let $K_{\mathbb{Z}_2}^J$ be the function Δ-set with $(K_{\mathbb{Z}_2}^J)^{(p)}$ the set of $\mathbb{Z}_2$-equivariant Δ-maps $J \wedge (\Delta^p)_+ \longrightarrow K$.

DEFINITION A9 Let $\mathbb{F}$ be an Ω-spectrum with an involution $T \colon \mathbb{F} \longrightarrow \mathbb{F}$, and let K be a locally finite Δ-set with a double cover $K^w \longrightarrow K$. The w-twisted $\left\{ \begin{array}{l} \mathbb{F}\text{-cohomology} \\ \mathbb{F}\text{-homology} \end{array} \right.$ Ω-spectrum of K is defined by

$$\left\{ \begin{array}{l} \mathbb{F}_{\mathbb{Z}_2}^{K_+} = \{ (\mathbb{F}_n)_{\mathbb{Z}_2}^{K_+} \mid n \in \mathbb{Z} \} \\ K_+^w \wedge_{\mathbb{Z}_2} \mathbb{F} = \{ \varinjlim_j \Omega^j (K_+^w \wedge_{\mathbb{Z}_2} \mathbb{F}_{n-j}) \mid n \in \mathbb{Z} \} \end{array} \right.$$

with homotopy groups the w-twisted $\left\{ \begin{array}{l} \mathbb{F}\text{-cohomology} \\ \mathbb{F}\text{-homology} \end{array} \right.$ groups of K

$$\left\{ \begin{array}{l} H^n(K, w; \mathbb{F}) = \pi_{-n}(\mathbb{F}_{\mathbb{Z}_2}^{K_+}) = [K_+^w, \mathbb{F}_{-n}]_{\mathbb{Z}_2} \\ H_n(K, w; \mathbb{F}) = \pi_n(K_+^w \wedge_{\mathbb{Z}_2} \mathbb{F}) = \varinjlim_j \pi_{n+j}(K_+^w \wedge_{\mathbb{Z}_2} \mathbb{F}_{-j}) \, . \end{array} \right.$$

□

In the untwisted case of the trivial double cover $K^w = K \sqcup K$

$$H_*(K, w; \mathbb{F}) = H_*(K; \mathbb{F}) \, , \quad H^*(K, w; \mathbb{F}) = H^*(K; \mathbb{F}) \, .$$

The w-twisted $\mathbb{F}$-cohomology group $H^n(K, w; \mathbb{F})$ of a locally finite simplicial complex K has a direct combinatorial description as the set of $\mathbb{Z}_2$-equivariant homotopy classes of $\mathbb{Z}_2$-equivariant Δ-maps $K_+^w \longrightarrow \mathbb{F}_{-n}$, which may be called '$w$-twisted $\mathbb{F}$-cocycles in K'. The w-twisted $\mathbb{F}$-homology group $H_n(K, w; \mathbb{F})$ has a similar description as the set of cobordism classes of 'w-twisted $\mathbb{F}$-cycles in K', by analogy with the untwisted case.

Construct a $\mathbb{Z}_2$-equivariant embedding of K^w in some $\partial \Delta^{2m+1}$, as follows. Let $T \colon K^w \longrightarrow K^w$ be the free involution defined by the covering translation, and let $\{v_0, v_1, \ldots, v_m\}$ be the vertices of K. Choose lifts $\{\tilde{v}_0, \tilde{v}_1, \ldots, \tilde{v}_m\}$ to half the vertices of K^w, so that the other half are given by $\{T\tilde{v}_0, T\tilde{v}_1, \ldots, T\tilde{v}_m\}$. Define a free action of $\mathbb{Z}_2$ on $\partial \Delta^{2m+1}$ by

$$T \colon \partial \Delta^{2m+1} \xrightarrow{\; \simeq \;} \partial \Delta^{2m+1} \; ; \; i \longrightarrow \begin{cases} i + m + 1 & \text{if } 0 \leq i \leq m \\ i - m - 1 & \text{if } m + 1 \leq i \leq 2m + 1, \end{cases}$$

and define a $\mathbb{Z}_2$-equivariant embedding

$$K^w \longrightarrow \partial \Delta^{2m+1} \; ; \; \tilde{v}_i \longrightarrow i \, , \; T\tilde{v}_i \longrightarrow i + m + 1 \, .$$

The simplicial complex Σ^{2m} and the supplement $\overline{K}^w \subseteq \Sigma^{2m}$ of $K^w \subseteq \partial \Delta^{2m+1}$ are defined as in the untwisted case. Σ^{2m} comes equipped with a free involution

$$T \colon \Sigma^{2m} \xrightarrow{\; \simeq \;} \Sigma^{2m} \; ; \; \sigma^* \longrightarrow T\sigma^* \, .$$

The inclusion $\overline{K}^w \longrightarrow \Sigma^{2m}$ is a $\mathbb{Z}_2$-equivariant map covering the inclusion
$$\overline{K} = \overline{K}^w/\mathbb{Z}_2 \longrightarrow \mathbb{RP}^{2m} = \Sigma^{2m}/\mathbb{Z}_2 \ .$$

DEFINITION A10 Given a simplicial complex K with a double cover $K^w \longrightarrow K$ and a $\mathbb{Z}_2$-equivariant embedding $K^w \subseteq \partial\Delta^{2m+1}$ let
$$\mathbb{H}.(K,w;\mathbb{F}) = \{\mathbb{H}_n(K,w;\mathbb{F}) \mid n \in \mathbb{Z}\}$$
be the Ω-spectrum defined by
$$\mathbb{H}_n(K,w;\mathbb{F}) = \varinjlim_J (\mathbb{F}_{n-2m},\emptyset)_{\mathbb{Z}_2}^{(\Sigma^{2m},\overline{J}^w)} \ ,$$
with the direct limit taken over the finite subcomplexes $J \subseteq K$ and using the canonical $\mathbb{Z}_2$-equivariant embedding $J^w \subseteq \partial\Delta^{2m+1}$, with homotopy groups
$$\pi_n(\mathbb{H}.(K,w;\mathbb{F})) = \varinjlim_J H^{2m-n}(\mathbb{RP}^{2m},\overline{J},w;\mathbb{F}) \quad (n \in \mathbb{Z}) \ .$$

$\square$

PROPOSITION A11 *The Ω-spectrum $\mathbb{H}.(K,w;\mathbb{F})$ is homotopy equivalent to the w-twisted $\mathbb{F}$-homology Ω-spectrum $K_+^w \wedge_{\mathbb{Z}_2} \mathbb{F}$, with homotopy groups*
$$\pi_n(\mathbb{H}.(K,w;\mathbb{F})) = \pi_n(K_+^w \wedge_{\mathbb{Z}_2} \mathbb{F}) = H_n(K,w;\mathbb{F}) \quad (n \in \mathbb{Z}) \ .$$
PROOF As for A4, using the w-twisted S-duality isomorphisms
$$\pi_*(\mathbb{H}.(K,w;\mathbb{F})) = H^{2m-*}(\mathbb{RP}^{2m},\overline{K},w;\mathbb{F})$$
$$\xrightarrow{\ \simeq\ } \pi_*(K_+^w \wedge_{\mathbb{Z}_2} \mathbb{F}) = H_*(K,w;\mathbb{F}) \ .$$

$\square$

There are also twisted assembly maps:

DEFINITION A12 Given an Ω-spectrum with involution $\mathbb{F}$, and a $\mathbb{Z}_2$-invariant subcomplex $K^w \subseteq \partial\Delta^{2m+1}$ define the *w-twisted assembly* to be the composite map of Ω-spectra
$$A : \mathbb{H}.(K,w;\mathbb{F}) \xrightarrow{\ w^!\ } \mathbb{H}.(K^w;\mathbb{F}) \xrightarrow{\ A\ } \mathbb{F}$$
inducing w-twisted assembly maps in the homotopy groups
$$A : \pi_n(\mathbb{H}.(K,w;\mathbb{F})) = H_n(K,w;\mathbb{F}) \longrightarrow \pi_n(\mathbb{F}) \quad (n \in \mathbb{Z}) \ ,$$
with $w^!$ the transfer map forgetting the $\mathbb{Z}_2$-equivariance
$$w^! : \mathbb{H}_n(K,w;\mathbb{F}) = (\mathbb{F}_{n-2m},\emptyset)_{\mathbb{Z}_2}^{(\Sigma^{2m},\overline{K}^w)}$$
$$\longrightarrow \mathbb{H}_n(K^w;\mathbb{F}) = (\mathbb{F}_{n-2m},\emptyset)^{(\Sigma^{2m},\overline{K}^w)}$$
and $A : \mathbb{H}.(K^w;\mathbb{F}) \longrightarrow \mathbb{F}$ the assembly map of 12.14.

$\square$

DEFINITION A13 Given a covariant functor
$$\mathbb{F} : \{\text{simplicial complexes with a double cover}\}$$
$$\longrightarrow \{\Omega\text{-spectra with involution}\} ; (K, w) \longrightarrow \mathbb{F}(K, w)$$
define the *local* $\{\mathbb{F}\}$-*coefficient homology* Ω-*spectrum* of (K, w)
$$\mathbb{H}.(K, w; \{\mathbb{F}\}) = \{\mathbb{H}_n(K, w; \{\mathbb{F}\}) \mid n \in \mathbb{Z}\}$$
by
$$\mathbb{H}_n(K, w; \{\mathbb{F}\}) = \varprojlim_{\tilde{\sigma} \in K^w} (\mathbb{F}_{n-2m}(D(\sigma, K), w'), \emptyset)_{\mathbb{Z}_2}^{(\overline{K^w}(\tilde{\sigma}), \overline{K^w})} ,$$
with $\sigma \in K$ the projection of $\tilde{\sigma} \in K^w$. The *local* $\{\mathbb{F}\}$-*coefficient homology groups of* (K, w) are the homotopy groups of $\mathbb{H}.(K, w; \{\mathbb{F}\})$
$$H_n(K, w; \{\mathbb{F}\}) = \pi_n(\mathbb{H}.(K, w; \{\mathbb{F}\})) \quad (n \in \mathbb{Z}) .$$
□

As in the untwisted case (12.6, 12.8) it is possible to express $H_n(K, w; \{\mathbb{F}\})$ as the cobordism group of n-dimensional $\{\mathbb{F}\}$-cycles in (K, w), which are collections
$$x = \{x(\tilde{\sigma}) \in \mathbb{F}_{n-2m}(D(\sigma, K), w')^{(2m-|\tilde{\sigma}|)} \mid \tilde{\sigma} \in K^w\}$$
such that

(i) $\partial_i x(\tilde{\sigma}) = \begin{cases} \Theta_i x(\delta_i \tilde{\sigma}) & \text{if } \delta_i \tilde{\sigma} \in K^w \\ \emptyset & \text{if } \delta_i \tilde{\sigma} \notin K^w \end{cases} \quad (0 \leq i \leq 2m - |\tilde{\sigma}|)$

(ii) $x(T\tilde{\sigma}) = Tx(\tilde{\sigma})$,

with $\Theta_i : \mathbb{F}(D(\delta_i\sigma, K), w') \longrightarrow \mathbb{F}(D(\sigma, K), w')$ induced by the inclusion $D(\delta_i\sigma, K) \subset D(\sigma, K)$.

DEFINITION A14 The *local* w-*twisted* $\{\mathbb{F}\}$-*coefficient assembly* is the map of Ω-spectra
$$A : \mathbb{H}.(K, w; \{\mathbb{F}\}) \longrightarrow \mathbb{F}(K', w')$$
given by the composite
$$A : \mathbb{H}.(K, w; \{\mathbb{F}\}) \longrightarrow \mathbb{H}.(K, w; \mathbb{F}(K', w')) \xrightarrow{A} \mathbb{F}(K', w')$$
of the forgetful map $\mathbb{H}.(K, w; \{\mathbb{F}\}) \longrightarrow \mathbb{H}.(K, w; \mathbb{F}(K', w'))$ induced by all the inclusions $D(\sigma, K) \subseteq K'$ ($\sigma \in K$) and the w-twisted assembly of A12
$$A : \mathbb{H}.(K, w; \mathbb{F}(K', w')) \longrightarrow \mathbb{F}(K', w') .$$
□

For a homotopy invariant functor
$$\mathbb{F} : \{\text{simplicial complexes with a double cover}\}$$
$$\longrightarrow \{\Omega\text{-spectra with involution}\} ; (K, w) \longrightarrow \mathbb{F}(K, w)$$

the forgetful map from local w-twisted $\{\mathbb{F}\}$-coefficient homology to constant w-twisted $\mathbb{F}(\{*\})$-coefficient homology is a homotopy equivalence

$$\mathbb{H}.(K,w;\{\mathbb{F}\}) \xrightarrow{\simeq} \mathbb{H}.(K,w;\{\mathbb{F}(\{*\})\}) \ .$$

DEFINITION A15 The *constant w-twisted $\mathbb{F}(\{*\})$-coefficient assembly* for a homotopy invariant functor $\mathbb{F}$ and a pair (K,w)

$$A : \mathbb{H}.(K,w;\mathbb{F}(\{*\})) \longrightarrow \mathbb{F}(K,w)$$

is given by the local w-twisted $\{\mathbb{F}\}$-coefficient assembly A of A14, using the homotopy equivalences

$$\mathbb{H}.(K,w;\{\mathbb{F}\}) \simeq \mathbb{H}.(K,w;\{\mathbb{F}(\{*\})\}) \ , \quad \mathbb{F}(K',w') \simeq \mathbb{F}(K,w) \ .$$

□

EXAMPLE A16 Let $\Omega_\cdot^{SO}(K,w) = \{\Omega_\cdot^{SO}(K,w)_n \mid n \in \mathbb{Z}\}$ be the Ω-spectrum consisting of the Kan Δ-sets $\Omega_\cdot^{SO}(K,w)_n$ with k-simplexes

$$\Omega_\cdot^{SO}(K,w)_n^{(k)} =$$

$\{\,(n+k)$-dimensional smooth oriented manifold k-ads

$(M;\partial_0 M,\partial_1 M,\ldots,\partial_k M)$ such that $\partial_0 M \cap \partial_1 M \cap \ldots \cap \partial_k M = \emptyset$,

with an orientation-reversing free involution $M \xrightarrow{\simeq} M$

and a $\mathbb{Z}_2$-equivariant map $f \colon M \longrightarrow |K^w| \,\}$

and base simplex the empty manifold k-ad $\emptyset$. Let

$$T : \Omega_\cdot^{SO}(K,w) \longrightarrow \Omega_\cdot^{SO}(K,w)$$

be the orientation-reversing involution. The homotopy groups

$$\pi_n(\Omega_\cdot^{SO}(K,w)) = \Omega_n^{SO}(K,w) \ \ (n \geq 0)$$

are the bordism groups of $\mathbb{Z}_2$-equivariant maps $M \longrightarrow |K^w|$ from closed oriented n-dimensional manifolds with an orientation-reversing free involution. The functor

$$\Omega_\cdot^{SO} : \{\text{simplicial complexes with double cover}\} \longrightarrow$$

$$\{\Omega\text{-spectra with involution}\} \ ; \ (K,w) \longrightarrow \Omega_\cdot^{SO}(K,w)$$

is homotopy invariant, and the assembly map of A15 is a homotopy equivalence

$$A : \mathbb{H}.(K,w;\Omega_\cdot^{SO}(\{*\})) \xrightarrow{\simeq} \Omega_\cdot^{SO}(K,w) \ ,$$

being a combinatorial version of the Pontrjagin–Thom isomorphism. (In fact, $\Omega_\cdot^{SO}(K,w)$ is just a combinatorial version of the Thom spectrum $|K^w|_+ \wedge_{\mathbb{Z}_2} \underline{MSO}$.) The assembly of an n-dimensional $\Omega_\cdot^{SO}(\{*\})$-coefficient cycle in (K,w)

$$x = \{M(\tilde{\sigma})^{n-|\tilde{\sigma}|} \mid \tilde{\sigma} \in K^w\}$$

is a $\mathbb{Z}_2$-equivariant map
$$A(x) : M^n = \bigcup_{\tilde{\sigma} \in K^w} M(\tilde{\sigma}) \longrightarrow |K^w| = |(K^w)'|$$
from a closed smooth oriented n-manifold with an orientation-reversing free involution, such that
$$A(x)^{-1}D(\tilde{\sigma}, K^w) = M(\tilde{\sigma}) \quad (\tilde{\sigma} \in K^w) .$$
In the untwisted case of the trivial double cover
$$K^w = K \sqcup K , \quad \Omega_.^{SO}(K, w) = \Omega_.^{SO}(K)$$
the spectrum is the oriented smooth bordism Ω-spectrum of 12.21.

□

The results of §13 concerning the algebraic $\mathbb{L}$-spectra also have twisted versions. Only the following special case of the twisted version of 13.7 is spelled out:

PROPOSITION A17 *The quadratic $\mathbb{L}$-spectrum of the twisted algebraic bordism category $\Lambda(R)_*(K, w)$ of A3 is the twisted generalized homology spectrum of (K, w) of A10*
$$\mathbb{L}.(\Lambda(R)_*(K, w)) = \mathbb{H}.(K, w; \mathbb{L}.(R)) ,$$
so that on the level of homotopy groups
$$L_n(\Lambda(R)_*(K, w)) = H_n(K, w; \mathbb{L}.(R)) \quad (n \in \mathbb{Z}) .$$

□

The algebraic surgery exact sequence of §14 also has a twisted version, with K replaced by (K, w). Define an assembly map
$$A : \mathbb{H}.(K, w; \mathbb{L}.(R)) \longrightarrow \mathbb{L}.(R[\pi_1(K)]^w)$$
by composing the forgetful map $\mathbb{H}.(K, w; \mathbb{L}.(R)) \longrightarrow \mathbb{L}.(R, K, w)$ with the homotopy equivalence $\mathbb{L}.(R, K, w) \simeq \mathbb{L}.(R[\pi_1(K)]^w)$ given by 10.6. Only the twisted version of 14.6 is spelled out:

DEFINITION A18 (i) *The twisted quadratic structure groups* of (R, K, w) are the cobordism groups
$$\mathbb{S}_n(R, K, w) = L_{n-1}(\mathbb{A}(R, K, w), \mathbb{C}(R, K, w), \mathbb{C}(R)_*(K, w)) \quad (n \in \mathbb{Z})$$
of $(n-1)$-dimensional quadratic complexes in $\mathbb{A}(R, K, w)$ which are globally contractible and locally Poincaré.
(ii) *The twisted quadratic structure spectrum* of (R, K, w) is the Ω-spectrum
$$\mathbb{S}.(R, K, w) = \Sigma \mathbb{L}.(\mathbb{A}(R, K, w), \mathbb{C}(R, K, w), \mathbb{C}(R)_*(K, w))$$
with homotopy groups
$$\pi_*(\mathbb{S}.(R, K, w)) = \mathbb{S}_*(R, K, w) .$$

(iii) The *twisted algebraic surgery exact sequence* is the exact sequence of homotopy groups

$$\ldots \longrightarrow H_n(K,w; \mathbb{L}.(R)) \overset{A}{\longrightarrow} L_n(R[\pi_1(K)]^w) \overset{\partial}{\longrightarrow}$$
$$\mathbb{S}_n(R,K,w) \longrightarrow H_{n-1}(K,w; \mathbb{L}.(R)) \longrightarrow \ldots$$

induced by the fibration sequence of spectra

$$\mathbb{H}.(K,w; \mathbb{L}.(R)) \longrightarrow \mathbb{L}.(R[\pi_1(K)]^w) \longrightarrow \mathbb{S}.(R,K,w) \ .$$

□

The results of §§15–26 extend to the nonorientable case in a straightforward manner.

Appendix B. Assembly via products

The quadratic L-theory assembly map of §14

$$A \: : \: H_*(K; \mathbb{L}.(\mathbb{Z})) \longrightarrow \mathbb{L}_*(\mathbb{Z}[\pi_1(K)])$$

will now be reconciled with the construction of A proposed in Ranicki [140] by means of a 'preassembly' map $A \colon K_+ \longrightarrow \mathbb{L}^0(\mathbb{Z}[\pi_1(K)])$ and a pairing of spectra

$$\otimes \: : \: \mathbb{L}^{\cdot}(\mathbb{Z}[\pi_1(K)]) \wedge \mathbb{L}.(\mathbb{Z}) \longrightarrow \mathbb{L}.(\mathbb{Z}[\pi_1(K)]) \: .$$

Although only the quadratic case is considered, there is an entirely analogous treatment for the symmetric L-theory assembly map.

B1. The cartesian product of Δ-sets X, Y is the Δ-set $X \times Y$ with

$$(X \times Y)^{(n)} \: = \: X^{(n)} \times Y^{(n)} \: , \quad \partial_i(x, y) \: = \: (\partial_i(x), \partial_i(y)) \: .$$

The Δ-map

$$X \times Y \longrightarrow X \otimes Y \: ; \: (\Delta^n \to X, \Delta^n \to Y) \longrightarrow (\Delta^n \to \Delta^n \otimes \Delta^n \to X \otimes Y)$$

is a homotopy equivalence for Kan Δ-sets X, Y (Rourke and Sanderson [152]), inducing a homotopy equivalence of the realizations

$$|X \times Y| \: \simeq \: |X \otimes Y| \: = \: |X| \times |Y| \: .$$

It follows that the cartesian smash product of pointed Kan Δ-sets X, Y

$$X \wedge Y \: = \: (X \times Y)/(X \times \emptyset_Y \cup \emptyset_X \times Y)$$

is homotopy equivalent to the geometric smash product, with $|X \wedge Y| \simeq |X| \wedge |Y|$.

B2. Let Δ^n have vertices $0, 1, \dots, n$. Define a cell structure on the realization $|\Delta^n|$ with one $(p + q)$-cell for each sequence $(j_0, j_1, \dots, j_p, k_0, \dots, k_q)$ of integers such that

$$0 \leq j_0 < j_1 < \dots < j_p \leq k_0 < k_1 < \dots < k_q \leq n \: ,$$

the convex hull of the vertices $\widehat{j_{p'} k_{q'}}$ $(0 \leq p' \leq p, \: 0 \leq q' \leq q)$ in the barycentric subdivision $(\Delta^n)'$. This is the *combinatorial diagonal approximation*.

B3. Write a geometric or chain complex n-ad as $C = \{C(\sigma) \,|\, \sigma \in \Delta^n\}$, with $C(\sigma) \subset C(\tau)$ for $\sigma < \tau \in \Delta^n$. Use the combinatorial diagonal approximation of B2 to define the product of n-ads C, D to be the n-ad $C \otimes D$ with

$$(C \otimes D)(01 \dots n) \: = \: \bigcup_{0 \leq j_0 < \dots < j_p \leq k_0 < \dots < k_q \leq n} C(j_0 \dots j_p) \otimes D(k_0 \dots k_q)$$

$$= \: \bigcup_{i=0}^{n} C(0 \dots i) \otimes D(i \dots n) \: .$$

There is one piece of the product for each cell in $|\Delta^n|$.

B4. For any rings with involution R, S use the chain complex n-ad product of B3 to define spectrum-level products
$$\otimes : \mathbb{L}^i(R) \wedge \mathbb{L}^j(S) \longrightarrow \mathbb{L}^{i+j}(R \otimes S) \ ,$$
$$\otimes : \mathbb{L}^i(R) \wedge \mathbb{L}_j(S) \longrightarrow \mathbb{L}_{i+j}(R \otimes S)$$
inducing the products of Ranicki [141, §8] on the level of homotopy groups
$$\otimes : L^i(R) \otimes L^j(S) \longrightarrow L^{i+j}(R \otimes S) \ ,$$
$$\otimes : L^i(R) \otimes L_j(S) \longrightarrow L_{i+j}(R \otimes S) \ .$$
(These products can also be defined using bisimplicial sets.) In particular, $\mathbb{L}^0(\mathbb{Z})$ is a ring spectrum, and $\mathbb{L}_0(\mathbb{Z})$ is an $\mathbb{L}^0(\mathbb{Z})$-module spectrum. For commutative rings R, S with the identity involution and a subcomplex $K \subseteq \partial \Delta^{m+1}$ the n-ad products can also be used to define spectrum-level products
$$\otimes : \mathbb{L}^i(R_*K) \wedge \mathbb{L}_j(S) \longrightarrow \mathbb{L}_{i+j}((R \otimes S)_*K) \ ,$$
$$\otimes : \mathbb{L}^i(R, K) \wedge \mathbb{L}_j(S) \longrightarrow \mathbb{L}_{i+j}(R \otimes S, K) \ .$$

B5. For any subcomplex $K \subseteq \partial \Delta^{m+1}$ define the framed (smooth or topological) framed Ω-spectrum
$$\Omega^{fr}_\bullet(K) = \{\Omega^{fr}_\bullet(K)_i = \Omega^{fr}_\bullet(\{*\})^{(\Sigma^m, \overline{K})}_{i-m} \, | \, i \in \mathbb{Z}\}$$
by analogy with
$$\mathbb{L}^\bullet(\mathbb{Z}_*K) = \{\mathbb{L}^i(\mathbb{Z}_*K) = \mathbb{L}^{i-m}(\mathbb{Z})^{(\Sigma^m, \overline{K})} \, | \, i \in \mathbb{Z}\} \ .$$
Use the construction of the symmetric signature to define a map
$$\sigma^* : \Omega^{fr}_\bullet(K) \longrightarrow \mathbb{L}^\bullet(\mathbb{Z}_*K) \ ; \quad M \longrightarrow (\Delta(M), \phi(M)) \ .$$
The framed bordism spectrum $\Omega^{fr}_\bullet(\{*\})$ of a point is an Ω-spectrum homotopy equivalent to the Ω-spectrum of the sphere spectrum $\underline{S}^0$, and
$$\Omega^{fr}_\bullet(K) \simeq K_+ \wedge \Omega^{fr}_\bullet(\{*\}) \ ,$$
$$\Omega^{fr}_\bullet(K)_i \simeq \varinjlim_j \Omega^{i+j} \Sigma^j K_+ \quad (\simeq_s \Sigma^{-i} K_+ \text{ for } i \le 0)$$
where $\simeq_s$ denotes stable homotopy equivalence. Use the products of B4 to define a product of Kan Δ-sets
$$\otimes : \Omega^{fr}_\bullet(\{*\})_i \wedge \mathbb{L}_j(\mathbb{Z}) \longrightarrow \mathbb{L}_{i+j}(\mathbb{Z}) \ ,$$
and define also products
$$\otimes : \Omega^{fr}_\bullet(K)_i \wedge \mathbb{L}_j(\mathbb{Z}) \xrightarrow{\sigma^* \wedge 1} \mathbb{L}^i(\mathbb{Z}_*K) \wedge \mathbb{L}_j(\mathbb{Z}) \xrightarrow{\otimes} \mathbb{L}_{i+j}(\mathbb{Z}_*K) \ .$$
By 13.7 there is an identification
$$\mathbb{L}_i(\mathbb{Z}_*K) = \mathbb{H}_i(K, \mathbb{L}.(\mathbb{Z})) \ ,$$
and for each $i \in \mathbb{Z}$ the products
$$\otimes : \Omega^{fr}_\bullet(K)_i \wedge \mathbb{L}_{-j}(\mathbb{Z}) \longrightarrow \mathbb{L}_{i-j}(\mathbb{Z}_*K) \quad (j \in \mathbb{Z})$$

induce a homotopy equivalence

$$\varinjlim_j \Omega^j(\Omega_{\bullet}^{fr}(K)_i \wedge \mathbb{L}_{-j}(\mathbb{Z})) \xrightarrow{\simeq}$$

$$\mathbb{H}_i(K, \mathbb{L}.(\mathbb{Z})) = \varinjlim_j \Omega^j \mathbb{L}_{i-j}(\mathbb{Z}_* K) = \mathbb{L}_i(\mathbb{Z}_* K) .$$

The inclusion $K_+ \longrightarrow \Omega_{\bullet}^{fr}(K)_0$ is a stable homotopy equivalence, and the ith space of the homology spectrum $\mathbb{H}.(K; \mathbb{L}.(\mathbb{Z}))$ is such that

$$\mathbb{H}_i(K; \mathbb{L}.(\mathbb{Z})) \simeq \varinjlim_j \Omega^j(\Omega_{\bullet}^{fr}(K)_i \wedge \mathbb{L}_{-j}(\mathbb{Z})) \simeq_s K_+ \wedge \mathbb{L}_i(\mathbb{Z}) .$$

B6. Given an algebraic bordism category Λ let

$$\mathbb{NL}^{\cdot}(\Lambda) = \{ \mathbb{NL}^i(\Lambda) \,|\, i \in \mathbb{Z} \}$$

be the normal symmetric $\mathbb{L}$-spectrum of 13.5. Let

$$\mathbb{NL}.(\Lambda) = \{ \mathbb{NL}_i(\Lambda) \,|\, i \in \mathbb{Z} \}$$

be the normal quadratic $\mathbb{L}$-spectrum defined using quadratic n-ads which are not required to be Poincaré. $\mathbb{NL}.(\Lambda)$ is contractible, since every quadratic complex (C, ψ) bounds the quadratic pair $(C \longrightarrow 0, (0, \psi))$. The normal L-spaces fit into fibration sequences

$$\mathbb{L}^i(\mathbb{Z}_* K) \longrightarrow \mathbb{NL}^i(\mathbb{Z}_* K) \xrightarrow{\partial} \mathbb{L}_{i-1}(\mathbb{Z}_* K) ,$$

$$\mathbb{L}^i(\mathbb{Z}, K) \longrightarrow \mathbb{NL}^i(\mathbb{Z}_* K) \xrightarrow{\partial} \mathbb{L}_{i-1}(\mathbb{Z}, K) ,$$

$$\mathbb{L}_i(\mathbb{Z}_* K) \longrightarrow \mathbb{NL}_i(\mathbb{Z}_* K) \xrightarrow{\partial} \mathbb{L}_{i-1}(\mathbb{Z}_* K) ,$$

$$\mathbb{L}_i(\mathbb{Z}, K) \longrightarrow \mathbb{NL}_i(\mathbb{Z}_* K) \xrightarrow{\partial} \mathbb{L}_{i-1}(\mathbb{Z}, K) .$$

B7. The normal symmetric $\mathbb{L}$-spectrum $\mathbb{NL}^{\cdot}(\mathbb{Z})$ is a ring spectrum with products

$$\otimes : \mathbb{NL}^i(\mathbb{Z}) \wedge \mathbb{NL}^j(\mathbb{Z}) \longrightarrow \mathbb{NL}^{i+j}(\mathbb{Z}) ,$$

acting on the normal quadratic $\mathbb{L}$-spectrum $\mathbb{NL}.(\mathbb{Z})$ by products

$$\otimes : \mathbb{NL}^i(\mathbb{Z}) \wedge \mathbb{NL}_j(\mathbb{Z}) \longrightarrow \mathbb{NL}_{i+j}(\mathbb{Z}) .$$

The products

$$\otimes : \mathbb{NL}^i(\mathbb{Z}_* K) \wedge \mathbb{NL}_j(\mathbb{Z}) \longrightarrow \mathbb{NL}_{i+j}(\mathbb{Z}_* K)$$

restrict to products

$$\otimes : \mathbb{L}^i(\mathbb{Z}, K) \wedge \mathbb{L}_j(\mathbb{Z}) \longrightarrow \mathbb{L}_{i+j}(\mathbb{Z}, K) ,$$

and there is defined a commutative diagram

$$
\begin{array}{ccc}
\Omega_.^{fr}(K)_i \wedge \mathbb{L}_j(\mathbb{Z}) & \xrightarrow{\ \otimes\ } & \mathbb{L}_{i+j}(\mathbb{Z}_* K) \\
\downarrow & & \Big\| \\
\mathbb{L}^i(\mathbb{Z}_* K) \wedge \mathbb{L}_j(\mathbb{Z}) & \xrightarrow{\ \otimes\ } & \mathbb{L}_{i+j}(\mathbb{Z}_* K) \\
\downarrow & & \Big\downarrow{\scriptstyle A} \\
\mathbb{L}^i(\mathbb{Z}, K) \wedge \mathbb{L}_j(\mathbb{Z}) & \xrightarrow{\ \otimes\ } & \mathbb{L}_{i+j}(\mathbb{Z}, K) \\
\downarrow & & \Big\downarrow{\scriptstyle \simeq} \\
\mathbb{L}^i(\mathbb{Z}[\pi_1(K)]) \wedge \mathbb{L}_j(\mathbb{Z}) & \xrightarrow{\ \otimes\ } & \mathbb{L}_{i+j}(\mathbb{Z}[\pi_1(K)]) \ .
\end{array}
$$

By B5 the products $\otimes : \Omega_.^{fr}(K)_i \wedge \mathbb{L}_j(\mathbb{Z}) \longrightarrow \mathbb{L}_{i+j}(\mathbb{Z}_* K)$ induce homotopy equivalences

$$
\varinjlim_j \Omega^j(\Omega_.^{fr}(K)_i \wedge \mathbb{L}_{-j}(\mathbb{Z}))
$$

$$
\simeq \ \mathbb{H}_i(K, \mathbb{L}.(\mathbb{Z})) \ = \ \varinjlim_j \Omega^j \mathbb{L}_{i-j}(\mathbb{Z}_* K) \ = \ \mathbb{L}_i(\mathbb{Z}_* K) \ .
$$

B8. The symmetric signature is a map of Ω-spectra

$$
\sigma^* \ = \ 1 \wedge U : \Omega_.^{fr}(K) \simeq_s K_+ \wedge \underline{S}^0 \longrightarrow \mathbb{L}^0(\mathbb{Z}_* K) \simeq_s K_+ \wedge \mathbb{L}^{\cdot} \ ,
$$

with

$$
U \ = \ (\mathbb{Z}, 1) : \Omega_.^{fr}(\{*\})_0 \simeq_s S^0 \longrightarrow \mathbb{L}^0(\mathbb{Z})
$$

the unit of the ring spectrum $\mathbb{L}^{\cdot}(\mathbb{Z})$, representing

$$
(\mathbb{Z}, 1) \ = \ 1 \in L^0(\mathbb{Z}) \ = \ \mathbb{Z} \ .
$$

Define the *preassembly* pointed Δ-map

$$
A : K_+ \xrightarrow{\ \sigma^*\ } \Omega_.^{fr}(K)_0 \xrightarrow{\ } \mathbb{L}^0(\mathbb{Z}_* K) \xrightarrow{\ A^*\ } \mathbb{L}^0(\mathbb{Z}[\pi_1(K)]) \ ;
$$

$$
(\Delta^n \longrightarrow K) \longrightarrow (C(\widetilde{\Delta}^n), \phi(\widetilde{\Delta}^n))
$$

by sending the characteristic map of an n-simplex $\Delta^n \longrightarrow K$ to the n-dimensional symmetric Poincaré n-ad over $\mathbb{Z}[\pi_1(K)]$ of the pullback $\widetilde{\Delta}^n \longrightarrow \widetilde{K}$ from the universal cover $\widetilde{K}$ of K, with A^* the symmetric L-theory assembly. The preassembly A is the composite

$$
A : K_+ \ = \ K_+ \wedge S^0 \xrightarrow{\ 1 \wedge U\ } K_+ \wedge \mathbb{L}^0(\mathbb{Z}) \simeq_s \mathbb{L}^0(\mathbb{Z}_* K) \xrightarrow{\ A^*\ } \mathbb{L}^0(\mathbb{Z}[\pi_1(K)]) \ .
$$

Thus the assembly map in quadratic L-theory factorizes as the composite

$$A : \mathbb{H}_0(K;\mathbb{L}.(\mathbb{Z})) = \mathbb{L}_0(\mathbb{Z}_* K) \simeq \varinjlim_j \Omega^j(\Omega_{\cdot}^{fr}(K)_0 \wedge \mathbb{L}_{-j}(\mathbb{Z}))$$

$$\xrightarrow{A} \dot{\mathbb{H}}_0(\mathbb{L}^0(\mathbb{Z}[\pi_1(K)]);\mathbb{L}.(\mathbb{Z})) \simeq \varinjlim_j \Omega^j(\mathbb{L}^0(\mathbb{Z}[\pi_1(K)]) \wedge \mathbb{L}_{-j}(\mathbb{Z}))$$

$$\xrightarrow{\otimes} \mathbb{L}_0(\mathbb{Z}[\pi_1(K)]) = \varinjlim_j \Omega^j \mathbb{L}_{-j}(\mathbb{Z}[\pi_1(K)]) \ .$$

On the level of homotopy groups this can be written as

$$A : H_*(K;\mathbb{L}.(\mathbb{Z})) = \dot{H}_*(\Omega_{\cdot}^{fr}(K)_0;\mathbb{L}.(\mathbb{Z}))$$

$$\xrightarrow{A} \dot{H}_*(\mathbb{L}^0(\mathbb{Z}[\pi_1(K)]);\mathbb{L}.(\mathbb{Z})) \xrightarrow{\otimes} L_*(\mathbb{Z}[\pi_1(K)]) \ .$$

This is the construction of assembly via products.

B9. From the multiplicative point of view the Sullivan–Wall factorization of the surgery map for an n-dimensional topological manifold M through bordism is given by

$$\theta = A : [M,G/TOP] = H^0(M;\mathbb{L}.) = H_n(M;\mathbb{L}.) = H_n(M;\mathbb{L}. \wedge \underline{S}^0)$$

$$\xrightarrow{1 \wedge U} H_n(M;\mathbb{L}. \wedge \underline{MSTOP}) = \dot{H}_n(M_+ \wedge \mathbb{L}_0; \underline{MSTOP})$$

$$= \Omega_n^{TOP}(M \times G/TOP, M \times \{*\})$$

$$\longrightarrow \Omega_n^{TOP}(B\pi \times G/TOP, B\pi \times \{*\}) \longrightarrow L_n(\mathbb{Z}[\pi]) \ (\pi = \pi_1(M)) \ ,$$

with

$$U : \underline{S}^0 \longrightarrow \underline{MSTOP} = \Omega_{\cdot}^{TOP}(\{*\})$$

the unit in the oriented topological bordism spectrum of a point. The map induced by $1 \wedge U$ is an injection, which is split by the map induced by the composite

$$\mathbb{L}. \wedge \underline{MSTOP} \xrightarrow{1 \wedge \sigma^*} \mathbb{L}. \wedge \mathbb{L}^{\cdot} \xrightarrow{\otimes} \mathbb{L}.$$

with $\sigma^*: \underline{MSTOP} \longrightarrow \mathbb{L}^{\cdot}$ the symmetric signature map. $\underline{MSTOP}$ is a ring spectrum, and $\mathbb{L}.(R)$ is an $\underline{MSTOP}$-module spectrum for any ring with involution R: see Taylor and Williams [169] for the homotopy theoretic consequences, such as the decomposition at 2 as a generalized Eilenberg–MacLane spectrum

$$\mathbb{L}.(R) \otimes \mathbb{Z}_{(2)} \simeq \bigvee_i K.(L_i(R)_{(2)}, i) \ .$$

B10. An automorphism $f:(M,\theta) \longrightarrow (M,\theta)$ of a nonsingular symmetric form (M,θ) over a ring with involution R determines a 1-dimensional symmetric Poincaré complex $A(f) = (C, \phi)$ by

$$d = 1 - f^* : C_1 = M^* \longrightarrow C_1 = M^* \, ,$$

$$\phi_0 = \begin{cases} f^*\theta : C^1 = M \longrightarrow C_0 = M^* \\ \theta : C^0 = M \longrightarrow C_1 = M^* \, , \end{cases}$$

$$\phi_1 = \theta : C^1 = M \longrightarrow C_1 = M^* \, ,$$

corresponding to the nonsingular symmetric formation over R

$$(M \oplus M, \theta \oplus -\theta; \Delta, (f \oplus 1)\Delta) \, ,$$

with $\Delta = \{(x, x) \in M \oplus M \mid x \in M\}$ the diagonal lagrangian in the non-singular symmetric form $(M \oplus M, \theta \oplus -\theta)$. For example, the 1-dimensional symmetric Poincaré complex of the circle S^1 is

$$\sigma^*(S^1) = A(z : (\mathbb{Z}[\mathbb{Z}], 1) \longrightarrow (\mathbb{Z}[\mathbb{Z}], 1)) \, ,$$

with the involution $\bar{z} = z^{-1}$ on $\mathbb{Z}[\mathbb{Z}] = \mathbb{Z}[z, z^{-1}]$, and

$$d = 1 - z^{-1} : C_1 = \mathbb{Z}[\mathbb{Z}] \longrightarrow C_0 = \mathbb{Z}[\mathbb{Z}] \, .$$

The preassembly map $A : B\pi_+ \longrightarrow \mathbb{L}^0(\mathbb{Z}[\pi])$ sends the 1-simplex

$$g \in (B\pi_+)^{(1)} = \pi \sqcup \{\emptyset\}$$

determined by an element $g \in \pi$ to the 1-dimensional symmetric Poincaré complex over $\mathbb{Z}[\pi]$

$$g_*\sigma^*(S^1) = A(g : (\mathbb{Z}[\pi], 1) \longrightarrow (\mathbb{Z}[\pi], 1)) = (C, \phi)$$

with

$$d = 1 - g^{-1} : C_1 = \mathbb{Z}[\pi] \longrightarrow C_0 = \mathbb{Z}[\pi] \, .$$

Loday [94] constructed the assembly map $A_\pi : H_*(B\pi; \mathbb{L}.) \longrightarrow L_*(\mathbb{Z}[\pi])$ away from 2, using products and the action of π on hermitian K-theory induced by the inclusion

$$\pi \longrightarrow \mathrm{Aut}(\mathbb{Z}[\pi], 1) = GL_1(\mathbb{Z}[\pi]) \; ; \; g \longrightarrow g \, .$$

The methods of this appendix show that this construction does indeed agree with the surgery assembly map, as conjectured in [94].

Appendix C. Assembly via bounded topology

The applications of algebraic L-theory to compact topological manifolds depend on the torus trick of Kirby and Siebenmann [84]. The controlled and bounded topology of non-compact manifolds subsequently developed by Chapman, Ferry and Quinn has led to an interesting hybrid of algebra and topology involving the lower K-groups of Bass [9] and the lower L-groups of Ranicki [137], [146], in which the algebraic operations are required to be small when measured in some metric space. The controlled surgery theories of Quinn [130]-[133], Yamasaki [187] and the bounded surgery theory of Ferry and Pedersen [50] have found wide applications to the structure theory of ANR homology manifolds, group actions, fibrations and rigidity. See Ferry, Hambleton and Pedersen [49] and Weinberger [181] for surveys of the applications.

Controlled and bounded topology offer an alternative construction of the 4-periodic algebraic L-theory assembly maps

$$A : H_*(X; \mathbb{L}.(\mathbb{Z})) \longrightarrow L_*(\mathbb{Z}[\pi_1(X)]) ,$$

using the lower L-groups and the Bass–Heller–Swan computation $K_{-i}(\mathbb{Z}) = 0$ $(i \geq 1)$ to express the $\mathbb{L}.(\mathbb{Z})$-coefficient generalized homology groups as the $\mathbb{R}^i$-bounded surgery obstruction groups for large $i \geq 1$. The 4-periodic algebraic L-theory assembly map will now be obtained using bounded topology, and some of the consequences of this approach will be explored. The generalized homology groups with L-theory coefficients arise as the cobordism groups of bounded algebraic Poincaré complexes, and the assembly maps are the forgetful maps to the unbounded cobordism groups. See Ranicki and Yamasaki [148] for a chain complex approach to assembly in controlled K-theory, which also applies to controlled L-theory.

C1. The projective L-groups $L_*^p(R)$ of Novikov [121] and Ranicki [136], [137] are defined for any ring with involution R, using quadratic structures on f.g. projective R-modules. The projective L-groups are related to the free L-groups $L_*^h(R) = L_*(R)$ by splittings

$$L_n(R[z, z^{-1}]) = L_n(R) \oplus L_{n-1}^p(R) \ (\bar{z} = z^{-1})$$

and a Rothenberg-type exact sequence

$$\dots \longrightarrow L_n(R) \longrightarrow L_n^p(R) \longrightarrow \widehat{H}^n(\mathbb{Z}_2 ; \widetilde{K}_0(R)) \longrightarrow L_{n-1}(R) \longrightarrow \dots .$$

The projective surgery theory of Pedersen and Ranicki [123] involves the projective $\mathbb{S}$-groups $\mathbb{S}_*^p(X)$ which are defined to fit into an exact sequence

$$\dots \longrightarrow H_n(X; \mathbb{L}.) \stackrel{A}{\longrightarrow} L_n^p(\mathbb{Z}[\pi_1(X)]) \longrightarrow \mathbb{S}_n^p(X) \longrightarrow H_{n-1}(X; \mathbb{L}.) \longrightarrow \dots$$

for any space X, and are such that

$$\mathbb{S}_n(X \times S^1) = \mathbb{S}_n(X) \oplus \mathbb{S}_{n-1}^p(X) .$$

The projective assembly map is the composite
$$A : H_n(X; \mathbb{L}) \xrightarrow{A} L_n(\mathbb{Z}[\pi_1(X)]) \longrightarrow L_n^p(\mathbb{Z}[\pi_1(X)]) .$$
If X is a finitely dominated n-dimensional geometric Poincaré complex then $X \times S^1$ is homotopy equivalent to a finite $(n+1)$-dimensional geometric Poincaré complex, by the Mather trick. The *projective total surgery obstruction* $s^p(X) \in \mathbb{S}_n^p(X)$ of [123] is such that
$$s(X \times S^1) = (0, s^p(X)) \in \mathbb{S}_{n+1}(X \times S^1) = \mathbb{S}_{n+1}(X) \oplus \mathbb{S}_n^p(X) .$$
Thus $s^p(X) = 0$ if (and for $n \geq 4$ only if) $X \times S^1$ is homotopy equivalent to a compact $(n+1)$-dimensional topological manifold.

The lower L-groups $L_*^{\langle -i \rangle}(R)$ $(i \geq 1)$ of Ranicki [137], [146] are the L-theoretic analogues of the lower K-groups $K_{-i}(R)$ of Bass [9, XII]. The free and projective L-groups
$$L_*(R) = L_*^h(R) = L_*^{\langle 1 \rangle}(R) \ , \ L_*^p(R) = L_*^{\langle 0 \rangle}(R)$$
are related to the lower L-groups $L_*^{\langle -i \rangle}(R)$ by splittings
$$L_n^{\langle 1-i \rangle}(R[z, z^{-1}]) = L_n^{\langle 1-i \rangle}(R) \oplus L_{n-1}^{\langle -i \rangle}(R) \ (i \geq 0)$$
and exact sequences
$$\ldots \longrightarrow L_n^{\langle 1-i \rangle}(R) \longrightarrow L_n^{\langle -i \rangle}(R) \longrightarrow \widehat{H}^n(\mathbb{Z}_2; \widetilde{K}_{-i}(R)) \longrightarrow L_{n-1}^{\langle 1-i \rangle}(R) \longrightarrow \ldots$$
with $\widetilde{K}_{-i}(R) = K_{-i}(R)$ for $i \geq 1$. For any space X the free and projective $\mathbb{S}$-groups
$$\mathbb{S}_*(X) = \mathbb{S}_*^h(X) = \mathbb{S}_*^{\langle 1 \rangle}(X) \ , \ \mathbb{S}_*^p(X) = \mathbb{S}_*^{\langle 0 \rangle}(X)$$
are related to the lower $\mathbb{S}$-groups $\mathbb{S}_*^{\langle -i \rangle}(X)$ by splittings
$$\mathbb{S}_n^{\langle 1-i \rangle}(X \times S^1) = \mathbb{S}_n^{\langle 1-i \rangle}(X) \oplus \mathbb{S}_{n-1}^{\langle -i \rangle}(X) \ (i \geq 0)$$
and a commutative braid of exact sequences

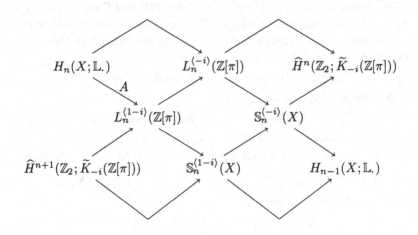

with $\pi = \pi_1(X)$. For $i \geq 0$ the *lower total surgery obstruction* of a finitely dominated n-dimensional geometric Poincaré complex X is the image of the projective total surgery obstruction

$$s^{\langle -i \rangle}(X) = [s^p(X)] \in \mathbb{S}_n^{\langle -i \rangle}(X) ,$$

and is such that $s^{\langle -i \rangle}(X) = 0$ if (and for $n + i \geq 4$ only if) $X \times T^{i+1}$ is homotopy equivalent to a compact $(n + i + 1)$-dimensional topological manifold.

C2. Given a metric space X and an additive category $\mathbb{A}$ let $\mathbb{C}_X(\mathbb{A})$ be the X-bounded additive category defined by Pedersen and Weibel [124]. The objects of $\mathbb{C}_X(\mathbb{A})$ are formal direct sums

$$M = \sum_{x \in X} M(x)$$

of objects $M(x)$ in $\mathbb{A}$. The morphisms $f : M \longrightarrow N$ in $\mathbb{C}_X(\mathbb{A})$ are collections of morphisms in $\mathbb{A}$

$$f = \{ f(y,x) : M(x) \longrightarrow N(y) \mid x, y \in X \}$$

such that there exists a number $b \geq 0$ with $f(y,x) = 0$ if $d(x,y) > b$. An involution $* : \mathbb{A} \longrightarrow \mathbb{A} ; A \longrightarrow A^*$ extends to an involution of $\mathbb{C}_X(\mathbb{A})$ by

$$* : \mathbb{C}_X(\mathbb{A}) \longrightarrow \mathbb{C}_X(\mathbb{A}) ;$$

$$M = \sum_{x \in X} M(x) \longrightarrow M^* = \sum_{x \in X} M^*(x) , \ M^*(x) = M(x)^* .$$

The $\begin{cases} \text{symmetric} \\ \text{quadratic} \end{cases}$ L-groups $\begin{cases} L^*(\mathbb{C}_X(\mathbb{A})) \\ L_*(\mathbb{C}_X(\mathbb{A})) \end{cases}$ are related by symmetrization maps

$$1 + T : L_*(\mathbb{C}_X(\mathbb{A})) \longrightarrow L^*(\mathbb{C}_X(\mathbb{A}))$$

which are isomorphisms modulo 8-torsion, since the ring $\widehat{L}^0(\mathbb{Z}) = \mathbb{Z}_8$ acts on the relative groups.

C3. Given a group π and an additive category $\mathbb{A}$ let $\mathbb{A}[\pi]$ be the additive category with one object $M[\pi]$ for each object M in $\mathbb{A}$, and one morphism

$$f = \sum_{g \in \pi} n_g f_g : M[\pi] \longrightarrow N[\pi] \quad (n_g \in \mathbb{Z})$$

for each formal linear combination of morphisms $f_g : M \longrightarrow N$ in $\mathbb{A}$, with $\{ g \in \pi \mid f_g \neq 0 \}$ finite. An involution on $\mathbb{A}$ is extended to an involution on $\mathbb{A}[\pi]$ by

$$* : \mathbb{A}[\pi] \longrightarrow \mathbb{A}[\pi] ;$$

$$M[\pi] \longrightarrow (M[\pi])^* = M^*[\pi] , \ f = \sum_{g \in \pi} n_g f_g \longrightarrow f^* = \sum_{g \in \pi} n_g (f_g)^* .$$

For any commutative ring R there is an identification
$$\mathbb{A}^h(R)[\pi] = \mathbb{A}^h(R[\pi])$$
with $\mathbb{A}^h(R)$ the additive category of based f.g. free R-modules. Write the category $\mathbb{C}_X(\mathbb{A}^h(R[\pi]))$ as $\mathbb{C}_X(R[\pi])$.

C4. The bounded surgery theory of Ferry and Pedersen [50] applies to geometric Poincaré complexes and manifolds which are 'X-bounded' for some metric space X, i.e. equipped with a proper map to X such that the diameters of cells are uniformly bounded in X. In the first instance the theory applies to 'allowable' metric spaces and Poincaré complexes with constant 'bounded fundamental group' π, and the same hypotheses will be in force here. The main construction of [50] associates to a normal map $(f,b): J \longrightarrow K$ from an n-dimensional X-bounded manifold J to an X-bounded geometric Poincaré complex K an X-bounded surgery obstruction
$$\sigma_*(f,b) \in L_n(\mathbb{C}_X(\mathbb{Z}[\pi]))$$
such that $\sigma_*(f,b) = 0$ if (and for $n \geq 5$ only if) (f,b) is normal bordant to an X-bounded homotopy equivalence. An n-dimensional X-bounded geometric Poincaré complex K has a Spivak normal fibration $\nu_K: K \longrightarrow BG$, such that the topological reductions $\tilde{\nu}_K: K \longrightarrow BTOP$ are in one–one correspondence with the bordism classes of normal maps $(f,b): (J, \nu_J) \longrightarrow (K, \tilde{\nu}_K)$ from n-dimensional X-bounded manifolds, as in the classical compact case $X = \{\text{pt.}\}$. There exists a topological reduction $\tilde{\nu}_K: K \longrightarrow BTOP$ such that $\sigma_*(f,b) = 0 \in L_n(\mathbb{C}_X(\mathbb{Z}[\pi]))$ if (and for $n \geq 5$ only if) K is X-bounded homotopy equivalent to an X-bounded topological manifold. For an X-bounded topological manifold K the X-bounded structure set $\mathbb{S}^b(K)$ fits into the bounded version of the Sullivan–Wall surgery exact sequence
$$\ldots \longrightarrow L_{n+1}(\mathbb{C}_X(\mathbb{Z}[\pi])) \longrightarrow \mathbb{S}^b(K) \longrightarrow [K, G/TOP] \longrightarrow L_n(\mathbb{C}_X(\mathbb{Z}[\pi])).$$
The X-bounded symmetric signature of an n-dimensional geometric X-bounded geometric Poincaré complex K is the algebraic Poincaré cobordism class
$$\sigma^*(K) = (C(\tilde{K}), \phi_K) \in L^n(\mathbb{C}_X(\mathbb{Z}[\pi])),$$
with $C(\tilde{K})$ the cellular chain complex in $\mathbb{C}_X(\mathbb{Z}[\pi])$ of the universal cover $\tilde{K}$ and $\phi_K = \Delta([K])$ the evaluation of an Alexander–Whitney–Steenrod diagonal chain approximation Δ on the locally finite fundamental class $[K] \in H_n^{lf}(K)$. The X-bounded symmetric signature is an X-bounded homotopy invariant. The X-bounded surgery obstruction of an n-dimensional X-bounded normal map $(f,b): J \longrightarrow K$ has symmetrization
$$(1+T)\sigma_*(f,b) = \sigma^*(J) - \sigma^*(K) \in L^n(\mathbb{C}_X(\mathbb{Z}[\pi])).$$

C5. Let K be a simplicial complex which is locally finite and finite-dimensional. Given an additive category $\mathbb{A}$ let $\mathbb{A}_*^{lf}(K)$ be the additive category of K-based objects in $\mathbb{A}$, the category with objects formal direct sums

$$M = \sum_{\sigma \in K} M(\sigma)$$

of objects $M(\sigma)$ in $\mathbb{A}$. A morphism $f: M \longrightarrow N$ in $\mathbb{A}_*^{lf}(K)$ is a collection of morphisms in $\mathbb{A}$

$$f = \{f(\tau, \sigma): M(\sigma) \longrightarrow N(\tau) \mid \sigma, \tau \in K\}$$

such that $f(\tau, \sigma) = 0: M(\sigma) \longrightarrow N(\tau)$ unless $\tau \geq \sigma$. (For finite K this is just the K-based category $\mathbb{A}_*(K)$ of §4.) Given an involution $*: \mathbb{A} \longrightarrow \mathbb{A}$; $A \longrightarrow A^*$ define a chain duality $T: \mathbb{A}_*^{lf}(K) \longrightarrow \mathbb{A}_*^{lf}(K)$ by the method of §5. The dual of an object M in $\mathbb{A}_*^{lf}(K)$ is a chain complex TM in $\mathbb{A}_*^{lf}(K)$ with

$$TM_r(\sigma) = \begin{cases} \sum_{\tau \geq \sigma} M(\tau)^* & \text{if } r = -|\sigma| \\ 0 & \text{otherwise} . \end{cases}$$

Working as in §14 it is possible to identify the algebraic L-groups of $\mathbb{A}_*^{lf}(K)$ with the locally finite generalized homology groups

$$L_n(\mathbb{A}_*^{lf}(K)) = H_n^{lf}(K; \mathbb{L}.(\mathbb{A})) \quad (n \in \mathbb{Z}) .$$

Assume that the diameters of the simplices of K are uniformly bounded, i.e. there exists a number $b \geq 0$ such that $d(x, y) \leq b$ if $x, y \in |\sigma|$ for any simplex $\sigma \in K$. Regard the polyhedron of K (also denoted by K) as a metric space using a proper embedding $K \subseteq \mathbb{R}^N$ for a sufficiently large $N \geq 0$, so that the K-bounded additive category with involution $\mathbb{C}_K(\mathbb{A})$ is defined as above. Let $\widetilde{K}$ be a regular covering of K with group of covering translations π. Define an assembly functor by forgetting all but the bounded aspects of the simplicial structure and passing to the cover

$$A : \mathbb{A}_*^{lf}(K) \longrightarrow \mathbb{C}_K(\mathbb{A}[\pi]) ; \quad M \longrightarrow \widehat{M} ,$$

sending an object M in $\mathbb{A}_*^{lf}(K)$ to the object $\widehat{M}$ in $\mathbb{C}_K(\mathbb{A}[\pi])$ defined by

$$\widehat{M}(x) = \begin{cases} M(\sigma)[\pi] & \text{if } x = \widehat{\sigma} \text{ is the barycentre of } \sigma \in K \\ 0 & \text{otherwise} . \end{cases}$$

Working as in 6.1 the chain duality on $\mathbb{A}_*^{lf}(K)$ is related to the involution on $\mathbb{C}_K(\mathbb{A})$ by a natural chain equivalence in $\mathbb{C}_K(\mathbb{A})$

$$T\beta : \widehat{TM} \xrightarrow{\simeq} (\widehat{M})^* .$$

The assembly functor of algebraic bordism categories

$$A : \Lambda(\mathbb{A}_*^{lf}(K)) \longrightarrow \Lambda(\mathbb{C}_K(\mathbb{A}[\pi]))$$

(with Λ as in 3.3) induces natural assembly maps of L-groups

$$A : L_n(\mathbb{A}_*^{lf}(K)) = H_n^{lf}(K; \mathbb{L}.(\mathbb{A})) \longrightarrow L_n(\mathbb{C}_K(\mathbb{A}[\pi])) \quad (n \in \mathbb{Z}) .$$

If $K = \tilde{J}$ is the universal cover of a finite simplicial complex J the simply connected assembly map

$$A : L_n(\mathbb{A}^{lf}_*(\tilde{J})) = H^{lf}_n(\tilde{J}; \mathbb{L}.(\mathbb{A})) \longrightarrow L_n(\mathbb{C}_{\tilde{J}}(\mathbb{A}))$$

(with $\pi = \{1\}$) is related to the universal assembly map of §9

$$A : H_n(J; \mathbb{L}.(\mathbb{A})) \longrightarrow L_n(\mathbb{A}[\rho]) \quad (\rho = \pi_1(J))$$

by a commutative diagram

$$
\begin{array}{ccc}
H_n(J; \mathbb{L}.(\mathbb{A})) & \xrightarrow{\;\;A\;\;} & L_n(\mathbb{A}[\rho]) \\
{\scriptstyle \text{trf}} \big\downarrow & & \big\downarrow {\scriptstyle \text{trf}} \\
H^{lf}_n(\tilde{J}; \mathbb{L}.(\mathbb{A})) & \xrightarrow{\;\;A\;\;} & L_n(\mathbb{C}_{\tilde{J}}(\mathbb{A})) \ .
\end{array}
$$

The infinite transfer map

$$\text{trf} : L_n(\mathbb{A}_*(J)) = H_n(J; \mathbb{L}.(\mathbb{A})) \longrightarrow L_n(\mathbb{A}^{lf}_*(\tilde{J})) = H^{lf}_n(\tilde{J}; \mathbb{L}.(\mathbb{A}))$$

is induced by the functor

$$\mathbb{A}_*(J) \longrightarrow \mathbb{A}^{lf}_*(\tilde{J}) \ ; \quad M = \sum_{\sigma \in J} M(\sigma) \longrightarrow \widetilde{M} = \sum_{\tilde{\sigma} \in \tilde{J}} M(p\tilde{\sigma})$$

with $p : \tilde{J} \longrightarrow J$ the covering projection. The infinite transfer map

$$\text{trf} : L_n(\mathbb{A}[\rho]) = L_n(\mathbb{C}_{\tilde{J}}(\mathbb{A})^\rho) \longrightarrow L_n(\mathbb{C}_{\tilde{J}}(\mathbb{A}))$$

is induced by the inclusion $\mathbb{C}_{\tilde{J}}(\mathbb{A})^\rho \longrightarrow \mathbb{C}_{\tilde{J}}(\mathbb{A})$ of the ρ-invariant subcategory, with objects the lifts $\widetilde{M}$ of objects M in $\mathbb{C}_J(\mathbb{A})$ and ρ-equivariant morphisms. The forgetful functor

$$\mathbb{C}_{\tilde{J}}(\mathbb{A})^\rho \longrightarrow \mathbb{A}[\rho] \ ; \quad \widetilde{M} \longrightarrow \widetilde{M}$$

is an equivalence of additive categories with involution, since J is finite.

C6. Given a metric space X and an X-bounded simplicial complex K with constant bounded fundamental group π let $\mathbb{S}^b_*(K)$ be the relative groups in the bounded algebraic surgery exact sequence

$$\ldots \longrightarrow H^{lf}_n(K; \mathbb{L}.) \xrightarrow{\;\;A\;\;} L_n(\mathbb{C}_X(\mathbb{Z}[\pi])) \longrightarrow \mathbb{S}^b_n(K)$$
$$\longrightarrow H^{lf}_{n-1}(K; \mathbb{L}.) \longrightarrow \ldots$$

with $\mathbb{L}. = \mathbb{L}.\langle 1 \rangle(\mathbb{Z})$ as in §17, and

$$A : H^{lf}_*(K; \mathbb{L}.) \xrightarrow{\;\;A\;\;} L_*(\mathbb{C}_K(\mathbb{Z}[\pi])) \longrightarrow L_*(\mathbb{C}_X(\mathbb{Z}[\pi])) \ .$$

The bordism group $\Omega^{bP}_n(K)$ of n-dimensional X-bounded geometric Poincaré complexes with a map to K fits into a commutative braid of exact sequences

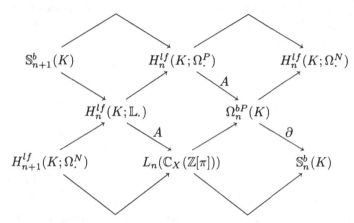

generalizing 19.6 (= the special case when X is compact). Define the *1/2-connective X-bounded visible symmetric L-groups* $VL_b^*(K)$ to be the cobordism groups of visible symmetric Poincaré complexes (C, ϕ) in $\mathbb{A}(\mathbb{Z})_*^{lf}(K)$ which are globally 0-connective and locally 1-Poincaré at ∞, by analogy with the 1/2-connective visible symmetric L-groups $VL^*(K)$ of §15. As in 15.18 (i) there is defined a commutative braid of exact sequences

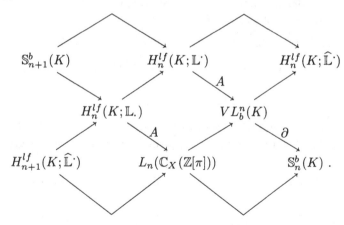

The *1/2-connective X-bounded visible symmetric signature* of an n-dimensional X-bounded geometric Poincaré complex K is

$$\sigma^*(K) = (C(K), \Delta[K]) \in VL_b^n(K) .$$

The *X-bounded total surgery obstruction* of K

$$s^b(K) = \partial \sigma^*(K) \in \mathbb{S}_n^b(K)$$

is such that $s^b(K) = 0$ if (and for $n \geq 5$ only if) K is X-bounded homotopy equivalent to an n-dimensional X-bounded topological manifold. The algebraic surgery exact sequence is related to the geometric surgery exact sequence of Ferry and Pedersen [50] for an n-dimensional X-bounded manifold K by an isomorphism

$$\cdots \to L_{n+1}(\mathbb{C}_X(\mathbb{Z}[\pi])) \longrightarrow \mathbb{S}^b(K) \longrightarrow [K, G/TOP] \longrightarrow L_n(\mathbb{C}_X(\mathbb{Z}[\pi]))$$

$$\Big\| \qquad\qquad s^b \Big\downarrow \cong \qquad\qquad t \Big\downarrow \cong \qquad\qquad \Big\|$$

$$\cdots \to L_{n+1}(\mathbb{C}_X(\mathbb{Z}[\pi])) \longrightarrow \mathbb{S}^b_{n+1}(K) \longrightarrow H^{lf}_n(K;\mathbb{L}_{\boldsymbol{.}}) \longrightarrow L_n(\mathbb{C}_X(\mathbb{Z}[\pi]))$$

with

$$s^b : \mathbb{S}^b(K) \xrightarrow{\;\cong\;} \mathbb{S}^b_{n+1}(K) \;;\; (f: J \longrightarrow K) \longrightarrow s^b(f) = s^b_\partial(W, J \sqcup -K)$$

given by the X-bounded rel ∂ total surgery obstruction of the mapping cylinder $W = J \times I \cup_f K$ of the X-bounded homotopy equivalence $f: J \longrightarrow K$, and

$$t = [K] \cap - : [K, G/TOP] = H^0(K;\mathbb{L}_{\boldsymbol{.}}) \xrightarrow{\;\cong\;} H^{lf}_n(K;\mathbb{L}_{\boldsymbol{.}})$$

the Poincaré duality isomorphism defined by cap product with the locally finite $\mathbb{L}_{\boldsymbol{.}}$-coefficient orientation $[K] \in H^{lf}_n(K;\mathbb{L}_{\boldsymbol{.}})$ ($\mathbb{L}_{\boldsymbol{.}} = \mathbb{L}_{\boldsymbol{.}}\langle 0 \rangle(\mathbb{Z}))$.

C7. Let $\mathbb{P}_X(\mathbb{A})$ denote the idempotent completion of $\mathbb{C}_X(\mathbb{A})$, the additive category in which an object is a pair

$$(M = \text{object of } \mathbb{C}_X(\mathbb{A}), \, p = p^2 : M \longrightarrow M)$$

and a morphism $f: (M, p) \longrightarrow (N, q)$ is a morphism $f: M \longrightarrow N$ in $\mathbb{C}_X(\mathbb{A})$ such that

$$qfp = f : M \longrightarrow N .$$

The algebraic K-theoretic methods of Pedersen and Weibel [124], Carlsson [31] and Ranicki [146] give an exact sequence for the algebraic K-groups of $\mathbb{C}_{X_1 \cup X_2}(\mathbb{A})$

$$\cdots \longrightarrow \varinjlim_b K_1(\mathbb{C}_{\mathcal{N}_b(X_1,X_2)}(\mathbb{A})) \longrightarrow K_1(\mathbb{C}_{X_1}(\mathbb{A})) \oplus K_1(\mathbb{C}_{X_2}(\mathbb{A}))$$

$$\longrightarrow K_1(\mathbb{C}_{X_1 \cup X_2}(\mathbb{A})) \longrightarrow \varinjlim_b K_0(\mathbb{P}_{\mathcal{N}_b(X_1,X_2)}(\mathbb{A})) \longrightarrow \cdots$$

with

$$\mathcal{N}_b(X_1, X_2) = \{x \in X_1 \cup X_2 \,|\, d(x, y_i) \le b \text{ for some } y_i \in X_i, \, i = 1, 2\} .$$

An involution $* : \mathbb{A} \longrightarrow \mathbb{A}; A \longrightarrow A^*$ is extended to an involution of $\mathbb{P}_X(\mathbb{A})$ by

$$* : \mathbb{P}_X(\mathbb{A}) \longrightarrow \mathbb{P}_X(\mathbb{A}) \;;\; (M, p) \longrightarrow (M, p)^* = (M^*, p^*) .$$

The quadratic L-groups of $\mathbb{C}_X(\mathbb{A})$ and $\mathbb{P}_X(\mathbb{A})$ are related by an exact sequence

$$\ldots \longrightarrow L_n(\mathbb{C}_X(\mathbb{A})) \longrightarrow L_n(\mathbb{P}_X(\mathbb{A})) \longrightarrow \widehat{H}^n(\mathbb{Z}_2 ; \widetilde{K}_0(\mathbb{P}_X(\mathbb{A})))$$
$$\longrightarrow L_{n-1}(\mathbb{C}_X(\mathbb{A})) \longrightarrow \ldots$$

involving the Tate $\mathbb{Z}_2$-cohomology groups of the duality involution on the reduced projective class group

$$\widetilde{K}_0(\mathbb{P}_X(\mathbb{A})) \;=\; \mathrm{coker}(K_0(\mathbb{C}_X(\mathbb{A})) \longrightarrow K_0(\mathbb{P}_X(\mathbb{A}))) \;.$$

The quadratic L-groups of $\mathbb{C}_{X_1 \cup X_2}(\mathbb{A})$ fit into the Mayer–Vietoris exact sequence of [146, 14.4]

$$\ldots \longrightarrow \varinjlim_b L_n(\mathbb{C}_{\mathcal{N}_b(X_1,X_2)}(\mathbb{A})) \longrightarrow L_n(\mathbb{C}_{X_1}(\mathbb{A})) \oplus L_n(\mathbb{C}_{X_2}(\mathbb{A}))$$
$$\longrightarrow L_n^Y(\mathbb{C}_{X_1 \cup X_2}(\mathbb{A})) \longrightarrow \varinjlim_b L_{n-1}(\mathbb{C}_{\mathcal{N}_b(X_1,X_2)}(\mathbb{A})) \longrightarrow \ldots$$

with

$$Y \;=\; \mathrm{im}(K_1(\mathbb{C}_{X_1}(\mathbb{A})) \oplus K_1(\mathbb{C}_{X_2}(\mathbb{A})) \longrightarrow K_1(\mathbb{C}_{X_1 \cup X_2}(\mathbb{A}))) \;.$$

Similarly for the symmetric L-groups L^*.

C8. Let X be a metric space with a K-dissection (4.14)

$$X \;=\; \bigcup_{\sigma \in K} X[\sigma]$$

for a finite simplicial complex K with fundamental group $\pi = \pi_1(K)$. Working as in 13.7 the generalized homology group $H_n(K; \{\mathbb{L}.(\mathbb{C}_{X[\sigma]}(\mathbb{A}))\})$ can be identified with the cobordism group of n-dimensional quadratic Poincaré cycles

$$(C,\psi) \;=\; \{(C(\sigma),\psi(\sigma)) \,|\, \sigma \in K\}$$

such that $(C(\sigma),\psi(\sigma))$ is defined in $\mathbb{C}_{X[\sigma]}(\mathbb{A})$, and there is defined an assembly map for any regular covering $p \colon \widetilde{K} \longrightarrow K$ with group of covering translations π

$$A : H_n(K; \{\mathbb{L}.(\mathbb{C}_{X[\sigma]}(\mathbb{A}))\}) \longrightarrow L_n(\mathbb{C}_X(\mathbb{A}[\pi])) \;;$$
$$(C,\psi) \longrightarrow (C(\widetilde{K}),\psi(\widetilde{K})) \;=\; \bigcup_{\tilde{\sigma} \in \widetilde{K}} (C(p\tilde{\sigma}),\psi(p\tilde{\sigma})) \;.$$

For any bound $b \geq 0$ and any n-simplex $\sigma = (v_0 v_1 \ldots v_n) \in K$ let

$$\mathcal{N}_b(X[\sigma]) \;=\; \{x \in X \,|\, d(x,y_i) \leq b \text{ for some } y_i \in X[v_i], 1 \leq i \leq n\} \;.$$

The algebraic transversality of [146, §14] shows that every n-dimensional quadratic complex in $\mathbb{C}_X(\mathbb{A})$ is homotopy equivalent to the assembly

$$A(C,\psi) \;=\; (C(K),\psi(K))$$

of an n-dimensional quadratic cycle (C, ψ) (although not necessarily one which is Poincaré) such that $(C(\sigma), \psi(\sigma))$ is defined in $\mathbb{C}_{\mathcal{N}_b(X[\sigma])}(\mathbb{A})$ for some bound $b \geq 0$. Working as in §13 the relative group $\mathbb{S}_n(K, X, \mathbb{A})$ in the bounded algebraic surgery exact sequence

$$\cdots \longrightarrow \varinjlim_b H_n(K; \{\mathbb{L}.(\mathbb{C}_{\mathcal{N}_b(X[\sigma])}(\mathbb{A}))\}) \overset{A}{\longrightarrow} L_n(\mathbb{C}_X(\mathbb{A}))$$

$$\overset{\partial}{\longrightarrow} \mathbb{S}_n(K, X, \mathbb{A}) \longrightarrow \varinjlim_b H_{n-1}(K; \{\mathbb{L}.(\mathbb{C}_{\mathcal{N}_b(X[\sigma])}(\mathbb{A}))\}) \longrightarrow \cdots$$

can be identified with the cobordism group of $(n-1)$-dimensional quadratic Poincaré cycles (C, ψ) such that $(C(\sigma), \psi(\sigma))$ is defined in $\mathbb{C}_{\mathcal{N}_b(X[\sigma])}(\mathbb{A})$ for some bound $b \geq 0$, and the assembly $C(K)$ is contractible in $\mathbb{C}_X(\mathbb{A})$. It follows from the Mayer–Vietoris exact sequences of [146, §14] that the groups $\mathbb{S}_*(K, X, \mathbb{A})$ are 2-primary torsion, and can be expressed in terms of the duality $\mathbb{Z}_2$-action on algebraic K-theory. In particular, for the case $K = \Delta^1 = \{0, 1, 01\}$ of a space X which is expressed as a union of two subspaces

$$X = X[0] \cup X[1] \ , \ \ X[0] \cap X[1] = X[01] \ ,$$

$$H_n(K; \{\mathbb{L}.(\mathbb{C}_{\mathcal{N}_b(X[\sigma])}(\mathbb{A}))\})$$

$$= L_n(\mathbb{C}_{\mathcal{N}_b(X[01])}(\mathbb{A}) \longrightarrow \mathbb{C}_{\mathcal{N}_b(X[0])}(\mathbb{A}) \times \mathbb{C}_{\mathcal{N}_b(X[1])}(\mathbb{A})) \ ,$$

$$\mathbb{S}_n(\Delta^1, X, \mathbb{A}) = \varinjlim_b \widehat{H}^n(\mathbb{Z}_2 ; I_b)$$

with

$$I_b = \ker(\widetilde{K}_0(\mathbb{P}_{\mathcal{N}_b(X[01])}(\mathbb{A})) \longrightarrow \widetilde{K}_0(\mathbb{P}_{\mathcal{N}_b(X[0])}(\mathbb{A})) \oplus \widetilde{K}_0(\mathbb{P}_{\mathcal{N}_b(X[1])}(\mathbb{A}))) \ .$$

C9. The *open cone* of a subspace $K \subseteq S^N$ is the metric space

$$O(K) = \{ tx \in \mathbb{R}^{N+1} \,|\, t \in [0, \infty) \, , \, x \in K \} \subseteq \mathbb{R}^{N+1} \ .$$

For a compact polyhedron $K \subset S^N$ define a K-dissection of $O(K^+)$ by

$$O(K^+)[\sigma] = O(D(\sigma, K)^+) \ \ (\sigma \in K) \ ,$$

with $K^+ = K \sqcup \{\mathrm{pt.}\}$. The assembly maps given by C5 and C8

$$A : H_*^{lf}(O(K^+); \mathbb{L}.(\mathbb{A})) \longrightarrow L_*(\mathbb{C}_{O(K^+)}(\mathbb{A})) \ ,$$

$$A : H_*(K; \{\mathbb{L}.(\mathbb{C}_{O(K^+)[\sigma]}(\mathbb{A}))\}) \longrightarrow L_*(\mathbb{C}_{O(K^+)}(\mathbb{A}))$$

are related as follows. Projections define homotopy equivalences of spectra,

$$\mathbb{L}.(\mathbb{C}_{O(K^+)[\sigma]}(\mathbb{A})) \overset{\simeq}{\longrightarrow} \mathbb{L}.(\mathbb{C}_{\mathbb{R}}(\mathbb{A})) \ \ (\sigma \in K) \ ,$$

and product with the generator

$$\sigma^*(\mathbb{R}) = 1 \in L^1(\mathbb{C}_{\mathbb{R}}(\mathbb{Z})) = L^0(\mathbb{Z}) = \mathbb{Z}$$

defines a homotopy equivalence

$$\sigma^*(\mathbb{R}) \otimes - : \Sigma \mathbb{L}.(\mathbb{P}_0(\mathbb{A})) \xrightarrow{\simeq} \mathbb{L}.(\mathbb{C}_{\mathbb{R}}(\mathbb{A}))$$

with $\mathbb{P}_0(\mathbb{A})$ the idempotent completion of $\mathbb{A}$. The assembly map of C8 factors through the assembly map of C5

$$A : H^{lf}_*(O(K^+); \mathbb{L}.(\mathbb{A})) = H_{*-1}(K; \mathbb{L}.(\mathbb{A})) \longrightarrow$$

$$H_*(K; \{\mathbb{L}.(\mathbb{C}_{O(K^+)[\sigma]}(\mathbb{A}))\}) = H_*(K; \mathbb{L}.(\mathbb{C}_{\mathbb{R}}(\mathbb{A}))) = H_{*-1}(K; \mathbb{L}.(\mathbb{P}_0(\mathbb{A})))$$

$$\xrightarrow{A} L_*(\mathbb{C}_{O(K^+)}(\mathbb{A})) ,$$

with both assembly maps isomorphisms modulo 2-primary torsion.

C10. Pedersen and Weibel [124] identified the torsion group of the $\mathbb{R}$-bounded category $\mathbb{C}_{\mathbb{R}}(\mathbb{A})$ of an additive category $\mathbb{A}$ with the class group of the idempotent completion $\mathbb{P}_0(\mathbb{A})$

$$K_1(\mathbb{C}_{\mathbb{R}}(\mathbb{A})) = K_0(\mathbb{P}_0(\mathbb{A})) ,$$

and expressed the lower K-groups of $\mathbb{A}$ as

$$K_{-i}(\mathbb{A}) = K_1(\mathbb{C}_{\mathbb{R}^{i+1}}(\mathbb{A})) = K_0(\mathbb{P}_{\mathbb{R}^i}(\mathbb{A})) \ (i \geq 1) .$$

The lower quadratic L-groups $L^{\langle -i \rangle}_*(\mathbb{A})$ of an additive category with involution $\mathbb{A}$ are defined in Ranicki [146], and shown to be such that

$$L^{\langle -i \rangle}_*(\mathbb{A}) = L_{*+i+1}(\mathbb{C}_{\mathbb{R}^{i+1}}(\mathbb{A})) = L_{*+i}(\mathbb{P}_{\mathbb{R}^i}(\mathbb{A})) ,$$

$$L^{\langle 1-i \rangle}_*(\mathbb{A}[z, z^{-1}]) = L^{\langle 1-i \rangle}_*(\mathbb{A}) \oplus L^{\langle -i \rangle}_{*-1}(\mathbb{A}) \ (i \geq 0)$$

with

$$\mathbb{A}[z, z^{-1}] = \mathbb{A}[\mathbb{Z}] \ , \quad L^{\langle 1 \rangle}_*(\mathbb{A}) = L_*(\mathbb{A}) \ , \quad L^{\langle 0 \rangle}_*(\mathbb{A}) = L_*(\mathbb{P}_0(\mathbb{A})) .$$

Also, there are defined exact sequences

$$\cdots \longrightarrow L^{\langle 1-i \rangle}_n(\mathbb{A}) \longrightarrow L^{\langle -i \rangle}_n(\mathbb{A}) \longrightarrow \hat{H}^n(\mathbb{Z}_2; \tilde{K}_{-i}(\mathbb{P}_0(\mathbb{A})))$$

$$\longrightarrow L^{\langle 1-i \rangle}_{n-1}(\mathbb{A}) \longrightarrow \cdots \ (i \geq 0)$$

with $\tilde{K}_{-i}(\mathbb{P}_0(\mathbb{A})) = K_{-i}(\mathbb{A})$ for $i \geq 1$. The lower L-groups of a ring with involution R are the special cases

$$L^{\langle -i \rangle}_*(\mathbb{A}^h(R)) = L^{\langle -i \rangle}_*(R) .$$

C11. For any compact polyhedron K there is defined an isomorphism of algebraic surgery exact sequences

$$\cdots \to H^{lf}_{n+i}(K \times \mathbb{R}^i; \mathbb{L}.) \xrightarrow{A} L_{n+i}(\mathbb{C}_{\mathbb{R}^i}(\mathbb{Z}[\pi])) \longrightarrow \mathbb{S}^b_{n+i}(K \times \mathbb{R}^i) \to \cdots$$

$$\downarrow{\simeq} \qquad\qquad \downarrow{\simeq} \qquad\qquad \downarrow{\simeq}$$

$$\cdots \longrightarrow H_n(K; \mathbb{L}.) \xrightarrow{A} L^{\langle 1-i \rangle}_n(\mathbb{Z}[\pi]) \longrightarrow \mathbb{S}^{\langle 1-i \rangle}_n(K) \longrightarrow \cdots$$

with $\pi = \pi_1(K)$. If K is an n-dimensional geometric Poincaré complex then for any $i \geq 1$ the lower total surgery obstruction (C1) to $K \times T^i$ being homotopy equivalent to a compact $(n + i)$-dimensional manifold coincides with the $\mathbb{R}^i$-bounded total surgery obstruction (C6) to $K \times \mathbb{R}^i$ being $\mathbb{R}^i$-bounded homotopy equivalent to an $\mathbb{R}^i$-bounded open $(n + i)$-dimensional manifold

$$s^{\langle 1-i\rangle}(K) \;=\; s^b(K \times \mathbb{R}^i) \in \mathbb{S}_n^{\langle 1-i\rangle}(K) \;=\; \mathbb{S}_{n+i}^b(K \times \mathbb{R}^i) \, .$$

A homotopy equivalence $f\colon M \longrightarrow K \times T^i$ from a compact $(n+i)$-dimensional manifold M lifts to a $\mathbb{Z}^i$-equivariant $\mathbb{R}^i$-bounded homotopy equivalence $\bar{f}\colon \overline{M} \longrightarrow K \times \mathbb{R}^i$. Conversely, if $n + i \geq 5$ an $\mathbb{R}^i$-bounded homotopy equivalence $g\colon L \longrightarrow K \times \mathbb{R}^i$ from an $\mathbb{R}^i$-bounded open $(n+i)$-dimensional manifold L can be 'wrapped up' to a $\mathbb{Z}^i$-equivariant lift $\bar{f}\colon L = \overline{M} \longrightarrow K \times \mathbb{R}^i$ of a homotopy equivalence $f\colon M \longrightarrow K \times T^i$ from a compact $(n+i)$-dimensional manifold M. See Hughes and Ranicki [76] for an algebraic treatment of wrapping up.

The $\mathbb{R}^i$-bounded geometric Poincaré complex bordism groups $\Omega_*^{bP}(K \times \mathbb{R}^i)$ (C6) fit into an exact sequence

$$\ldots \longrightarrow L_{n+i}(\mathbb{C}_{\mathbb{R}^i}(\mathbb{Z}[\pi])) \longrightarrow \Omega_{n+i}^{bP}(K \times \mathbb{R}^i)$$
$$\longrightarrow H_{n+i}^{lf}(K \times \mathbb{R}^i; \Omega_.^N) \longrightarrow L_{n+i-1}(\mathbb{C}_{\mathbb{R}^i}(\mathbb{Z}[\pi])) \longrightarrow \ldots$$

with

$$L_{n+i}(\mathbb{C}_{\mathbb{R}^i}(\mathbb{Z}[\pi])) \;=\; L_n^{\langle 1-i\rangle}(\mathbb{Z}[\pi]) \;\;,\;\; H_{n+i}^{lf}(K \times \mathbb{R}^i; \Omega_.^N) \;=\; H_n(K; \Omega_.^N) \, .$$

In particular, for $i = 1$ the $\mathbb{R}$-bounded geometric Poincaré complex bordism groups $\Omega_{*+1}^{bP}(K \times \mathbb{R})$ coincide with the finitely dominated geometric Poincaré complex bordism groups $\Omega_*^p(K)$ of Pedersen and Ranicki [123]

$$\Omega_{n+1}^{bP}(K \times \mathbb{R}) \;=\; \Omega_n^p(K) \, ,$$

and there is defined an isomorphism of exact sequences

$$\ldots \to L_{n+1}(\mathbb{C}_{\mathbb{R}}(\mathbb{Z}[\pi])) \longrightarrow \Omega_{n+1}^{bP}(K \times \mathbb{R}) \longrightarrow H_{n+1}^{lf}(K \times \mathbb{R}; \Omega_.^N) \to \ldots$$
$$\Big\downarrow \simeq \qquad\qquad \Big\downarrow \simeq \qquad\qquad \Big\downarrow \simeq$$
$$\ldots \longrightarrow L_n^p(\mathbb{Z}[\pi]) \longrightarrow \Omega_n^p(K) \longrightarrow H_n(K; \Omega_.^N) \longrightarrow \ldots \, .$$

The ultimate lower quadratic L-groups and L-spectrum of an additive category $\mathbb{A}$ are defined by

$$L_*^{\langle -\infty\rangle}(\mathbb{A}) \;=\; \varinjlim_i L_*^{\langle -i\rangle}(\mathbb{A}) \;\;,\;\; \mathbb{L}_.^{\langle -\infty\rangle}(\mathbb{A}) \;=\; \varinjlim_i \mathbb{L}_.^{\langle -i\rangle}(\mathbb{A}) \, ,$$

with
$$\pi_*(\mathbb{L}_.^{\langle-\infty\rangle}(\mathbb{A})) = L_*^{\langle-\infty\rangle}(\mathbb{A}) .$$
For any $i \geq 0$ products with the generator
$$\sigma^*(\mathbb{R}^i) = 1 \in L^i(\mathbb{C}_{\mathbb{R}^i}(\mathbb{Z})) = L^0(\mathbb{Z}) = \mathbb{Z}$$
define homotopy equivalences of the non-connective quadratic $\mathbb{L}$-spectra
$$\sigma^*(\mathbb{R}^i) \otimes - : \mathbb{L}.(\mathbb{Z}) = \{ \mathbb{L}_{-i}(\mathbb{Z}) \mid i \geq 0 \}$$
$$\xrightarrow{\simeq} \mathbb{L}_.^{\langle-\infty\rangle}(\mathbb{Z}) = \{ \mathbb{L}_0(\mathbb{C}_{\mathbb{R}^i}(\mathbb{Z})) \mid i \geq 0 \} ,$$
since they induce isomorphisms in the homotopy groups
$$\sigma^*(\mathbb{R}^\infty) \otimes - : \pi_*(\mathbb{L}.(\mathbb{Z})) = L_*(\mathbb{Z})$$
$$\xrightarrow{\simeq} \pi_*(\mathbb{L}_.^{\langle-\infty\rangle}(\mathbb{Z})) = L_{*+\infty}(\mathbb{C}_{\mathbb{R}^\infty}(\mathbb{Z}))$$
(using $K_{-i}(\mathbb{Z}) = 0$ for $i \geq 1$). Thus the deloopings by lower L-theory correspond to the deloopings by dimension shift.

C13. Pedersen and Weibel [124] identify the algebraic K-theory of $\mathbb{P}_{O(K)}(\mathbb{A})$ for a compact polyhedron $K \subseteq S^N$ with the reduced generalized homology groups of K with coefficients in the algebraic K-theory spectrum $\mathbb{K}(\mathbb{P}_0(\mathbb{A}))$ of the idempotent completion $\mathbb{P}_0(\mathbb{A})$
$$K_*(\mathbb{P}_{O(K)}(\mathbb{A})) = \dot{H}_{*-1}(K; \mathbb{K}(\mathbb{P}_0(\mathbb{A}))) .$$
The Mayer–Vietoris exact sequences of Ranicki [146, §14] show that the assembly maps in the ultimate lower quadratic L-groups are isomorphisms
$$A: H_*^{lf}(O(K); \mathbb{L}_.^{\langle-\infty\rangle}(\mathbb{A})) = \dot{H}_{*-1}(K; \mathbb{L}_.^{\langle-\infty\rangle}(\mathbb{A})) \xrightarrow{\simeq} L_*^{\langle-\infty\rangle}(\mathbb{C}_{O(K)}(\mathbb{A})) .$$
The simply connected assembly maps are isomorphisms
$$A: H_*^{lf}(O(K); \mathbb{L}.(\mathbb{Z})) = \dot{H}_{*-1}(K; \mathbb{L}.(\mathbb{Z}))$$
$$\xrightarrow{\simeq} L_*^{\langle-\infty\rangle}(\mathbb{C}_{O(K)}(\mathbb{Z})) = L_*(\mathbb{C}_{O(K)}(\mathbb{Z}))$$
since $K_{-i}(\mathbb{Z}) = 0$ for $i \geq 1$.

C14. For any pair of metric spaces $(X, Y \subseteq X)$ and any additive category $\mathbb{A}$ let $\mathbb{C}_{X,Y}(\mathbb{A})$ be the additive category with the objects M of $\mathbb{C}_X(\mathbb{A})$ and morphisms $[f]: M \longrightarrow N$ the equivalence classes of morphisms $f: M \longrightarrow N$ in $\mathbb{C}_X(\mathbb{A})$ which agree more than a bounded distance away from Y. A morphism in $\mathbb{C}_{X,Y}(\mathbb{A})$ is thus a 'germ' of morphisms in $\mathbb{C}_X(\mathbb{A})$ which agree far away from $\mathbb{C}_Y(\mathbb{A})$, by analogy with the germs at Y of functions defined on X. The germ category $\mathbb{C}_{X,Y}(\mathbb{A})$ was introduced by Munkholm in the special case $(X, Y) = (\mathbb{R}^k, \{0\})$ (Anderson and Munkholm [3, VII.3]). See Ferry, Hambleton and Pedersen [49] for a survey of the applications of the germ categories. The X-bounded topology away from Y is measured by the algebraic K- and L-groups of $\mathbb{C}_{X,Y}(\mathbb{A})$. See Ranicki [146, 4.1, 14.2] for the exact sequences

$$\ldots \longrightarrow K_1(\mathbb{C}_Y(\mathbb{A})) \longrightarrow K_1(\mathbb{C}_X(\mathbb{A})) \longrightarrow K_1(\mathbb{C}_{X,Y}(\mathbb{A}))$$
$$\longrightarrow K_0(\mathbb{P}_Y(\mathbb{A})) \longrightarrow K_0(\mathbb{P}_X(\mathbb{A})) \longrightarrow \ldots$$
$$\ldots \longrightarrow L_n^J(\mathbb{P}_Y(\mathbb{A})) \longrightarrow L_n(\mathbb{C}_X(\mathbb{A})) \longrightarrow L_n(\mathbb{C}_{X,Y}(\mathbb{A}))$$
$$\longrightarrow L_{n-1}^J(\mathbb{P}_Y(\mathbb{A})) \longrightarrow L_{n-1}(\mathbb{C}_X(\mathbb{A})) \longrightarrow \ldots$$

with $J = \ker(\widetilde{K}_0(\mathbb{P}_Y(\mathbb{A})) \longrightarrow \widetilde{K}_0(\mathbb{P}_X(\mathbb{A})))$. For a compact subspace $K \subseteq S^N$ the forgetful map

$$\mathbb{C}_{O(K^+)}(\mathbb{A}) \longrightarrow \mathbb{C}_{O(K^+),O(S^0)}(\mathbb{A}) = \mathbb{C}_{O(K),\{0\}}(\mathbb{A})$$

induces isomorphisms in algebraic K- and L-theory

$$K_*(\mathbb{C}_{O(K^+)}(\mathbb{A})) \cong K_*(\mathbb{C}_{O(K),\{0\}}(\mathbb{A}))$$
$$L_*(\mathbb{C}_{O(K^+)}(\mathbb{A})) \cong L_*(\mathbb{C}_{O(K),\{0\}}(\mathbb{A}))$$

so that $O(K^+)$-bounded surgery and $(O(K), \{0\})$-bounded surgery are essentially the same, namely $O(K)$-bounded surgery at ∞ (= away from $\{0\}$). Ferry and Pedersen [50] use $O(K)$-bounded surgery at ∞ and the controlled end theory of Quinn [131], [132] as a substitute for K-controlled surgery. Since $K_{-i}(\mathbb{Z}) = 0$ for $i \geq 1$

$$L_*(\mathbb{C}_{O(K),\{0\}}(\mathbb{Z})) = L_*(\mathbb{C}_{O(K^+)}(\mathbb{Z}))$$
$$= H_*^{lf}(O(K^+); \mathbb{L}.(\mathbb{Z})) = H_{*-1}(K; \mathbb{L}.(\mathbb{Z})) .$$

Similarly for the bounded symmetric L-groups, and also for the bounded visible symmetric L-groups.

For any subspace $K \subseteq S^N$ with the homotopy type of a compact n-dimensional polyhedron the bounded L-theory braid of C6

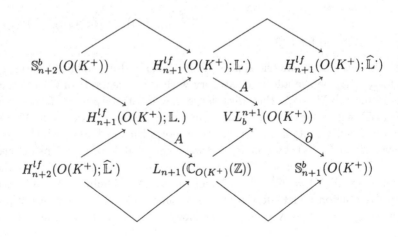

can be written as

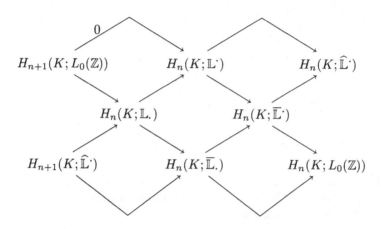

with $\overline{\mathbb{L}}^{\cdot} = K.(L_0(\mathbb{Z}), 0) \vee \mathbb{L}^{\cdot}$ as in §25. Similarly, the bounded geometric Poincaré bordism braid of C6

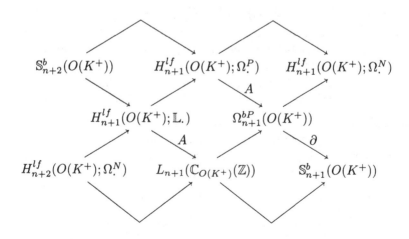

can be written as

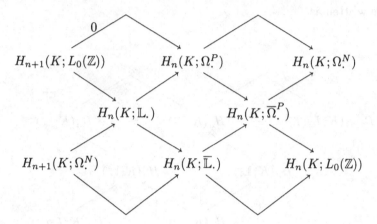

with $\overline{\Omega}^P_{.} = K.(L_0(\mathbb{Z}), 0) \vee \Omega^P_{.}$. The $O(K^+)$-bounded visible symmetric signature of an $O(K^+)$-bounded $(n + 1)$-dimensional geometric Poincaré complex X is a cobordism class

$$\sigma^*(X) = (C(X), \Delta[X])$$

$$\in VL^{n+1}_b(O(K^+)) = H_n(K; \overline{\mathbb{L}}^{\cdot}) = H_n(K; L_0(\mathbb{Z})) \oplus H_n(K; \mathbb{L}^{\cdot})$$

with components the $O(K^+)$-bounded total surgery obstruction

$$s^b(X) = \partial\sigma^*(X) \in \mathbb{S}^b_{n+1}(O(K^+)) = H_n(K; L_0(\mathbb{Z}))$$

and the $O(K^+)$-bounded symmetric signature

$$\sigma^*(X) = (C(X), \Delta[X]) \in L^{n+1}(\mathbb{C}_{O(K^+)}(\mathbb{Z})) = H_n(K; \mathbb{L}^{\cdot}) .$$

The $O(K^+)$-bounded total surgery obstruction can be expressed as the difference of local and global codimension n signatures at ∞, by analogy with the expression in 24.20 of the total surgery obstruction $s(B\pi) \in \mathbb{S}_n(B\pi) = H_n(B\pi; L_0(\mathbb{Z}))$ of the classifying space $B\pi$ of an n-dimensional Novikov group π as the difference of local and global codimension n signatures.

C15.   A compact n-dimensional ANR homology manifold X is an X-controlled Poincaré complex (Quinn [132]), and $X \times \mathbb{R}$ has the $O(X^+)$-bounded homotopy type of an $(n+1)$-dimensional $O(X^+)$-bounded Poincaré complex via the projection map

$$X \times \mathbb{R} \longrightarrow O(X^+) = O(X) \vee (-\infty, 0] ; (x, t) \longrightarrow \begin{cases} tx & \text{if } t \geq 0 \\ t & \text{if } t < 0 \end{cases}$$

(Ferry and Pedersen [50]). The $O(X^+)$-bounded total surgery obstruction $s^b(X \times \mathbb{R}) \in \mathbb{S}^b_{n+1}(O(X^+))$ is identified in [50] with the resolution obstruction $i(X) \in L_0(\mathbb{Z})$ of Quinn [133]

$$s^b(X \times \mathbb{R}) = i(X) \in \mathbb{S}^b_{n+1}(O(X^+)) = H_n(X; L_0(\mathbb{Z})) = L_0(\mathbb{Z}) .$$

A resolution of X corresponds to an $O(X^+)$-bounded homotopy equivalence $f: M \longrightarrow X \times \mathbb{R}$ from an $O(X^+)$-bounded open $(n+1)$-dimensional manifold M. There exists a resolution of X if and only if the $O(X^+)$-bounded Poincaré duality chain equivalence

$$[X \times \mathbb{R}] \cap - : C(X \times \mathbb{R})^{n+1-*} \longrightarrow C(X \times \mathbb{R})$$

is sufficiently close to being 'cell-like'. Let $(f, b): M \longrightarrow X$ be an n-dimensional normal map from a topological manifold M determined as in 25.8 by the canonical topological reduction $\nu_X: X \longrightarrow BTOP$ of the Spivak normal fibration. The canonical $\mathbb{L}$-homology fundamental class of M is

$$[M]_{\mathbb{L}} = \sigma^*(M \times \mathbb{R}) \in L^{n+1}(\mathbb{C}_{O(M^+)}(\mathbb{Z})) = H_n(M; \mathbb{L}^{\cdot})$$

with codimension n signature

$$B[M]_{\mathbb{L}} = 1 \in H_n(M; L^0(\mathbb{Z})) = L^0(\mathbb{Z}) = \mathbb{Z} .$$

The canonical $\mathbb{L}$-homology fundamental class of X is the image

$$[X]_{\mathbb{L}} = f_*[M]_{\mathbb{L}} = \sigma^*(M \times \mathbb{R}) \in L^{n+1}(\mathbb{C}_{O(X^+)}(\mathbb{Z})) = H_n(X; \mathbb{L}^{\cdot}) ,$$

with codimension n signature

$$B[X]_{\mathbb{L}} = 1 \in H_n(X; L^0(\mathbb{Z})) = L^0(\mathbb{Z}) = \mathbb{Z} .$$

The canonical $\overline{\mathbb{L}}$-homology fundamental class of X (25.10) is given by

$$[X]_{\overline{\mathbb{L}}} = (i(X), [X]_{\mathbb{L}})$$

$$\in VL_b^{n+1}(O(X^+)) = H_n(X; \overline{\mathbb{L}}^{\cdot}) = H_n(X; L_0(\mathbb{Z})) \oplus H_n(X; \mathbb{L}^{\cdot}) ,$$

with codimension n signature

$$B[X]_{\overline{\mathbb{L}}} = 8i(X) + 1 \in H_n(X; L^0(\mathbb{Z})) = L^0(\mathbb{Z}) = \mathbb{Z} .$$

The $(n+1)$-dimensional $O(X^+)$-bounded normal map $(f, b) \times 1: M \times \mathbb{R} \longrightarrow X \times \mathbb{R}$ has $O(X^+)$-bounded surgery obstruction

$$\sigma_*((f, b) \times 1) = (-i(X), 0)$$

$$\in L_{n+1}(\mathbb{C}_{O(X^+)}(\mathbb{Z})) = H_n(X; \overline{\mathbb{L}}_{\cdot}) = H_n(X; L_0(\mathbb{Z})) \oplus H_n(X; \mathbb{L}_{\cdot}) .$$

The $O(X^+)$-bounded symmetric signature of $X \times \mathbb{R}$

$$\sigma^*(X \times \mathbb{R}) = \sigma^*(M \times \mathbb{R}) - (1 + T)\sigma_*((f, b) \times 1) = [X]_{\mathbb{L}} + (1 + T)i(X)$$

$$\in L^{n+1}(\mathbb{C}_{O(X^+)}(\mathbb{Z})) = H_n(X; \mathbb{L}^{\cdot}) = H_n(X; L^0(\mathbb{Z})) \oplus H_n(X; \mathbb{L}^{\cdot}\langle 1 \rangle(\mathbb{Z})) .$$

is thus the image of $[X]_{\overline{\mathbb{L}}} \in H_n(X; \overline{\mathbb{L}}^{\cdot})$ under the map

$$\begin{pmatrix} 1 + T & 1 & 0 \\ 0 & 0 & 1 \end{pmatrix} :$$

$$H_n(X; \overline{\mathbb{L}}^{\cdot}) = H_n(X; L_0(\mathbb{Z})) \oplus H_n(X; L^0(\mathbb{Z})) \oplus H_n(X; \mathbb{L}^{\cdot}\langle 1 \rangle(\mathbb{Z}))$$

$$\longrightarrow H_n(X; \mathbb{L}^{\cdot}) = H_n(X; L^0(\mathbb{Z})) \oplus H_n(X; \mathbb{L}^{\cdot}\langle 1 \rangle(\mathbb{Z})) .$$

C16. If $h: M' \longrightarrow M$ is a homeomorphism of compact n-dimensional ANR homology manifolds then $h \times 1: M' \times \mathbb{R} \longrightarrow M \times \mathbb{R}$ is an $O(M^+)$-bounded homotopy equivalence of $(n + 1)$-dimensional $O(M^+)$-bounded geometric Poincaré complexes. The $1/2$-connective $O(M^+)$-bounded visible symmetric signature of $M \times \mathbb{R}$

$$\sigma^*(M \times \mathbb{R}) = ([M]_{\mathbb{L}}, i(M))$$

$$\in VL_b^{n+1}(O(M^+)) = H_n(M; \mathbb{L}^{\cdot}) \oplus H_n(M; L_0(\mathbb{Z}))$$

is an $O(M^+)$-bounded homotopy invariant of $M \times \mathbb{R}$, and hence a topological invariant of M. The topological invariance of the canonical $\mathbb{L}^{\cdot}$-homology fundamental class $[M]_{\mathbb{L}} \in H_n(M; \mathbb{L}^{\cdot})$ is an integral version of the topological invariance of the rational Pontrjagin classes due to Novikov [120].

Rationally, the $\mathbb{L}^{\cdot}$-orientation of a compact oriented n-dimensional topological manifold M is the Poincaré dual of the $\mathcal{L}$-genus $\mathcal{L}(M) = \mathcal{L}(\tau_M) \in H^{4*}(M; \mathbb{Q})$

$$[M]_{\mathbb{L}} \otimes 1 = [M]_{\mathbb{Q}} \cap \mathcal{L}(M) \in H_{n-4*}(M; \mathbb{Q}) ,$$

with $[M]_{\mathbb{Q}} \in H_n(M; \mathbb{Q})$ the $\mathbb{Q}$-coefficient fundamental class. The usual Hirzebruch L-polynomial relations

$$\mathcal{L}_k(M) = L_k(p_1, p_2, \ldots, p_k) \in H^{4k}(M; \mathbb{Q}) \ (k \geq 0) ,$$

express the $\mathcal{L}$-genus in terms of the rational Pontrjagin classes $p_* = p_*(\tau_M) \in H^{4*}(M; \mathbb{Q})$ of the stable tangent bundle $\tau_M = -\tilde{\nu}_M : M \longrightarrow BSTOP$. Conversely, the rational Pontrjagin classes are determined by the $\mathcal{L}$-genus, for example $p_1 = 3L_1 \in H^4(M; \mathbb{Q})$. Originally, the expression for the $\mathcal{L}$-genus in terms of the signatures of submanifolds was obtained for differentiable manifolds, but successive developments have shown that it also applies for PL, topological and ANR homology manifolds (taking account of the resolution obstruction, as in 25.17).

For a compact oriented n-dimensional topological manifold M^n the $4k$-dimensional component $\mathcal{L}_k(M) \in H^{4k}(M; \mathbb{Q})$ of the $\mathcal{L}$-genus is detected by the signatures of compact $4k$-dimensional submanifolds $N^{4k} \subset M^n \times \mathbb{R}^j$ (j large) with trivial normal bundle

$$\langle \mathcal{L}_k(M), i_*[N]_{\mathbb{Q}} \rangle = \text{signature}(N) \in L^{4k}(\mathbb{Z}) = \mathbb{Z} ,$$

since every element in $H_{4k}(M; \mathbb{Q})$ is a rational multiple of an element of the form

$$x = i_*[N]_{\mathbb{Q}} = [M]_{\mathbb{Q}} \cap g^*(1)$$

$$\in H_{4k}(M; \mathbb{Q}) = H_{4k}(M \times \mathbb{R}^j; \mathbb{Q}) \ (1 \in H_{lf}^m(\mathbb{R}^m) = \mathbb{Z})$$

with

$$g : M^n \times \mathbb{R}^j \longrightarrow \mathbb{R}^m \ (m = n + j - 4k)$$

a proper map transverse regular at $0 \in \mathbb{R}^m$ and

$$i = \text{inclusion} : N^{4k} = g^{-1}(0) \longrightarrow M^n \times \mathbb{R}^j .$$

The topological invariance of the rational Pontrjagin classes is thus a direct consequence of topological transversality for high-dimensional manifolds, which was established subsequently by Kirby and Siebenmann [84]. However, it is instructive to interpret the original argument of Novikov [120] for the topological invariance of the rational Pontrjagin classes of differentiable and PL manifolds in terms of bounded topology, as follows.

Let $h: M' \longrightarrow M$ be a homeomorphism of compact n-dimensional oriented PL manifolds. Let x, g, N be as above, so that

$$\langle \mathcal{L}_k(M), x \rangle = \text{signature}(N) \in L^{4k}(\mathbb{Z}) = \mathbb{Z},$$

and let

$$x' = (h^{-1})_*(x) \in H_{4k}(M'; \mathbb{Q}).$$

It is required to prove that

$$\langle \mathcal{L}_k(M'), x' \rangle = \text{signature}(N) \in L^{4k}(\mathbb{Z}) = \mathbb{Z}.$$

The inverse image of an open regular neighbourhood $N^{4k} \times \mathbb{R}^m \subset M^n \times \mathbb{R}^j$ of N in $M \times \mathbb{R}^j$ is an open codimension 0 PL submanifold

$$W^{n+j} = (h \times 1_{\mathbb{R}^j})^{-1}(N \times \mathbb{R}^m) \subseteq M' \times \mathbb{R}^j$$

with a homeomorphism

$$H = (h \times 1_{\mathbb{R}^j})| : W \longrightarrow N \times \mathbb{R}^m.$$

Making H PL transverse regular at $N \times \{0\} \subset N \times \mathbb{R}^m$ there is obtained a normal map of closed $4k$-dimensional PL manifolds

$$(f, b) = H| : N'^{4k} = H^{-1}(N \times \{0\}) \longrightarrow N$$

with simply-connected surgery obstruction

$$\sigma_*(f, b) = (\text{signature}(N') - \text{signature}(N))/8$$

$$= (\langle \mathcal{L}_k(M'), x' \rangle - \langle \mathcal{L}_k(M), x \rangle)/8 \in L_{4k}(\mathbb{Z}) = \mathbb{Z}.$$

Approximate the homeomorphism H by an $\mathbb{R}^m$-bounded homotopy equivalence $W \simeq N \times \mathbb{R}^m$ of $\mathbb{R}^m$-bounded open $(4k+m)$-dimensional PL manifolds with $\mathbb{R}^m$-bounded symmetric signature

$$\sigma^*(W) = \sigma^*(N \times \mathbb{R}^m)$$

$$= \text{signature}(N') = \text{signature}(N)$$

$$\in L^{4k+m}(\mathbb{C}_{\mathbb{R}^m}(\mathbb{Z})) = L^{4k}(\mathbb{Z}) = \mathbb{Z}.$$

Equivalently, identify

$$\sigma_*(H) = \sigma_*(f, b) = 0 \in L_{4k+m}(\mathbb{C}_{\mathbb{R}^m}(\mathbb{Z})) = L_{4k}(\mathbb{Z}) = \mathbb{Z}.$$

Equivalently, use geometric 'wrapping up' to identify W with the pullback cover $\overline{V} = e^*(N \times \mathbb{R}^m)$ of a compact $(4k + m)$-dimensional PL manifold V along a homeomorphism $e: V \longrightarrow N \times T^m$ with a lift to a $\mathbb{Z}^m$-equivariant homeomorphism

$$\bar{e} = H : \overline{V} = W \longrightarrow N \times \mathbb{R}^m ,$$

and

$$\sigma^*(V) = \sigma^*(N \times T^m)$$
$$= (\text{signature}(N'), 0) = (\text{signature}(N), 0)$$
$$\in L^{4k+m}(\mathbb{Z}[\mathbb{Z}^m]) = L^{4k}(\mathbb{Z}) \oplus \left(\sum_{i=1}^{m} \binom{m}{i} L^{4k+i}(\mathbb{Z}) \right) .$$

The evaluation of $\mathcal{L}(M') \in H^{4*}(M'; \mathbb{Q})$ on $x' = (h^{-1})_*(x) \in H_{4k}(M'; \mathbb{Q})$ is thus given by

$$\langle \mathcal{L}_k(M'), x' \rangle = \text{signature}(N') = \text{signature}(N)$$
$$= \langle \mathcal{L}_k(M), x \rangle = \langle h^* \mathcal{L}_k(M), x' \rangle \in \mathbb{Z} ,$$

and

$$\mathcal{L}(M') = h^* \mathcal{L}(M) \in H^{4*}(M'; \mathbb{Q}) .$$

See Sullivan and Teleman [167] and Weinberger [180] for analytic proofs of the topological invariance of the rational Pontrjagin classes $p_*(\tau_M) \in H^{4*}(M; \mathbb{Q})$ of a compact oriented topological manifold M. The most systematic way of obtaining the topological invariance of the $\mathbb{L}^{\cdot}$-orientation $[M]_{\mathbb{L}} \in H_n(M; \mathbb{L}^{\cdot})$ is to follow up the proposal in the Introduction of developing the sheaf-theoretic versions of the methods of this text, allowing the construction of $[M]_{\mathbb{L}}$ directly from the local homology sheaf.

Bibliography

[1] S. AKBULUT and J. D. MCCARTHY *Casson's invariant for oriented homology 3-spheres.* Mathematical Notes **36**, Princeton University Press (1990)

[2] J. ALEXANDER, G. HAMRICK and J. VICK Linking forms and maps of odd order. *Trans. Am. Math. Soc.* **221**, 169–185 (1976)

[3] D. R. ANDERSON and H. J. MUNKHOLM *Boundedly Controlled Topology.* Lecture Notes in Mathematics **1323**, Springer (1988)

[4] D. W. ANDERSON Chain functors and homology theories. *Proceedings 1971 Seattle Algebraic Topology Symposium,* Lecture Notes in Mathematics **249**, Springer, 1–12 (1971)

[5] M. A. ARMSTRONG, G. E. COOKE and C. P. ROURKE *The Princeton notes on the Hauptvermutung.* Warwick University notes (1972) K-monographs in Mathematical Sciences, Roderer (to appear)

[6] M. ATIYAH The signature of fibre bundles. *Papers in the honour of Kodaira,* Tokyo University Press, 73–84 (1969)

[7] M. ATIYAH and I. M. SINGER The index of elliptic operators III. *Ann. Math.* **87**, 546–604 (1968)

[8] A. BAK and M. KOLSTER The computation of odd-dimensional projective surgery groups for finite groups. *Topology* **21**, 35–63 (1982)

[9] H. BASS *Algebraic K-theory.* Benjamin (1968)

[10] A. L. BLAKERS and W. S. MASSEY The homotopy groups of a triad II. *Ann. Math.* **55**, 192–201 (1952)

[11] A. BOREL and J. MOORE Homology theory for locally compact spaces. *Michigan Math. J.* **7**, 137–159 (1960)

[12] R. BOTT and L. TU *Differential forms in algebraic topology.* Springer (1982)

[13] A. K. BOUSFIELD and D. M. KAN *Homotopy limits, completions and localizations.* Lecture Notes in Mathematics **304**, Springer (1972)

[14] W. BROWDER Torsion in H-spaces. *Ann. Math.* **74**, 24–51 (1961)

[15] W. BROWDER Homotopy type of differentiable manifolds. *Proceedings Arhus Colloquium,* 42–46 (1962)

[16] W. BROWDER *Surgery on simply connected manifolds.* Springer (1972)

[17] W. BROWDER Poincaré spaces, their normal fibrations and surgery. *Inventiones Math.* **17**, 191–202 (1972)

[18] W. BROWDER and G. LIVESAY Fixed point free involutions on homotopy spheres. *Tohôku J. Math.* **25**, 69–88 (1973)

[19] W. BROWDER and F. QUINN A surgery theory for G-manifolds and stratified sets. *Proceedings 1973 Tokyo Conference on Manifolds,* Tokyo University Press, 27–36 (1975)

[20] G. BRUMFIEL and J. MORGAN The homotopy–theoretic consequences

of N. Levitt's obstruction theory to transversality for spherical
fibrations. *Pacific J. Math.* **67**, 1–100 (1976)

[21] S. BUONCRISTIANO, C. P. ROURKE and B. J. SANDERSON *A geometric
approach to homology theory.* Lond. Math. Soc. Lecture Notes
18, Cambridge University Press (1976)

[22] S. CAPPELL Manifolds with fundamental group a generalized free product. *Bull. Am. Math. Soc.* **80**, 1193–1198 (1974)

[23] S. CAPPELL On connected sums of manifolds. *Topology* **13**, 395–400
(1974)

[24] S. CAPPELL On homotopy invariance of higher signatures. *Inventiones
Math.* **33**, 171–179 (1976)

[25] S. CAPPELL and J. SHANESON The codimension two placement problem, and homology equivalent manifolds. *Ann. Math.* **99**, 277–348
(1974)

[26] S. CAPPELL and J. SHANESON Pseudo-free actions I. *Proceedings 1978
Arhus Topology Conference*, Lecture Notes in Mathematics **763**,
Springer, 395–447 (1979)

[27] S. CAPPELL and J. SHANESON Singular spaces, characteristic classes,
and intersection homology. *Ann. Math.* **134**, 325–374 (1991)

[28] S. CAPPELL and S. WEINBERGER A geometric interpretation of Siebenmann's periodicity phenomenon. *Proceedings 1985 Georgia Conference on Geometry and Topology*, Dekker, 47–52 (1987)

[29] S. CAPPELL and S. WEINBERGER Which *H*-spaces are manifolds? I.
Topology **27**, 377–386 (1988)

[30] S. CAPPELL and S. WEINBERGER Classification de certains espaces stra-
tifiés *C. R. Acad. Sci. Paris* **313**, *Série I*, 399-401 (1991)

[31] G. CARLSSON Homotopy fixed points in the algebraic *K*-theory of certain infinite discrete groups. *Advances in Homotopy Theory (Cor-
tona, 1988)*, Lond. Math. Soc. Lecture Notes **139**, Cambridge
University Press, 5–10 (1989)

[32] G. CARLSSON and J. MILGRAM The structure of odd *L*-groups. *Proceedings 1978 Waterloo Algebraic Topology Conference*, Lecture
Notes in Mathematics **741**, Springer, 1–72 (1979)

[33] T. A. CHAPMAN and S. FERRY Approximating homotopy equivalences
by homeomorphisms. *Amer. J. Math.* **101**, 583–607 (1979)

[34] S. S. CHERN, F. HIRZEBRUCH and J. P. SERRE On the index of a fibered
manifold. *Proc. Am. Math. Soc.* **8**, 587–596 (1957)

[35] C. CIBILS Groupe de Witt d'une algèbre avec involution. *l'Enseignement Math.* **29**, 27–43 (1983)

[36] M. COHEN Simplicial structures and transverse cellularity. *Ann. Math.*
85, 218–245 (1967)

[37] M. COHEN Homeomorphisms between homotopy manifolds and their resolutions. *Inventiones Math.* **10**, 239–250 (1970)

[38] P. CONNER and F. RAYMOND A quadratic form on the quotient of a periodic map. *Semigroup Forum* **7**, 310–333 (1974)

[39] A. CONNES and H. MOSCOVICI Cyclic homology, the Novikov conjecture, and hyperbolic groups. *Topology* **29**, 345–388 (1990)

[40] C. W. CURTIS and I. REINER *Methods of representation theory, with applications to finite groups and orders.* Wiley, Vol. I (1981), Vol. II (1987)

[41] R. DAVERMAN *Decompositions of manifolds.* Academic Press (1986)

[42] J. DAVIS and J. MILGRAM *A survey of the spherical space form problem.* Mathematical Reports 2, Harwood (1984)

[43] M. DAVIS Coxeter groups and aspherical manifolds. *Proceedings 1982 Arhus Algebraic Topology Conference*, Lecture Notes in Mathematics **1051**, Springer, 197–221 (1984)

[44] K. H. DOVERMANN $\mathbb{Z}_2$-surgery theory. *Michigan Math. J.* **28**, 267–287 (1981)

[45] F. T. FARRELL and W. C. HSIANG Manifolds with $\pi_1 = G \times_\alpha T$. *Amer. J. Math.* **95**, 813–845 (1973)

[46] F. T. FARRELL and W. C. HSIANG On Novikov's conjecture for nonpositively curved manifolds. *Ann. Math.* **113**, 197–209 (1981)

[47] F. T. FARRELL and L. E. JONES A topological analogue of Mostow's rigidity theorem. *J. Am. Math. Soc.* **2**, 257–370 (1989)

[48] F. T. FARRELL and L. E. JONES *Classical aspherical manifolds* CBMS Regional Conference Series in Mathematics **75**, American Mathematical Society (1990)

[49] S. FERRY, I. HAMBLETON and E. K. PEDERSEN A survey of bounded surgery theory and applications. *Math. Pub. MSRI* (to appear)

[50] S. FERRY and E. K. PEDERSEN Epsilon surgery I. *Math. Gott.* **17** (1990)

[51] S. FERRY and S. WEINBERGER Curvature, tangentiality and controlled topology. *Inventiones Math.* **105**, 401–415 (1991)

[52] D. FRANK The first exotic class of a manifold. *Trans. Am. Math. Soc.* **146**, 387–395 (1969)

[53] M. FREEDMAN and F. QUINN *The topology of 4-manifolds.* Princeton University Press (1990)

[54] A. FRÖHLICH and A. MCEVETT The representation of groups by automorphisms of forms. *J. Algebra* **12**, 114–133 (1969)

[55] D. E. GALEWSKI and R. J. STERN The relationship between homology and topological manifolds via homology transversality. *Inventiones Math.* **39**, 277–292 (1977)

[56] D. E. GALEWSKI and R. J. STERN Classification of simplicial triangulations of topological manifolds. *Ann. Math.* **111**, 1–34 (1980)

[57] I. M. GELFAND and A. S. MISHCHENKO Quadratic forms over commutative group rings and K-theory. *Funct. Anal. Appl.* **2**, 277–281 (1969)

[58] S. GITLER and J. STASHEFF The first exotic class of *BF*. *Topology* **4**, 257–266 (1965)

[59] M. GORESKY and R. MACPHERSON Intersection homology theory. *Topology* **19**, 135–162 (1980)

[60] M. GORESKY and R. MACPHERSON Intersection homology II. *Inventiones Math.* **71**, 77–129 (1983)

[61] M. GORESKY and P. SIEGEL Linking pairings on singular spaces. *Comm. Math. Helv.* **58**, 96–110 (1983)

[62] I. HAMBLETON Projective surgery obstructions on closed manifolds. *Proceedings 1980 Oberwolfach Algebraic K-theory Conference, Vol. II*, Lecture Notes in Mathematics **967**, Springer, 101–131 (1982)

[63] I. HAMBLETON and J. C. HAUSMANN Acyclic maps and Poincaré spaces. *Proceedings 1982 Arhus Algebraic Topology Conference*, Lecture Notes in Mathematics **1051**, Springer, 222–245 (1984)

[64] I. HAMBLETON and I. MADSEN On the computation of the projective surgery obstruction groups. (*preprint*)

[65] I. HAMBLETON and J. MILGRAM Poincaré transversality for double covers. *Canad. J. Math.* **30**, 1319–1330 (1978)

[66] I. HAMBLETON, J. MILGRAM, L. TAYLOR and B. WILLIAMS Surgery with finite fundamental group. *Proc. Lond. Math. Soc.* **56** (3), 349–379 (1988)

[67] I. HAMBLETON, A. RANICKI and L. TAYLOR Round L-theory. *J. Pure App. Alg.* **47**, 131–154 (1987)

[68] I. HAMBLETON, L. TAYLOR and B. WILLIAMS An introduction to maps between surgery obstruction groups. *Proceedings 1982 Arhus Algebraic Topology Conference*, Lecture Notes in Mathematics **1051**, Springer, 49–127 (1984)

[69] A. HARSILADZE Hermitian K-theory and quadratic extensions. *Trudy Moskov. Math. Obshch.* **41**, 3–36 (1980)

[70] H. HASSE Äquivalenz quadratischer Formen in einem beliebigen algebraischen Zahlkörper. *J. reine u. angew. Math.* **153**, 158–162 (1924)

[71] A. HATCHER Higher simple homotopy theory. *Ann. Math.* **102**, 101–137 (1975)

[72] J. C. HAUSMANN and P. VOGEL *Geometry on Poincaré spaces*. Mathematical Notes, Princeton University Press (to appear)

[73] F. HIRZEBRUCH *Topological methods in algebraic geometry.* Springer (1966)

[74] F. HIRZEBRUCH Involutionen auf Mannigfaltigkeiten. *Proceedings Conference on Transformation Groups, New Orleans 1967*, Springer, 148–166 (1968)

[75] F. HIRZEBRUCH and D. ZAGIER *The Atiyah–Singer theorem and elementary number theory.* Publish or Perish (1974)

[76] B. HUGHES and A. RANICKI *Ends of complexes.* (to appear)

[77] L. E. JONES Patch spaces: a geometric representation for Poincaré spaces. *Ann. Math.* **97**, 276–306 (1973) Corrigendum: *ibid.* **102**, 183–185 (1975)

[78] J. KAMINKER and J. G. MILLER A comment on the Novikov conjecture. *Proc. Am. Math. Soc.* **83**, 656–658 (1981)

[79] J. KAMINKER and J. G. MILLER Homotopy invariance of the analytic index of signature operators over C^*-algebras. *J. Operator Theory* **14**, 113–127 (1985)

[80] G. KASPAROV On the homotopy invariance of rational Pontrjagin numbers. *Dokl. Akad. Nauk SSSR* **190**, 1022–1025 (1970)

[81] G. KASPAROV Equivariant KK-theory and the Novikov conjecture. *Inventiones Math.* **91**, 147–202 (1988)

[82] M. KERVAIRE A manifold which does not admit a differentiable structure. *Comm. Math. Helv.* **34**, 257–270 (1960)

[83] M. KERVAIRE and J. MILNOR Groups of homotopy spheres I. *Ann. Math.* **77**, 504–537 (1963)

[84] R. KIRBY and L. SIEBENMANN *Foundational essays on topological manifolds, smoothings, and triangulations.* Ann. Math. Stud. **88**, Princeton University Press (1977)

[85] M. KOLSTER Even-dimensional projective surgery groups of finite groups. *Proceedings 1980 Oberwolfach Algebraic K-theory Conference, Vol. II*, Lecture Notes in Mathematics **976**, Springer, 239–279 (1982)

[86] W. LANDHERR Äquivalenz Hermitescher Formen über einen beliebigen algebraischen Zahlkörper. *Abh. Math. Sem. Univ. Hamburg* **11**, 245–248 (1935)

[87] R. LEE Semicharacteristic classes. *Topology* **12**, 183–199 (1973)

[88] J. LEVINE Lectures on groups of homotopy spheres. *Algebraic and geometric topology, Proc. 1983 Rutgers Topology Conference*, Lecture Notes in Mathematics **1126**, Springer, 62–95 (1985)

[89] N. LEVITT Poincaré duality cobordism. *Ann. Math.* **96**, 211–244 (1972)

[90] N. LEVITT and J. MORGAN Transversality structures and *PL* struc-

tures on spherical fibrations. *Bull. Am. Math. Soc.* **78**, 1064–1068 (1972)

[91] N. LEVITT and A. RANICKI Intrinsic transversality structures. *Pacific J. Math.* **129**, 85–144 (1987)

[92] D. W. LEWIS Forms over real algebras and the multisignature of a manifold. *Adv. Math.* **23**, 272–284 (1977)

[93] D. W. LEWIS Exact sequences of Witt groups of equivariant forms. *l'Enseignement Math.* **29**, 45–51 (1983)

[94] J.-L. LODAY K-théorie algébrique et représentations de groupes. *Ann. scient. Éc. Norm. Sup.* (4) **9**, 309–377 (1976)

[95] S. LOPEZ DE MEDRANO *Involutions on manifolds* Springer (1971)

[96] W. LÜCK and A. RANICKI Surgery transfer. *Proceedings 1987 Göttingen Conference on Algebraic Topology*, Lecture Notes in Mathematics **1361**, Springer, 167–246 (1989)

[97] W. LÜCK and A. RANICKI Surgery obstructions of fibre bundles. *J. Pure App. Alg.* (to appear)

[98] G. LUSZTIG Novikov's signature and families of elliptic operators. *J. Diff. Geo.* **7**, 229–256 (1971)

[99] I. MADSEN and J. MILGRAM *The classifying spaces for surgery and cobordism of manifolds.* Ann. Math. Stud. **92**, Princeton University Press (1979)

[100] I. MADSEN, C. B. THOMAS and C. T. C. WALL The topological space form problem. *Topology* **15**, 375–382 (1976)

[101] C. MCCRORY Cone complexes and *PL* transversality. *Trans. Am. Math. Soc.* **207**, 269–291 (1975)

[102] C. MCCRORY A characterization of homology manifolds. *J. Lond. Math. Soc.* **16** (2), 149–159 (1977)

[103] C. MCCRORY Zeeman's filtration of homology. *Trans. Am. Math. Soc.* **250**, 147–166 (1979)

[104] W. MEYER *Die Signatur von lokalen Koeffizientensystemen und Faserbündeln.* Bonner Math. Schriften **53** (1972)

[105] J. MILGRAM Orientations for Poincaré duality spaces and applications. *Proceedings 1986 Arcata Conference on Algebraic Topology*, Lecture Notes in Mathematics **1370**, Springer, 293–324 (1989)

[106] J. MILGRAM Surgery with finite fundamental group. *Pacific J. Math.* **151**, I. 65–115, II. 117–150 (1991)

[107] J. MILGRAM and A. RANICKI The *L*-theory of Laurent polynomial extensions and genus 0 function fields. *J. reine u. angew. Math.* **406**, 121–166 (1990)

[108] J. MILNOR On manifolds homeomorphic to the 7–sphere. *Ann. Math.* **64**, 399–405 (1956)

[109] J. MILNOR *Introduction to algebraic K-theory.* Ann. Math. Stud. **72**, Princeton University Press (1971)

[110] J. MILNOR and D. HUSEMOLLER *Symmetric bilinear forms.* Springer (1973)

[111] J. MILNOR and J. STASHEFF *Characteristic classes.* Ann. Math. Stud. **76**, Princeton University Press (1974)

[112] A. S. MISHCHENKO Homotopy invariants of non–simply connected manifolds. III. Higher signatures. *Izv. Akad. Nauk SSSR, ser. mat.* **35**, 1316–1355 (1971)

[113] A. S. MISHCHENKO Hermitian K-theory. The theory of characteristic classes. The method of functional analysis. *Uspeki Mat.* **31**, 69–134 (1976) English translation: *Russian Math. Surv.* **31**:2, 71–138 (1976)

[114] A. S. MISHCHENKO and A. T. FOMENKO The index of elliptic operators over C^*-algebras. *Izv. Akad. Nauk SSSR, ser. mat.* **43**, 831–859 (1979) English translation: *Math. USSR - Izv.* **15**, 87 (1980)

[115] A. S. MISHCHENKO and Y. P. SOLOVEV A classifying space for hermitian K-theory. *Trudy Sem. Vect. and Tensor Anal.* **18**, 140–168 (1976)

[116] J. MORGAN and D. SULLIVAN The transversality characteristic class and linking cycles in surgery theory. *Ann. Math.* **99**, 463–544 (1974)

[117] W. NEUMANN Signature related invariants of manifolds I. Monodromy and γ-invariants. *Topology* **18**, 147–172 (1979)

[118] A. NICAS *Induction theorems for groups of homotopy manifold structures.* Memoirs Am. Math. Soc. **267** (1982)

[119] S. P. NOVIKOV Homotopy equivalent smooth manifolds I. *Izv. Akad. Nauk SSSR, ser. mat.* **28**, 365–474 (1965)

[120] S. P. NOVIKOV On manifolds with free abelian fundamental group and applications (Pontrjagin classes, smoothings, high–dimensional knots). *Izv. Akad. Nauk SSSR, ser. mat.* **30**, 208–246 (1966)

[121] S. P. NOVIKOV The algebraic construction and properties of hermitian analogues of K-theory for rings with involution, from the point of view of the hamiltonian formalism. Some applications to differential topology and the theory of characteristic classes. *Izv. Akad. Nauk SSSR, ser. mat.* **34**, 253–288, 478–500 (1970)

[122] W. PARDON Intersection homology Poincaré spaces and the characteristic variety theorem. *Comm. Math. Helv.* **65**, 198–233 (1990)

[123] E. K. PEDERSEN and A. RANICKI Projective surgery theory. *Topology* **19**, 239–254 (1980)

[124] E. K. PEDERSEN and C. WEIBEL K-theory homology of spaces. *Pro-*

ceedings 1986 Arcata Conference on Algebraic Topology, Lecture Notes in Mathematics **1370**, Springer, 346–361 (1989)

[125] T. PETRIE The Atiyah–Singer invariant, the Wall groups $L_n(\pi, 1)$, and the function $(te^x + 1)/(te^x - 1)$. *Ann. Math.* **92**, 174–187 (1970)

[126] D. QUILLEN Higher algebraic K-theory. *Proceedings 1972 Battelle Seattle Algebraic K-theory Conference, Vol. I*, Lecture Notes in Mathematics **341**, Springer, 85–147 (1973)

[127] F. QUINN A geometric formulation of surgery. *Topology of manifolds, Proceedings 1969 Georgia Topology Conference*, Markham Press, 500–511 (1970)

[128] F. QUINN $B_{(TOP_n)}\sim$ and the surgery obstruction. *Bull. Am. Math. Soc.* **77**, 596–600 (1971)

[129] F. QUINN Surgery on Poincaré and normal spaces. *Bull. Am. Math. Soc.* **78**, 262–267 (1972)

[130] F. QUINN Ends of maps I. *Ann. Math.* **110**, 275–331 (1979)

[131] F. QUINN Ends of maps II. *Inventiones Math.* **68**, 353–424 (1982)

[132] F. QUINN Resolutions of homology manifolds, and the topological characterization of manifolds. *Inventiones Math.* **72**, 267–284 (1983) Corrigendum: *ibid.* **85**, 653 (1986)

[133] F. QUINN An obstruction to the resolution of homology manifolds. *Michigan Math. J.* **34**, 284–291 (1987)

[134] F. QUINN Assembly maps in bordism-type theories. (*preprint*)

[135] A. RANICKI *An algebraic formulation of surgery*. Trinity College, Cambridge fellowship dissertation (1972)

[136] A. RANICKI Algebraic L-theory I. Foundations. *Proc. Lond. Math. Soc.* **27** (3), 101–125 (1973)

[137] A. RANICKI Algebraic L-theory II. Laurent extensions. *Proc. Lond. Math. Soc.* **27** (3), 126–158 (1973)

[138] A. RANICKI On the algebraic L-theory of semisimple rings. *J. Algebra* **50**, 242–243 (1978)

[139] A. RANICKI Localization in quadratic L-theory. *Proceedings 1978 Waterloo Algebraic Topology Conference*, Lecture Notes in Mathematics **741**, Springer, 102–157 (1979)

[140] A. RANICKI The total surgery obstruction. *Proceedings 1978 Arhus Topology Conference*, Lecture Notes in Mathematics **763**, Springer, 275–316 (1979)

[141] A. RANICKI The algebraic theory of surgery, I. Foundations. *Proc. Lond. Math. Soc.* **40** (3), 87–192 (1980)

[142] A. RANICKI The algebraic theory of surgery, II. Applications to topology. *Proc. Lond. Math. Soc.* **40** (3), 193–287 (1980)

[143] A. RANICKI *Exact sequences in the algebraic theory of surgery.* Math-

ematical Notes **26**, Princeton University Press (1981)

[144] A. RANICKI The L-theory of twisted quadratic extensions. *Canad. J. Math.* **39**, 345–364 (1987)

[145] A. RANICKI Additive L-theory. *K-theory* **3**, 163–195 (1989)

[146] A. RANICKI *Lower K- and L-theory.* Lond. Math. Soc. Lecture Notes **178**, Cambridge University Press (1992)

[147] A. RANICKI and M. WEISS Chain complexes and assembly. *Math. Z.* **204**, 157–185 (1990)

[148] A. RANICKI and M. YAMASAKI Controlled K-theory. (*preprint*)

[149] V. A. ROHLIN The Pontrjagin–Hirzebruch class in codimension 2. *Izv. Akad. Nauk SSSR, ser. mat.* **30**, 705–718 (1966)

[150] J. ROSENBERG C^*-algebras, positive scalar curvature, and the Novikov conjecture III. *Topology* **25**, 319–336 (1987)

[151] C. P. ROURKE and B. J. SANDERSON On topological neighbourhoods. *Compositio Math.* **22**, 387–424 (1970)

[152] C. P. ROURKE and B. J. SANDERSON Δ-sets I: Homotopy theory. *Qu. J. Math. Oxford* (2) **22**, 321–338 (1971)

[153] W. SCHARLAU *Quadratic and hermitian forms.* Springer (1985)

[154] J. P. SERRE *Linear representations of finite groups.* Springer (1977)

[155] J. SHANESON Wall's surgery obstruction groups for $G \times \mathbb{Z}$. *Ann. Math.* **90**, 296–334 (1969)

[156] L. SIEBENMANN Are nontriangulable manifolds triangulable? *Proceedings 1969 Georgia Topology Conference*, Markham Press, 77–84 (1970)

[157] L. SIEBENMANN Topological manifolds. *Proceedings 1970 Nice I. C. M.*, Gauthier–Villars, Vol. 2, 133–163 (1971) (also in [84], 307–337)

[158] L. SIEBENMANN Approximating cellular maps by homeomorphisms. *Topology* **11**, 271–294 (1972)

[159] P. SIEGEL Witt spaces: a geometric cycle theory for KO homology at odd primes. *Am. J. Math.* **105**, 1067–1105 (1983)

[160] E. SPANIER *Algebraic topology* McGraw–Hill (1966)

[161] M. SPIVAK Spaces satisfying Poincaré duality. *Topology* **6**, 77–102 (1967)

[162] C. W. STARK L-theory and graphs of free abelian groups. *J. Pure App. Alg.* **47**, 299–309 (1987)

[163] D. SULLIVAN *Triangulating homotopy equivalences.* Princeton Ph.D. thesis (1965)

[164] D. SULLIVAN *Geometric Topology I. Localization, Periodicity and Galois symmetry.* M. I. T. notes (1970)

[165] D. SULLIVAN Geometric periodicity and the invariants of manifolds. *Proceedings 1970 Amsterdam Conference on Manifolds*, Lecture

Notes in Mathematics **197**, Springer, 44-75 (1971)

[166] D. SULLIVAN Singularities in spaces. *Proceedings Liverpool Singularities Symposium (1969/70), Vol. II*, Lecture Notes in Mathematics **209**, Springer, 196-206 (1971)

[167] D. SULLIVAN and N. TELEMAN An analytic proof of Novikov's theorem on rational Pontrjagin classes. *Pub. Math. I. H. E. S.* **58**, 291-293 (1983)

[168] R. SWAN Periodic resolutions for finite groups. *Ann. Math.* **72**, 267-291 (1960)

[169] L. TAYLOR and B. WILLIAMS Surgery spaces: formulae and structure. *Proceedings 1978 Waterloo Algebraic Topology Conference*, Lecture Notes in Mathematics **741**, Springer, 170-195 (1979)

[170] P. VOGEL Une nouvelle famille de groupes en *L*-théorie algébrique. *Proceedings 1982 Bielefeld Algebraic K-theory Conference*, Lecture Notes in Mathematics **1046**, Springer, 385-421 (1984)

[171] F. WALDHAUSEN Algebraic *K*-theory of generalized free products. *Ann. Math.* **108**, 135-256 (1978)

[172] C. T. C. WALL Surgery of non–simply connected manifolds. *Ann. Math.* **84**, 217-276 (1966)

[173] C. T. C. WALL Non–additivity of the signature. *Inventiones Math.* **7**, 269-274 (1969)

[174] C. T. C. WALL Poincaré complexes. *Ann. Math.* **86**, 213-245 (1970)

[175] C. T. C. WALL On the axiomatic foundations of the theory of Hermitian forms. *Proc. Camb. Phil. Soc.* **67**, 243-250 (1970)

[176] C. T. C. WALL *Surgery on compact manifolds*. Academic Press (1971)

[177] C. T. C. WALL On the classification of hermitian forms II. Semisimple rings. *Inventiones Math.* **18**, 119-141 (1972)

[178] C. T. C. WALL Formulae for surgery obstructions. *Topology* **15**, 189-210 (1976)

[179] S. WEINBERGER Aspects of the Novikov conjecture. *Contemp. Math.* **105**, 281-297 (1990)

[180] S. WEINBERGER An analytic proof of the topological invariance of rational Pontrjagin classes. (*preprint*)

[181] S. WEINBERGER *The topological classification of stratified spaces*. (preprint)

[182] M. WEISS Surgery and the generalized Kervaire invariant. *Proc. Lond. Math. Soc.* **51** (3), I. 146-192, II. 193-230 (1985)

[183] M. WEISS Visible *L*-theory. *Forum Math.* **4**, 465-498 (1992)

[184] M. WEISS and B. WILLIAMS Automorphisms of manifolds and algebraic *K*-theory II. *J. Pure App. Alg.* **62**, 47-107 (1989)

[185] G. W. WHITEHEAD Generalized homology theories. *Trans. Am. Math.*

Soc. **102**, 227–283 (1962)

[186] J. H. C. WHITEHEAD A certain exact sequence. *Ann. Math.* **52**, 51–110 (1950)

[187] M. YAMASAKI *L*-groups of crystallographic groups. *Inventiones Math.* **88**, 571–602 (1987)

[188] E. C. ZEEMAN Dihomology III. A generalization of the Poincaré duality for manifolds. *Proc. Lond. Math. Soc.* **13** (3), 155–183 (1963)

Index